computational mathematics in engineering and applied science: ODEs, DAEs, and PDEs

William E. Schiesser
Lehigh University
Bethlehem, Pennsylvania

CRC Press
Boca Raton Ann Arbor London Tokyo

Library of Congress Cataloging-in-Publication Data

Schiesser, W. E.
Computational mathematics in engineering and applied science: ODEs, DAEs, and PDEs / William E. Schiesser.
p. cm.
Includes bibliographical references and index.
ISBN 0-8493-7373-5
1. Engineering mathematics. 2. Differential equations.
3. Mathematical models. I. Title
TA347.D45S34 1993
515'.35'0285–dc20 93-17599
CIP

Direct all inquiries to CRC Press, Inc., 2000 Corporate Blvd., N.W., Boca Raton, Florida, 33431.

International Standard Book Number 0-8493-7373-5

Library of Congress Card Number 93-17599

Printed in the United States of America 3 4 5 6 7 8 9 0

Printed on acid-free paper

contents

preface

Mathematicians, scientists, and engineers have produced mathematical descriptions of physical systems that originate in several ways:

1. Precise mathematical relationships, e.g., the velocity of a point mass is the integral of its acceleration with respect to time and the position is the time integral of the velocity; the Gauss and Stoke theorems which relate various volume, surface, and line integrals.

2. Empirical relationships that are so well established they have the status of "laws," e.g., Fourier's first law for heat conduction which expresses the undeniable fact that heat flows from regions of higher temperature to regions of lower temperature.

3. Mathematical expressions of principles that are also undeniably correct in the sense that they apply almost universally, e.g., the conservation of mass, charge, momentum, and energy.

4. Systems of equations that have been so successful in describing physical phenomena, their validity and utility under certain circumstances is now unquestioned, e.g., Newton's equations of motion, Maxwell's equations for electromagnetic fields, the Euler and Navier-Stokes equations in fluid mechanics.

5. Continuing development, extension, and generalization of mathematical descriptions to encompass still broader classes of physical systems.

These sources of mathematical descriptions for physical systems are, of course, often related. For example, the Euler equations of fluid mechanics are a statement of the conservation of mass and momentum.

Many scientific and engineering studies lead to a system of equations, often called a "mathematical model," that can, in principle, be used to describe the behavior of the problem system. These descriptions of physical systems are often expressed as differential equations that are among the most widely used mathematical forms in science and engineering.

Thus, scientists and engineers have available a spectrum of mathematical representations of physical systems; the extension of these descriptions, and the development of new descriptions, are active areas of research. The mathematical description (or mathematical modeling) of physical systems is therefore well developed, and continues to advance, usually in combination with experimental studies that test the validity of the descriptive mathematical relationships. The ultimate test of a mathematical model is whether it reflects the behavior of the physical system that it purports to describe. The ultimate use of a mathematical model, particularly in engineering, is the design of a proposed system with prescribed performance characteristics.

However, before the mathematical model can be used in this way, the solution of the model equations is required, since the solution indicates how the modeled system will perform; also, computational experiments can be run with the model to search for unexpected characteristics of the problem system, and possibly to optimize its design.

The second step (the solution in the two-step model formulation/solution sequence) is often the more difficult step to accomplish. Analytical methods of solutions are essentially limited to systems of a few linear equations. Numerical methods, in principle, can accommodate systems of equations of any size and complexity. However, numerical methods are "approximate," and the accuracy of the computed results must be assessed through some form of error analysis. The word "approximate" should not be interpreted to mean that the solutions are necessarily of poor accuracy (in the sense that the computed results depart significantly from the actual or correct solution), and therefore any conclusions drawn from the mathematical model may be questionable or unreliable. Rather, in principle, the numerical solution can be raised to any required level of accuracy through the selection of the numerical algorithms and the completion of enough numerical calculations to ensure convergence of the numerical solution to any required accuracy.

This book is devoted primarily to the second step in the use of mathematical models, that is, the solution stage, with emphasis on numerical methods. The general approach to the solution of the model equations is therefore through "computational mathematics." The organization of the book is according to equation types, that is, ordinary differential equations (ODEs), differential algebraic equations (DAEs), and partial differential equations (PDEs).

Chapter 1 is a summary of the various types of ODEs, DAEs, and PDEs, including classification of ODEs/DAEs as initial or boundary value problems, and geometric classification of PDEs as elliptic, hyperbolic, parabolic, and combinations of these three types. Chapter 2 then covers the numerical integration of ODEs and DAEs with emphasis on the use of library routines. Library integrators are recommended as the most reliable and efficient way for scientists and engineers to produce numerical solutions to ODEs and DAEs. This is illustrated with an example of an initial ODE problem, which, when "approximated" by finite differences that appear to be quite reasonable, produces completely incorrect solutions; the proposed remedy is to use a quality library ODE integrator. Chapter 2 concludes with a series of examples of rather unconventional uses of ODE integrators, e.g., the evaluation of one- and two-dimensional integrals, the solution of integro-differential equations, and systems of nonlinear algebraic and transcendental equations.

PDEs are the most difficult problem conceptually and computationally, if for no other reason than the types and forms of PDEs are essentially unlimited, e.g., linear and nonlinear; in one, two, and three dimensions plus time; with initial and boundary value independent variables; having constant and variable coefficients; in various coordinate systems; with Dirichlet, Neumann, and type-three boundary conditions (both linear and nonlinear); over finite, semi-infinite, and infinite, multiply connected domains; with irregular geometrics, etc. Thus, in the case of PDEs, the first major step in developing a numerical solution is a classification that will direct the selection of a numerical method or algorithm.

The approach to classification of PDEs in this book is to consider the type of independent variables, and the order of the derivatives in which the independent variables appear. For example, do the PDEs have an initial value variable (e.g., are they parabolic or hyperbolic), or do they have only boundary value independent variables (e.g., are they elliptic)? If they have an initial value variable, what is the highest order derivative for this variable? If it is first order, and (a) the highest order spatial (boundary value) derivative is first order, then we have first-order hyperbolic PDEs, (b) the highest order spatial derivative is second order, then we have parabolic PDEs. Chapters 3 and 4 discuss numerical methods for PDEs first order in an initial value independent variable, including solutions of PDEs in different coordinate systems, of mixed type, with mixed partial derivatives, and defined over multiple, connected spatial domains. The numerical method of lines is used for the solution of PDEs in Chapters 3 and 4, which builds naturally on the discussion of methods for ODEs and DAEs covered in Chapter 2. Most examples are in terms of finite difference approximations for the spatial derivatives; Chapter 4 concludes with a discussion of weighted residual, finite element, and finite volume methods within the context of the method of lines.

PDEs that are zeroth order in an initial value variable (i.e., elliptic) and second order in an initial value variable (i.e., second-order hyperbolic) are discussed in Chapter 5, again, with the solution by the method of lines, presented as a natural extension of the material in Chapters 3 and 4. Thus, the overall organization of the book is by equation type, rather than by applications of the differential equations, or by solution methods. This organization is intended to serve the needs of scientists and engineers who have a problem in ODEs, DAEs, and/or PDEs, and who wish to study an approach to the numerical solution of the problem system according to the type of differential equations to be solved rather than following an approach based on a specific numerical method or applications of a numerical method.

In other words, the focus of the book is on computational methods directed by the form and structure of the problem system differential equations. The intention with this approach is to accommodate the general needs of scientists and engineers within the presentation of one book. This is admittedly a rather ambitious goal, and the author would welcome the comments of readers concerning this point (or concerning any of the material in the book). Analytical solutions are presented when available for comparison with the numerical solutions; the parallel discussion of analytical and numerical methods indicates how they can be used together for added insights and understanding, verification and error analysis of solutions.

The computational methods are presented by examples from science and engineering. The examples illustrate a rather diverse collection of differential equations taken largely from problems given to the author. In particular, I wish to acknowledge the contributions of the following colleagues: Don Greenspan for the orbit problem in Chapter 2, Larry Biegler for the system of two nonlinear equations solved by Davidenko's differential equation in Chapter 2, Mike McAshan for the Bateman problem in Chapter 3, Shaul Abramovich for the Maxwell's equation problem in Chapter 3 and Mike McAshan for boundary condition (3.93) of this problem, Fernando Aguirre for the two-region problem in Chapter 4, and Shiv Prasher for the aquifier model in Chapter 5.

Also, my appreciation is extended to several colleagues who provided numerical methods and mathematical software: Mike Carver for the five-point biased upwind approximations, Alan Hindmarsh for the ODEPACK integrators (e.g., LSODE), Linda Petzold for DASSL, Larry Shampine and Buddy Watts for RKF45, and Gilbert Stengle for assistance in the derivation of the second derivative formulas (used in DSS042 and DSS044). All of the problems and library routines discussed in the book are available as Fortran source code from the author on 3.5 inch, 1.44 mb, DOS-formatted diskettes; requests can be directed to: W. E. Schiesser, Iacocca Hall, D307, Lehigh University, 111 Research Drive, Bethlehem, PA 18015, USA, (215) 758-4264 (phone), (215) 758-5057 (fax), wesl@lehigh.edu (Internet).

Other problems and numerical methods taken from the literature are acknowledged by reference to the original papers and books. My intention is to appropriately acknowledge all of the contributors to this book, and I extend sincere apologies for any oversights. Careful editing and formatting by Lori Pickert is gratefully acknowledged. Finally, I welcome any communication from the readers and users of this book; if past experience is any indication for the future, I will receive totally unexpected questions, suggestions, and applications that are an intellectually rewarding aspect of this work, and the basis for new developments in computational mathematics applied to differential systems.

I dedicate this book to my wife Dolores with my love and appreciation for her patience, encouragement, and support.

William E. Schiesser
Lehigh University
Bethlehem, PA
and
SSC Laboratory*
Dallas, TX
September 1, 1993

*Operated by the Universities Research Association, Inc., for the U.S. Department of Energy under contract No. DE-AC35-89ER40486.

To Dolores

computational mathematics in engineering and applied science: ODEs, DAEs, and PDEs

chapter one

The General Problems in Ordinary, Differential Algebraic and Partial Differential Equations

Ordinary differential equations (ODEs), *differential algebraic equations* (DAEs), and *partial differential equations* (PDEs) are mathematical forms that have broad application in science and engineering. The formulation of mathematical models expressed as sets of ODEs/DAEs/PDEs is frequently the starting point for quantitative studies of the behavior and performance of scientific and engineering systems. Mathematical models, however, have limited utility unless solutions to the models can be produced and studied with reasonable effort. Before the general purpose scientific computer became available, models were typically manipulated analytically, often with major assumptions made during the analysis so that the mathematics would be tractable. Practically, this meant that models consisting of at most a few ODEs/DAEs/PDEs could be solved, and this process often required great ingenuity, particularly in the case of nonlinear equations.

With the availability of the scientific computer, this general requirement for models to be highly simplified was completely circumvented; now large sets of nonlinear ODEs/DAEs/PDEs, in principle, can be integrated numerically. In practice, the coding (programming) of a numerical algorithm to solve a particular ODE/DAE/PDE problem system can appear daunting to the scientist or engineer who has limited knowledge and experience in numerical mathematics, and who wishes primarily to arrive at a useful solution without a major investment of time and effort to learn the details of numerical analysis and programming. Additionally, even if the analyst is willing to make this investment of time and effort, the direction this effort should take is often far from clear. The numerical analysis literature typically conveys the impression that highly specialized methods are reported which can be applied only to very specific problems; just finding one or more candidate numerical algorithms can be a major undertaking, not to mention the effort to code these algorithms, test the code with problems for which independent solutions are available, program the problem of interest, and finally evaluate the numerical solution with respect to criteria such as stability, accuracy, convergence, and consistency.

In this book, we describe a methodology we believe relieves the scientist or engineer of much of this effort in computing a numerical solution to an ODE/DAE/PDE problem. The approach is to illustrate the methodology through detailed examples, including complete codes written in transportable Fortran 77. The underlying mathematics of the methods are not discussed in detail for several reasons: (1) these details are available elsewhere; (2) the mathematical concepts that must be understood to ensure the general applicability of the methods and the validity of the computed results have been considered by recognized experts in numerical mathematics and have been integrated into the algorithms and software discussed in the book; and (3) the error analysis of computed solutions is presented numerically as part of the example applications (rather than mathematically as theorems and proofs). We start with a general statement of the ODE/DAE/PDE problem that can be addressed by the numerical methods. This statement serves as the basic framework for the discussion in subsequent chapters.

1.1 The ODE Problem

We now consider several forms of the ODE problem which arise in physical applications. As we will see in the remainder of the book, these forms are quite general in the sense that they can be applied to a broad spectrum of problems in science and engineering.

1.1.1 Explicit Initial Value ODEs

Consider the following system of ODEs

$$\begin{aligned} dy_1/dt &= f_1(y_1, y_2, \ldots, y_n, t) \\ dy_2/dt &= f_2(y_1, y_2, \ldots, y_n, t) \\ \vdots \quad & \quad \vdots \\ dy_n/dt &= f_n(y_1, y_2, \ldots, y_n, t) \end{aligned} \tag{1.1}$$

$$y_1(t_0) = y_{1,0}, y_2(t_0) = y_{2,0}, \ldots, y_n(t_0) = y_{n,0} \tag{1.2}$$

where $y_1, y_2, \ldots, y_n$ is the *vector of dependent variables*, with n components (also termed an n vector); t is the *initial value independent variable*, typically time in a physical problem; $f_1, f_2, \ldots, f_n$ is the *vector of derivatives*, also termed the *derivative vector*; t_0 is the initial value of t; and $y_{1,0}, y_{2,0}, \ldots, y_{n,0}$ are the initial values of the dependent variables at $t = t_0$, generally termed the *initial condition vector*.

The definition of an ODE problem consists in the specification of (1) the derivative vector, $f_1, f_2, \ldots, f_n$, (2) the value of t_0, and (3) the initial condition vector, $y_{1,0}$, $y_{2,0}, \ldots, y_{n,0}$.

For example, if we wish to write the ODE for a simple harmonic oscillator in the format of equations (1.1) and (1.2)

$$d^2y/dt^2 + \omega^2 y = 0 \tag{1.3}$$

$$y(0) = 1 \tag{1.4}$$

$$dy(0)/dt = 0 \tag{1.5}$$

we first define two *state variables*, $y_1 = y, y_2 = dy/dt$. Then equations (1.3) to (1.5) can be written in the form

$$dy_2/dt = -\omega^2 y_1 \tag{1.6}$$

$$dy_1/dt = y_2 \tag{1.7}$$

$$y_1(0) = 1 \tag{1.8}$$

$$y_2(0) = 0 \tag{1.9}$$

Thus, equations (1.6) to (1.9) can be written in the general format of equations (1.1) and (1.2) as

$$f_1(y_1, y_2, t) = y_2$$
$$f_2(y_1, y_2, t) = -\omega^2 y_1$$
$$t_0 = 0$$
$$y_{1,0} = 1$$
$$y_{2,0} = 0$$

Observe that the specification of this problem consisted of specifiying: (1) the derivative vector, $f_1(= y_2)$, $f_2(= -\omega^2 y_1)$; (2) the value of $t_0(= 0)$; and (3) the initial condition vector, $y_{1,0}$ $(= 1)$, $y_{2,0}$ $(= 0)$. We note that although equations (1.1) are first order, (do not contain higher order derivatives in t, e.g., d^2y_1/dt^2), they can easily be applied to higher order ODEs, in this case a second-order ODE, equation (1.3), by first defining two state variables, y_1 and y_2. In general, we can apply numerical methods for first-order ODEs to an nth order ODE, i.e., an ODE in which the highest order derivative is nth order, by rewriting it as n first-order ODEs. This procedure is general (can be applied to essentially all higher order ODEs), and therefore we can infer that equations (1.1) and (1.2) are generally applicable to higher order ODEs. Also, equations (1.8) and (1.9) indicate that the initial conditions for higher order ODEs can be naturally included in the formulation of equations (1.2).

Finally, we would naturally assume that if we integrate equations (1.6) to (1.9) numerically, we would obtain a numerical solution which closely approximates the analytical (exact or closed form) solution, $y_1 = \cos(\omega t)$, $y_2 = -\omega \sin(\omega t)$. To restate this idea in other words, we can use analytical solutions, when they are available, to evaluate numerical solutions, and more importantly, to evaluate the integration algorithm(s) that produced the numerical solutions.

To reiterate, equations (1.1) and (1.2) are quite general, and have only the limitations that they are: (a) first order (which, as we just observed, is not a significant limitation); (b) initial value, i.e., the initial conditions are all evaluated at the same value of the independent variable, $t = t_0$; and (3) explicit ODEs because each equation contains only one derivative that is explicitly defined mathematically. As we will see, such systems of ODEs and their associated numerical integration algorithms are broadly applicable to scientific and engineering systems. In particular, we should note that the dependent variables, $y_1, y_2, \ldots, y_n$, do not have to be to the first power or first degree in the derivative functions, $f_1, f_2, \ldots, f_n$, i.e., the derivative functions can be nonlinear. Thus equations (1.1) are first order, but not necessarily first degree; the distinction between order and degree is quite important and often confused.

Also, equations (1.1) and (1.2) can be stated in *matrix form* as

$$d\bar{y}/dt = \bar{f}(\bar{y}, t) \tag{1.10}$$

$$\bar{y}(t_0) = \bar{y}_0 \tag{1.11}$$

where

$$\bar{y} = \begin{bmatrix} y_1 & y_2 & \cdots & y_n \end{bmatrix}^T$$
$$\bar{f} = \begin{bmatrix} f_1 & f_2 & \cdots & f_n \end{bmatrix}^T$$
$$\bar{y}_0 = \begin{bmatrix} y_{1,0} & y_{2,0} & \cdots & y_{n,0} \end{bmatrix}^T$$

(a superscript T denotes a transpose, which in this case, changes a row vector to a column vector, and is used here only for convenience in writing vectors). Equations (1.1) and (1.2) or (1.10) and (1.11) are referred to as (a) a system of n first-order ODEs or (b) an nth order system of ODEs.

1.1.2 Semi-Implicit Initial Value ODEs

As we noted previously, equations (1.1) and (1.2) are broadly applicable in science and engineering. They can, however, be generalized to encompass an even broader spectrum of problems. For example, if we write the ODEs and initial conditions as

$$\begin{aligned}
a_{11}dy_1/dt + a_{12}dy_2/dt + \cdots + a_{1n}dy_n/dt &= f_1(y_1, y_2, \ldots, y_n, t)\\
a_{21}dy_1/dt + a_{22}dy_2/dt + \cdots + a_{2n}dy_n/dt &= f_2(y_1, y_2, \ldots, y_n, t)\\
\vdots \qquad\qquad & \qquad \vdots\\
a_{n1}dy_1/dt + a_{n2}dy_2/dt + \cdots + a_{nn}dy_n/dt &= f_n(y_1, y_2, \ldots, y_n, t)
\end{aligned} \tag{1.12}$$

$$y_1(t_0) = y_{1,0}, y_2(t_0) = y_{2,0}, \ldots, y_n(t_0) = y_{n,0} \tag{1.13}$$

the coefficients a_{ij}, $i = 1, 2, \ldots, n$, $j = 1, 2, \ldots, n$, constitute an $n \times n$ *coupling matrix;* a_{ij} is the coefficient for derivative dy_j/dt in equation i. Thus we now have more than one derivative in each ODE (equations (1.12) reduce to equation (1.1) if the coupling matrix is the *identity matrix*). Equations (1.12) are termed *linearly coupled* or *semi-implicit* in contrast with explicit ODEs, equations (1.1).

Within this class of ODEs, three important special cases can be considered: (1) the coupling coefficients a_{ij} are constants or functions of t only; (2) the a_{ij} are functions of the dependent variables, $y_1, y_2, \ldots, y_n$ (as well as possibly of t); and (3) all of the a_{ij} in a given equation, say equation i, are zero, in which case the ODE reduces to the algebraic equation

$$0 = f_i(y_1, y_2, \ldots, y_n, t)$$

All of these special cases are important in applications. For example, the first and second cases result from *weighted residual methods* and finite element methods applied to PDEs which are reduced to systems of ODEs via the method of lines [Schiesser (1991)], as will be demonstrated in subsequent examples. Also, the third case of one or more algebraic equations leads to DAEs [Brenan et al. (1989)]. Integrators to accommodate the class of semi-implicit ODEs defined by equations (1.12) and (1.13) have been developed, notably `LSODI`, which is in the library of integrators, `ODEPACK`, developed by A. C. Hindmarsh [Byrne and Hindmarsh (1987)].

We should also note that when initial conditions (1.13) are applied to equations (1.12), we obtain a set of n equations for the n initial derivatives, $dy_1(t_0)/dt$, $dy_2(t_0)/dt$, $\ldots, dy_n(t_0)/dt$. Thus both the initial dependent variable vector of equations (1.13) and the *vector of initial derivatives* are generally required to start the integration of equations (1.12). In other words, this is the requirement for a consistent set of initial conditions to start the numerical integration.

Equations (1.12) and (1.13) can also be stated in matrix form as

$$\bar{A}d\bar{y}/dt = \bar{f}(\bar{y}, t) \tag{1.14}$$

$$\bar{y}(t_0) = \bar{y}_0 \tag{1.15}$$

where $\bar{y}$, $\bar{f}$, and $\bar{y}_0$ have the same definitions as for equations (1.10) and (1.11) and

$$\bar{A} = \begin{bmatrix} a_{11} & a_{12} & \cdots & a_{1n}\\ a_{21} & a_{22} & \cdots & a_{2n}\\ \vdots & \vdots & \vdots & \vdots\\ a_{n1} & a_{n2} & \ldots & a_{nn} \end{bmatrix} \tag{1.16}$$

1.1.3 Implicit Initial Value ODEs

The form of equations (1.12) and (1.13) can be generalized still further to

$$\begin{aligned} f_1(y_1, y_2, \ldots, y_n, dy_1/dt, dy_2/dt, \ldots, dy_n/dt, t) &= 0 \\ f_2(y_1, y_2, \ldots, y_n, dy_1/dt, dy_2/dt, \ldots, dy_n/dt, t) &= 0 \\ \vdots \qquad\qquad & \vdots \\ f_n(y_1, y_2, \ldots, y_n, dy_1/dt, dy_2/dt, \ldots, dy_n/dt, t) &= 0 \end{aligned} \tag{1.17}$$

$$\begin{aligned} f_1(y_1(t_0), y_2(t_0), \ldots, y_n(t_0), dy_1(t_0)/dt, dy_2(t_0)/dt, \ldots, dy_n(t_0)/dt, t_0) &= 0 \\ f_2(y_1(t_0), y_2(t_0), \ldots, y_n(t_0), dy_1(t_0)/dt, dy_2(t_0)/dt, \ldots, dy_n(t_0)/dt, t_0) &= 0 \\ \vdots \qquad\qquad & \vdots \\ f_n(y_1(t_0), y_2(t_0), \ldots, y_n(t_0), dy_1(t_0)/dt, dy_2(t_0)/dt, \ldots, dy_n(t_0)/dt, t_0) &= 0 \end{aligned} \tag{1.18}$$

or in matrix form

$$\bar{f}(\bar{y}, d\bar{y}/dt, t) = \bar{0} \tag{1.19}$$

$$\bar{f}(\bar{y}(t_0), d\bar{y}(t_0)/dt, t_0) = \bar{0} \tag{1.20}$$

Equations (1.17) are termed *fully implicit* since the derivative vector dy_1/dt, $dy_2/dt, \ldots$ dy_n/dt is defined implicitly through the arguments of the function vector $f_1, f_2, \ldots, f_n$ (which, in turn, defines the problem along with the initial conditions, equations (1.18)). Two other features of equations (1.17) and (1.18), or (1.19) and (1.20), should be noted:

1. If any function, say f_i, does not include at least one derivative in its arguments, it is an algebraic equation, i.e.,

$$f_i(y_1, y_2, \ldots, y_n, t) = 0 \tag{1.21}$$

   Thus equations (1.17) and (1.18) are a system of DAEs.

2. The initial conditions, equations (1.18) or (1.20), require initial values of the derivatives, $dy_1(t_0)/dt$, $dy_2(t_0)/dt, \ldots, dy_n(t_0)/dt$, as well as initial values of the dependent variables, $y_1(t_0), y_2(t_0), \ldots, y_n(t_0)$. This requirement for a consistent set of initial conditions is generally more difficult to accomplish than for the explicit ODEs (for which only the initial dependent variable vector, $y_1(t_0), y_2(t_0), \ldots,$ $y_n(t_0)$, has to be specified). The difficulty of computing a numerical solution to the DAE system of equations (1.19) and (1.20) is also closely tied to the index of the particular problem [Brenan et al. (1989)]. We will not get into a discussion of the index in detail, but merely point out that it is a measure of the singularity of the algebraic part of a DAE system. In general, the higher the index, the more difficult the calculation of a numerical solution to a DAE system. Index 0 and 1 problems can be handled rather routinely, while index 2 and higher problems are difficult to solve numerically; these higher index problems are at the forefront of research in numerical methods for DAE systems.

Thus, we might expect that implicit ODEs (equations (1.19) and (1.20)) are inherently more difficult to integrate numerically than explicit and semi-implicit ODEs, because of their greater generality. This is indeed the case, but the increased difficulty of solution is additionally due to the properties of implicit ODEs not shared by the simpler forms of ODEs, particularly in the case of DAEs. The theoretical and numerical aspects of implicit ODEs and DAEs have been explored by Brenan et al. (1989), and a computer code, DASSL developed by L. R. Petzold [Brenan et al. (1989)], is available for this class of problems.

The three preceding sets of initial value ODEs, i.e., explicit, semi-implicit, and implicit, can be generalized further to boundary value ODEs. The distinction between initial and boundary value ODEs has to do with the values of the independent variable, t, at which the auxiliary conditions are specified. In the case of initial value problems, the auxiliary conditions, i.e., the initial conditions, are all specified at the same initial value of the independent variable, e.g., $t = t_0$, and the solution evolves from this initial point. For boundary value problems, the auxiliary conditions are specified at two or more values of the independent variable, e.g., $t = t_0$ and $t = t_1$, which often in applications correspond to physical boundaries of the problem system; hence the name boundary conditions for these auxiliary conditions. We now restate the preceding initial value problems as boundary value problems where the single initial value of the independent variable, t_0, is merely replaced with two or more boundary values which are represented as the components of the vector $\bar{t}_b$.

1.1.4 Explicit Boundary Value ODEs

$$\begin{aligned} dy_1/dt &= f_1(y_1, y_2, \ldots, y_n, t) \\ dy_2/dt &= f_2(y_1, y_2, \ldots, y_n, t) \\ \vdots \quad & \qquad \vdots \\ dy_n/dt &= f_n(y_1, y_2, \ldots, y_n, t) \end{aligned} \tag{1.22}$$

$$y_1(\bar{t}_b) = y_{1,0}, y_2(\bar{t}_b) = y_{2,0}, \ldots, y_n(\bar{t}_b) = y_{n,0} \tag{1.23}$$

As an example of the application of equations (1.22) and (1.23), we return to equations (1.3) to (1.5), but now restated as a boundary value ODE

$$d^2y/dt^2 + \omega^2 y = 0 \tag{1.24}$$

$$y(0) = 1 \tag{1.25}$$

$$dy(\pi/\omega)/dt = 0 \tag{1.26}$$

or, in terms of the two state variables, $y_1 = y$, $y_2 = dy/dt$

$$dy_2/dt = -\omega^2 y_1 \tag{1.27}$$

$$dy_1/dt = y_2 \tag{1.28}$$

$$y_1(0) = 1 \tag{1.29}$$

$$y_2(\pi/\omega) = 0 \tag{1.30}$$

Note in particular that the boundary conditions, equations (1.25) and (1.26) or equations (1.29) and (1.30), are evaluated at two different values of the independent variable, $t = 0$ and π/ω, in contrast with initial conditions (1.4) and (1.5) or (1.8) and (1.9), which are both evaluated at $t = 0$.

The solution in this case is the same, i.e., $y_1(t) = \cos(\omega t)$, $y_2(t) = -\omega \sin(\omega t)$. Thus, we might conclude that the distinction between initial and boundary value problems is essentially trivial. This is not the case, however; generally, boundary value problems are inherently more difficult to solve than initial value problems. In fact, a standard approach to the solution of boundary value problems is to solve an associated initial value problem. For example, since equations (1.6) to (1.9) and equations (1.27) to (1.30) have the same solutions, we could solve equations (1.27) to (1.30) by solving equations (1.6) to (1.9). This would require replacing boundary condition (1.30) with initial condition (1.9). The catch in this procedure is that generally we will not know beforehand the correct initial condition(s) to use in place of the boundary condition(s), so usually a trial and error procedure for the missing initial conditions is required, i.e., we would assume the required number of initial conditions, then integrate to the other boundary value(s) of the independent variable and observe whether the boundary conditions are satisfied; if they are not, the procedure is repeated with new assumed initial conditions until the boundary conditions are satisfied. This is the so-called *shooting method.* This trial and error can be difficult, particularly as the number of missing initial conditions increases (e.g., with increasing order of the ODE boundary value problem). Fortunately, alternative algorithms have been developed which are much more effective than this shooting method, and they have been implemented in robust, general purpose computer codes such as COLSYS [Ascher et al. (1981)].

1.1.5 Semi-Implicit Boundary Value ODEs

Equations (1.12) and (1.13) can be stated in boundary value form as

$$\begin{aligned}
a_{11}dy_1/dt + a_{12}dy_2/dt + \cdots + a_{1n}dy_n/dt &= f_1(y_1, y_2, \ldots, y_n, t) \\
a_{21}dy_1/dt + a_{22}dy_2/dt + \cdots + a_{2n}dy_n/dt &= f_2(y_1, y_2, \ldots, y_n, t) \\
\vdots \qquad\qquad & \qquad \vdots \\
a_{n1}dy_1/dt + a_{n2}dy_2/dt + \cdots + a_{nn}dy_n/dt &= f_n(y_1, y_2, \ldots, y_n, t)
\end{aligned} \tag{1.31}$$

$$y_1(\bar{t}_b) = y_{1,0}, y_2(\bar{t}_b) = y_{2,0}, \ldots, y_n(\bar{t}_b) = y_{n,0} \tag{1.32}$$

In matrix form

$$\bar{A}d\bar{y}/dt = \bar{f}(\bar{y}, t) \tag{1.33}$$

$$\bar{y}(\bar{t}_b) = \bar{y}_0 \tag{1.34}$$

To the best of the author's knowledge, semi-implicit boundary value ODEs have not been studied systematically. Also, the following implicit boundary value ODEs are now receiving attention, and some theoretical and numerical analysis results have been reported by Ascher and Petzold (1992). Since semi-implicit ODEs are a special case of implicit ODEs, these results could also, in principle, be applied to semi-implicit ODEs.

1.1.6 Implicit Boundary Value ODEs

Equations (1.17) and (1.18) can be written as boundary value DAEs

$$\begin{aligned}
f_1(y_1, y_2, \ldots, y_n, dy_1/dt, dy_2/dt, \ldots, dy_n/dt, t) &= 0 \\
f_2(y_1, y_2, \ldots, y_n, dy_1/dt, dy_2/dt, \ldots, dy_n/dt, t) &= 0 \\
\vdots \qquad\qquad & \qquad \vdots \\
f_n(y_1, y_2, \ldots, y_n, dy_1/dt, dy_2/dt, \ldots, dy_n/dt, t) &= 0
\end{aligned} \tag{1.35}$$

$$\begin{aligned}
f_1(y_1(\bar{t}_b), y_2(\bar{t}_b), \ldots, y_n(\bar{t}_b), dy_1(\bar{t}_b)/dt, dy_2(\bar{t}_b)/dt, \ldots, dy_n(\bar{t}_b)/dt, \bar{t}_b) &= 0 \\
f_2(y_1(\bar{t}_b), y_2(\bar{t}_b), \ldots, y_n(\bar{t}_b), dy_1(\bar{t}_b)/dt, dy_2(\bar{t}_b)/dt, \ldots, dy_n(\bar{t}_b)/dt, \bar{t}_b) &= 0 \\
\vdots \qquad\qquad & \qquad \vdots \\
f_n(y_1(\bar{t}_b), y_2(\bar{t}_b), \ldots, y_n(\bar{t}_b), dy_1(\bar{t}_b)/dt, dy_2(\bar{t}_b)/dt, \ldots, dy_n(\bar{t}_b)/dt, \bar{t}_b) &= 0
\end{aligned} \tag{1.36}$$

In matrix form

$$\bar{f}(\bar{y}, d\bar{y}/dt, t) = \bar{0} \tag{1.37}$$

$$\bar{f}(\bar{y}(\bar{t}_b), d\bar{y}(\bar{t}_b)/dt, \bar{t}_b) = \bar{0} \tag{1.38}$$

The reader is referred to Ascher and Petzold (1992) for an analysis of these implicit boundary value DAEs. Also, Ascher and Spiteri (1992) have reported two codes, `COLNEW` and `COLDAE`, for the solution of implicit boundary value DAEs.

We have now covered most of the major classes of ODEs that commonly occur in scientific and engineering applications. We now proceed to a classification of PDEs in a format that is tied closely to the preceding classification of ODEs through the *numerical method of lines.*

1.2 The PDE Problem

PDEs are distinguished from ODEs by the number of independent variables; ODEs have only one independent variable (t in the preceding discussion), while PDEs have more than one independent variable. We will represent the additional independent variables with the vector $\bar{x}$, which typically has three components representing three-dimensional space. Some representative elements of $\bar{x}$ include:

$$\bar{x} = \begin{bmatrix} x & y & z \end{bmatrix}^T \quad \text{(Cartesian coordinates)} \tag{1.39}$$

$$\bar{x} = \begin{bmatrix} r & \theta & z \end{bmatrix}^T \quad \text{(cylindrical coordinates)} \tag{1.40}$$

$$\bar{x} = \begin{bmatrix} r & \theta & \phi \end{bmatrix}^T \quad \text{(spherical coordinates)} \tag{1.41}$$

The general PDE problem can now be stated as

$$\begin{aligned} u_{1t} &= f_1(\bar{x}, t, u_1, u_2, \ldots, u_n, u_{1\bar{x}}, u_{2\bar{x}}, \ldots u_{n\bar{x}}, u_{1\bar{x}\bar{x}}, u_{2\bar{x}\bar{x}}, \ldots, u_{n\bar{x}\bar{x}}, \ldots) \\ u_{2t} &= f_2(\bar{x}, t, u_1, u_2, \ldots, u_n, u_{1\bar{x}}, u_{2\bar{x}}, \ldots u_{n\bar{x}}, u_{1\bar{x}\bar{x}}, u_{2\bar{x}\bar{x}}, \ldots, u_{n\bar{x}\bar{x}}, \ldots) \\ &\vdots \qquad\qquad\qquad\qquad\qquad \vdots \\ u_{nt} &= f_n(\bar{x}, t, u_1, u_2, \ldots, u_n, u_{1\bar{x}}, u_{2\bar{x}}, \ldots u_{n\bar{x}}, u_{1\bar{x}\bar{x}}, u_{2\bar{x}\bar{x}}, \ldots, u_{n\bar{x}\bar{x}}, \ldots) \end{aligned} \tag{1.42}$$

where $u_1, u_2, \ldots, u_n$ is the vector of dependent variables of length n to be computed by the method of lines; t is the initial value independent variable, typically time; $f_1, f_2, \ldots f_n$ is the vector of RHS functions defined for a particular PDE problem; and $\bar{x}$ is the *vector of boundary value* (spatial) *independent variables*, e.g., $[x\,y\,z]^T$ for Cartesian coordinates, $[r\,\theta\,z]^T$ for cylindrical coordinates, $[r\,\theta\,\phi]^T$ for spherical coordinates.

A subscript with respect to t or $\bar{x}$ indicates a partial derivative with respect to t or $\bar{x}$. For example,

$$u_{1t} = \frac{\partial u_1}{\partial t}$$

$$u_{2\bar{x}} = \frac{\partial u_2}{\partial x}, \frac{\partial u_2}{\partial y}, \frac{\partial u_2}{\partial z}$$

(Cartesian coordinates)

$$u_{3\bar{x}\bar{x}} = \frac{\partial^2 u_3}{\partial r^2}, \frac{\partial^2 u_3}{\partial \theta^2}, \frac{\partial^2 u_3}{\partial \phi^2}$$

(spherical coordinates)

Equation (1.42) also requires an initial condition vector

$$u_1(\bar{x}, t_0) = g_1(\bar{x}), u_2(\bar{x}, t_0) = g_2(\bar{x}), \ldots, u_n(\bar{x}, t_0) = g_n(\bar{x}) \tag{1.43}$$

and a vector of boundary conditions

$$\begin{aligned} &h_1(\bar{x}_b, t, u_1(\bar{x}_b, t), u_2(\bar{x}_b, t) \ldots, u_n(\bar{x}_b, t), u_{1\bar{x}}(\bar{x}_b, t), u_{2\bar{x}}(\bar{x}_b, t), \\ &\qquad \ldots, u_{n\bar{x}}(\bar{x}_b, t), \ldots) = 0 \\ &h_2(\bar{x}_b, t, u_1(\bar{x}_b, t), u_2(\bar{x}_b, t) \ldots, u_n(\bar{x}_b, t), u_{1\bar{x}}(\bar{x}_b, t), u_{2\bar{x}}(\bar{x}_b, t), \\ &\qquad \ldots, u_{n\bar{x}}(\bar{x}_b, t), \ldots) = 0 \\ &\qquad \vdots \qquad\qquad\qquad\qquad \vdots \end{aligned} \tag{1.44}$$

where t_0 is the initial value of t; $g_1, g_2, \ldots, g_n$ is the vector of initial condition functions; $h_1, h_2, \ldots$ is the vector of boundary condition functions; and $\bar{x}_b$ is the boundary value of $\bar{x}$.

The length of the boundary condition vector $[h_1\, h_2\, \ldots]^T$ cannot be stated generally for equation (1.44) since it will depend on the number and order of the spatial derivatives in equation (1.42). Also, $\bar{x}_b$, which generally denotes $\bar{x}$ including the boundary values, cannot be stated more explicitly since it will depend on the number of boundary value independent variables in equations (1.42) and (1.43) (typically, one, two, or three for each PDE).

Equations (1.42) through (1.44) can be stated in a more concise vector form as

$$\bar{u}_t = \bar{f}(\bar{x}, t, \bar{u}, \bar{u}_{\bar{x}}, \bar{u}_{\bar{x}\bar{x}}, \ldots) \tag{1.45}$$

$$\bar{u}(\bar{x}, t_o) = \bar{g}(\bar{x}) \tag{1.46}$$

$$\bar{h}(\bar{x}_b, t, \bar{u}(\bar{x}_b, t), \bar{u}_{\bar{x}}(\bar{x}_b, t), \ldots) = \bar{0} \tag{1.47}$$

1.2.1 The Numerical Method of Lines

The numerical integration of PDEs can be approached in many ways, and the extensive PDE numerical analysis literature reflects this diversity of approaches. In fact, the prospect of surveying the literature, selecting a candidate method (numerical algorithm), coding the method and testing the code, and finally, applying the code to the problem system of interest can be daunting for a scientist or engineer with only limited experience in these areas. We present here a general approach to the computer solution of PDEs, the numerical method of lines, which has proven to be relatively easy to understand, yet comprehensive in the class of problems that it can accommodate. Further, the method of lines is based on the integration of systems of ODEs that approximate the PDEs of interest, and therefore the recent progress in algorithms and computer codes for ODEs can be applied to PDEs as well. We consider here the basic features of the method of lines. Detailed applications are discussed in subsequent chapters and elsewhere [Schiesser (1991)].

The basic idea in the method of lines is to replace the boundary value derivatives with respect to $\bar{x}$ (also termed spatial derivatives since $\bar{x}$ frequently represents the spatial coordinates of the PDE problem) in equations (1.42) and (1.44) or (1.45) and (1.47) with algebaic approximations. This in effect leaves only derivatives with respect to the initial value variable, t, i.e., the PDE problem reverts to a set of approximating

ODEs in t. Since $\bar{x}$ is a boundary value (spatial) independent variable, it is defined over a spatial grid which runs between the boundary values of $\bar{x}$; approximating ODEs are defined at each of the spatial grid points. This set of ODEs defined over the entire spatial grid is then integrated simultaneously along lines of constant $\bar{x}$; hence the name the numerical method of lines. Admittedly, this description is general and vague. The examples in subsequent chapters will demonstrate concretely the elements of a method of lines solution. As in the case of ODEs, we now classify PDEs, which facilitates the selection of a specific method of lines implementation; the classification is geometric, i.e., we characterize PDEs as *elliptic*, *parabolic*, or *hyperbolic* (or a combination for PDEs which share the properties of these three equation types and for systems of PDEs which have individual equations of these three types).

1.2.2 Elliptic PDEs

If all of the derivatives with respect to t in equation (1.45) are identically zero, then we have a problem which is entirely of the boundary value type in $\bar{x}_b$

$$0 = \bar{f}(\bar{x}, \bar{u}, \bar{u}_{\bar{x}}, \bar{u}_{\bar{x}\bar{x}}, \ldots) \tag{1.48}$$

$$\bar{h}(\bar{x}_b, \bar{u}(\bar{x}_b), \bar{u}_{\bar{x}}(\bar{x}_b), \ldots) = \bar{0} \tag{1.49}$$

We classify this problem as elliptic; examples include *Laplace's*, *Poisson's*, and *Helmholtz's* equations

Laplace's equation:

$$\frac{\partial^2 u}{\partial x^2} + \frac{\partial^2 u}{\partial y^2} = 0 \quad \text{or} \quad u_{xx} + u_{yy} = 0 \tag{1.50}$$

Poisson's equation:

$$\frac{\partial^2 u}{\partial x^2} + \frac{\partial^2 u}{\partial y^2} = f(x,y) \quad \text{or} \quad u_{xx} + u_{yy} = f(x,y) \tag{1.51}$$

Helmholtz's equation:

$$\frac{\partial^2 u}{\partial x^2} + \frac{\partial^2 u}{\partial y^2} + u = 0 \quad \text{or} \quad u_{xx} + u_{yy} + u = 0 \tag{1.52}$$

For example, equation (1.50) follows from equation (1.42) with:

$$n = 1 \text{ (one PDE and one dependent variable, } u_1)$$

$$u_{1t} = 0 \text{ (for an elliptic problem)}$$

$$f_1 = \frac{\partial^2 u}{\partial x^2} + \frac{\partial^2 u}{\partial y^2} = u_{xx} + u_{yy}.$$

Equations (1.50) through (1.52) each require two boundary conditions in x and two boundary conditions in y. Generally, the rule for the required number of boundary conditions is: for each boundary value independent variable, the number of required boundary conditions equals the order of the highest order derivative for that variable. For example, since equation (1.50) is second order in x, it requires two boundary conditions in x, and since it is second order in y, it also requires two boundary conditions in y. Thus, we might specify as the boundary conditions in x for equation (1.50)

$$u(0,y) = f_0(y), \qquad u(x_1,y) = f_1(y) \tag{1.53)(1.54}$$

and the boundary conditions in y as

$$\frac{\partial u}{\partial y}(x,0) = \frac{\partial u}{\partial y}(x,y_1) = 0 \qquad (1.55)(1.56)$$

or in subscript notation

$$u_y(x,0) = u_y(x,y_1) = 0$$

The notation in these boundary conditions should be reviewed. The arguments of u are x and y, i.e., $u(x,y)$. Then specific values of these arguments are given to specify the boundary conditions, e.g., $u(0,y)$ is the value of u at $x = 0$ for all y.

Also, equations (1.53) to (1.56) correspond to special cases of boundary conditions (1.44) with the following interpretation

$$\begin{aligned} h_1 &= u(0,y) - f_0(y) \\ \bar{x}_b &= x_b(=0), y \end{aligned} \qquad (1.53)$$

$$\begin{aligned} h_2 &= u(x_1,y) - f_1(y) \\ \bar{x}_b &= x_b(=x_1), y \end{aligned} \qquad (1.54)$$

$$\begin{aligned} h_3 &= u_y(x,0) \\ \bar{x}_b &= y_b(=0), x \end{aligned} \qquad (1.55)$$

$$\begin{aligned} h_4 &= u_y(x,y_1) \\ \bar{x}_b &= y_b(=y_1), x \end{aligned} \qquad (1.56)$$

Equations (1.53) and (1.54) are examples of *Dirichlet boundary conditions* since they specify the dependent variable, u, at the boundary values of x, $x = 0$ and x_1. Equations (1.55) and (1.56) are examples of *Neumann boundary conditions* since they specify the normal derivative of u with respect to y at the boundary values of y, $y = 0$ and y_1. A third possbility is a combination of Dirichlet and Neumann boundary conditions, i.e., a boundary condition of the third type; an example would be

$$u_x(0,y) + u(0,y) = 0$$

(note that u and its derivative, u_x, are both included).

Equations (1.50) and (1.53) through (1.56) constitute a well-posed problem, and we can now proceed to compute a solution. Before we do that, however, we should answer the question: What do we mean by a solution to this problem or any other problem in PDEs? For the present problem, we seek a function $u(x,y)$ that satisfies the PDE (equation (1.50)) and all of the auxiliary conditions (equations (1.53) to (1.56)). In general, for a system of n PDEs, we seek n functions which satisfy the PDEs and all of their auxiliary conditions. Ideally, we would like to have analytical solutions, but this is rarely possible (because of the nonlinearity and number of the PDEs which result from the analysis of a realistic scientific or engineering problem). Rather, we will compute numerically the solution $u(x,y)$ with the objective of having the numerical solution close to the (unknown) analytical or exact solution. An essential part of this numerical approach then is to specify what is meant by "close," and then to try to establish that we have, in fact, met this requirement (of close agreement between the numerical and exact solutions). Finally, to illustrate a PDE solution, what is the solution to equations (1.50) and (1.53) to (1.56) for $f_0 = f_1 = 0$ or $f_0 = f_1 = 1$?

The procedure for solving elliptic problems within the method of lines is to add a vector of "time" derivatives to equation (1.48) and include an assumed initial condition vector, then numerically integrate the resulting system of ODEs in t to equilibrium where the time derivatives are essentially zero, the so-called *method of false transients*. For example, in the case of equation (1.50) we add a derivative in t

$$\frac{\partial^2 u}{\partial x^2}+\frac{\partial^2 u}{\partial y^2}=\frac{\partial u}{\partial t} \tag{1.57}$$

then integrate equation (1.57) until $\partial u/\partial t \approx 0$, in which case equation (1.57) reduces to equation (1.50). In order to start the solution of equation (1.57), we need an initial condition (because of the derivative $\partial u/\partial t$), which can be chosen arbitrarily since t is not part of the original problem (equations (1.50) and (1.53) to (1.56)); thus, the assumed initial condition will be inconsequential as the solution approaches the condition $\partial u/\partial t \approx 0$. Typically, a constant is used as the initial condition, i.e.,

$$u(x,y,0)=u_0 \tag{1.58}$$

In conclusion, the method of lines accommodates elliptic PDEs indirectly by converting them to "time"-dependent PDEs which are then integrated until the time derivatives essentially vanish. These modified PDEs, such as equation (1.57), are parabolic PDEs to be considered next.

Finally, as a special case, if $\bar{x}_b$ is one-dimensional (has one component), then this procedure gives a solution to a system of boundary value ODEs. For example, if in equations (1.42) to (1.44),

$$n=1 \text{ (one equation or dependent variable)}$$

$$\bar{x}=x \text{ (one spatial dimension)}$$

$$f_1=u_{xx}(u_1=u)$$

$$g_1(x)=0, t_0=0$$

$$h_1=u(0,t)-1$$

$$h_2=u(1,t)$$

then the corresponding PDE problem is

$$u_t=u_{xx} \tag{1.59}$$

$$u(x,0)=0 \tag{1.60}$$

$$u(0,t)=1 \tag{1.61}$$

$$u(1,t)=0 \tag{1.62}$$

We can now integrate equation (1.59) until $u_t \approx 0$, in which case, equations (1.59) to (1.62) reduce to the boundary value ODE problem (t essentially drops out of the problem)

$$\frac{d^2u}{dx^2}=0 \tag{1.63}$$

$$u(0)=1 \tag{1.64}$$

$$u(1)=0 \tag{1.65}$$

What is the analytical solution to equations (1.63) to (1.65)?

1.2.3 Parabolic PDEs

If each PDE in equation (1.45) has only a first-order derivative in t and a second-order derivative in $\bar{x}$

$$\bar{u}_t = \bar{f}(\bar{x}, t, \bar{u}, \bar{u}_{\bar{x}\bar{x}}) \tag{1.66}$$

$$\bar{u}(\bar{x}, t_o) = \bar{g}(\bar{x}) \tag{1.67}$$

$$\bar{h}(\bar{x}_b, t, \bar{u}(\bar{x}_b, t), \bar{u}_{\bar{x}}(\bar{x}_b, t)) = \bar{0} \tag{1.68}$$

we classify this problem as *parabolic*; examples include equation (1.59) which is *Fourier's second law in one dimension* (or the one-dimensional diffusion or heat conduction equation in Cartesian coordinates), and equation (1.57), which is *Fourier's second law in two dimensions* (or the two-dimensional diffusion or heat conduction equation in Cartesian coordinates).

Other examples include

$$\frac{\partial u}{\partial t} = \frac{1}{r}\frac{\partial}{\partial r}\left(r\frac{\partial u}{\partial r}\right) + \frac{1}{r^2}\frac{\partial^2 u}{\partial \theta^2} + \frac{\partial^2 u}{\partial z^2} \tag{1.69}$$

$$\frac{\partial u}{\partial t} = \frac{1}{r^2}\frac{\partial}{\partial r}\left(r^2\frac{\partial u}{\partial r}\right) + \frac{1}{r^2 \sin^2\phi}\frac{\partial^2 u}{\partial \theta^2} + \frac{1}{r^2 \sin\phi}\frac{\partial}{\partial \phi}\left(\sin\phi\frac{\partial u}{\partial \phi}\right) \tag{1.70}$$

Equations (1.69) and (1.70) are *Fourier's second law in cylindrical and spherical coordinates*, respectively. Note that in each case the PDE is first order in t and second order in the spatial independent variables which, again, are the characteristics of a parabolic PDE. Also, these equations have variable coefficients that are functions of the independent variables; for example

$$\frac{1}{r^2}$$

is a variable coefficient in the term

$$\frac{1}{r^2}\frac{\partial^2 u}{\partial \theta^2}$$

of equation (1.69). Similarly, the term

$$\frac{1}{r^2 \sin^2\phi}$$

is a variable coefficient in the term

$$\frac{1}{r^2 \sin^2\phi}\frac{\partial^2 u}{\partial \theta^2}$$

of equation (1.70). However, equations (1.69) and (1.70) are linear since all of the terms contain u and its various derivatives to the first degree.

1.2.4 First-Order Hyperbolic PDEs

If each PDE in equation (1.45) has only a first-order derivative in t and a first-order derivative in $\bar{x}$,

$$\bar{u}_t = \bar{f}(\bar{x}, t, \bar{u}, \bar{u}_{\bar{x}}) \tag{1.71}$$

$$\bar{u}(\bar{x}, t_o) = \bar{g}(\bar{x}) \tag{1.72}$$

$$\bar{h}(\bar{x}_b, t, \bar{u}(\bar{x}_b, t)) = \bar{0} \tag{1.73}$$

we classify this problem as *first-order hyperbolic*. As examples, we have

$$u_t = -vu_x \tag{1.74}$$

$$u_t = -uu_x = -\left(\frac{u^2}{2}\right)_x \tag{1.75}$$

Equations (1.74) and (1.75) are deceptively simple in appearance; in fact, they are among the most difficult of the three types of PDEs (elliptic, parabolic, hyperbolic) to integrate numerically, as we will observe in Chapter 3. Equation (1.74) is the *advection equation*, so named because it models flow, in this case with constant velocity v. Equation (1.75) is a nonlinear variant of (1.74). Why is it nonlinear or not first degree?

Equation (1.74) is first order in t and x. It can be converted to a PDE second order in t and x by differentiation. If we differentiate with respect to t we obtain

$$u_{tt} = -vu_{xt}$$

Similarly, differentiation of equation (1.74) with respect to x gives

$$u_{tx} = -vu_{xx}$$

If we now assume the mixed partials are equal (i.e., the order of differentiation does not matter, so $u_{xt} = u_{tx}$), these two equations can be combined to give

$$u_{tt} = v^2 u_{xx} \tag{1.76}$$

which is a second-order hyperbolic PDE or the wave equation. We therefore consider this class of PDEs next.

1.2.5 Second-Order Hyperbolic PDEs

If each PDE in equation (1.42) has only a first-order derivative in t and a second-order derivative in x but the derivatives in t are related so that when two PDEs are combined, a second-order derivative in t results,

$$u_{1t} = u_2$$

$$u_{2t} = f_2(\bar{x}, t, u_1, u_2, \ldots, u_n, , u_{1\bar{x}\bar{x}}, u_{2\bar{x}\bar{x}}, \ldots, u_{n\bar{x}\bar{x}},) \tag{1.77}$$
$$\vdots \qquad\qquad \vdots$$

$$u_1(\bar{x}, t_o) = g_1(\bar{x}), u_2(\bar{x}, t_o) = g_2(\bar{x}), \ldots \tag{1.78}$$

$$\begin{aligned} &h_1(\bar{x}_b, t, u_1(\bar{x}_b, t), u_2(\bar{x}_b, t) \ldots, u_n(\bar{x}_b, t), u_{1\bar{x}}(\bar{x}_b, t), u_{2\bar{x}}(\bar{x}_b, t), \ldots, u_{n\bar{x}}(\bar{x}_b, t), \ldots) = 0 \\ &h_2(\bar{x}_b, t, u_1(\bar{x}_b, t), u_2(\bar{x}_b, t) \ldots, u_n(\bar{x}_b, t), u_{1\bar{x}}(\bar{x}_b, t), u_{2\bar{x}}(\bar{x}_b, t), \ldots, u_{n\bar{x}}(\bar{x}_b, t), \ldots) = 0 \\ &\vdots \qquad\qquad \vdots \end{aligned} \tag{1.79}$$

we classify this problem as *second-order hyperbolic*; the wave equation, equation (1.76), is an example since it can be written as two first-order PDEs in t in the format of equations (1.76)

$$u_{1t} = u_2 \tag{1.80}$$

$$u_{2t} = v^2 u_{1xx} \tag{1.81}$$

As another example, the *wave equation in three dimensions* (Cartesian coordinates) is

$$u_{tt} = c^2(u_{xx} + u_{yy} + u_{zz}) \tag{1.82}$$

where again c is a velocity.

The geometric classification of Sections 1.2.2 to 1.2.5 is not presented in the conventional way for two reasons:

1. The conventional classification involves a single, two-dimensional linear PDE, for which special cases lead to mathematical definitions of the elliptic, parabolic, and hyperbolic cases [Strang (1986), pp. 246–247]. While this approach is interesting mathematically, it is too restrictive to be of much use in many realistic PDE applications.

2. The approach in Sections 1.2.2 to 1.2.5 is less rigorous mathematically but more useful in classifying PDEs to serve as a guide in their numerical solution. Thus systems of nonlinear PDEs with an initial value independent variable (time-like variable) and one, two, and three boundary value independent variables (spatial dimensions) can be classified. Also, PDEs with mixed characteristics can be included in this classification.

We now consider some examples of PDEs of mixed characteristics.

1.2.6 Mixed PDEs

If each PDE in equation (1.45) has only a first-order derivative in t and first- and second-order derivatives in $\bar{x}$, we classify this problem as *hyperbolic parabolic*. Examples are the one-dimensional, linear PDE

$$u_t = -vu_x + Du_{xx} \tag{1.83}$$

and the one-dimensional, nonlinear PDE

$$u_t = -uu_x + Du_{xx} \tag{1.84}$$

Equation (1.83) is a *convective diffusion* PDE since it models flow with velocity v and diffusion with diffusivity D. Equation (1.84) is *Burgers' equation* in one dimension; it is widely used as a test problem for numerical integration algorithms since it (1) is nonlinear, (2) has an exact solution which can be used to evaluate numerical solutions, and (3) can be changed through a spectrum of hyperbolic and parabolic characteristics by varying D. If $D = 0$, equations (1.83) and (1.84) reduce to hyperbolic equations (1.74) and (1.75), respectively. If the diffusion term is much larger than the convective term, equations (1.83) and (1.84) are predominantly parabolic, and in the limit of no convection, become equation (1.59). In fact, equation (1.83) can be transformed to equation (1.59) with dependent variable w through the change of variable $u = we^{(v/(2D)x)}$ [Myint-U and Debnath (1987), p. 49].

The following are other examples of mixed PDEs

$$u_{tt} = au_{xx} + bu_{xt} + cu \tag{1.85}$$

$$u_{tt} + u_t = u_{xx} + bu \tag{1.86}$$

$$u_t = au_x + bu_{xxx} \tag{1.87}$$

Equation (1.85) is the *Klein-Gordon equation* for $b = 0$. Equation (1.86) is the *time-dependent Maxwell equation* or the *damped wave equation* with $b = 0$, and the *telegraph equation* with $b \neq 0$. Equation (1.87) is the *linearized Korteweg-deVries equation* [Myint-U and Debnath (1987), pp. 395–396]. Thus, an essentially unlimited variety of PDEs is possible. Additional variations are possible through different combinations of boundary conditions, both linear and nonlinear, and through mixed systems of ODEs/DAEs/PDEs. The method of lines is a methodology that accommodates most of the possible forms of ODEs/DAEs/PDEs; we will attempt to demonstrate its versatility and flexibility in subsequent chapters.

We conclude this chapter with one final example, the *cubic Schrödinger equation*, which is complex, both mathematically and somewhat figuratively [Sanz-Serna and Christie (1986)]

$$iu_t + u_{xx} + q|u|^2 u = 0 \tag{1.88}$$

$i = \sqrt{-1}$ and therefore $u(x,t)$ is complex. Equation (1.88) has the solution (for $q = 1$)

$$u(x,t) = \sqrt{2}e^{i(0.5x+0.75t)}sech(x-t) \tag{1.89}$$

and has the interesting characteristc that $|u|$ is a wave of height $\sqrt{2}$ initially centered at $x = 0$ which travels to the right at speed 1 without changing shape, i.e., *a soliton*. We will develop a method of lines solution to equation (1.88) in Chapter 4 and compare it with the analytical solution, equation (1.89).

References

Ascher, U. M., J. Christensen and R. D. Russell (1981), "COLSYS: Collocation Software for Boundary Value ODEs", ACM Trans. Math. Software, June.

Ascher, U. M. and L. R. Petzold (1992), "Numerical Methods for Boundary Value Problems in Differential-Algebraic Equations", in *Recent Developments in Numerical Methods and Software for ODEs/DAEs/PDEs*, G. D. Byrne and W. E. Schiesser (eds.), World Scientific, Teaneck, NJ.

Ascher, U. M. and R. I. Spiteri (1992), "Collocation Software for Boundary Value Differential-Algebraic Equations," Technical Report 92-18, Department of Computer Science, University of British Columbia, Vancouver, B.C., V6T 1Z2, Canada.

Brenan, K. E., S. L. Campbell and L. R. Petzold (1989), *Numerical Solution of Initial-Value Problems in Differential-Algebraic Equations*, North-Holland, New York.

Byrne, G. D. and A. C. Hindmarsh (1987), "Stiff ODE Solvers: A Review of Current and Coming Attractions", *J. Comput. Phys.*, **70**, pp. 1–62.

Myint-U, Tyn and L. Debnath (1987), *Partial Differential Equations for Scientists and Engineers*, Third Edition, North Holland, New York.

Sanz-Serna, J. M. and I. Christie (1986), "A Simple Adaptive Technique for Nonlinear Wave Problems", *J. Comput. Phys.*, **67**, pp. 348–360.

Schiesser, W. E. (1991), *The Numerical Method of Lines Integration of Partial Differential Equations*, Academic Press, San Diego.

Strang, G. (1986), *Introduction to Applied Mathematics*, Wellesley-Cambridge Press, Wellesley, MA.

chapter two

The Numerical Integration of Initial Value Ordinary Differential Equations

In Chapter 1, Section 1, we reviewed the mathematical forms of the ODEs commonly used in scientific and engineering applications, i.e., explicit, semi-implicit, and implicit initial and boundary value ODEs. ODEs are a natural way to express many physical problems, and we routinely write sets of ODEs which can number in the hundreds or thousands. Prior to the development of digital computers, this would have been an academic exercise since we would have had no way to solve such large sets of equations (recall again that the most we can do with analytical methods is about four linear, constant coefficient ODEs). Now, however, with the digital computer this bottleneck has been eliminated, and we are essentially limited only by our imagination in writing ODE models. In principle we can solve any ODE model using *numerical integration*. Obviously, then, we require some knowledge of numerical integration of ODEs to use digital computers effectively; this is topic of this chapter.

First, we state our preference in approaching the general problem of the numerical integration of initial value ODEs. One approach would be to select one of the classical integration algorithms, e.g., Euler's method, or the modified Euler method, then code (program) the algorithm, test the code, and finally, apply it to the ODE problem system of interest. While this approach has the apparent advantages that the analyst will understand the details of the code and therefore presumably be able to use it effectively, it also has several disadvantages which we think outweigh the advantages: (1) the process of selecting an algorithm and coding it requires some knowledge of numerical analysis and programming which might not be part of the background of the scientist or engineer seeking a solution to a specific ODE problem; (2) the ODE problem may have characteristics which the analyst does not appreciate, e.g., stiffness, so that an ineffective algorithm is selected, e.g., a nonstiff integrator for a stiff problem; (3) even if an appropriate algorithm is selected, e.g., the backward differentiation formulas (BDF) for a stiff problem, the algorithm may be relatively complicated and require insights for effective implementation (coding) that the analyst does not have, e.g., understanding the details of step size and order selection for the BDFs to start a solution and maintain user prescribed error tolerances, the exploitation of the ODE Jacobian matrix structure; (4) the analyst will generally not have the experience to anticipate what can go wrong during a numerical solution and how to program the algorithm to deal with such problems, e.g., inappropriate error tolerances and excessive computational work due to stiffness, which should at least be flagged with error messages.

The preceding comments imply that these issues of the complexity of numerical integration have been dealt with, and in fact, this is the case. Extensive research on numerical methods for the integration of ODEs has been carried out and reported in the numerical analysis literature (see, for example, the reviews by Byrne and Hindmarsh (1987) and Shampine et al. (1976)). This research represents many years of work by leading numerical analysts, and it would be, to say the least, presumptuous to think that an ODE integrator of the quality of the current state-of-the-art integrators could be produced with reasonable effort by starting with a search for an appropriate algorithm, with the intention of developing an ODE integration code. In other words, what we are proposing is the use of the current quality library integrators, which have achieved the status of international standards.

Scientists and engineers have a tendency to think they can solve a problem better than others, if for no other reason than they formulated the problem and therefore they best understand (possibly) the problem. While this may be true with regard to problem formulation, it is almost certainly not true with regard to the computer solution, which, in general, requires in-depth knowledge of integration algorithms and their computer implementation. In other words, it is generally far more efficient to assume that others have this knowledge, and therefore we should make use of their work. This is the approach we will follow in this book, i.e., we will consider briefly some of the characteristics of ODEs and their integration, primarily so we can understand and more effectively use library integrators. Please note that we are not advocating the uninformed use of such integrators. Rather, some knowledge of ODE equation characteristics and numerical methods will invariably lead to more effective use of existing integrators. We will also find that the use of quality integrators will enhance this knowledge and contribute significantly to our experience by reporting results that we would not normally expect and look for, e.g., computational statistics like the number of derivative evaluations and Jacobian matrix updates, algorithm order, and integration step size. We now proceed to some basic notions about ODE numerical integration.

2.1 *An Example of What Can Go Wrong*

We begin the discussion of the numerical integration of ODEs by considering an example of a finite difference approximation for a system of four ODEs which seems completely reasonable, but which, in fact, gives incorrect solutions. The following equations of motion for a simple orbit problem are discussed by Feynman (1963)

$$m\frac{d^2x}{dt^2} = F_x, \qquad m\frac{d^2y}{dt^2} = F_y \tag{2.1)(2.2}$$

$$F_x = -F\cos\theta, \qquad F_y = -F\sin\theta \tag{2.3)(2.4}$$

$$F = \frac{GMm}{r^2} \tag{2.5}$$

$$\sin\theta = \frac{y}{r} = \frac{y}{\sqrt{x^2+y^2}}, \qquad \cos\theta = \frac{x}{r} = \frac{x}{\sqrt{x^2+y^2}} \tag{2.6)(2.7}$$

The combination of equations (2.1) to (2.7) then gives the ODEs to be integrated numerically

$$\frac{d^2x}{dt^2} = \frac{-x}{\{x^2+y^2\}^{3/2}}, \qquad \frac{d^2y}{dt^2} = \frac{-y}{\{x^2+y^2\}^{3/2}} \tag{2.8)(2.9}$$

where we take $GM = 1$ for simplicity.

Equations (2.8) and (2.9) are two second-order ODEs which can be written as four first-order ODEs in x, y, v_x, and v_y

$$\frac{dx}{dt} = v_x, \qquad \frac{dy}{dt} = v_y \qquad (2.10)(2.11)$$

$$\frac{dv_x}{dt} = \frac{-x}{\{x^2+y^2\}^{3/2}}, \qquad \frac{dv_y}{dt} = \frac{-y}{\{x^2+y^2\}^{3/2}} \qquad (2.12)(2.13)$$

The analytical solution of equations (2.10) to (2.13) is not known. We therefore consider the numerical integration of these ODEs by finite difference approximations as suggested by Greenspan (1990). For example, equation (2.10) can be approximated as

$$\frac{x_{k+1} - x_k}{\Delta t} = \frac{v_{k+1,x} + v_{k,x}}{2} \qquad (2.14)$$

where the subscript k denotes a point along the numerical solution at t; the numerical solution at $k+1$ corresponds to $t + \Delta t$ where Δt is the integration step of the independent variable t. Note that in the limit of $\Delta t \to 0$, equation (2.14) reverts to equation (2.10).

Similarly, equations (2.11) to (2.13) can be approximated as

$$\frac{y_{k+1} - y_k}{\Delta t} = \frac{v_{k+1,y} + v_{k,y}}{2} \qquad (2.15)$$

$$\frac{v_{k+1,x} - v_{k,x}}{\Delta t} = \frac{1}{(x_k^2+y_k^2)^{1/2}(x_{k+1}^2+y_{k+1}^2)^{1/2}} \cdot \frac{-0.5\,\{x_{k+1}+x_k\}}{0.5\,\{(x_k^2+y_k^2)^{1/2}+(x_{k+1}^2+y_{k+1}^2)^{1/2}\}} \qquad (2.16)$$

$$\frac{v_{k+1,y} - v_{k,y}}{\Delta t} = \frac{1}{(x_k^2+y_k^2)^{1/2}(x_{k+1}^2+y_{k+1}^2)^{1/2}} \cdot \frac{-0.5\,\{y_{k+1}+y_k\}}{0.5\,\{(x_k^2+y_k^2)^{1/2}+(x_{k+1}^2+y_{k+1}^2)^{1/2}\}} \qquad (2.17)$$

Again, in the limit of $\Delta t \to 0$, equations (2.15) to (2.17) reduce to equations (2.11) to (2.13), respectively.

Alternatively, we can express equations (2.10) to (2.13) in terms of the state variables y_1, y_2, y_3, and y_4, which will facilitate the subsequent programming

$$y_1 = x, \quad y_2 = dx/dt, \quad y_3 = y, \quad y_4 = dy/dt$$

Thus,

$$dy_1/dt = y_2, \qquad dy_3/dt = y_4 \qquad (2.18)(2.19)$$

$$\frac{dy_2}{dt} = \frac{-y_1}{\left(y_1{}^2+y_3{}^2\right)^{3/2}}, \qquad \frac{dy_4}{dt} = \frac{-y_3}{\left(y_1{}^2+y_3{}^2\right)^{3/2}} \qquad (2.20)(2.21)$$

and we now specify the initial conditions proposed by Feynman (1963)

$$y_1(0) = 0.5, \quad y_2(0) = 0, \quad y_3(0) = 0, \quad y_4(0) = 1.63 \qquad (2.22)\text{–}(2.25)$$

The finite difference approximations of equations (2.18) to (2.21) are

$$\frac{y_{1,k+1} - y_{1,k}}{\Delta t} = \frac{y_{2,k+1} + y_{2,k}}{2} \tag{2.26}$$

$$\frac{y_{3,k+1} - y_{3,k}}{\Delta t} = \frac{y_{4,k+1} + y_{4,k}}{2} \tag{2.27}$$

$$\frac{y_{2,k+1} - y_{2,k}}{\Delta t} = \frac{1}{(y_{1,k}^2 + y_{3,k}^2)^{1/2}(y_{1,k+1}^2 + y_{3,k+1}^2)^{1/2}} \cdot \frac{-0.5\left\{y_{1,k+1} + y_{1,k}\right\}}{0.5\left\{(y_{1,k}^2 + y_{3,k}^2)^{1/2} + (y_{1,k+1}^2 + y_{3,k+1}^2)^{1/2}\right\}} \tag{2.28}$$

$$\frac{y_{4,k+1} - y_{4,k}}{\Delta t} = \frac{1}{(y_{1,k}^2 + y_{3,k}^2)^{1/2}(y_{1,k+1}^2 + y_{3,k+1}^2)^{1/2}} \cdot \frac{-0.5\left\{y_{3,k+1} + y_{3,k}\right\}}{0.5\left\{(y_{1,k}^2 + y_{3,k}^2)^{1/2} + (y_{1,k+1}^2 + y_{3,k+1}^2)^{1/2}\right\}} \tag{2.29}$$

with the initial conditions

$$y_{1,0} = 0.5, \quad y_{2,0} = 0, \quad y_{3,0} = 0, \quad y_{4,0} = 1.63 \qquad (2.30)\text{–}(2.33)$$

Equations (2.26) to (2.33) constitute a system of nonlinear algebraic equations which will be solved numerically for a series of values of t starting at $t = 0$ (i.e., a series of values of k, $k = 0, 1, 2, \ldots$).

Since we do not have an analytical solution to the ODE problem system, equations (2.18) to (2.25), we seek other means for checking the numerical solution produced by equations (2.26) to (2.33). A frequently used criterion is to check that the numerical solution conserves energy. The purpose of the following excerise is to demonstrate that this is a necessary, but not sufficient, condition, i.e., the numerical solution from equations (2.26) to (2.33) does conserve energy, but can, in fact, be seriously in error. The total energy of the system defined by equations (2.8) and (2.9) (or equivalently, equations (2.18) to (2.21)) is obtained by first multiplying equations (2.8) and (2.9) by dx/dt and dy/dt, respectively

$$\frac{dx}{dt}\frac{d^2x}{dt^2} = \frac{-\dfrac{dx}{dt}x}{\left(x^2+y^2\right)^{3/2}}, \qquad \frac{dy}{dt}\frac{d^2y}{dt^2} = \frac{-\dfrac{dy}{dt}y}{\left(x^2+y^2\right)^{3/2}} \qquad (2.34)(2.35)$$

Adding equations (2.34) and (2.35) gives

$$\frac{dx}{dt}\frac{d^2x}{dt^2} + \frac{dy}{dt}\frac{d^2y}{dt^2} = -\left\{\frac{\dfrac{dx}{dt}x}{\left(x^2+y^2\right)^{3/2}} + \frac{\dfrac{dy}{dt}y}{\left(x^2+y^2\right)^{3/2}}\right\} \tag{2.36}$$

If equation (2.36) is integrated with respect to t,

$$\int_0^t \left\{\frac{dx}{dt}\frac{d^2x}{dt^2} + \frac{dy}{dt}\frac{d^2y}{dt^2}\right\} dt = -\int_0^t \left\{\frac{\dfrac{dx}{dt}x}{\left(x^2+y^2\right)^{3/2}} + \frac{\dfrac{dy}{dt}y}{\left(x^2+y^2\right)^{3/2}}\right\} dt$$

or

$$\int_0^t \frac{d}{dt}\left\{(1/2)\left(\frac{dx}{dt}\right)^2 + (1/2)\left(\frac{dy}{dt}\right)^2\right\}dt = \int_0^t \frac{d}{dt}\left\{\frac{1}{(x^2+y^2)^{1/2}}\right\}dt \tag{2.37}$$

If the integrations in equation (2.37) are expressed in terms of a constant of integration, E, we have

$$\left\{(1/2)\left(\frac{dx}{dt}\right)^2 + (1/2)\left(\frac{dy}{dt}\right)^2\right\} - \left\{\frac{1}{(x^2+y^2)^{1/2}}\right\} = E$$

or

$$\left\{(1/2)\left(\frac{dx}{dt}\right)^2 + (1/2)\left(\frac{dy}{dt}\right)^2\right\} - (1/r) = E$$

or in terms of y_1, y_2, y_3, and y_4,

$$(1/2)\left\{{y_2}^2 + {y_4}^2\right\} - (1/r) = E \tag{2.38}$$

where $r = \sqrt{x^2+y^2}$. The left-hand side is the sum of the kinetic energy, $\frac{1}{2}\left\{{y_2}^2 + {y_4}^2\right\}$, and potential energy, $-(1/r)$, of the system, and this total energy must be constant, i.e., the system is *conservative*. We now consider the solution of equations (2.26) to (2.33) which does conserve energy according to equation (2.38).

To compute a solution to equations (2.26) to (2.29), subject to initial conditions (2.30) to (2.33), we must develop a method for the solution of systems of nonlinear algebraic equations. Generally, this is not an easy problem, which is really an understatement, i.e., the solution of systems of nonlinear algebraic and transcendental equations is generally considered to be one of the most difficult problems in the field of scientific computation. Much effort has been devoted to the solution of systems of nonlinear equations, and fortunately, this effort has produced quality library subroutines that can easily be called; the alternative is to develop a computer code, perhaps starting with Newton's method or a variant. This is not recommended for the reasons we discussed at the beginning of the chapter, i.e., the use of quality library routines written by experts in numerical analysis is usually far more efficient and effective than attempting to write a code starting with a basic numerical algorithm.

We therefore turn to subroutine `SNSQE` for systems of nonlinear equations [Kahaner et al. (1989)] which is based on a variant of Newton's method called *Powell's hybrid method*. `SNSQE` is actually a combination of two routines taken from the `MINPACK` library (see Kahaner et al. for details). Since Powell's hybrid method is based on Newton's method, it requires the Jacobian matrix of the system of nonlinear equations. The Jacobian matrix can be provided either analytically in a subroutine called by `SNSQE`, or the user can specify that the Jacobian matrix should be computed by `SNSQE` numerically using forward finite differences as approximations to the partial derivatives of the Jacobian matrix; this latter approach (numerical Jacobian) is almost always selected in realistic scientific and engineering problems since programming of the analytical Jacobian is usually impractical (recall that for a system of n equations, the Jacobian matrix has n^2 elements, and this number grows very rapidly with n). A main program, `ORBIT`, to call subroutine `SNSQE` for the solution of equations (2.26) to (2.33) is listed in Program 2.1a.

The following points should be noted about Program 2.1a:

1. The call to SNSQE is inside DO loop 3. This loop has the primary purpose of stepping the solution of equations (2.26) to (2.33) through successive values of t, starting with $t = 0$, i.e., $k = 0, 1, 2, \ldots$ in equations (2.26) to (2.29). The arguments of SNSQE are discussed below after the other essential details of Program 2.1a are covered.

2. The PARAMETER statement

```
      PARAMETER (N=4,LWA=(3*N**2+13*N)/2)
```

   specifies four equations are to be solved (N=4), in this case, equations (2.26) to (2.29); then, the length of the real work array, WORK, required by SNSQE, is computed, LWA=(3*N**2+13*N)/2. Work arrays are frequently used in library routines such as SNSQE to provide generality in the routines, usually so that the coding in the routines is essentially independent of the problem size, in this case, the number of equations to be solved, N. Note also that LWA is proportional to N**2 to accommodate the Jacobian matrix and the associated linear algebra of Powell's hybrid method, as discussed previously, i.e., the Jacobian matrix is of size n^2.

3. A COMMON block follows the PARAMETER statement in which some of the arrays are sized by the PARAMETER statement:

```
C...
C...  VARIABLES AND PARAMETERS ARE IN COMMON FOR USE IN
C...  SUBORDINATE SUBROUTINES
      COMMON /T/        T,  NSTOP,  NORUN
     +       /Y/    Y(N)
     +       /F/   YT(N)
     +       /C/     Y10,      Y20,      Y30,      Y40,      TP
     +       /I/      IP,       NP
```

   This COMMON block may seem excessively complicated, but it has a format with a well-defined purpose, and this format will be used consistently throughout the remainder of the book. The contents of the various COMMON areas are briefly described below:

   (a) COMMON/T/ contains the independent variable, t, of equations (2.8) and (2.9); NSTOP and NORUN will be discussed in subsequent examples.

   (b) COMMON/Y/ contains the vector of dependent variables in equations (2.26) to (2.29), Y(N), i.e., y_1 to y_4.

   (c) COMMON/F/ contains the *residuals* of equations (2.26) to (2.29), that is, the functions of y_1 to y_4 which result from taking all of the terms in equations (2.26) to (2.29) to one side of the equations; these residuals are programmed in a subroutine called by SNSQE, and the purpose of SNSQE is to find values of y_1 to y_4 that make these residuals zero, which will then be the solutions to equations (2.26) to (2.29). COMMON/T/, /Y/, and /F/ are the essential COMMON areas in all of the examples to follow; other COMMON areas can be added as required, but they must not have any of the reserved names /T/, /Y/, or /F/.

```
      PROGRAM ORBIT
C...
C...  GREENSPAN ORBIT PROBLEM
C...
C...  A SYSTEM OF FOUR NONLINEAR ALGEBRAIC EQUATIONS PROPOSED BY
C...  GREENSPAN, DONALD, A COUNTEREXAMPLE OF THE USE OF ENERGY AS
C...  A MEASURE OF COMPUTATIONAL EFFICIENCY, J. COMPUTATIONAL
C...  PHYSICS, V. 91, 490-494 (1990).  THE NUMERICAL SOLUTION IS
C...  BY NONLINEAR EQUATION SOLVER SNSQE TAKEN FROM KAHANER, DAVID,
C...  CLEVE MOLER AND STEPHEN NASH, NUMERICAL METHODS AND SOFTWARE,
C...  CHAPTER 7, PRENTICE HALL, ENGLEWOOD CLIFFS, NJ, 1989
C...
C...  DECLARATIVE STATEMENTS (NOTE - N SHOULD EQUAL NEQN READ FROM
C...  THE DATA FILE VIA FORMAT 902)
      PARAMETER (N=4,LWA=(3*N**2+13*N)/2)
C...
C...  VARIABLES AND PARAMETERS ARE IN COMMON FOR USE IN SUBORDINATE
C...  SUBROUTINES
      COMMON /T/      T,  NSTOP,  NORUN
     +       /Y/   Y(N)
     +       /F/  YT(N)
     +       /C/    Y10,     Y20,     Y30,     Y40,      TP
     +       /I/     IP,      NP
C...
C...  WORK ARRAY FOR SNSQE
      REAL WA(LWA)
C...
C...  FILE TO HOLD DOCUMENTATION TITLE
      CHARACTER TITLE(20)*4
C...
C...  EXTERNALS CALLED BY SNSQE
      EXTERNAL DERV, JAC
C...
C...  INIITIALIZE THE RUN COUNTER
      DATA NORUN/0/
C...
C...  INPUT/OUTPUT UNIT NUMBERS AND FILES
         NI=5
         NO=6
         OPEN(NI,FILE='DATA'  ,STATUS='OLD')
         OPEN(NO,FILE='OUTPUT',STATUS='NEW')
C...
C...  START THE NEXT RUN
4     NORUN=NORUN+1
C...
C...  READ DATA FILE, STARTING WITH DOCUMENTATION TITLE
      READ(NI,900)(TITLE(I),I=1,20)
C...
C...  CHECK FOR END OF RUNS (IN THE DATA FILE)
      IF((TITLE(1).EQ.'END ').AND.
```

Program 2.1a Main Program ORBIT to Call Subroutine SNSQE for the Solution of Equations (2.26) to (2.33). *Continued next pages.*

```
     +    (TITLE(2).EQ.'OF R').AND.
     +    (TITLE(3).EQ.'UNS '))THEN
C...
C...      END OF RUNS WAS READ, SO TERMINATE EXECUTION
          STOP
C...
C...      CONTINUE RUN WITH CURRENT SET OF DATA
      END IF
C...
C...  READ INITIAL TIME, FINAL TIME AND PRINT INTERVAL OF TIME
      READ(NI,901)T0,TF,TP
C...
C...  READ NUMBER OF EQUATIONS, ERROR TOLERANCE
      READ(NI,902)NEQN,TOL
C...
C...  PRINT DATA SUMMARY
      WRITE(NO,903)NORUN,(TITLE(I),I=1,20),T0,TF,TP,NEQN,TOL
C...
C...  INITIAL CONDITIONS
         T=T0
         CALL INITAL
C...
C...  PRINT THE INITIAL CONDITIONS
      CALL PRINT(NI,NO)
C...
C...  PARAMETERS FOR SNSQE
         IOPT=2
         NPRINT=0
C...
C...  NUMBER OF STEPS AND POINTS ALONG THE SOLUTION
         NSTEPS=INT((TF-T0)/TP)
         NP=NSTEPS+1
C...
C...  STEP THROUGH THE NSTEPS STEPS
      DO 3 NSTEP=1,NSTEPS
C...
C...  TAKE A STEP ALONG THE SOLUTION
      CALL SNSQE(DERV,JAC,IOPT,N,Y,YT,TOL,NPRINT,INFO,WA,LWA)
C...
C...  UPDATE TIME
      T=T+TP
C...
C...  PRINT THE SOLUTION
      CALL PRINT(NI,NO)
C...
C...  UPDATE THE EQUATIONS
         Y10=Y(1)
         Y20=Y(2)
         Y30=Y(3)
         Y40=Y(4)
```

Program 2.1a *Continued.*

```
C...
C...   TAKE THE NEXT TIME STEP
3      CONTINUE
C...
C...   ALL TIME STEPS HAVE BEEN TAKEN (RUN IS COMPLETE), SO GO ON TO
C...   THE NEXT RUN BY READING THE NEXT SET OF DATA
       GO TO 4
900    FORMAT(20A4)
901    FORMAT(3E10.0)
902    FORMAT(I5,20X,E10.0)
903    FORMAT(1H1,/,' RUN NO. ',I3,' - ',20A4,//,
     + '   INITIAL VALUE OF TIME = ',E10.3,//,
     + '     FINAL VALUE OF TIME = ',E10.3,//,
     + ' PRINT INTERVAL OF TIME = ',E10.3,//,
     + ' SOLUTION BY SUBROUTINE SNSQE',//,
     + '    NUMBER OF EQUATIONS = ',I3,//,
     + '    ERROR TOLERANCE = ',E10.3,//)
       END
```

Program 2.1a *Continued.*

(d) `COMMON/C/` has the initial values of y_1 to y_4 at $t = 0$ $(k = 0)$, `Y10` to `Y40`, as defined by initial conditions (2.30) to (2.33), which are also the base values of y_1 to y_4 (at t or k) when the next step is taken along the solution (to $t+\Delta t$ or $k+1$) according to equations (2.26) to (2.29); `TP` in the interval in t at which the solution is to printed.

(e) `COMMON/I/` contains two integer counters that control the operation of program `ORBIT`, particularly the output. Again, `COMMON/C/` and `/I/` are optional.

4. The `EXTERNAL` statement defines two subroutines `DERV` and `JAC`. `DERV` defines the equation residuals as discussed in (c) above, and is the first argument of `SNSQE` (the `EXTERNAL` statement permits a subroutine (`DERV` is this case), to be passed as an argument of another subroutine (`SNSQE`), and has the advantage that the name of the argument subroutine can be selected by the programmer, e.g., `DERV`). `JAC` defines the analytical Jacobian matrix of the nonlinear equation system, and is also an argument of `SNSQE`. In the present case, the option for computation of the numerical Jacobian matrix by `SNSQE` is selected so that `JAC` is not actually called by `SNSQE`; however, a three-line dummy subroutine `JAC` is provided to satisfy the loader of the computer when main program `ORBIT` is executed (this use of a dummy routine may not be required for a particular computer and is presented here to indicate what may be required).

5. Input and output unit numbers are defined as 5 and 6 respectively, and the corresponding files are `DATA` and `OUTPUT` as defined in the `OPEN` statements.

6. Program `ORBIT` is set up for multiple runs (solutions) of equations (2.26) to (2.29), and the first run is initiated with `NORUN=1` at statement 4. A documentation title is read from file `DATA` into array `TITLE` via `FORMAT 900`, and this documentation title is then checked against the characters `END OF RUNS`. If these characters have been read from file `DATA` the `STOP` statement is executed; otherwise, the title is printed as the first line of output in file `OUTPUT` and the run continues by

reading the initial, final, and print interval values of t (the independent variable in equations (2.8) and (2.9)) via `FORMAT 901`. Then the number of equations and the error tolerance to be used by `SNSQE` in finding a solution to equations (2.26) to (2.29) is read from `DATA` via `FORMAT 902`. All of these data from file `DATA` are summarized by a write statement via `FORMAT 903`.

7. Independent variable t is then initialized to the starting value read from file `DATA` (`T=T0`), and the stepping along the solution at successive values of t begins, that is, for $t = \Delta t$, $t = t + \Delta t$, $t = t + 2\Delta t, \ldots$ Subroutine `INITAL` is called to set the initial conditions of equations (2.30) to (2.33) and subroutine `PRINT` is called to print these initial conditions (subroutines `INITAL`, `DERV`, `PRINT`, and `JAC` will be considered subsequently). The communication between the various subroutines is through the `COMMON` areas discussed previously, and through arguments discussed subsequently.

8. Two input parameters for `SNSQE` are defined, `IOPT=2` indicating a numerical Jacobian will be used, and `NPRINT=0` indicating `SNSQE` will not generate any printed output. A detailed discussion of the arguments of `SNSQE` is given in Kahaner et al. (1989) and in the prologue of subroutine `SNSQE`.

9. The number of steps along the solution, corresponding to $k = 1, 2, \ldots$ in equations (2.26) to (2.29) is computed as the difference between the final and initial values of t, divided by the print (output) interval of t. This number of steps, `NSTEPS`, is then the upper limit of `DO` loop 3 which contains the call to `SNSQE`. Thus, each pass through `DO` loop 3 corresponds to another step along the solution to equations (2.26) to (2.29).

10. Upon the return from `SNSQE`, the solution to equations (2.26) to (2.29) is contained in array `Y`. After updating t corresponding to these new solution values, the solution is printed by a call to `PRINT`; the next step along the solution is then initiated by setting the new current solution to the base values (the values of y_1 to y_4 at $k+1$ in equations (2.26) to (2.29) now become the values at k) and cycling through `DO` loop 3 again. In this way, the complete solution is computed from $t = 0$ to the final value of t.

This program structure will be used throughout the examples in this book. In particular, subroutines `INITAL`, `DERV`, and `PRINT` will be used to initialize the calculations, define the problem system equations and print the solution, respectively; communication between these subroutines will be through `COMMON/T/`, `/Y/`, and `/F/`. Overall control of the calculations will be by data read from file `DATA`, and the output from the calculations will be on file `OUTPUT`.

Subroutine `INITAL` to define the initial conditions, equations (2.30) to (2.33) is listed in Program 2.1b.

Of particular importance is the necessity to define initial values for all of the elements of array `Y(N)` which is in `COMMON/Y/`. These values are then passed to main program `ORBIT` (through `COMMON/Y/`) to start the calculation of the solution of equations (2.26) to (2.29) as explained previously.

The residuals of equations (2.26) to (2.29) are computed in subroutine `DERV`, called as the first argument of subroutine `SNSQE` and listed in Program 2.1c.

Note that subroutine `DERV` has four arguments, and the overall requirement is to program the four residuals in array `YT` from the four input values of y (y_1 to y_4) in array `Y` (which are set by subroutine `SNSQE`). Generally, `SNSQE` will provide values of y_1 to y_4 that will decrease the residuals, and ideally, will eventually find values of y_1 to y_4 that will reduce the residuals to values less than the tolerance read from the third

```
      SUBROUTINE INITAL
C...
C...  DECLARATIVE STATEMENTS
      PARAMETER (N=4)
C...
C...  COMMON AREAS DEFINED IN MAIN PROGRAM ORBIT
      COMMON /T/       T,  NSTOP,  NORUN
     +       /Y/   Y(N)
     +       /F/  YT(N)
     +       /C/    Y10,    Y20,    Y30,    Y40,     TP
     +       /I/     IP,     NP
C...
C...  INITIAL CONDITIONS
         Y10=0.5
         Y20=0.
         Y30=0.
         Y40=1.63
C...
C...  ESTIMATES OF THE SOLUTION FOR SUBROUTINE SNSQE AT T = 0
         Y(1)=Y10
         Y(2)=Y20
         Y(3)=Y30
         Y(4)=Y40
C...
C...  INITIALIZE A COUNTER FOR THE PLOTTED SOLUTION
      IP=0
      RETURN
      END
```

Program 2.1b Subroutine INITAL to Define Initial Conditions (2.30) to (2.33).

line of data in main program ORBIT. In order to find these solution values of y_1 to y_4, SNSQE must have the values of the residuals computed by DERV (and returned to SNSQE from DERV through the third argument of DERV). Note also that COMMON/C/ provides the base values of y_1 to y_4 at k in equations (2.26) to (2.29) as set in main program ORBIT, and SNSQE attempts to compute the values of y_1 to y_4 at $k+1$ in equations (2.26) to (2.29).

Subroutine PRINT is listed in Program 2.1d. First, the energy of the solution is computed according to equation (2.38) by a call to FUNCTION ENERGY. The four equation residuals are then computed by a call to DERV, and the sum of squares of these residuals is computed as variable FNORM. Then t and the solution in array Y is printed along with the energy and the sum of squares of the residuals. Finally a call to subroutine PLOT plots the numerical solution as y_3 vs. y_1.

Function ENERGY, subroutine PLOT and dummy subroutine JAC are listed in Program 2.1e.

Subroutine PLOT writes file ORBIT.TOP which was then sent to a plotting system to produce Figure 2.1. Generally, plotting is hardware and software dependent, and we present subroutine PLOT just as an example of how plotting can be done; some conversion will most likely be required for different hardware and software configurations. Note also that JAC is a do-nothing routine provided to satisfy the EXTERNAL requirements of subroutine SNSQE; it is not actually called because IOPT=2 in main program ORBIT specifies a numerical Jacobian matrix computed internally by SNSQE.

```
      SUBROUTINE DERV(N,Y,YT,IFLAG)
      REAL Y(N),  YT(N)
C...
C...  COMMON AREA FOR THE SOLUTION AT THE PRECEDING TIME STEP
      COMMON /C/    Y10,    Y20,    Y30,    Y40,    DT

C...  EQUATION (2.26)
         YT(1)=(Y(1)-Y10)/DT-(Y(2)+Y20)/2.0
C...
C...  EQUATION (2.27)
         YT(2)=(Y(3)-Y30)/DT-(Y(4)+Y40)/2.0
C...
C...  EQUATION (2.28)
         TERM0=SQRT( Y10**2+ Y30**2)
         TERM1=SQRT(Y(1)**2+Y(3)**2)
         TERM2=TERM0*TERM1
         TERM3=0.5*(TERM0+TERM1)
         YT(3)=(Y(2)-Y20)/DT+(1.0/TERM2)*(0.5*(Y(1)+Y10))/TERM3
C...
C...  EQUATION (2.29)
         YT(4)=(Y(4)-Y40)/DT+(1.0/TERM2)*(0.5*(Y(3)+Y30))/TERM3
      RETURN
      END
```

Program 2.1c Subroutine DERV to Calculate the Residuals of Equations (2.26) to (2.29).

The data file read by main program ORBIT is listed in Program 2.1f. Note that three runs are programmed for error tolerances of 0.1, 0.001, and 0.00001 for the solution of equations (2.26) to (2.29) computed by SNSQE. The independent variable t runs from $t = 0$ to $t = 10$ with printing of the solution at intervals of 0.5. Also, four equations are specified, as expected. The final line is END OF RUNS which terminates execution of main program ORBIT as explained previously.

Abbreviated output produced by program 2.1 is listed in Table 2.1, and the plotted solution for the third run produced by subroutine PLOT is given in Figure 2.1.

The reduction in the equation residuals and the sum of squares of the residuals with decreasing error tolerance is apparent from comparing the solutions from the three runs. In other words, SNSQE computed successively more accurate solutions to equations (2.26) to (2.29) with decreasing error tolerance, as expected. Note also that the energy of the solutions appears to have converged to a value of -0.671550, which is correct to six figures (as we will indicate in the subsequent discussion).

In fact, this correct value of the energy to six figures is the point of this problem. The plot of the orbit in Figure 2.1 indicates that it is highly inaccurate since the orbit defined by equations (2.8) and (2.9) is closed and periodic. The drift away from a periodic orbit and into an annulus, as reported by Greenspan (1990) is evident, yet the energy is correct. In other words, the conclusion that a numerical solution is accurate because its energy is accurate is incorrect. The correct energy is a necessary, but insufficient, condition for the solution to be correct.

We might then ask why the solution to equations (2.26) to (2.29) is such a poor approximation to the solution of equations (2.8) and (2.9). The answer is essentially that the integration step, Δt, in equations (2.26) to (2.29) is too large, i.e., $\Delta t = 0.5$

```
      SUBROUTINE PRINT(NI,NO)
C...
C...  DECLARATIVE STATEMENTS
      PARAMETER (N=4)
C...
C...  COMMON AREAS DEFINED IN MAIN PROGRAM ORBIT
      COMMON /T/       T,  NSTOP,  NORUN
     +       /Y/   Y(N)
     +       /F/  YT(N)
     +       /C/     Y10,     Y20,     Y30,     Y40,      TP
     +       /I/      IP,      NP
C...
C...  PRINT THE NUMERICAL SOLUTION
C...
C...     ENERGY, EQUATION (2.38)
         EN=ENERGY(N,Y)
C...
C...     L2 NORM USING SUBROUTINE DERV
         FNORM=0.
         CALL DERV(N,Y,YT,IFLAG)
         DO 2 I=1,N
            FNORM=FNORM+YT(I)**2
2        CONTINUE
         FNORM=SQRT(FNORM)
C...
C...     PRINT NUMERICAL SOLUTION, RESIDUAL, ENERGY AND NORM
         WRITE(NO,1)T,(I,Y(I),YT(I),I=1,N),EN,FNORM
1        FORMAT('     T = ',F7.3,//,
     +      4(2X,  '   I = ',I2,     '    Y = ',F9.6,2X,
     +             '  RE = ',E11.3,/),/,
     +        2X,  '  EN = ',F9.6,7X,'   FNORM = ',E11.3,///)
C...
C...  STORE THE NUMERICAL SOLUTION FOR PLOTTING
         IP=IP+1
         CALL PLOT
      RETURN
      END
```

Program 2.1d Subroutine `PRINT` to Print the Solution of Equations (2.26) to (2.29).

read from the data file in Program 2.1f. We could therefore improve the accuracy by decreasing the integration step. This was done in a second set of three runs, with $\Delta t = 0.05$ as defined by the data file in Program 2.1g.

All of the other programming in Program 2.1 remained the same. Abbreviated output from the third run for this case of $\Delta t = 0.05$ and a plot of the numerical solution are given in Table 2.2 and Figure 2.2, respectively.

The plotted orbit in Figure 2.2 now closely approximates the periodic orbit of equations (2.8) and (2.9) as will be demonstrated in the subsequent discussion. Also, as before, the energy of the numerical solution is correct to six figures, i.e., -0.671550. Thus, the choice of the integration step, Δt, is clearly an important issue in computing an accurate numerical solution to a system of differential equations. Ideally, the computer program would select an integration step, and in fact, in the next section we

```
      FUNCTION ENERGY(N,Y)
C...
C...  FUNCTION ENERGY COMPUTES THE TOTAL ENERGY OF THE SYSTEM
C...
      REAL Y(N)
C...
C...  ENERGY, EQUATION (2.38)
         R=SQRT(Y(1)**2+Y(3)**2)
         ENERGY=0.5*(Y(2)**2+Y(4)**2)-1.0/R
      RETURN
      END

      SUBROUTINE PLOT
C...
C...  DECLARATIVE STATEMENTS
      PARAMETER (N=4)
C...
C...  COMMON AREAS DEFINED IN MAIN PROGRAM ORBIT
      COMMON /T/      T,  NSTOP,  NORUN
     +       /Y/   Y(N)
     +       /F/  YT(N)
     +       /C/    Y10,    Y20,     Y30,     Y40,      TP
     +       /I/     IP,     NP
C...
C...  INITIATE PLOTTING
      IF(IP.EQ.1)THEN
C...
C...     OPEN FILE FOR TOP DRAWER PLOTTING
         IT=4
         OPEN(IT,FILE= 'ORBIT.TOP',STATUS='UNKNOWN')
C...
C...     SCALE THE AXES
         WRITE(IT,7)
7        FORMAT(' SET LIMITS X FROM -1 TO 1 Y FROM -1 TO 1',/,
     1          ' SET FONT DUPLEX')
         WRITE(IT,4)
4        FORMAT(' SET WINDOW X 2 TO 8 Y 2 TO 8')
      END IF
C...
C...  WRITE Y(3) VS Y(1) FOR PLOTTING
         WRITE(IT,5)Y(1),Y(3)
5        FORMAT(2F10.4)
C...
C...  COMPLETE PLOTTING
      IF(IP.EQ.NP)THEN
C...
C...     CONNECT POINTS
         WRITE(IT,6)
6        FORMAT(' JOIN 1')
```

Program 2.1e Function ENERGY, Subroutines PLOT and JAC to Compute the Energy and Plot the Solution of Equations (2.26) to (2.33), and Meet the EXTERNAL Requirements of SNSQE. *Continued next page.*

```
C...
C...      LABEL THE AXES
          WRITE(IT,8)NP
8         FORMAT(
     1    ' TITLE 2.75 8.5 "',I3,'-point orbit from SNSQE - SP"'
     2    ,/,' TITLE LEFT "        y3(t)"'
     3    ,/,' TITLE BOTTOM "y1(t)"'
     4    ,/,' TITLE 2.5 0.7 "Greenspan/Feynman Orbit Problem"')
C...
C...      END PLOT
          WRITE(IT,10)
10        FORMAT(' NEW FRAME')
       END IF
       RETURN
       END

       SUBROUTINE JAC
C...
C...   JAC IS A DUMMY ROUTINE TO SATISFY THE LOADER.  IT IS NOT CALLED
C...   BY SNSQE SINCE IOPT = 2
       RETURN
       END
```

Program 2.1e *Continued.*

```
GREENSPAN ORBIT PROBLEM - TOL = 0.1,      COARSE T INTEGRATION
0.         10.0       0.5
    4                      0.1
GREENSPAN ORBIT PROBLEM - TOL = 0.001,    COARSE T INTEGRATION
0.         10.0       0.5
    4                      0.001
GREENSPAN ORBIT PROBLEM - TOL = 0.00001, COARSE T INTEGRATION
0.         10.0       0.5
    4                      0.00001
END OF RUNS
```

Program 2.1f Data File `DATA` Read by Main Program `ORBIT`.

consider a library routine, `RKF45`, that automatically adjusts Δt to meet a user-specified error tolerance for the numerical solution.

2.2 *The Solution—Use a Quality ODE Integrator*

We observed in the preceding section that a solution to the finite difference equations which approximate a system of ODEs can be inaccurate if the integration step is not selected carefully. The question naturally arises then of how we can select the integration step to ensure a numerical solution of acceptable accuracy, e.g., to a given number of significant figures or percent accuracy. One approach would be to compute a series of numerical solutions at successively smaller integration steps, then compare the solutions. If this process converges, we should observe that the numerical so-

Table 2.1 Abbreviated File `OUTPUT` produced by Program 2.1. *Continued next pages.*

```
RUN NO. 1 - GREENSPAN ORBIT PROBLEM - TOL = 0.1, COARSE T INTEGRATION

 INITIAL VALUE OF TIME =  0.000E+00

   FINAL VALUE OF TIME =  0.100E+02

PRINT INTERVAL OF TIME =  0.500E+00

SOLUTION BY SUBROUTINE SNSQE

   NUMBER OF EQUATIONS =   4

   ERROR TOLERANCE =  0.100E+00


    T =    0.000

    I =  1   Y =  0.500000    RE =   0.000E+00
    I =  2   Y =  0.000000    RE =  -0.163E+01
    I =  3   Y =  0.000000    RE =   0.400E+01
    I =  4   Y =  1.630000    RE =   0.000E+00

   EN = -0.671550           FNORM =   0.432E+01


    T =    0.500

    I =  1   Y =  0.257927    RE =   0.298E-07
    I =  2   Y = -0.968291    RE =   0.000E+00
    I =  3   Y =  0.617728    RE =  -0.183E-03
    I =  4   Y =  0.840911    RE =   0.327E-04

   EN = -0.671486           FNORM =   0.185E-03


    T =    1.000

    I =  1   Y = -0.230197    RE =  -0.101E-05
    I =  2   Y = -0.984204    RE =  -0.209E-06
    I =  3   Y =  0.834933    RE =  -0.676E-03
    I =  4   Y =  0.027908    RE =   0.577E-02

   EN = -0.669903           FNORM =   0.581E-02
             .                      .
             .                      .
             .                      .

    T =    9.000

    I =  1   Y =  0.983693    RE =   0.000E+00
    I =  2   Y = -0.072228    RE =  -0.238E-06
    I =  3   Y =  0.151216    RE =  -0.703E-03
    I =  4   Y =  0.818117    RE =   0.681E-05
```

Table 2.1 *Continued.*

```
   EN = -0.667508          FNORM =   0.703E-03


    T =   9.500

    I =  1   Y =  0.829397    RE =   0.000E+00
    I =  2   Y = -0.544958    RE =   0.298E-06
    I =  3   Y =  0.516737    RE =  -0.311E-03
    I =  4   Y =  0.643965    RE =  -0.107E-03

   EN = -0.667499          FNORM =   0.328E-03


    T =  10.000

    I =  1   Y =  0.454121    RE =   0.179E-06
    I =  2   Y = -0.956147    RE =   0.209E-06
    I =  3   Y =  0.738221    RE =  -0.510E-03
    I =  4   Y =  0.241973    RE =  -0.403E-03

   EN = -0.667397          FNORM =   0.650E-03



RUN NO. 2 - GREENSPAN ORBIT PROBLEM - TOL = 0.001, COARSE T INTEGRATION

 INITIAL VALUE OF TIME =  0.000E+00

   FINAL VALUE OF TIME =  0.100E+02

PRINT INTERVAL OF TIME =  0.500E+00

SOLUTION BY SUBROUTINE SNSQE

   NUMBER OF EQUATIONS =   4

   ERROR TOLERANCE =  0.100E-02


    T =   0.000

    I =  1   Y =  0.500000    RE =   0.000E+00
    I =  2   Y =  0.000000    RE =  -0.163E+01
    I =  3   Y =  0.000000    RE =   0.400E+01
    I =  4   Y =  1.630000    RE =   0.000E+00

   EN = -0.671550          FNORM =   0.432E+01


    T =   0.500

    I =  1   Y =  0.257954    RE =   0.000E+00
    I =  2   Y = -0.968182    RE =   0.119E-06
```

Table 2.1 *Continued.*

```
   I =  3   Y =  0.617734     RE =    0.321E-04
   I =  4   Y =  0.840935     RE =    0.364E-04

  EN = -0.671535           FNORM =    0.485E-04


   T =   1.000

   I =  1   Y = -0.230060     RE =    0.000E+00
   I =  2   Y = -0.983875     RE =    0.298E-07
   I =  3   Y =  0.834028     RE =   -0.484E-07
   I =  4   Y =  0.024242     RE =    0.119E-06

  EN = -0.671535           FNORM =    0.132E-06
            .                       .
            .                       .
            .                       .
   T =   9.000

   I =  1   Y =  0.967912     RE =   -0.149E-07
   I =  2   Y = -0.127421     RE =    0.000E+00
   I =  3   Y =  0.194368     RE =    0.271E-04
   I =  4   Y =  0.816433     RE =    0.932E-06

  EN = -0.671531           FNORM =    0.271E-04

   T =   9.500

   I =  1   Y =  0.785449     RE =    0.000E+00
   I =  2   Y = -0.602432     RE =    0.000E+00
   I =  3   Y =  0.552032     RE =    0.817E-05
   I =  4   Y =  0.614223     RE =    0.337E-05

  EN = -0.671531           FNORM =    0.883E-05


   T =  10.000

   I =  1   Y =  0.383559     RE =    0.000E+00
   I =  2   Y = -1.005125     RE =   -0.596E-07
   I =  3   Y =  0.747251     RE =    0.190E-04
   I =  4   Y =  0.166652     RE =    0.176E-04

  EN = -0.671535           FNORM =    0.259E-04


RUN NO. 3 - GREENSPAN ORBIT PROBLEM - TOL = 0.00001, COARSE T INTEGRATION

 INITIAL VALUE OF TIME =  0.000E+00

   FINAL VALUE OF TIME =  0.100E+02

PRINT INTERVAL OF TIME =  0.500E+00
```

Table 2.1 *Continued.*

```
SOLUTION BY SUBROUTINE SNSQE

   NUMBER OF EQUATIONS =   4

   ERROR TOLERANCE =  0.100E-04


    T =    0.000

    I =  1    Y =  0.500000     RE =    0.000E+00
    I =  2    Y =  0.000000     RE =   -0.163E+01
    I =  3    Y =  0.000000     RE =    0.400E+01
    I =  4    Y =  1.630000     RE =    0.000E+00

   EN = -0.671550           FNORM =    0.432E+01

    T =    0.500

    I =  1    Y =  0.257948     RE =   -0.298E-07
    I =  2    Y = -0.968209     RE =    0.000E+00
    I =  3    Y =  0.617727     RE =   -0.119E-06
    I =  4    Y =  0.840909     RE =   -0.119E-06

   EN = -0.671550           FNORM =    0.171E-06


    T =    1.000

    I =  1    Y = -0.230077     RE =   -0.596E-07
    I =  2    Y = -0.983889     RE =    0.000E+00
    I =  3    Y =  0.834002     RE =   -0.224E-07
    I =  4    Y =  0.024188     RE =    0.119E-06

   EN = -0.671550           FNORM =    0.135E-06
              .                   .
              .                   .
              .                   .

    T =    9.000

    I =  1    Y =  0.967813     RE =    0.000E+00
    I =  2    Y = -0.127867     RE =    0.000E+00
    I =  3    Y =  0.194683     RE =    0.000E+00
    I =  4    Y =  0.816383     RE =    0.522E-07

   EN = -0.671550           FNORM =    0.522E-07


    T =    9.500

    I =  1    Y =  0.785125     RE =    0.298E-07
    I =  2    Y = -0.602886     RE =    0.000E+00
```

Table 2.1 *Continued.*

```
 I =  3   Y =  0.552271     RE =    0.000E+00
 I =  4   Y =  0.613970     RE =    0.596E-07

EN = -0.671550            FNORM =    0.666E-07

 T =  10.000

 I =  1   Y =  0.383020     RE =    0.000E+00
 I =  2   Y = -1.005533     RE =   -0.596E-07
 I =  3   Y =  0.747272     RE =   -0.119E-06
 I =  4   Y =  0.166032     RE =    0.000E+00

EN = -0.671550            FNORM =    0.133E-06
```

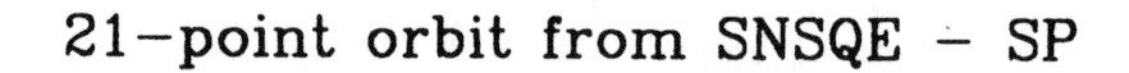

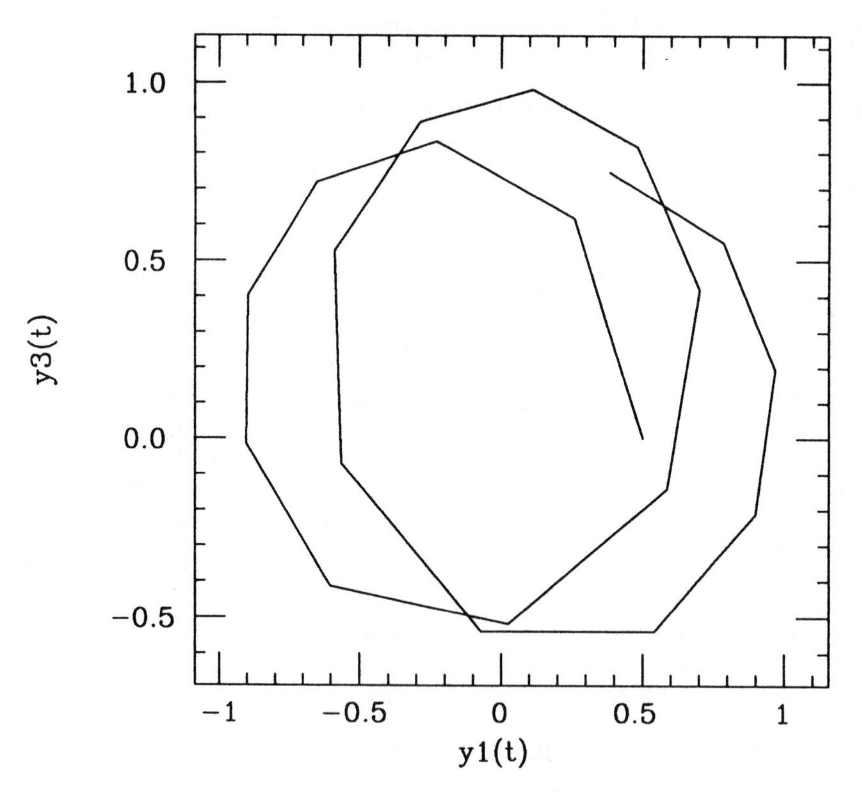

Figure 2.1 21-point Plot of Solution produced by Program 2.1 from the Third Run with Error Tolerance $= 0.00001$ for $\Delta t = 0.5$.

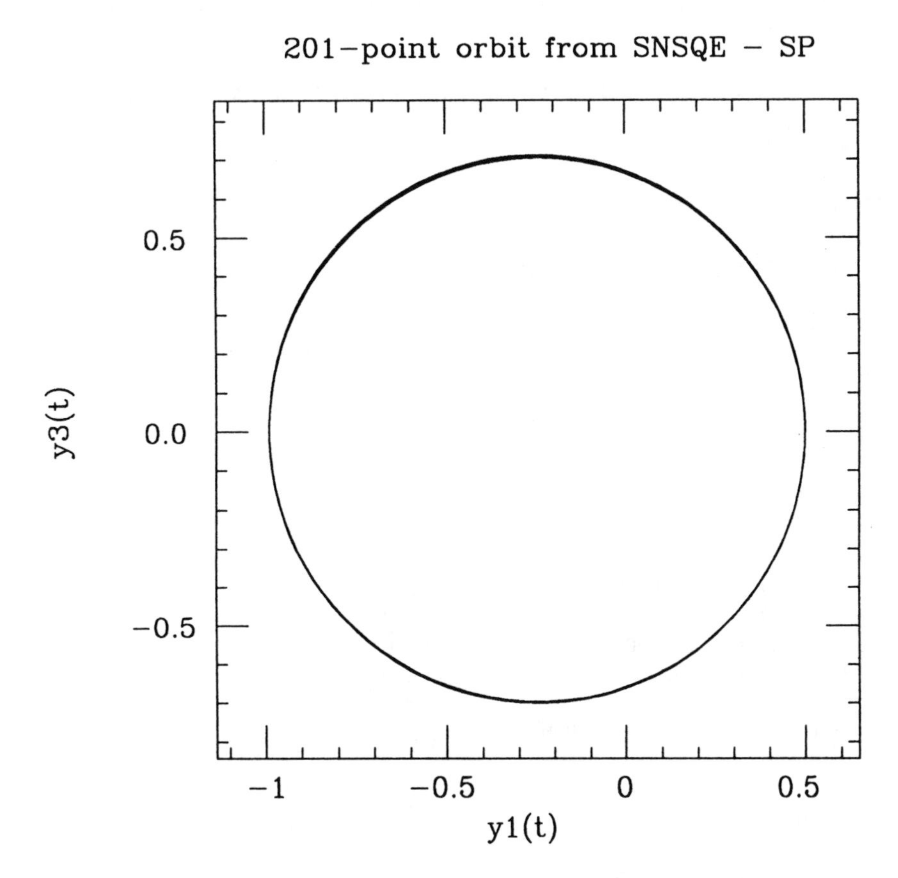

Figure 2.2 201-point Plot of Solution produced by Program 2.1 from the Third Run with Error Tolerance = 0.00001 for $\Delta t = 0.05$.

lutions are approaching a limit with increasing numbers of significant figures. This is a rather cumbersome procedure, particularly if a large number of numbers must

```
GREENSPAN ORBIT PROBLEM - TOL = 0.1,     FINE T INTEGRATION
0.         10.0        0.05
    4                       0.1
GREENSPAN ORBIT PROBLEM - TOL = 0.001,   FINE T INTEGRATION
0.         10.0        0.05
    4                       0.001
GREENSPAN ORBIT PROBLEM - TOL = 0.00001, FINE T INTEGRATION
0.         10.0        0.05
    4                       0.00001
END OF RUNS
```

Program 2.1g Data File DATA Read by Main Program ORBIT with $\Delta t = 0.05$.

***Table* 2.2** Abbreviated File OUTPUT produced by Program 2.1 for $\Delta t = 0.05$. *Continued next page.*

```
RUN NO.   3 - GREENSPAN ORBIT PROBLEM - TOL = 0.00001, FINE T INTEGRATION

  INITIAL VALUE OF TIME =  0.000E+00

    FINAL VALUE OF TIME =  0.100E+02

 PRINT INTERVAL OF TIME =  0.500E-01

 SOLUTION BY SUBROUTINE SNSQE

    NUMBER OF EQUATIONS =   4

    ERROR TOLERANCE =  0.100E-04


     T =   0.000

     I =  1   Y =  0.500000    RE =   0.000E+00
     I =  2   Y =  0.000000    RE =  -0.163E+01
     I =  3   Y =  0.000000    RE =   0.400E+01
     I =  4   Y =  1.630000    RE =   0.000E+00

    EN = -0.671550           FNORM =   0.432E+01


     T =   0.050

     I =  1   Y =  0.495049    RE =  -0.283E-06
     I =  2   Y = -0.198030    RE =  -0.119E-06
     I =  3   Y =  0.081097    RE =   0.238E-06
     I =  4   Y =  1.613861    RE =   0.101E-05

    EN = -0.671550           FNORM =   0.109E-05


     T =   0.100

     I =  1   Y =  0.480388    RE =  -0.298E-07
     I =  2   Y = -0.388415    RE =  -0.119E-06
     I =  3   Y =  0.160610    RE =   0.000E+00
     I =  4   Y =  1.566684    RE =  -0.715E-06

    EN = -0.671550           FNORM =   0.726E-06
              .                   .
              .                   .
              .                   .

     T =   9.900

     I =  1   Y = -0.966197    RE =  -0.238E-06
     I =  2   Y = -0.214721    RE =  -0.596E-07
```

Table 2.2 *Continued.*

```
  I =  3    Y =  0.184643    RE =   0.119E-06
  I =  4    Y = -0.802480    RE =  -0.313E-06

 EN = -0.671550            FNORM =   0.415E-06


  T =   9.950


  I =  1    Y = -0.975663    RE =   0.179E-06
  I =  2    Y = -0.163919    RE =   0.119E-06
  I =  3    Y =  0.144304    RE =  -0.119E-06
  I =  4    Y = -0.811086    RE =   0.447E-06

 EN = -0.671550            FNORM =   0.510E-06


  T =  10.000

  I =  1    Y = -0.982586    RE =  -0.551E-06
  I =  2    Y = -0.113027    RE =  -0.596E-07
  I =  3    Y =  0.103588    RE =   0.000E+00
  I =  4    Y = -0.817528    RE =  -0.447E-06

 EN = -0.671550            FNORM =   0.712E-06
```

be examined for this convergence, e.g., this would not be especially difficult in the case of only four ODEs as in the orbit problem, but we could possibly be numerically integrating several thousand ODEs, which is not uncommon in applications. We could automate this process to some extent by storing the successive numerical solutions, then have the computer compare the solutions, perhaps by computing a sum of squares of differences between successive solutions.

However, it would be much more desirable to have a numerical algorithm included in the solution procedure which would automatically adjust the integration step Δt as the solution evolves to produce a solution with a user-specified accuracy. At first, this might seem like an impossibility because it implies that we (or the computer) know what the error is at each point along the solution so that the integration step can be adjusted accordingly, and this in turn implies that we know the exact solution (in order to know the error in the numerical solution); if we know the exact solution, there would be no need to compute a numerical solution. The resolution of this apparent dilemma, however, is that we will estimate the error in the numerical solution, rather than know it precisely. The estimate of the error does not require a knowledge of the exact solution, but is accurate enough to determine whether the integration step Δt should be changed, e.g., reduced if the estimated error is above the user-specified error criterion. This procedure of estimating the integration error is built into all quality ODE integrators, and we now consider the use of one of these integrators, subroutine RKF45 [Forsythe et al. (1977)].

The name `RKF45` denotes the following characteristics of the integrator: (a) a Runge-Kutta algorithm is used which is named after the two German mathematicians who first developed this approach in the 1890s, (b) the particular version of the Runge-Kutta algorithm was developed by Fehlberg (1970), and (c) the special feature of the RKF algorithm is the use of a fourth- and fifth-order method to compute solutions to the ODE problem system; subtraction of the two solutions gives an estimate of the integration error at each point along the solution which can then be used to adjust the integration step Δt. The details of this algorithm will be discussed subsequently. We now consider the solution of the orbit problem using RKF45.

First, as in the case of SNSQE, we require a main program to call RKF45, which is listed in Program 2.2a.

The following points should be noted about main program ORBIT2:

1. The structure is very similar to main program ORBIT (Program 2.1a). The essential difference is that RKF45 is called rather than SNSQE. Note that RKF45 requires a real work array, WORK, and an integer work array, IWORK, which are sized according to the formulas given in the documentation comments in RKF45. These arrays have been sized to accommodate up to 250 ODEs (note the arrays in COMMON/Y/ and /F/), thus the main program ORBIT2 can accommodate larger ODE problems than the orbit problem we are now considering. Also, a user-supplied subroutine, FCN, defines the ODEs; it is the first argument of RKF45 and therefore is declared an external.

2. In programming subroutine FCN, we made the decision to again define the problem system equations in a subroutine called DERV. Therefore, FCN is really just a small interface routine between RKF45 and DERV (it is listed at the end of main program ORBIT2). This choice is somewhat arbitrary, and was made for the following reasons: (a) we use DERV consistently throughout all of the examples for the programming of the problem system equations and (b) in subsequent examples, DERV will not have arguments; rather, its input and output information will be passed through COMMON/T/, /Y/, and /F/, but again, arguments could have been used rather than COMMON.

3. A data file, named DATA, is again used with three lines: (a) a documentation title; (b) the initial, final, and print or output intervals of t; and (c) the number of ODEs and the error tolerance to be used by RKF45 to compute a solution. As before, we have four ODEs, equations (2.26) to (2.29). The data are then printed in a summary at the beginning of each run, i.e., multiple runs are possible by using sets of the three basic lines of data in file DATA.

4. Subroutine INITAL is called to define initial conditions (2.30) to (2.33). Then the input parameters for RKF45 are set so that the calculation can proceed from the initial conditions. In particular, the absolute and relative error tolerances to be used by RKF45 must be specified. In the present problem, these are selected as zero and TOL respectively, where TOL is the error tolerance read from the third line of data in file DATA. Parenthetically, experience has indicated that the choice of error tolerances must be done carefully. If inappropriate error tolerances are selected, an excessively long computer run or a solution with excessive error may result, or in the worst case, the ODE integrator may be unable to compute a solution. Unfortunately, general guidelines for the selection of error tolerances are difficult to state since appropriate error tolerances are often problem dependent or specific. RKF45 has two basic modes of operation as defined by the initial value of IFLAG; if IFLAG=1 in the first call to RKF45, a "multistep" integration will be performed which is unconstrained by the output interval (TP read from the second data line) while with IFLAG=-1 in the first call to RKF45, a "one-step"

```
      PROGRAM ORBIT2
C...
C...  FEYNMAN ORBIT PROBLEM - SOLUTION BY SUBROUTINE RKF45
C...
C...  SEE SUBROUTINE INITAL FOR A DISCUSSION OF THE PROBLEM.
C...
C...  THIS MAIN PROGRAM ESSENTIALLY CALLS SUBROUTINE RKF45 FOR THE
C...  SOLUTION OF A SYSTEM OF ORDINARY DIFFERENTIAL EQUATIONS (ODES).
C...
C...  THE MODEL INITIAL CONDITIONS ARE SET IN SUBROUTINE INITAL, AND
C...  THE MODEL DERIVATIVES ARE PROGRAMMED IN SUBROUTINE DERV.  THE
C...  NUMERICAL SOLUTION IS PRINTED AND PLOTTED IN SUBROUTINE PRINT.
C...
C...  SUBROUTINES INITAL, DERV AND PRINT.  THE FOLLOWING CODING IS FOR
C...  250 ORDINARY DIFFERENTIAL EQUATIONS (ODES).  IF MORE ODES ARE TO
C...  BE INTEGRATED, ALL OF THE 250*S SHOULD BE CHANGED TO THE
C...  REQUIRED NUMBER
      COMMON/T/          T,      NSTOP,     NORUN
     1      /Y/    Y(250)
     2      /F/    F(250)
C...
C...  THE NUMBER OF DIFFERENTIAL EQUATIONS IS IN COMMON/N/ FOR USE IN
C...  SUBROUTINE FCN
      COMMON/N/      NEQN
C...
C...  COMMON AREA TO PROVIDE THE INPUT/OUTPUT UNIT NUMBERS TO OTHER
C...  SUBROUTINES
      COMMON/IO/       NI,        NO
C...
C...  ABSOLUTE DIMENSIONING OF THE ARRAYS REQUIRED BY RKF45
      DIMENSION YV(250), WORK(1503), IWORK(5)
C...
C...  EXTERNAL THE DERIVATIVE ROUTINE CALLED BY RKF45
      EXTERNAL FCN
C...
C...  ARRAY FOR THE TITLE (FIRST LINE OF DATA), CHARACTERS  END OF
C...  RUNS
      CHARACTER TITLE(20)*4, ENDRUN(3)*4
C...
C...  DEFINE THE CHARACTERS  END OF RUNS
      DATA ENDRUN/'END ','OF R','UNS '/
C...
C...  DEFINE THE INPUT/OUTPUT UNIT NUMBERS
         NI=5
         NO=6
C...
C...  DEFINE INPUT/OUTPUT FILES
         OPEN(NI,FILE='DATA'   ,STATUS='OLD')
         OPEN(NO,FILE='OUTPUT',STATUS='NEW')
C...
C...  INITIALIZE THE RUN COUNTER
```

Program 2.2a Main Program ORBIT2 to Call Subroutine RKF45 for the Solution of Equations (2.26) to (2.33). *Continued next pages.*

```
      NORUN=0
C...
C...  BEGIN A RUN
1     NORUN=NORUN+1
C...
C...  INITIALIZE THE RUN TERMINATION VARIABLE
      NSTOP=0
C...
C...  READ THE FIRST LINE OF DATA
      READ(NI,1000,END=999)(TITLE(I),I=1,20)
C...
C...  TEST FOR  END OF RUNS  IN THE DATA
      DO 2 I=1,3
         IF(TITLE(I).NE.ENDRUN(I))GO TO 3
2     CONTINUE
C...
C...  AN END OF RUNS HAS BEEN READ, SO TERMINATE EXECUTION
999   STOP
C...
C...  READ THE SECOND LINE OF DATA
3     READ(NI,1001,END=999)T0,TF,TP
C...
C...  READ THE THIRD LINE OF DATA
      READ(NI,1002,END=999)NEQN,TOL
C...
C...  PRINT A DATA SUMMARY
      WRITE(NO,1003)NORUN,(TITLE(I),I=1,20),T0,TF,TP,NEQN,TOL
C...
C...  INITIALIZE TIME
      T=T0
C...
C...  SET THE INITIAL CONDITIONS
      CALL INITAL
C...
C...  SET THE INITIAL CONDITIONS FOR SUBROUTINE RKF45
      DO 5 I=1,NEQN
         YV(I)=Y(I)
5     CONTINUE
C...
C...  SET THE PARAMETERS FOR SUBROUTINE RKF45
C...
C...     INITIAL TIME
         TV=T0
C...
C...     ERROR TOLERANCE
         ABSERR=0.
         RELERR=TOL
C...
C...     INITIALIZATION FOR MULTI-STEP MODE (RKF45 NOT CONSTRAINED
C...     BY OUTPUT INTERVAL TP)
         IFLAG=1
C...
```

Program 2.2a *Continued.*

```
C...      INITIALIZATION FOR ONE-STEP MODE (RKF45 CONSTRAINED BY
C...      OUTPUT INTERVAL TP)
C...      IFLAG=-1
C...
C...      FINAL TIME
          TEND=T0
C...
C...   CALL SUBROUTINE RKF45 TO START THE SOLUTION FROM THE INITIAL
C...   CONDITION (IFLAG = 1 OR -1) OR COMPUTE THE SOLUTION TO THE
C...   NEXT PRINT POINT (IFLAG = 2 OR -2)
4      CALL RKF45(FCN,NEQN,YV,TV,TEND,RELERR,ABSERR,IFLAG,WORK,IWORK)
C...
C...   PRINT THE INITIAL CONDITION (IFLAG = 1 OR -1) OR THE SOLUTION
C...   AT THE NEXT PRINT POINT (IFLAG = 2 OR -2)
       T=TV
       DO 6 I=1,NEQN
          Y(I)=YV(I)
6      CONTINUE
       CALL PRINT(NI,NO)
C...
C...   TEST FOR AN ERROR CONDITION
       IF((IFLAG.NE.2).AND.(IFLAG.NE.-2))THEN
C...
C...      PRINT A MESSAGE INDICATING AN ERROR CONDITION
          WRITE(NO,1004)IFLAG
C...
C...      GO ON TO THE NEXT RUN
          GO TO 1
       END IF
C...
C...   CHECK FOR A RUN TERMINATION
       IF(NSTOP.NE.0)GO TO 1
C...
C...   CHECK FOR THE END OF THE RUN
          TEND=TV+TP
          IF(TV.LT.(TF-0.5*TP))GO TO 4
C...
C...   THE CURRENT RUN IS COMPLETE, SO GO ON TO THE NEXT RUN
       GO TO 1
C...
C...   ****************************************************************
C...
C...   FORMATS
C...
1000   FORMAT(20A4)
1001   FORMAT(3E10.0)
1002   FORMAT(I5,20X,E10.0)
1003   FORMAT(1H1,
     1 ' RUN NO. - ',I3,2X,20A4,//,
     2 ' INITIAL T - ',E10.3,//,
     3 '   FINAL T - ',E10.3,//,
     4 '   PRINT T - ',E10.3,//,
```

Program 2.2a *Continued.*

```
     5 ' NUMBER OF DIFFERENTIAL EQUATIONS - ',I3,//,
     6 ' INTEGRATION ALGORITHM - RKF45 ',//,
     7 ' MAXIMUM INTEGRATION ERROR - ',E10.3,//,
     8 1H1)
1004  FORMAT(1H ,//,' IFLAG = ',I3,//,
     1 ' INDICATING AN INTEGRATION ERROR, SO THE CURRENT RUN'      ,/,
     2 ' IS TERMINATED.  PLEASE REFER TO THE DOCUMENTATION FOR'   ,/,
     3 ' SUBROUTINE',//,25X,'RKF45',//,
     4 ' FOR AN EXPLANATION OF THESE ERROR INDICATORS'             )
      END

      SUBROUTINE FCN(TV,YV,YDOT)
C...
C...  SUBROUTINE FCN IS AN INTERFACE ROUTINE BETWEEN SUBROUTINES RKF45
C...  AND DERV
C...
C...  ODE COMMON AREA
      COMMON/T/          T,      NSTOP,      NORUN
     1      /Y/       Y(1)
     2      /F/       F(1)
C...
C...  THE NUMBER OF DIFFERENTIAL EQUATIONS IS AVAILABLE THROUGH COMMON
C...  /N/
      COMMON/N/       NEQN
C...
C...  ABSOLUTE DIMENSION THE DEPENDENT, DERIVATIVE VECTORS
      DIMENSION YV(250), YDOT(250)
C...
C...  TRANSFER THE INDEPENDENT VARIABLE, DEPENDENT VARIABLE VECTOR
C...  FOR USE IN SUBROUTINE DERV
      T=TV
      DO 1 I=1,NEQN
         Y(I)=YV(I)
1     CONTINUE
C...
C...  EVALUATE THE DERIVATIVE VECTOR
      CALL DERV
C...
C...  TRANSFER THE DERIVATIVE VECTOR FOR USE BY SUBROUTINE RKF45
      DO 2 I=1,NEQN
         YDOT(I)=F(I)
2     CONTINUE
      RETURN
      END
```

Program 2.2a *Continued.*

integration will occur constrained by the output interval. Generally, the multistep option should be selected unless many output points are specified as, for example, in a detailed plotting of the solution.

5. RKF45 is then called in a loop to step through the successive values of t, starting with $t = 0$. This loop is not based on the use of a DO loop as was done in main

program ORBIT, but rather through a test of whether the final value of t has been reached

```
IF(TV.LT.(TF-0.5*TP))GO TO 4
```

If this statement is true, the final value of t (TF) has not been reached, and the calculation continues by returning to statement 4, which is the call to RKF45. Otherwise, the solution is complete and main program ORBIT2 goes on to the next run by reading the next set of data from file DATA.

6. Immediately after each call to RKF45, the solution is printed by a call to PRINT (which will also print the initial conditions after the first call to RKF45). Then a check is made for any error conditions that may have occurred during the call to RKF45. In general, a normal return from RKF45 will correspond to IFLAG=2 if IFLAG=1 initially, or IFLAG=-2 if IFLAG=-1 initially. Other values of IFLAG indicate various error conditions which are described in the documentation comments in RKF45, and which result in termination of the current run in main program ORBIT2. As might be expected, these possible error conditions include the inability of RKF45 to calculate a solution with the required user-specified error (TOL read from the third line of data) and excessive numbers of steps along the solution (perhaps because the error tolerance is too stringent); other possibilities include inappropriate error tolerances as sensed by RKF45 and even incorrect input parameters. This checking and flagging of error conditions is a hallmark of a quality library routine, and indicates that much thought has gone into the writing and testing of the routine. In other words, we make the point again that in writing an ODE integrator, the analyst will generally not be able to anticipate the error conditions that can occur and build safeguards into the code, at least in an initial effort; we can, however, take advantage of the experiences of others who have written quality routines such as RKF45.

7. The next run is initiated by a return to statement 1 if (a) the current run is complete (see (5) above), (b) an error condition occurred (IFLAG not equal to 2 or -2 after the return from RKF45), or (c) NSTOP, an integer variable in the second position in COMMON/T/ is set to a nonzero value in subroutines INITAL, DERV, or PRINT; this might be done, for example, if the solution is clearly violating some physical principle such as the calculation of a negative absolute temperature, density, or concentration, in which case NSTOP is set to a nonzero value, possibly with the printing of an error message to explain the error condition.

The initial conditions, equations (2.22) to (2.25), are set in subroutine INITAL, listed in Program 2.2b.

The following points should be noted about subroutine INITAL:

1. Initial conditions (2.22) to (2.25) are set, which is the essential function of INITAL, i.e., the dependent variables in COMMON/Y/ must be initialized to start the integration of the ODEs. Then a call to DERV confirms that the code in this subroutine for the problem ODEs, equations (2.18) to (2.21), executes (this does not guarantee that a correct solution will be computed, but does ensure that the code in DERV will compile and execute at least once; this call to DERV also provides initial values of the derivatives in COMMON/F/ in case these initial derivatives are to be printed by a subsequent call to PRINT).

2. The coding in subroutine INITAL includes an initial condition for a fifth ODE, Y(5)=0., and also, the PARAMETER statement at the beginning of INITAL specifies

```
      SUBROUTINE INITAL
C...
C...  THE FOLLOWING SYSTEM OF ORDINARY DIFFERENTIAL EQUATIONS
C...  (ODES) FOR PLANETARY MOTION IS DISCUSSED BY FEYNMAN ET AL
C...  (1963)
C...
C...       2    2
C...      D X/DT = - X/R**3                               (1)
C...
C...       2    2
C...      D Y/DT = - Y/R**3                               (2)
C...
C...      R = (X**2 + Y**2)**0.5                          (3)
C...
C...      X(0) = 0.5, DX(0)/DT =  0                       (4)(5)
C...
C...      Y(0) = 0,   DY(0)/DT = 1.63                     (6)(7)
C...
C...  THE TOTAL ENERGY FOR THE SYSTEM, WHICH IS CONSTANT, IS GIVEN
C...  BY
C...
C...      E = (1/2)*(V **2 + V **2) - 1/R                 (8)
C...                  X       Y
C...  WHERE
C...
C...      V  = DX/DT, V  = DY/DT
C...       X           Y
C...
C...  E FROM EQUATION (8) IS COMPUTED AND PRINTED IN SUBROUTINE
C...  PRINT TO TEST THE OPERATION OF THE INTEGRATOR.
C...
C...  EQUATIONS (1) AND (2) CAN BE STATED IN TERMS OF NEW VARIABLES
C...  DEFINED AS
C...
C...      Y1 = X, Y2 = DX/DT
C...
C...      Y3 = Y, Y4 = DY/DT
C...
C...  THE FOLLOWING PROGRAMMING IS IN TERMS OF X1, X2, Y1 AND Y2.
C...
C...  EQUATIONS (1) TO (7) HAVE AN EXACT SOLUTION (GREENSPAN, 1990),
C...  WHICH, IF WRITTEN IN PARAMETRIC FORM (STENGLE, 1991), IS
C...
C...      X(T) = -C + A*COS(THETA(T))                     (9)
C...
C...      Y(T) = B*SIN(THETA(T))                          (10)
C...
C...  WHERE THETA(T) IS GIVEN BY THE ODE (2)
C...
C...      DTHETA/DT = SQRT((1/TERM1)*TERM2)               (11)
C...
```

Program 2.2b Subroutine `INITAL` to Define Initial Conditions (2.22) to (2.25). *Continued next pages.*

```
C...      TERM1 = (1/2)*((A*SIN(THETA))**2 + (B*COS(THETA))**2)    (12)
C...
C...      TERM2 = -1/(2*A) + 1/SQRT(TERM3)                         (13)
C...
C...      TERM3 = SQRT((A*COS(THETA) - C)**2 + (B*SIN(THETA))**2) (14)
C...
C...   THE INITIAL CONDITION FOR EQUATION (11) IS
C...
C...      THETA(0) = 0                                             (15)
C...
C...   THE SOLUTION TO EQUATIONS (1), (2) AND (11) IS BY ODE SOLVER
C...   RKF45 DISCUSSED IN FORSYTHE (1976).
C...
C...   REFERENCES:
C...
C...   FEYNMAN, RICHARD P., ROBERT B. LEIGHTON AND MATTHEW SANDS, LEC-
C...   TURES ON PHYSICS, ADDISON-WESLEY PUBLISHING COMPANY, READING,
C...   MA, PP 9-6 TO 9-9 (1963)
C...
C...   FORSYTHE, GEORGE E., MICHAEL A. MALCOLM, AND CLEVE B. MOLER,
C...   COMPUTER METHODS FOR MATHEMATICAL COMPUTATIONS, PRENTICE-HALL,
C...   ENGLEWOOD CLIFFS, NJ, PP 129-147 (1977)
C...
C...   GREENSPAN, DONALD, A COUNTEREXAMPLE OF THE USE OF ENERGY AS A
C...   MEASURE OF COMPUTATIONAL EFFICIENCY, J. COMPUTATIONAL PHYSICS,
C...   V. 91, 490-494 (1990)
C...
C...   STENGLE, GILBERT, PRIVATE COMMUNICATION, OCTOBER 8, 1991
C...
      PARAMETER (N=5)
      COMMON/T/          T,      NSTOP,      NORUN
     1      /Y/       Y(N)
     2      /F/    DYDT(N)
     3      /C/          R,          A,          B,          C
     4      /I/         IP,         NP
C...
C...   MODEL PARAMETERS
         A=10000.0/13431.0
         B=SQRT(664225.0/1343100.0)
         C=6569.0/26862.0
C...
C...   INITIAL CONDITIONS, EQUATIONS (4) TO (7), (11)
         Y(1)=0.5
         Y(2)=0.
         Y(3)=0.
         Y(4)=SQRT(2.0*(-1.0/(2.0*A)+1.0/(A-C)))
         Y(5)=0.
C...
```

Program 2.2b Subroutine `INITAL` to Define Initial Conditions (2.22) to (2.25). *Continued next pages.*

```
C...  COMPUTE THE INITIAL DERIVATIVES
      CALL DERV
C...
C...  INITIALIZE INTEGER VARIABLES TO CONTROL THE PRINTING
         IP=0
         NP=201
      RETURN
      END
```

***Program 2.2b** Continued.*

five ODES. This additional ODE is discussed in the comments in `INITAL`, particularly the comments pertaining to equations (9) to (15). We now consider this fifth ODE in some detail. The solution to equations (2.8) and (2.9) can be written in *parametric form* [Stengle (1991)] as

$$x(t) = -c + a\cos(\theta(t)), y(t) = b\sin(\theta(t)) \qquad (2.39)(2.40)$$

where $\theta(t+\tau) = \theta(t)+2\pi$. The parameters of the problem are given by Greenspan (1990) as:

$$a = \frac{10000}{13431}, \qquad b = \sqrt{\frac{664225}{1343100}}, \qquad c = \frac{6569}{26862} \qquad (2.41)\text{–}(2.43)$$

$$E = -\frac{1}{2a} = -0.67155, \qquad \tau = \frac{2\pi}{\sqrt{(-2E)^3}} = \frac{2\pi}{(1.3431)^{3/2}} \qquad (2.44)\text{–}(2.46)$$

In order to compute $x(t)$ and $y(t)$ from equations (2.39) and (2.40), we require $\theta(t)$, for which an ODE can be derived in the following way. From equations (2.38) and (2.44)

$$\left\{(1/2)\left(\frac{dx}{dt}\right)^2 + (1/2)\left(\frac{dy}{dt}\right)^2\right\} - \frac{1}{(x^2+y^2)^{1/2}} = -\frac{1}{2a} = E \qquad (2.47)$$

From equations (2.39) and (2.40),

$$dx/dt = -a\sin(\theta(t))d\theta/dt, \qquad dy/dt = b\cos(\theta(t))d\theta/dt \qquad (2.48)(2.49)$$

Substitution of equations (2.39), (2.40), (2.48), and (2.49) in equation (2.47) gives

$$(d\theta/dt)^2(1/2)\left\{(a\sin\theta)^2 + (b\cos\theta)^2\right\} - \frac{1}{\{(-c+a\cos\theta)^2 + (b\sin\theta)^2\}^{1/2}} = -\frac{1}{2a} = E$$

or

$$d\theta/dt = \sqrt{\frac{1}{(1/2)\left\{(a\sin\theta)^2 + (b\cos\theta)^2\right\}}\left\{-\frac{1}{2a} + \frac{1}{\{(a\cos\theta - c)^2 + (b\sin\theta)^2\}^{1/2}}\right\}} \qquad (2.50)$$

Equation (2.50) can be integrated numerically, subject to the initial condition

$$\theta(0) = 0 \tag{2.51}$$

Also, from equations (2.50) and (2.51) we have

$$d\theta(0)/dt = \sqrt{\frac{1}{(1/2)b^2}\left\{-\frac{1}{2a} + \frac{1}{(a-c)}\right\}} \tag{2.52}$$

Then, from equations (2.49) and (2.52)

$$dy(0)/dt = bd\theta(0)/dt = \sqrt{2\left\{-\frac{1}{2a} + \frac{1}{(a-c)}\right\}} \tag{2.53}$$

From equations (2.41), (2.43), and (2.53)

$$dy(0)/dt = bd\theta(0)/dt = \sqrt{-\frac{1}{\frac{10000}{13431}} + \frac{2}{\frac{10000}{13431} - \frac{6569}{26862}}} = 1.63000 \tag{2.54}$$

Equation (2.51) is the fifth initial condition in subroutine INITAL, Y(5) = 0.; that is, we now have the following correspondences: $\theta(t) \rightarrow Y(5)$ (in COMMON/Y/), $d\theta/dt$ (from equation (2.50)) $\rightarrow$ YT(5) (in COMMON/F/). Also, equation (2.53), or equivalently, equation (2.54), is the basis for the fourth initial condition in INITAL, Y(4) = 1.63. In summary, we can use equations (2.39), (2.40), (2.50), and (2.51) to compute a solution to equations (2.8) and (2.9), i.e., $x(t)$ and $y(t)$, which can be compared with the numerical solution computed by RKF45, i.e., Y(1) and Y(3); this second solution, in turn, requires the numerical integration of equation (2.50), which is the fifth ODE programmed in INITAL, DERV, and PRINT.

The five ODEs, equations (2.18) to (2.21) and (2.50), are programmed in subroutine DERV (the equation numbers in the comments refer to the equation numbers in subroutine INITAL), listed in Program 2.2c.

The coding in DERV is a straightforward representation of equations (2.18) to (2.21) and equation (2.50). In other words, a point we can note about this coding is that it follows directly from the ODEs, in contrast with the finite difference equations (2.26) to (2.29) in which the appearance departs significantly from the ODEs, equations (2.18) to (2.21), and therefore complicates the coding of a solution.

Subroutine PRINT to print and plot the numerical solution is listed in Program 2.2d. Subroutine PRINT first computes the solution given by equations (2.39) and (2.40) using $\theta(t)$ computed from equation (2.50), which we will call the "exact" solution even though it involves a numerical integration of equation (2.50). Then the difference between the numerical and exact solutions is computed. Finally, t, $\theta(t)$, the numerical and exact solutions and the difference between the two solutions, and the energy of the numerical solution are printed with FORMAT 1. Also, subroutine PLOT is called to plot the solution.

Function ENERGY and subroutine PLOT are listed in Program 2.2e, and as might be expected, are similar to those in Program 2.1e. Again, subroutine PLOT produces a file of numbers and characters to be plotted which is sent to a plotting system; as indicated previously for Program 2.1e. This plotting is machine specific, and will most likely have to be changed for different hardware and software configurations. One essential difference between Programs 2.1e and 2.2e is that the latter does not have a subroutine JAC because RKF45 does not use the Jacobian matrix of the ODE system since it is an explicit integrator. We will get into these mathematical details later.

```
      SUBROUTINE DERV
      PARAMETER (N=5)
      COMMON/T/          T,       NSTOP,       NORUN
     1      /Y/       Y(N)
     2      /F/    DYDT(N)
     3      /C/          R,           A,           B,           C
     4      /I/         IP,          NP
C...
C...  RADIUS (WITH CUBING TO MINIMIZE CALCULATIONS)
         R=SQRT(Y(1)**2+Y(3)**2)**3
C...
C...  EQUATION (1)
         DYDT(1)=Y(2)
         DYDT(2)=-Y(1)/R
C...
C...  EQUATION (2)
         DYDT(3)=Y(4)
         DYDT(4)=-Y(3)/R
C...
C...  EQUATION (11)
         SINE=SIN(Y(5))
         COSE=COS(Y(5))
         TERM1=0.5*((A*SINE)**2+(B*COSE)**2)
         TERM2=-1.0/(2.0*A)+1.0/SQRT((A*COSE-C)**2+(B*SINE)**2)
         DYDT(5)=SQRT((1.0/TERM1)*TERM2)
      RETURN
      END
```

Program 2.2c Subroutine DERV to Calculate the Derivatives of Equations (2.18) to (2.21), and (2.50).

Finally, file DATA read by main program ORBIT2 is listed in Program 2.2f. These data define a 21-point orbit since t runs from 0 to 10 in steps of 0.5 (counting the initial condition). As expected, five ODEs are specified, and they are integrated with an error tolerance of 0.0000001; this tolerance is read by main program ORBIT2 (Program 2.2a) and is used as the relative error tolerance (RELERR) when RKF45 is called. Ordinarily, this error tolerance would be excessively stringent for scientific applications (1 part in 10^7). However, we intentionally chose such a small tolerance in computing a numerical solution for comparison with the exact solution, as implemented in subroutine PRINT. In other words, we wanted to avoid having the comparison between the numerical and exact solutions confounded by integration errors resulting from a loose error tolerance.

Abbreviated output from Program 2.2 is listed in Table 2.3 and plotted in Figure 2.3.

Table 2.3 indicates that the numerical and exact solutions agree to six figures at the beginning of the solution, but by the end of the solution, the two solutions agree to only about four figures. This would suggest some accumulation of integration error as the solution evolves, and a possible explanation is given subsequently. However, the accuracy is still much better than for the numerical solution in Table 2.1 which was substantially in error; this latter point can be further confirmed by comparing the exact solution of Table 2.3 with the numerical solution of Table 2.1. RKF45 computed this relatively accurate numerical solution without any specification of the integration step; rather, RKF45 automatically selected the integration step as it moved along the solution in an attempt to satisfy the specified error criterion of 1.0×10^{-7} (although,

```
      SUBROUTINE PRINT(NI,NO)
      PARAMETER (N=5)
      COMMON/T/          T,      NSTOP,      NORUN
     1      /Y/      Y(N)
     2      /F/   DYDT(N)
     3      /C/          R,          A,          B,          C
     4      /I/         IP,         NP
      REAL YE(N),   ER(N)
C...
C...  PRINT THE NUMERICAL SOLUTION
C...
C...     ENERGY, EQUATION (8)
         EN=ENERGY(N,Y)
C...
C...     EXACT SOLUTION AND ERROR IN NUMERICAL SOLUTION
C...     (THETA = Y(5), DTHETA/DT = DYDT(5) IN EQUATION (11))
         THETA=   Y(5)
         DTHDT=DYDT(5)
         YE(1)=-C+A*COS(THETA)
         YE(2)=-A*SIN(THETA)*DTHDT
         YE(3)=   B*SIN(THETA)
         YE(4)= B*COS(THETA)*DTHDT
         DO 2 I=1,N-1
            ER(I)=Y(I)-YE(I)
2        CONTINUE
C...
C...     PRINT NUMERICAL AND EXACT SOLUTIONS, NUMERICAL ERROR AND
C...     ENERGY
         WRITE(NO,1)T,THETA,(I,Y(I),YE(I),ER(I),I=1,N-1),EN
1        FORMAT('     T = ',F6.3,'    THETA = ',F6.3,//,
     +     4(2X,   '     I = ',I2,        '     Y = ',F9.6,
     +             '    YE = ',F9.6,      '    ER = ',E9.3,/),/,2X,
     +             '    EN = ',F9.6,///)
C...
C...  STORE THE NUMERICAL SOLUTION FOR PLOTTING
         IP=IP+1
         CALL PLOT
      RETURN
      END
```

Program 2.2d Subroutine `PRINT` to Print and Plot the Solution of Equations (2.18) to (2.21).

again, this error criterion was not met toward the end of the solution). This feature of automatic step size adjustment is a hallmark of a quality ODE integrator.

In addition, as we noted before, it was not necessary to write the ODEs, equations (2.18) to (2.21), in some approximate form suitable for programming (e.g., in terms of finite differences); rather, they could be programmed directly in subroutine `DERV`, which greatly facilitates the preparation of new problems for computer solution, particularly as the number and complexity of the ODEs increases.

The plot of the solution in Figure 2.3 suggests that the orbit is again not closed and periodic as in Figure 2.1. However, this is not the correct conclusion to be drawn from

```
      FUNCTION ENERGY(N,Y)
C...
C...  FUNCTION ENERGY COMPUTES THE TOTAL ENERGY OF THE SYSTEM
C...
      REAL Y(N)
C...
C...  ENERGY, EQUATION (8)
         R=SQRT(Y(1)**2+Y(3)**2)
         ENERGY=0.5*(Y(2)**2+Y(4)**2)-1.0/R
      RETURN
      END

      SUBROUTINE PLOT
      PARAMETER (N=5)
      COMMON/T/          T,      NSTOP,      NORUN
     1      /Y/      Y(N)
     2      /F/   DYDT(N)
     3      /C/         R,          A,          B,          C
     4      /I/        IP,         NP
C...
C...  INITIATE PLOTTING
      IF(IP.EQ.1)THEN
C...
C...     OPEN FILE FOR TOP DRAWER PLOTTING
         IT=4
         OPEN(IT,FILE= 'ORBIT.TOP',STATUS='UNKNOWN')
C...
C...     SCALE THE AXES
         WRITE(IT,7)
7        FORMAT(' SET LIMITS X FROM -1 TO 1 Y FROM -1 TO 1',/,
     1          ' SET FONT DUPLEX')
         WRITE(IT,4)
4        FORMAT(' SET WINDOW X 2 TO 8 Y 2 TO 8')
      END IF
C...
C...  WRITE Y(3) VS Y(1) FOR PLOTTING
         WRITE(IT,5)Y(1),Y(3)
5        FORMAT(2F10.4)
C...
C...  COMPLETE PLOTTING
      IF(IP.EQ.NP)THEN
C...
C...     CONNECT POINTS
         WRITE(IT,6)
6        FORMAT(' JOIN 1')
C...
C...     LABEL THE AXES
         WRITE(IT,8)NP
8        FORMAT(
     1   ' TITLE 2.75 8.5 "',I3,'-point orbit from RKF45 - SP"'
     2   ,/,' TITLE LEFT "       y3(t)"'
```

Program 2.2e Function ENERGY and Subroutine PLOT to Compute the Energy and Plot the Solution of Equations (2.18) to (2.25). *Continued next page.*

```
     3   ,/,' TITLE BOTTOM "y1(t)"'
     4   ,/,' TITLE 2.5 0.7 "Greenspan/Feynman Orbit Problem"')
C...
C...       END PLOT
           WRITE(IT,10)
10         FORMAT(' NEW FRAME')
        END IF
        RETURN
        END
```

Program 2.2e *Continued.*

```
FOURTH ORDER ORBIT PROBLEM - RKF45, SINGLE PRECISION
0.          10.0       0.5
    5                        0.0000001
END OF RUNS
```

Program 2.2f Data File DATA Read by Main Program ORBIT2 for an Output Interval of 0.5.

Figure 2.3. The apparent lack of a closed periodic orbit is due to the coarse grid in t of only 21 points, in combination with the fact that equal spacing in t of 0.5 is not the same as equal spacing in $\theta(t)$ given by the integration of equation (2.50). To substantiate this conclusion, the calculation was repeated with the data file of programming (all other coding remained the same).

The abbreviated output and plot produced by the data file in Program 2.2g are in Table 2.4 and Figure 2.4. We see from Table 2.4 that again the numerical and exact solutions are in six figure agreement at the beginning of the solution and in four figure agreement at the end of the solution. However, these differences are imperceptible in the plot of Figure 2.4 which indicates the orbit is closed and periodic.

We might then inquire why the solutions in Tables 2.3 and 2.4 lost accuracy as the numerical integration proceeded. As we will observe in the discussion in the next section, the automatic adjustment of the integration step to achieve the user-specified error tolerance is based on an estimate of the truncation error as the numerical integration proceeds, i.e., the error resulting from the truncation of the underlying Taylor series of the integration algorithm. However, another principal source of error which the integration algorithm cannot monitor is *round-off* due to the *finite word length* of the computer. The preceding calculations were performed with 32-bit arithmetic with a maximum accuracy of about one part in 10^7, which is roughly the accuracy of the numerical solution at the beginning of the solution. However, by the end of the solution, the *round-off error* had apparently accumulated to the extent that an accuracy of only one part in 10^4 was achieved; recall that the integrator might take hundreds, or even thousands, of steps in computing a solution; thus even a small amount of round-off error at each step can accumulate to produce a rather substantial total error toward the end of the solution.

If this explanation is correct, then executing Program 2.2 in double precision, which uses 64-bit arithmetic corresponding to an accuracy of approximately one part in 10^{14}

Table 2.3 Abbreviated File OUTPUT produced by Program 2.2 for an Output Interval of 0.5. *Continued next page.*

```
RUN NO. -    1  FOURTH ORDER ORBIT PROBLEM - RKF45, SINGLE PRECISION

INITIAL T -   0.000E+00

  FINAL T -   0.100E+02

  PRINT T -   0.500E+00

NUMBER OF DIFFERENTIAL EQUATIONS -    5

INTEGRATION ALGORITHM - RKF45

MAXIMUM INTEGRATION ERROR -   0.100E-06

    T =   0.000    THETA =   0.000

    I =   1    Y =   0.500000  YE =   0.500000  ER = 0.298E-07
    I =   2    Y =   0.000000  YE =   0.000000  ER = 0.000E+00
    I =   3    Y =   0.000000  YE =   0.000000  ER = 0.000E+00
    I =   4    Y =   1.630000  YE =   1.630000  ER = -.119E-06

   EN = -0.671550


    T =   0.500    THETA =   1.066

    I =   1    Y =   0.115726  YE =   0.115726  ER = -.104E-06
    I =   2    Y = -1.205860  YE = -1.205860  ER = -.358E-06
    I =   3    Y =   0.615429  YE =   0.615429  ER = 0.000E+00
    I =   4    Y =   0.629758  YE =   0.629757  ER = 0.596E-07

   EN = -0.671550


    T =   1.000    THETA =   1.870

    I =   1    Y = -0.464271  YE = -0.464271  ER = -.358E-06
    I =   2    Y = -1.009460  YE = -1.009460  ER = 0.119E-06
    I =   3    Y =   0.671919  YE =   0.671919  ER = 0.596E-07
    I =   4    Y = -0.294493  YE = -0.294493  ER = 0.000E+00

   EN = -0.671550
                  .                                  .
                  .                                  .
                  .                                  .
    T =   9.000    THETA =  14.331

    I =   1    Y = -0.388097  YE = -0.388118  ER = 0.214E-04
    I =   2    Y = -1.069453  YE = -1.069438  ER = -.149E-04
    I =   3    Y =   0.690048  YE =   0.690041  ER = 0.715E-05
    I =   4    Y = -0.198473  YE = -0.198506  ER = 0.325E-04
```

Table 2.3 *Continued.*

```
EN = -0.671547


 T =  9.500   THETA = 15.001

 I =  1   Y = -0.810463  YE = -0.810474  ER = 0.110E-04
 I =  2   Y = -0.602653  YE = -0.602629  ER = -.244E-04
 I =  3   Y =  0.456991  YE =  0.456971  ER = 0.193E-04
 I =  4   Y = -0.665785  YE = -0.665803  ER = 0.176E-04

EN = -0.671547


 T = 10.000   THETA = 15.601

 I =  1   Y = -0.984812  YE = -0.984810  ER = -.280E-05
 I =  2   Y = -0.093600  YE = -0.093570  ER = -.302E-04
 I =  3   Y =  0.075345  YE =  0.075320  ER = 0.251E-04
 I =  4   Y = -0.820409  YE = -0.820415  ER = 0.596E-05

EN = -0.671547
```

```
FOURTH ORDER ORBIT PROBLEM - RKF45, SINGLE PRECISION
0.        10.0       0.05
    5                     0.0000001
END OF RUNS
```

Program 2.2g Data File `DATA` Read by Main Program `ORBIT2` for an Output Interval of 0.05.

should show an improvement in the accuracy of the numerical solution. To test this idea, Program 2.2 was executed with double-precision arithmetic. The abbreviated output for the 21-point oribit, which can be compared with Table 2.3, is listed in Table 2.5.

Note that now the agreement between the numerical and exact solutions is better than seven figures, so that the use of double-precision arithmetic was clearly effective in improving the accuracy of the numerical solution to this problem. This agreement is consistent with the error tolerance specified for `RKF45` of 1.0×10^{-7}; also, the energy of the numerical solution is correct to six figures.

Just to complete the story, we also list in Table 2.6 the abbreviated output for the 201-point orbit. As in the case of the double-precision, 21-point orbit, the numerical and exact solutions are in agreement to better than seven figures, which is consistent with the error tolerance specified for `RKF45` of 1.0×10^{-7}; also, the energy of the numerical solution is correct to six figures.

Thus, in summary, we conclude that `RKF45` has produced an accurate solution with a minimum of programming and concern about solution accuracy. It is for these reasons that we recommend the use of quality ODE library integrators. However, we do not recommend the uninformed use of library routines, and in the present case,

Table 2.4 Abbreviated File OUTPUT produced by Program 2.2 for an Output Interval of 0.05. *Continued next page.*

```
RUN NO. -    1  FOURTH ORDER ORBIT PROBLEM - RKF45, SINGLE PRECISION

INITIAL T -  0.000E+00

  FINAL T -  0.100E+02

  PRINT T -  0.500E-01

NUMBER OF DIFFERENTIAL EQUATIONS -   5

INTEGRATION ALGORITHM - RKF45

MAXIMUM INTEGRATION ERROR -  0.100E-06


    T =  0.000   THETA =  0.000

    I =  1   Y =  0.500000  YE =  0.500000  ER = 0.298E-07
    I =  2   Y =  0.000000  YE =  0.000000  ER = 0.000E+00
    I =  3   Y =  0.000000  YE =  0.000000  ER = 0.000E+00
    I =  4   Y =  1.630000  YE =  1.630000  ER = -.119E-06

   EN = -0.671550


    T =  0.050   THETA =  0.116

    I =  1   Y =  0.495016  YE =  0.495016  ER = 0.298E-07
    I =  2   Y = -0.198686  YE = -0.198686  ER = 0.000E+00
    I =  3   Y =  0.081229  YE =  0.081229  ER = 0.745E-08
    I =  4   Y =  1.613807  YE =  1.613807  ER = 0.000E+00

   EN = -0.671550


    T =  0.100   THETA =  0.231

    I =  1   Y =  0.480260  YE =  0.480260  ER = 0.000E+00
    I =  2   Y = -0.389696  YE = -0.389696  ER = 0.000E+00
    I =  3   Y =  0.160860  YE =  0.160860  ER = 0.000E+00
    I =  4   Y =  1.566471  YE =  1.566471  ER = 0.238E-06

   EN = -0.671550

                  .                               .
                  .                               .
                  .                               .
    T =  9.900   THETA = 15.483

    I =  1   Y = -0.970343  YE = -0.970347  ER = 0.352E-05
    I =  2   Y = -0.195777  YE = -0.195745  ER = -.315E-04
    I =  3   Y =  0.156835  YE =  0.156810  ER = 0.250E-04
```

Table 2.4 *Continued.*

```
 I =  4   Y = -0.808267  YE = -0.808273  ER = 0.602E-05

EN = -0.671548


 T =  9.950   THETA = 15.542

 I =  1   Y = -0.978855  YE = -0.978857  ER = 0.197E-05
 I =  2   Y = -0.144683  YE = -0.144651  ER = -.318E-04
 I =  3   Y =  0.116233  YE =  0.116208  ER = 0.253E-04
 I =  4   Y = -0.815427  YE = -0.815431  ER = 0.483E-05

EN = -0.671548


 T = 10.000   THETA = 15.601

 I =  1   Y = -0.984811  YE = -0.984812  ER = 0.417E-06
 I =  2   Y = -0.093580  YE = -0.093548  ER = -.319E-04
 I =  3   Y =  0.075328  YE =  0.075303  ER = 0.254E-04
 I =  4   Y = -0.820413  YE = -0.820416  ER = 0.346E-05

EN = -0.671548
```

we should have some understanding of how RKF45 functions. We therefore consider in the next section some of the general characteristics of single-step methods such as those in RKF45, and then go on to multistep methods in the following section.

We conclude this discussion with two additional details:

1. In the preceding discussion, we considered the accuracy of single- and double-precision Fortran, and stated that they correspond to approximately one part in 10^7 and one part in 10^{14}, respectively. Another way to state this conclusion is in terms of the *unit roundoff* or *machine epsilon* of the computer. In the present case, these are approximately 10^{-7} and 10^{-14} for single- and double-precision Fortran, respectively. An estimate of the machine epsilon can easily be computed with the following small programs

```
      PROGRAM EPSILON
      EPS=1.0E0
1     EPS=2.0E-01*EPS
      IF((EPS+1.0E0).GT.1.0E0)GO TO 1
      WRITE(*,2)EPS
2     FORMAT(' THE APPROXIMATE MACHINE EPSILON = ',E12.3)
      STOP
      END
```

   for single precision and

```
      PROGRAM EPSILON
      IMPLICIT DOUBLE PRECISION (A-H,O-Z)
```

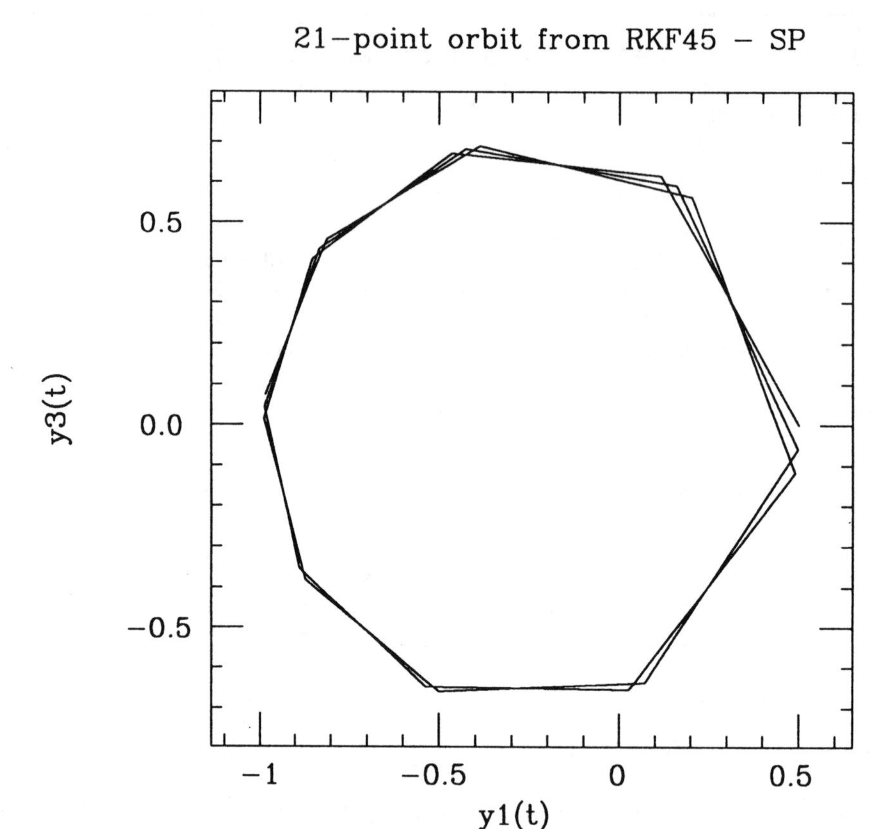

Figure 2.3 21-point Plot of Solution produced by Program 2.2.

```
      EPS=1.0D0
1     EPS=2.0D-01*EPS
      IF((EPS+1.0D0).GT.1.0D0)GO TO 1
      WRITE(*,2)EPS
2     FORMAT(' THE APPROXIMATE MACHINE EPSILON = ',D12.3)
      STOP
      END
```

in double precision. These programs indicate that the machine epsilon is the smallest number the computer can detect relative to the number one. The reader should execute these programs on one or more local computers to observe the values of the machine epsilon.

2. The double-precision versions of Programs 2.1 and 2.2 are included on the diskette available from the author along with the single-precision versions.

Table 2.5 Abbreviated File `OUTPUT` produced by Program 2.2 in Double Precision for an Output Interval of 0.5. *Continued next page.*

```
RUN NO. -    1  FOURTH ORDER ORBIT PROBLEM - RKF45, DOUBLE PRECISION

INITIAL T -  0.000E+00

  FINAL T -  0.100E+02

  PRINT T -  0.500E+00

NUMBER OF DIFFERENTIAL EQUATIONS -   5

INTEGRATION ALGORITHM - RKF45

MAXIMUM INTEGRATION ERROR -  0.100E-07

    T =  0.000   THETA =  0.000

    I =  1   Y =  0.500000  YE =  0.500000  ER = 0.000E+00
    I =  2   Y =  0.000000  YE =  0.000000  ER = 0.000E+00
    I =  3   Y =  0.000000  YE =  0.000000  ER = 0.000E+00
    I =  4   Y =  1.630000  YE =  1.630000  ER = -.278E-16

   EN = -0.671550


    T =  0.500   THETA =  1.066

    I =  1   Y =  0.115726  YE =  0.115726  ER = -.575E-08
    I =  2   Y = -1.205860  YE = -1.205860  ER = -.125E-07
    I =  3   Y =  0.615429  YE =  0.615429  ER = -.236E-08
    I =  4   Y =  0.629758  YE =  0.629758  ER = -.105E-08

   EN = -0.671550


    T =  1.000   THETA =  1.870

    I =  1   Y = -0.464271  YE = -0.464271  ER = -.103E-07
    I =  2   Y = -1.009460  YE = -1.009460  ER = -.332E-08
    I =  3   Y =  0.671919  YE =  0.671919  ER = -.512E-08
    I =  4   Y = -0.294493  YE = -0.294493  ER = -.217E-08

   EN = -0.671550

                  .                                .
                  .                                .
                  .                                .

    T =  9.000   THETA = 14.331

    I =  1   Y = -0.388121  YE = -0.388121  ER = -.209E-07
    I =  2   Y = -1.069436  YE = -1.069436  ER = -.400E-07
```

Table 2.5 *Continued.*

```
 I =  3   Y =  0.690041  YE =  0.690041  ER = -.239E-07
 I =  4   Y = -0.198509  YE = -0.198509  ER = 0.580E-07

EN = -0.671550


 T =  9.500   THETA = 15.001

 I =  1   Y = -0.810476  YE = -0.810476  ER = -.390E-07
 I =  2   Y = -0.602625  YE = -0.602625  ER = -.417E-07
 I =  3   Y =  0.456969  YE =  0.456969  ER = 0.797E-08
 I =  4   Y = -0.665805  YE = -0.665805  ER = 0.833E-07

EN = -0.671550


 T = 10.000   THETA = 15.601

 I =  1   Y = -0.984810  YE = -0.984810  ER = -.736E-07
 I =  2   Y = -0.093566  YE = -0.093566  ER = -.104E-06
 I =  3   Y =  0.075317  YE =  0.075317  ER = 0.538E-07
 I =  4   Y = -0.820415  YE = -0.820415  ER = 0.938E-07

EN = -0.671550
```

2.3 Single-Step Methods

As noted previously, the "45" in RKF45 denotes the use of fourth- and fifth-order formulas for the integration of ODEs. Basically, this means that these formulas match a *Taylor series* up to fourth- and fifth-order terms, respectively, in taking a step along the solution. We will subsequently consider the use of the Taylor series as the basis for ODE integration. For now, we simply list the integration formulas in RKF45. First, we compute six so-called Runge-Kutta constants, k_1 to k_6 (which, in fact, are not constant but change with the numerical solution)

$$k_1 = hf(y_i, t_i) \tag{2.55}$$

$$k_2 = hf(y_i + k_1/4, t_i + h/4) \tag{2.56}$$

$$k_3 = hf(y_i + (3/32)k_1 + (9/32)k_2, t + (3/8)h) \tag{2.57}$$

$$k_4 = hf(y_i + (1932/2197)k_1 - (7200/2197)k_2 + (7296/2197)k_3, t_i + (12/13)h) \tag{2.58}$$

$$k_5 = hf(y_i + (439/216)k_1 - 8k_2 + (3680/513)k_3 - (845/4104)k_4, t_i + h) \tag{2.59}$$

$$k_6 = hf(y_i - (8/27)k_1 + 2k_2 - (3544/2565)k_3 + (1859/4104)k_4 - (11/40)k_5, t_i + (1/2)h) \tag{2.60}$$

where we have now switched to the index "i" to denote a position along the solution rather than "k" (as in equations (2.26) to (2.29)) to avoid confusion with the Runge-Kutta constants, k_1 to k_6; also, the integration step is now denoted with h rather than Δt, in accordance with the usual convention. Equation (2.55) defines k_1 in terms of the derivative function to be integrated

$$dy/dt = f(y, t) \tag{2.61}$$

for which the initial condition is

$$y(t_0) = y_0 \tag{2.62}$$

Note that equations (2.61) and (2.62) are just a special case of equations (1.1) and (1.2) with $n = 1$ (one first-order ODE). However, the integration algorithms we will discuss apply to systems of first-order ODEs such as equations (1.1) and (1.2) as well as to a single ODE such as equations (2.61) and (2.62).

Thus we can compute k_1 at the initial point using equations (2.55), (2.61), and (2.62) with $i = 0$ and a given value of h

$$k_1 = hf(y_0, t_0) \tag{2.63}$$

Recall again that $f(y, t)$ is a given function defined by the ODE problem, as illustrated by the examples which follow equations (1.1) and (1.2).

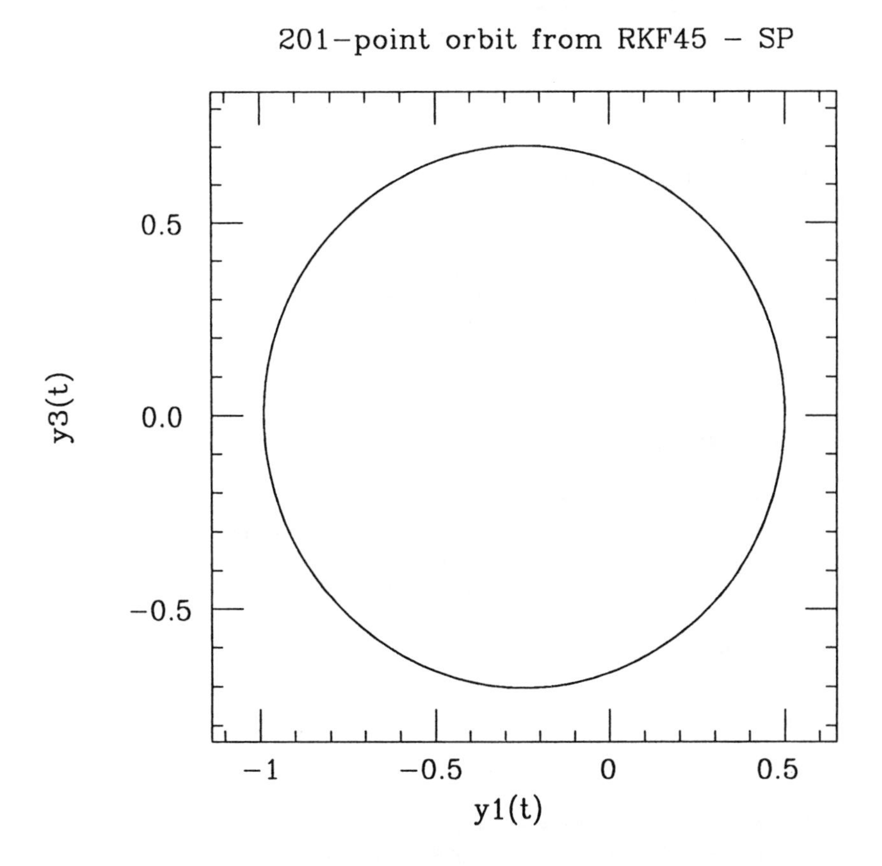

Figure 2.4 201-point Plot of Solution produced by Program 2.2.

Table 2.6 Abbreviated File OUTPUT produced by Program 2.2 in Double Precision for an Output Interval of 0.05. *Continued next page.*

```
RUN NO. -    1  FOURTH ORDER ORBIT PROBLEM - RKF45, DOUBLE PRECISION

INITIAL T -   0.000E+00

  FINAL T -   0.100E+02

  PRINT T -   0.500E-01

NUMBER OF DIFFERENTIAL EQUATIONS -    5

INTEGRATION ALGORITHM - RKF45

MAXIMUM INTEGRATION ERROR -  0.100E-07

    T =  0.000   THETA =  0.000

    I =  1   Y =  0.500000  YE =  0.500000  ER = 0.000E+00
    I =  2   Y =  0.000000  YE =  0.000000  ER = 0.000E+00
    I =  3   Y =  0.000000  YE =  0.000000  ER = 0.000E+00
    I =  4   Y =  1.630000  YE =  1.630000  ER = -.278E-16

   EN = -0.671550


    T =  0.050   THETA =  0.116

    I =  1   Y =  0.495016  YE =  0.495016  ER = -.236E-10
    I =  2   Y = -0.198686  YE = -0.198686  ER = -.252E-10
    I =  3   Y =  0.081229  YE =  0.081229  ER = -.455E-12
    I =  4   Y =  1.613807  YE =  1.613807  ER = 0.757E-10

   EN = -0.671550

    T =  0.100   THETA =  0.231

    I =  1   Y =  0.480260  YE =  0.480260  ER = -.125E-09
    I =  2   Y = -0.389696  YE = -0.389696  ER = -.213E-09
    I =  3   Y =  0.160860  YE =  0.160860  ER = -.182E-10
    I =  4   Y =  1.566471  YE =  1.566471  ER = 0.368E-09

   EN = -0.671550
                 .                                     .
                 .                                     .
                 .                                     .
    T =  9.900   THETA = 15.483

    I =  1   Y = -0.970343  YE = -0.970343  ER = -.259E-07
    I =  2   Y = -0.195763  YE = -0.195763  ER = -.343E-07
    I =  3   Y =  0.156824  YE =  0.156824  ER = 0.189E-07
    I =  4   Y = -0.808270  YE = -0.808270  ER = 0.392E-07
```

Table 2.6 *Continued.*

```
EN = -0.671550


 T =  9.950    THETA = 15.542

 I =  1    Y = -0.978854  YE = -0.978854  ER = -.277E-07
 I =  2    Y = -0.144668  YE = -0.144668  ER = -.375E-07
 I =  3    Y =  0.116222  YE =  0.116222  ER = 0.209E-07
 I =  4    Y = -0.815429  YE = -0.815429  ER = 0.388E-07

EN = -0.671550


 T = 10.000    THETA = 15.601

 I =  1    Y = -0.984810  YE = -0.984810  ER = -.297E-07
 I =  2    Y = -0.093566  YE = -0.093566  ER = -.407E-07
 I =  3    Y =  0.075317  YE =  0.075317  ER = 0.228E-07
 I =  4    Y = -0.820415  YE = -0.820415  ER = 0.382E-07

EN = -0.671550
```

Then k_2 can be calculated from equations (2.56) and (2.63)

$$k_2 = hf(y_0 + k_1/4, t_0 + h/4) \tag{2.64}$$

Proceeding in this way, using equations (2.57) to (2.60), we can calculate k_3, k_4, k_5, and k_6. Then we can step to the first point along the solution after the initial condition using the fourth-order *stepping formula* (with $i = 0$)

$$y_{i+1,4} = y_i + (25/216)k_1 + (1408/2565)k_3 + (2197/4104)k_4 - (1/5)k_5 \tag{2.65}$$

and the fifth-order stepping formula formula

$$\begin{aligned} y_{i+1,5} = y_i &+ (16/135)k_1 + (6656/12825)k_3 + (28561/56430)k_4 \\ &- (9/50)k_5 + (2/55)k_6 \end{aligned} \tag{2.66}$$

Since for the first point, $i = 0$, we are calculating $y_{1,4}$ and $y_{1,5}$ from the initial condition point y_0 (which, of course, is known according to initial condition (2.62)). $y_{1,4}$ and $y_{1,5}$ are the solutions from the fourth- and fifth-order formulas that match the Taylor series approximation to the solution to fourth- and fifth-order terms, respectively, that is, up to and including terms $O(h^4)$ and $O(h^5)$, where "$O(\ )$" denotes "of order". Therefore, subtraction of these two solutions, i.e., $y_{1,4}$ from $y_{1,5}$, should give an estimate of the $O(h^5)$ term in the Taylor series, which can be taken as an estimate of the integration error for the fourth-order result, $\epsilon_{i+1,4}$

$$\epsilon_{i+1,4} = y_{i+1,5} - y_{i+1,4} \tag{2.67}$$

(with $i = 0$ for the first point). This error, which is termed the *truncation error* since it results from a truncated Taylor series approximation of the solution, can then be used to adjust the integration step, h. For example, if this estimated error is larger than the user-specified error, h can be reduced, and the calculation from $i = 0$ to $i = 1$ can be repeated.

Once a step is taken along the solution with the user-prescribed accuracy, the procedure is repeated for the next step, using equations (2.55) to (2.60) and (2.65) to (2.67) with $i = 1$ to compute $y_{2,4}$ and $y_{2,5}$, then $y_{3,4}$ and $y_{3,5}$, etc. until the final value of t is reached. This briefly is the operation of subroutine `RKF45` as illustrated by Program 2.2. To further elucidate the details of the calculations, subroutine `FS_RKF45`, a fixed-step counterpart of `RKF45`, is listed in Program 2.3a.

The arguments of `FS_RKF45` are the number of first order ODEs to be integrated, `N`, and the fixed integration step `H`; all of the other communication with this subroutine is through the three basic `COMMON` blocks, `/T/`, `/Y/`, and `/F/`. The relationship of the coding to equations (2.55) to (2.60) and (2.65) to (2.67) is apparent. For example, `DO` loops 1 to 6 compute the six Runge-Kutta constants, k_1 to k_6, according to equations (2.55) to (2.60). Whenever a derivative function is required in these equations, i.e., $f(y, t)$ in equation (2.61), a call to `DERV` provides the derivative function (or derivative vector for a system of `N` ODEs) through `COMMON/F/`. Once all six Runge-Kutta constants are computed, `DO` loop 6 implements equations (2.65) and (2.66), and `DO` loop 7 implements equation (2.67). Note that the estimated error is stored in array `E` and is available to other subroutines through `COMMON/RK7/`. Also, as the final step in `FS_RKF45`, the estimated error is added to the fourth-order result (in `DO` loop 7) to produce a fifth-order result in array `Y` which is then available as the solution of the ODEs through `COMMON/Y/`.

Also, we should note that the `RKF45` formulas require six derivative evaluations, i.e., six calls to `DERV`, for each step along the solution. This computational effort is the major disadvantage of the Runge-Kutta method since it can be the dominant portion of the total computation of a numerical ODE system solution, particularly if the calculation of derivatives is computationally intensive, e.g., if the calculations in `DERV` require substantial computer time. The calculation of derivatives is not a major concern for small ODE problems, but when the number of ODEs is large, e.g., several thousand ODEs is not uncommon in scientific and engineering applications, then the number of derivative evaluations becomes a major consideration, and this number should be minimized. Again, `RKF45` requires six derivative evaluations per step along the solution; we will subsequently consider the Adams-Bashforth formulas which have accuracy comparable to the `RKF45` formulas, yet require only one derivative evaluation per step along the solution.

To illustrate the use of `FS_RKF45`, a main calling program, `FS`, and subroutines `INITAL`, `DERV`, and `PRINT` are listed in Program 2.3b for the model problem (a single, first-order linear ODE with eigenvalue λ)

$$dy/dt = \lambda y \tag{2.68}$$

$$y(t_0) = y_0 \tag{2.69}$$

where for simplicity, we take $t_0 = 0$. Main program `FS` is similar to main program `ORBIT2` in Program 2.2a except that the variable step features required by the call to `RKF45` in `ORBIT2` are now replaced by the fixed-step features for the call to `FS_RKF45` in `FS`; also, `FS` is coded in double-precision format to minimize the effect of computer word length (precision) on the error in the numerical ODE solutions. Subroutines `INITAL`, `DERV`, and `PRINT` follow directly from equations (2.68) and (2.69). Also, in subroutine `PRINT`, the exact solution to equations (2.68) and (2.69), with $y_0 = 1$, is programmed

$$y(t) = e^{\lambda t} \tag{2.70}$$

```
      SUBROUTINE FS_RKF45(N,H)
C...
C...  SUBROUTINE FS_RKF45 IS A FIXED STEP IMPLEMENTATION OF THE RKF45
C...  FORMULAS
C...
C...  ARGUMENT LIST
C...
C...     N     NUMBER OF FIRST ORDER ODES TO BE INTEGRATED (INPUT)
C...
C...     H     (FIXED) INTEGRATION STEP (INPUT)
C...
C...  THE ODE SOLUTION IS RETURNED THROUGH COMMON/Y/
C...
      IMPLICIT DOUBLE PRECISION (A-H,O-Z)
C...
C...  DECLARE CERTAIN VARIABLES STARTING WITH I TO N DOUBLE PRECISION
      DOUBLE PRECISION K1,K2,K3,K4,K5,K6
C...
C...  COMMON AREAS FOR LINKAGE TO INITAL, DERV AND PRINT (NOTE THAT
C...  THE UNIT DIMENSIONS, E.G., Y(1), REQUIRE THAT THE CALLING
C...  PROGRAM MUST BE LOADED FIRST TO DEFINE THE SIZE OF THE COMMON
C...  BLOCKS)
      COMMON/T/T,NSTOP,NORUN
      COMMON/Y/Y(1)
      COMMON/F/F(1)
C...
C...  COMMON AREA TO SIZE WORKING ARRAYS FOR A MAXIMUM OF NMAX ODES
      PARAMETER (NMAX=250)
      COMMON/RK1/K1(NMAX)
      COMMON/RK2/K2(NMAX)
      COMMON/RK3/K3(NMAX)
      COMMON/RK4/K4(NMAX)
      COMMON/RK5/K5(NMAX)
      COMMON/RK6/K6(NMAX)
      COMMON/RK7/ E(NMAX)
      COMMON/RK8/Y0(NMAX)
      COMMON/RK9/F0(NMAX)
C...
C...  STORE THE DEPENDENT VARIABLE AND DERIVATIVE VECTORS, AND THE
C...  INDEPENDENT VARIABLE AT THE BASE POINT FOR THE NEXT STEP ALONG
C...  THE SOLUTION
      CALL DERV
      DO 8 I=1,N
      Y0(I)=Y(I)
8     F0(I)=F(I)
      T0=T
C...
C...  COMPUTE K1
      DO 1 I=1,N
      K1(I)=H*F0(I)
```

Program 2.3a Subroutine FS_RKF45 for the Fixed-Step Integration of a System of First-Order ODEs. *Continued next pages.*

```
1     Y(I)=Y0(I)+0.25D+00*K1(I)
      T=T0+0.25D+00*H
      CALL DERV
C...
C...  COMPUTE K2
      DO 2 I=1,N
      K2(I)=H*F(I)
2     Y(I)=Y0(I)+(3.0D+00/32.0D+00)*K1(I)
     1          +(9.0D+00/32.0D+00)*K2(I)
      T=T0+(3.0D+00/8.0D+00)*H
      CALL DERV
C...
C...  COMPUTE K3
      DO 3 I=1,N
      K3(I)=H*F(I)
3     Y(I)=Y0(I)+(1932.0D+00/2197.0D+00)*K1(I)
     1          -(7200.0D+00/2197.0D+00)*K2(I)
     2          +(7296.0D+00/2197.0D+00)*K3(I)
      T=T0+(12.0D+00/13.0D+00)*H
      CALL DERV
C...
C...  COMPUTE K4
      DO 4 I=1,N
      K4(I)=H*F(I)
4     Y(I)=Y0(I)+( 439.0D+00/ 216.0D+00)*K1(I)
     1          -(   8.0D+00            )*K2(I)
     2          +(3680.0D+00/ 513.0D+00)*K3(I)
     3          -( 845.0D+00/4104.0D+00)*K4(I)
      T=T0+H
      CALL DERV
C...
C...  COMPUTE K5
      DO 5 I=1,N
      K5(I)=H*F(I)
5     Y(I)=Y0(I)-(   8.0D+00/  27.0D+00)*K1(I)
     1          +(   2.0D+00            )*K2(I)
     2          -(3544.0D+00/2565.0D+00)*K3(I)
     3          +(1859.0D+00/4104.0D+00)*K4(I)
     4          -(  11.0D+00/  40.0D+00)*K5(I)
      T=T0+0.5D+00*H
      CALL DERV
C...
C...  COMPUTE K6
      DO 6 I=1,N
      K6(I)=H*F(I)
C...
C...  STEP ALONG THE SOLUTION FROM T = T0 TO T = T0 + H USING THE
C...  FOURTH ORDER RKF45 STEPPING FORMULA
      Y(I)=Y0(I)+(  25.0D+00/ 216.0D+00)*K1(I)
     1          +(1408.0D+00/2565.0D+00)*K3(I)
     2          +(2197.0D+00/4104.0D+00)*K4(I)
     3          -(   1.0D+00/   5.0D+00)*K5(I)
```

Program 2.3a *Continued.*

```
C...
C...  ESTIMATE THE TRUNCATION ERROR FOR EACH DEPENDENT VARIABLE
C...  (THE RHS OF SUM2 IS THE FIFTH ORDER RKF45 STEPPING FORMULA)
      SUM2=Y0(I)+(   16.0D+00/  135.0D+00)*K1(I)
     1          +( 6656.0D+00/12825.0D+00)*K3(I)
     2          +(28561.0D+00/56430.0D+00)*K4(I)
     3          -(    9.0D+00/   50.0D+00)*K5(I)
     4          +(    2.0D+00/   55.0D+00)*K6(I)
6     E(I)=SUM2-Y(I)
C...
C...  INTEGRATION CONTINUES AFTER TRUNCATION ERROR CORRECTION IS
C...  APPLIED
      DO 7 I=1,N
7     Y(I)=Y(I)+E(I)
      T=T0+H
      RETURN
      END
```

Program 2.3a *Continued.*

Since the exact solution is known, the exact error in the numerical solution is also known, and can be compared with the estimated solution from equation (2.67) in `COMMON/RK7/`.

Main Program 2.3b reads a data file which is very similar to the data files read by main programs 2.1a and 2.2a. A sample data file is listed in Program 2.3c.

The only difference between data files Programs 2.1f and 2.2f, and Program 2.3c is the interpretation of the error tolerance read as the second number of the third line. For the former, this error tolerance is used by `SNSQE` and `RKF45`, ideally to achieve the accuracy in the numerical solution specified by the error tolerance. In the case of Program 2.3c, the integration step is fixed through the integration, so the tolerance is interpreted as the ratio of the fixed integration step to the output step or interval. For example, this ratio is one for the first set of data in Program 2.3c, meaning that the fixed integration step equals the output interval of 0.1 (the third number in the second line). The use of this tolerance, which has the Fortran name `TOL` in Program 2.3b is clear from the coding in Program 2.3b.

We now have a complete program consisting of Programs 2.3a, 2.3b, and 2.3c for the fixed-step integration of equations (2.68) and (2.69), with evaluation of the numerical solution using the exact solution of equation (2.70). The output is listed in Table 2.7. The following points can be noted about this output:

1. The numerical and exact solutions agree to five figures.

2. For the first run, the estimated and exact errors are in order-of-magnitude agreement with the estimated error exceeding the exact error; the latter is desirable since the error estimate is then conservative (recall that for most problems we do not have the exact error, and therefore we can only rely on the estimated error to give an indication of the accuracy of the numerical solution).

3. The second run gives the same output as the first run, and was included just to illustrate the change in the output interval and tolerance from 0.1 and 1.0 (first run) to 1.0 and 0.1 (second run).

4. For the third run, the estimated error is less than the exact error, but experience

```
      PROGRAM FS
C...
C...  MAIN PROGRAM TO TEST SUBROUTINE FS_RKF45
C...
C...  THIS MAIN PROGRAM ESSENTIALLY CALLS SUBROUTINE FS_RKF45 FOR THE
C...  SOLUTION OF A SYSTEM OF ORDINARY DIFFERENTIAL EQUATIONS (ODES).
C...
C...  THE MODEL INITIAL CONDITIONS ARE SET IN SUBROUTINE INITAL, AND
C...  THE MODEL DERIVATIVES ARE PROGRAMMED IN SUBROUTINE DERV.  THE
C...  NUMERICAL SOLUTION IS PRINTED AND PLOTTED IN SUBROUTINE PRINT.
C...
C...  DOUBLE PRECISION USED TO MINIMIZE THE EFFECTS OF ROUND-OFF
      IMPLICIT DOUBLE PRECISION (A-H,O-Z)
C...
C...  SUBROUTINES INITAL, DERV AND PRINT.  THE FOLLOWING CODING IS FOR
C...  250 ORDINARY DIFFERENTIAL EQUATIONS (ODES).  IF MORE ODES ARE TO
C...  BE INTEGRATED, ALL OF THE 250*S SHOULD BE CHANGED TO THE
C...  REQUIRED NUMBER
      COMMON/T/          T,      NSTOP,      NORUN
     1      /Y/    Y(250)
     2      /F/    F(250)
C...
C...  COMMON AREA TO PROVIDE THE INPUT/OUTPUT UNIT NUMBERS TO OTHER
C...  SUBROUTINES
      COMMON/IO/       NI,        NO
C...
C...  ARRAY FOR THE TITLE (FIRST LINE OF DATA), CHARACTERS  END OF
C...  RUNS
      CHARACTER TITLE(20)*4, ENDRUN(3)*4
C...
C...  DEFINE THE CHARACTERS  END OF RUNS
      DATA ENDRUN/'END ','OF R','UNS '/
C...
C...  DEFINE THE INPUT/OUTPUT UNIT NUMBERS
         NI=5
         NO=6
C...
C...  DEFINE INPUT/OUTPUT FILES
         OPEN(NI,FILE='DATA'  ,STATUS='OLD')
         OPEN(NO,FILE='OUTPUT',STATUS='NEW')
C...
C...  INITIALIZE THE RUN COUNTER
      NORUN=0
C...
C...  BEGIN A RUN
1     NORUN=NORUN+1
C...
C...  INITIALIZE THE RUN TERMINATION VARIABLE
      NSTOP=0
C...
C...  READ THE FIRST LINE OF DATA
```

Program 2.3b Main Calling Program FS for Subroutine FS_RKF45 and Subroutines INITAL, DERV, and PRINT for Equations (2.68) and (2.69). *Continued next pages.*

```
      READ(NI,1000,END=999)(TITLE(I),I=1,20)
C...
C...  TEST FOR  END OF RUNS  IN THE DATA
      DO 2 I=1,3
         IF(TITLE(I).NE.ENDRUN(I))GO TO 3
2     CONTINUE
C...
C...  AN END OF RUNS HAS BEEN READ, SO TERMINATE EXECUTION
999   STOP
C...
C...  READ THE SECOND LINE OF DATA
3     READ(NI,1001,END=999)T0,TF,TP
C...
C...  READ THE THIRD LINE OF DATA
      READ(NI,1002,END=999)NEQN,TOL
C...
C...  PRINT A DATA SUMMARY
      WRITE(NO,1003)NORUN,(TITLE(I),I=1,20),T0,TF,TP,NEQN,TOL
C...
C...  INITIALIZE TIME
      T=T0
C...
C...  SET THE INITIAL CONDITIONS
      CALL INITAL
C...
C...  PRINT THE INITAL CONDITIONS
      CALL PRINT(NI,NO)
C...
C...  SET THE INTEGRATION STEP, NUMBER OF STEPS FROM T TO T + TP
      H=TOL*TP
      NS=INT(REAL(0.99D0/TOL))+1
C...
C...  STEP THROUGH NS STEPS ALONG THE SOLUTION, EACH OF LENGTH H
4     DO 5 I=1,NS
      CALL FS_RKF45(NEQN,H)
5     CONTINUE
C...
C...  PRINT THE SOLUTION
      CALL PRINT(NI,NO)
C...
C...  CHECK FOR A RUN TERMINATION
      IF(NSTOP.NE.0)GO TO 1
C...
C...  CHECK FOR THE END OF THE RUN
         IF(T.LT.(TF-0.5D0*TP))GO TO 4
C...
C...  THE CURRENT RUN IS COMPLETE, SO GO ON TO THE NEXT RUN
      GO TO 1
C...
```

Program 2.3b *Continued.*

```
C...  **************************************************************
C...
C...  FORMATS
C...
1000  FORMAT(20A4)
1001  FORMAT(3D10.0)
1002  FORMAT(I5,20X,D10.0)
1003  FORMAT(1H1,//,
     1 ' RUN NO. - ',I3,2X,20A4,//,
     2 ' INITIAL T - ',D10.3,//,
     3 '   FINAL T - ',D10.3,//,
     4 '   PRINT T - ',D10.3,//,
     5 ' NUMBER OF DIFFERENTIAL EQUATIONS - ',I3,//,
     6 ' INTEGRATION ALGORITHM - FS_RKF45 ',//,
     7 ' INTEGRATION STEP/PRINT INTERVAL = ',D10.3,//
     8   1H1)
      END

      SUBROUTINE INITAL
      IMPLICIT DOUBLE PRECISION (A-H,O-Z)
      COMMON/T/      T,  NSTOP,  NORUN
     +       /Y/   Y(1)
     +       /F/  YT(1)
C...
C...  INITIAL CONDITION
      Y(1)=1.0D0
      RETURN
      END

      SUBROUTINE DERV
      IMPLICIT DOUBLE PRECISION (A-H,O-Z)
      COMMON/T/      T,  NSTOP,  NORUN
     +       /Y/   Y(1)
     +       /F/  YT(1)
C...
C...  ODE
      YT(1)=-Y(1)
      RETURN
      END

      SUBROUTINE PRINT(NI,NO)
      IMPLICIT DOUBLE PRECISION (A-H,O-Z)
      COMMON/T/      T,  NSTOP,  NORUN
     +       /Y/   Y(1)
     +       /F/  YT(1)
C...
C...  COMMON AREA TO ACCESS THE ESTIMATED ERRORS FROM SUBROUTINE
C...  FS_RKF45
      COMMON/RK7/ E(250)
C...
C...  PRINT HEADING FOR SOLUTION
      IF(T.LT.0.001D0)WRITE(NO,2)
```

Program 2.3b *Continued.*

```
2     FORMAT(//,6X,'T',8X,'Y(1)',6X,'YEXACT',5X,'E (EST)',3X'E (EXACT)',
     +          5X,'RATIO')
C...
C...  PRINT NUMERICAL AND EXACT SOLUTIONS, ESTIMATED AND EXACT ERRORS,
C...  RATIO OF ERROR
C...
C...     EXACT SOLUTION
         YE=DEXP(-T)
C...
C...     EXACT ERROR
         EE=Y(1)-YE
C...
C...     RATIO OF ERRORS
         IF(T.LT.0.001D0)THEN
            E(1)=0.0D0
            RATIO=1.0D0
         ELSE
            RATIO=E(1)/EE
         END IF
C...
C...     PRINT RESULTS
         WRITE(NO,1)T,Y(1),YE,E(1),EE,RATIO
1        FORMAT(F7.3,2F12.5,2D12.3,F10.3)
      RETURN
      END
```

Program 2.3b *Continued.*

```
DY/DT = -Y, Y(0) = 1, FS_RKF45
0.        1.0       0.1
   1                    1.0
DY/DT = -Y, Y(0) = 1, FS_RKF45
0.        1.0       1.0
   1                    0.1
DY/DT = -Y, Y(0) = 1, FS_RKF45
0.        10.0      1.0
   1                    0.1
END OF RUNS
```

Program 2.3c Data File Read by Main Program FS.

has indicated that generally the estimated error is reliable enough to be used to adjust the integration step as in subroutine RKF45 to meet the user-specified tolerance; the success of automatic step size adjustment also depends to some extend on the details of implementation.

5. The estimated and exact errors are opposite in sign, which is due to the way they are computed (or defined). Note that in subroutine FS_RKF45 (Program 2.3a), the estimated error is added as a correction in DO loop 7. Therefore it should have the correct sign to improve the accuracy of the fourth-order solution, which therefore will be opposite in sign to the error of the numerical solution.

Table 2.7 Output of Program 2.3. *Continued next page.*

```
RUN NO. -    1  DY/DT = -Y, Y(0) = 1, FS_RKF45

INITIAL T -  0.000D+00

  FINAL T -  0.100D+01

  PRINT T -  0.100D+00

NUMBER OF DIFFERENTIAL EQUATIONS -   1

INTEGRATION ALGORITHM - FS_RKF45

INTEGRATION STEP/PRINT INTERVAL =  0.100D+01

      T          Y(1)        YEXACT       E (EST)    E (EXACT)       RATIO
  0.000       1.00000       1.00000     0.000D+00    0.000D+00       1.000
  0.100       0.90484       0.90484     0.133D-07   -0.889D-09     -14.970
  0.200       0.81873       0.81873     0.120D-07   -0.161D-08      -7.485
  0.300       0.74082       0.74082     0.109D-07   -0.218D-08      -4.990
  0.400       0.67032       0.67032     0.985D-08   -0.263D-08      -3.743
  0.500       0.60653       0.60653     0.892D-08   -0.298D-08      -2.994
  0.600       0.54881       0.54881     0.807D-08   -0.323D-08      -2.495
  0.700       0.49659       0.49659     0.730D-08   -0.341D-08      -2.139
  0.800       0.44933       0.44933     0.661D-08   -0.353D-08      -1.871
  0.900       0.40657       0.40657     0.598D-08   -0.359D-08      -1.663
  1.000       0.36788       0.36788     0.541D-08   -0.361D-08      -1.497


RUN NO. -    2  DY/DT = -Y, Y(0) = 1, FS_RKF45

INITIAL T -  0.000D+00

  FINAL T -  0.100D+01

  PRINT T -  0.100D+01

NUMBER OF DIFFERENTIAL EQUATIONS -   1

INTEGRATION ALGORITHM - FS_RKF45

INTEGRATION STEP/PRINT INTERVAL =  0.100D+00

      T          Y(1)        YEXACT       E (EST)    E (EXACT)       RATIO
  0.000       1.00000       1.00000     0.000D+00    0.000D+00       1.000
  1.000       0.36788       0.36788     0.541D-08   -0.361D-08      -1.497


RUN NO. -    3  DY/DT = -Y, Y(0) = 1, FS_RKF45

INITIAL T -  0.000D+00

  FINAL T -  0.100D+02
```

***Table 2.7** Continued.*

```
  PRINT T -   0.100D+01

NUMBER OF DIFFERENTIAL EQUATIONS -    1

INTEGRATION ALGORITHM - FS_RKF45

INTEGRATION STEP/PRINT INTERVAL =   0.100D+00
```

T	Y(1)	YEXACT	E (EST)	E (EXACT)	RATIO
0.000	1.00000	1.00000	0.000D+00	0.000D+00	1.000
1.000	0.36788	0.36788	0.541D-08	-0.361D-08	-1.497
2.000	0.13534	0.13534	0.199D-08	-0.266D-08	-0.749
3.000	0.04979	0.04979	0.732D-09	-0.147D-08	-0.499
4.000	0.01832	0.01832	0.269D-09	-0.719D-09	-0.374
5.000	0.00674	0.00674	0.990D-10	-0.331D-09	-0.299
6.000	0.00248	0.00248	0.364D-10	-0.146D-09	-0.250
7.000	0.00091	0.00091	0.134D-10	-0.627D-10	-0.214
8.000	0.00034	0.00034	0.493D-11	-0.264D-10	-0.187
9.000	0.00012	0.00012	0.181D-11	-0.109D-10	-0.166
10.000	0.00005	0.00005	0.667D-12	-0.446D-11	-0.150

Three other important characteristics of the RKF45 formulas, and Runge-Kutta integrators in general, should be noted:

1. They proceed to the next point along the solution using only the current solution point, i.e., they are so-called "one-step methods". In fact, Runge-Kutta methods and single-step methods are terms that are used interchangeably.

2. As a consequence of (1), the Runge-Kutta methods are self starting, that is, they can proceed directly from the initial condition. This contrasts with multistep methods which require more than one point to take the next step along the solution and are therefore not self-starting.

3. Lower order formulas can be embedded in higher order formulas. For example, in the case of RKF45, a fourth-order methed is embedded in a fifth-order method. This property is very useful in formulating Runge-Kutta integrators with explicit error estimates, since these estimates are simply the difference between the high and low order formulas, as we observed in the case of the RKF45 formulas (as stated in equation (2.67)).

To illustrate these properties in a second example, we consider the classical fourth-order method originally proposed by Runge and Kutta

$$k_1 = hf(y_i, t) \tag{2.71}$$

$$k_2 = hf(y_i + k_1/2, t_i + h/2) \tag{2.72}$$

$$k_3 = hf(y_i + k_2/2, t_i + h/2) \tag{2.73}$$

$$k_4 = hf(y_i + k_3, t_i + h) \tag{2.74}$$

$$y_{i+1} = y_i + (1/6)(k_1 + 2k_2 + 2k_3 + k_4) \tag{2.75}$$

The stepping formula, equation (2.75), is fourth-order correct, i.e., it fits the Taylor series,

$$y_{i+1} = y_i + \frac{dy_i}{dt}\frac{h}{1!} + \frac{d^2y_i}{dt^2}\frac{h^2}{2!} + \frac{d^3y_i}{dt^3}\frac{h^3}{3!} + \frac{d^4y_i}{dt^4}\frac{h^4}{4!} + \cdots \tag{2.76}$$

term by term, up to and including the fourth-order term, $(d^4y_i/dt^4)(h^4/4!)$.

However, a second-order Runge-Kutta is also available from the preceding equations. In particular, if k_1 and k_2 are defined by equations (2.71) and (2.72), the stepping formula for the second-order Runge-Kutta is

$$y_{i+1} = y_i + k_2 \tag{2.77}$$

which fits the Taylor series, equation (2.76), up to and including the term $(d^2y_i/dt^2)\times(h^2/2!)$.

Now, we can rewrite equation (2.75) as

$$y_{i+1} = y_i + k_2 + (1/6)(k_1 - 4k_2 + 2k_3 + k_4) \tag{2.78}$$

so that the term $(1/6)(k_1 - 4k_2 + 2k_3 + k_4)$ becomes an error estimate for the second-order method of equation (2.77). In other words, the second-order method of equation (2.77) is embedded in the fourth-order method of equation (2.75). We can then use this fact to produce an error estimate for automatic step size adjustment as we did with the `RKF45` formulas.

Thus, the error estimate for the second-order formula, equation (2.77), $\epsilon_{i+1,2}$, is

$$\begin{aligned}\epsilon_{i+1,2} &= y_{i+1,4} - y_{i+1,2} \\ &= (1/6)(k_1 + 2k_2 + 2k_3 + k_4) - k_2 \\ &= (1/6)(k_1 - 4k_2 + 2k_3 + k_4)\end{aligned} \tag{2.79}$$

Note in particular that the subtraction of the terms involving k_2 is possible since it is defined the same way for both the second- and fourth-order formulas according to equation (2.72), i.e., k_2 is the same in both cases. Note also that k_1 is defined the same way for both methods according to equation (2.71) which is necessary to give the same k_2 of equation (2.72). Of course, k_3 and k_4 appear in only the fourth-order stepping formula; thus subtraction involving them is not a problem.

Note that since $\epsilon_{i+1,2}$ of equation (2.79) is the difference between a second- and fourth-order method, it is a two-term error estimate, i.e., it estimates the terms $(d^3y_i/dt^3)(h^3/3!)$ + $(d^4y_i/dt^4)(h^4/4!)$ in the Taylor series of equation (2.76). Also, this error estimate can be added as a correction to the second-order solution of equation (2.77)

$$y_{i+1,2} = y_{i,2} + k_2 + \epsilon_{i+1,2} \tag{2.80}$$

to produce the fourth-order result of equation (2.75).

Finally, we should note the ease with which the error estimate of equation (2.79) was derived, i.e., a simple subtraction of two Runge-Kutta stepping formulas, which is possible if the lower order method is embedded in the higher order method, i.e., the Runge-Kutta constants for the two methods are defined in the same way. This general approach can then be applied to any two existing Runge-Kutta methods to produce error estimates with varying numbers of terms from the underlying Taylor series, e.g., the one-term estimate of the `RKF45` formulas, the two-term estimate of equation (2.79). For example, this use of the embedding concept has been applied by Silebi et al. (1992)

to produce Runge-Kutta methods with three-term error estimates, i.e., from second-order methods embedded in fifth-order methods; two- and three-term error estimates have been found to be very reliable and therefore to give accurate numerical solutions to ODEs.

We conclude this section by deriving the general formulas for the second-order Runge-Kutta method, of which equation (2.77) is a specific example. Also, this derivation will demonstrate that the Runge-Kutta formulas result from a term-by-term matching of the underlying Taylor series, equation (2.76). If we consider a general stepping formula for the Runge-Kutta method [Forsythe et al. (1977)]

$$y_{i+1} = y_i + \sum_{j=1}^{n} \gamma_j k_j \tag{2.81}$$

where

$$k_j = hf\left(y_i + \sum_{m=1}^{j-1} \beta_{jm} k_m, t_i + \alpha_j h\right) \tag{2.82}$$

For a second-order method, we take $n = 2$; also, in general (for a Runge-Kutta method of any order), $\beta_{11} = \alpha_1 = 0$. Equations (2.81) and (2.82) then become

$$y_{i+1} = y_i + \gamma_1 k_1 + \gamma_2 k_2 \tag{2.83}$$

$$k_1 = hf(y_i, t_i) \tag{2.84}$$

$$k_2 = hf(y_i + \beta_{21} k_1, t_i + \alpha_2 h) \tag{2.85}$$

k_2 in equation (2.85) can be expanded in a Taylor series

$$k_2 = hf(y_i, t_i) + \frac{\partial f(y_i, t_i)}{\partial y} \beta_{21} k_1 + \frac{\partial f(y_i, t_i)}{\partial t} \alpha_2 h + \cdots \tag{2.86}$$

Substitution of equation (2.86), truncated after the α_2 term, and equation (2.84) in equation (2.83) gives

$$y_{i+1} = y_i + \gamma_1 hf(y_i, t_i) + \gamma_2 hf(y_i, t_i) + \gamma_2 h^2 \frac{\partial f(y_i, t_i)}{\partial y} f(y_i, t_i) \beta_{21} + \gamma_2 h^2 \frac{\partial f(y_i, t_i)}{\partial t} \alpha_2 \tag{2.87}$$

We now compare equation (2.87) with the Taylor series, equation (2.76), up to and including second-order terms

$$y_{i+1} = y_i + \frac{dy_i}{dt} \frac{h}{1!} + \frac{d^2 y_i}{dt^2} \frac{h^2}{2!}$$

or

$$y_{i+1} = y_i + f(y_i, t_i) \frac{h}{1!} + \left\{ \frac{\partial f(y_i, t_i)}{\partial y} \frac{dy_i}{dt} + \frac{\partial f(y_i, t_i)}{\partial t} \right\} \frac{h^2}{2!} \tag{2.88}$$

A term-by-term matching of equations (2.87) and (2.88) then gives

$$\gamma_1 + \gamma_2 = 1 \tag{2.89}$$

$$\gamma_2 \beta_{21} = 1/2, \gamma_2 \alpha_2 = 1/2 \qquad (2.90)(2.91)$$

where we have made use of $dy_i/dt = f(y_i, t_i)$, i.e., the ODE to be integrated numerically. Equations (2.89) to (2.91) define three relations between four parameters, γ_1, γ_2, β_{21}, and α_1. Thus, we can choose one parameter arbitrarily, which means there are an infinite number of possible second-order Runge-Kutta formulas. For example, if we choose $\gamma_1 = 0$, then $\gamma_2 = 1$, $\beta_{21} = \alpha_2 = 1/2$ and we have equations from (2.83) to (2.85)

$$y_{i+1} = y_i + k_2 \tag{2.92}$$

$$k_1 = hf(y_i, t_i)$$

$$k_2 = hf(y_i + (1/2)k_1, t_i + (1/2)h)$$

which is just the second-order Runge-Kutta considered previously, equations (2.71), (2.72), and (2.77). Equation (2.92) is also called the *midpoint formula* since it is based on the derivative at $t_i + (1/2)h$.

Another choice of the parameters corresponds to $\gamma_1 = 1/2$, for which $\gamma_2 = 1/2$, $\beta_{21} = \alpha_2 = 1$, and we have from equations (2.83) to (2.85)

$$y_{i+1} = y_i + (1/2)k_1 + (1/2)k_2 \tag{2.93}$$

$$k_1 = hf(y_i, t_i) \tag{2.94}$$

$$k_2 = hf(y_i + k_1, t_i + h) \tag{2.95}$$

Equations (2.93) to (2.95) are called the *modified Euler method*, and are based on the average of the derivatives at t_i and t_i+h (note the term $(k_1+k_2)/2$ in equation (2.93)). The name modified Euler method suggests that equations (2.93) to (2.95) are an extension of the Euler method, which is *the* (one and only) first-order Runge-Kutta method

$$y_{i+1} = y_i + k_1 \tag{2.96}$$

$$k_1 = hf(y_i, t_i) \tag{2.97}$$

We note also that since k_1 is defined the same way for the Euler method and the modified Euler method (according to equations (2.94) and (2.97)), the former is embedded in the latter. In other words, equation (2.93) can be written as

$$y_{i+1} = y_i + k_1 + \{(1/2)k_2 - (1/2)k_1\} \tag{2.98}$$

or

$$y_{i+1} = y_i + k_1 + \epsilon_i \tag{2.99}$$

with

$$\epsilon_1 = \{(1/2)k_2 - (1/2)k_1\} \tag{2.100}$$

where ϵ_1 of equation (2.100) is the estimated error for the Euler method which can be used to adjust the integration step and also can be applied as a correction to the solution from the Euler method to give a second-order result (according to equation (2.99)). The algorithm defined by equations (2.94), (2.95), (2.99), and (2.100) has been implemented in a general purpose ODE integrator [Silebi et al. (1992)].

The derivation of higher order Runge-Kutta algorithms, in principle, follows in the same way from equations (2.81) and (2.82), in combination with the Taylor series, equation (2.76). In practice, the mathematical manipulations become quite complicated for higher order Runge-Kutta methods. The details are given by Gear (1971), including the derivation of the classical fourth-order Runge-Kutta method, equations (2.71) to (2.75), as well as other higher Runge-Kutta methods that have an optimized choice of the free parameters.

Thus far, we have been concerned with the accuracy of one-step (Runge-Kutta) ODE integrators. There is, however, another major consideration in using these integrators, their stability. We now briefly consider this second basic characteristic. The stability of ODE integrators is usually studied via the model problem of equations (2.68) and (2.69), which has the analytical solution

$$y(t) = y_0 e^{\lambda t} \tag{2.101}$$

For a stable problem, $\lambda < 0$ and equation (2.101) indicates an exponential decay with increasing t. However, we can easily observe that the numerical solution to equations (2.68) and (2.69) may depart radically from an exponential decay. To demonstrate this sharp departure of the numerical solution from the analytical solution, consider the integration of equation (2.68) by the modified Euler method, equations (2.93) to (2.95). For equation (2.68) $f(y,t) = \lambda y$, which, when substituted in equations (2.93) to (2.95) gives

$$k_1 = hf(y_i, t_i) = h\lambda y_i \tag{2.102}$$

$$k_2 = hf(y_i + k_1, t_i + h) = h\left\{\lambda(y_i + h\lambda y_i)\right\} \tag{2.103}$$

$$\begin{aligned} y_{i+1} &= y_i + (1/2)k_1 + (1/2)k_2 = y_i + (1/2)h\lambda y_i \\ &\quad + (1/2)h\left\{\lambda(y_i + h\lambda y_i)\right\} && (2.104) \\ &= \left\{1 + \lambda h + (\lambda h)^2/2!\right\} y_i && (2.105) \end{aligned}$$

This result seems reasonable if we consider t_i and t_{i+1} $(= t_i + h)$, for which the corresponding values of $y(t)$ are y_i and y_{i+1}, respectively, and we write equation (2.101) as

$$\begin{aligned} y_{i+1} &= y_0 e^{\lambda t_{i+1}} = y_0 e^{\lambda(t_i + h)} = y_0 e^{\lambda h} e^{\lambda t_i} = e^{\lambda h} y_i \\ &= \left\{1 + \lambda h + (\lambda h)^2/2! + (\lambda h)^3/3! + \cdots\right\} y_i && (2.106) \end{aligned}$$

A comparison of equations (2.105) and equation (2.106) indicates that the modified Euler method approximates the solution to equation (2.68) up to and including the h^2 terms; this is expected since the modified Euler method is a second-order Runge-Kutta method, and according to the derivation of equations (2.89) to (2.91), a second-order Runge-Kutta matches the Taylor series of equation (2.88) (or equivalently, equation (2.106)) up to and including the h^2 terms.

The same result (equation (2.105)) is obtained for the midpoint formula, equation (2.92)

$$k_1 = hf(y_i, t_i) = h\lambda y_i \tag{2.107}$$

$$k_2 = hf(y_i + (1/2)k_1, t_i + (1/2)h) = h\lambda\left\{y_i + (1/2)h\lambda y_i\right\} \tag{2.108}$$

$$\begin{aligned} y_{i+1} &= y_i + k_2 = y_i + h\lambda\left\{y_i + (1/2)h\lambda y_i\right\} \\ &= \left\{1 + \lambda h + (\lambda h)^2/2!\right\} y_i && (2.109) \end{aligned}$$

The term $\left\{1 + \lambda h + (\lambda h)^2/2! + (\lambda h)^3/3! + \cdots\right\}$ in equation (2.106) converges to $e^{\lambda h}$ for all finite λh. However, we now observe an important and unexpected property of the term $\left\{1 + \lambda h + (\lambda h)^2/2!\right\}$ in equations (2.105) and (2.109). Consider the case $\lambda < 0$

(the stable case of equation (2.68)) for which $e^{\lambda h} \leq 1$ (we take the integration step $h \geq 0$). In other words, from equation (2.106)

$$y_{i+1} = e^{\lambda h} y_i$$

$$y_{i+1} \leq y_i \tag{2.110}$$

which we would expect for an exponential decay ($\lambda < 0$). However, the term $\{1 + \lambda h + (\lambda h)^2/2!\}$ in equations (2.105) and (2.109), which is called the *amplification factor* for the modified Euler method or the midpoint formula, can be greater than one in absolute value, depending on the value of λh, so that from either equation (2.105) or (2.109)

$$y_{i+1} \geq y_i \tag{2.111}$$

In other words, the numerical solution actually increases with t (i.e., increasing i) and is unstable (again, even though the analytical solution decays exponentially with t). This unexpected behavior is numerical instability and is a concern with any numerical integration algorithm.

We see from equations (2.105) and (2.109) that the condition at which instability in the numerical solution develops is

$$|1 + \lambda h + (\lambda h)^2/2!| = 1 \tag{2.112}$$

which defines the critical value of λh for stability (for the modified Euler method or midpoint formula). This value is $\lambda h = -2$ since

$$1 + (-2) + (-2)^2/2! = 1$$

In other words, for stability, we require

$$|\lambda h| < 2 \tag{2.113}$$

for which the amplification factor $\{1 + \lambda h + (\lambda h)^2/2!\}$ will be less than one in absolute value.

We also note that condition (2.113) is the stability limit for the Euler method which has the amplification factor $\{1 + \lambda h\}$. If $\lambda h = -2$, $\{1 + (-2)\} = -1$ and therefore the Euler solution of equation (2.68) (for $\lambda h = -2$) will oscillate between y_0 and $-y_0$, i.e., it will be just on the border of instability, according to the stepping formula for the Euler method

$$y_{i+1} = \{1 + \lambda h\} y_i, \qquad i = 0, 1, 2, \ldots \tag{2.114}$$

To further illustrate the critical value $\lambda h = -2$ for the Euler method (the first-order Runge-Kutta method), the modified Euler method, and the midpoint formula (or the explicit second-order Runge-Kutta method in general), Program 2.4 calculates the amplification factor for an ODE integration algorithm of order n (`N` in the Fortran), as a function of λh. Program 2.4 merely computes the amplification factor in brackets in equation (2.106) for n terms in the Taylor series. The output for $N = 1$ is listed in Table 2.8.

Note again the critical value $\lambda h = -2$. The output of Program 2.4 for $N = 2$ (e.g., the modified Euler method and the midpoint formula) is listed in Table 2.9. The critical value $\lambda h = -2$ is clear.

The comparison of the critical values of λh for the first- and second-order Runge-Kutta methods suggests that the stability interval, in this case $-2 \leq \lambda h \leq 0$, cannot be increased by going to a higher order method, e.g., from first to second order. This is not quite the case, but the increase is small, and as we will see, does not substantially

```
      PROGRAM RK_STABILITY
C...
C...  PROGRAM RK_STABILITY COMPUTES THE STABILITY INTERVAL ALONG
C...  THE NEGATIVE REAL AXIS OF AN EXPLICIT RUNGE KUTTA METHOD OF
C...  ORDER N (THE PRODUCT LAMBDA*H IS PLOTTED ALONG THE NEGATIVE
C...  REAL AXIS WHERE LAMBDA IS THE EIGENVALUE OF THE FIRST ORDER
C...  ODE DY/DT = LAMBDA*Y AND H IS THE INTEGRATION STEP USED IN
C...  THE RUNGE KUTTA METHOD).
C...
C...  REAL SELECTED VARIABLES (LAMH IS LAMBDA*H)
      REAL LAMH
C...
C...  OPEN AN OUTPUT FILE
      OPEN(6,FILE='OUTPUT',STATUS='NEW')
C...
C...  SET THE ORDER OF THE RK METHOD
      N=1
C...
C...  PRINT A HEADING FOR THE NUMERICAL OUTPUT
         WRITE(6,4)N
4        FORMAT(' ORDER OF THE EXPLICIT RUNGE KUTTA METHOD = ',I2,//,
     +          10X,' LAMBDA*H',5X,'Y(I+1)/Y(I)',/)
C...
C...  STEP THROUGH A SERIES OF VALUES OF LAMBDA*H ALONG THE NEGATIVE
C...  REAL AXIS FROM LAMDA*H = 0 TO LAMBDA*H = -4.  AT EACH VALUE OF
C...  LAMBDA*H, COMPUTE THE AMPLIFICATION FACTOR, I.E., Y(I+1)/Y(I),
C...  FOR THE RK METHOD
C...
      STEP=0.25
      LAMH=STEP
      DO 1 J=1,17
         LAMH=LAMH-STEP
         AMP=1.0
         DO 2 I=1,N
            AMP=AMP+TERM(I,LAMH)
2        CONTINUE
C...
C...  PRINT LAMBDA*H AND THE AMPLIFICATION FACTOR FOR THE RK METHOD
         WRITE(6,3)LAMH,AMP
3        FORMAT(2X,F15.2,F15.4)
C...
C...  GO ON TO THE NEXT LAMBDA*H
1     CONTINUE
C...
C...  ALL OF THE LAMBDA*H VALUES HAVE BEEN USED
      STOP
      END

      REAL FUNCTION TERM(I,LAMH)
C...
C...  FUNCTION TERM COMPUTES THE ITH TERM IN THE TAYLOR SERIES
C...  EXPANSION OF EXP(LAMBDA*H)
```

Program 2.4 Calculation of the Amplification Factor as a Function of λh for Explicit Runge-Kutta Methods of Different Orders. *Continued next page.*

```
C...
      REAL LAMH
      TERM=LAMH**I/FACT(I)
      RETURN
      END

      REAL FUNCTION FACT(I)
C...
C...  FUNCTION FACT COMPUTES THE ITH FACTORIAL IN THE TAYLOR SERIES
C...  EXPANSION OF EXP(LAMBDA*H)
C...
      FACT=1.0
      DO 1 J=1,I
         FACT=FACT*FLOAT(J)
1     CONTINUE
      RETURN
      END
```

Program 2.4 *Continued.*

Table 2.8 Amplification Factor vs. λh for the First-Order Runge-Kutta Method.

```
ORDER OF THE EXPLICIT RUNGE KUTTA METHOD =  1

           LAMBDA*H      Y(I+1)/Y(I)

             0.00           1.0000
            -0.25           0.7500
            -0.50           0.5000
            -0.75           0.2500
            -1.00           0.0000
            -1.25          -0.2500
            -1.50          -0.5000
            -1.75          -0.7500
            -2.00          -1.0000
            -2.25          -1.2500
            -2.50          -1.5000
            -2.75          -1.7500
            -3.00          -2.0000
            -3.25          -2.2500
            -3.50          -2.5000
            -3.75          -2.7500
            -4.00          -3.0000
```

improve the stability characteristics of the explicit Runge-Kutta methods. To illustrate this point, we have in Table 2.10 the output of Program 2.4 for $n = 4$, e.g., the classical fourth-order Runge-Kutta method of equations (2.71) to (2.75).

We see that the critical value of λh is between -2.75 and -3.00 (more precisely, it is -2.785); this is also the case for the fourth-order formulas in subroutine `RKF45`. Similarly, we would find the stability interval for fifth-order Runge-Kutta methods is not much larger (for example, for the fifth-order formulas in subroutine `RKF45`).

Table 2.9 Amplification Factor vs. λh for the Second-Order Runge-Kutta Method.

ORDER OF THE EXPLICIT RUNGE KUTTA METHOD = 2

LAMBDA*H	Y(I+1)/Y(I)
0.00	1.0000
-0.25	0.7813
-0.50	0.6250
-0.75	0.5313
-1.00	0.5000
-1.25	0.5313
-1.50	0.6250
-1.75	0.7813
-2.00	1.0000
-2.25	1.2813
-2.50	1.6250
-2.75	2.0313
-3.00	2.5000
-3.25	3.0313
-3.50	3.6250
-3.75	4.2813
-4.00	5.0000

Table 2.10 Amplification Factor vs. λh for the Fourth-Order Runge-Kutta Method.

ORDER OF THE EXPLICIT RUNGE KUTTA METHOD = 4

LAMBDA*H	Y(I+1)/Y(I)
0.00	1.0000
-0.25	0.7788
-0.50	0.6068
-0.75	0.4741
-1.00	0.3750
-1.25	0.3075
-1.50	0.2734
-1.75	0.2788
-2.00	0.3333
-2.25	0.4507
-2.50	0.6484
-2.75	0.9481
-3.00	1.3750
-3.25	1.9585
-3.50	2.7318
-3.75	3.7319
-4.00	5.0000

For many ODE problems, the integration step, h, is limited by the accuracy of the integration algorithm, rather than by the stability (e.g., according to stability condition (2.113)). However, there is an important class of ODE problems for which the stability of the numerical integration determines the maximum integration step (rather than accuracy), so-called stiff ODEs. This name came from the ODE for a second-order mass-spring-damper mechanical system,

$$m\frac{d^2x}{dt^2} + c\frac{dx}{dt} + kx = f(t) \tag{2.115}$$

where x is the displacement of the mass from an equilibrium position; t is time; m is mass; c is the damping coefficient; k is the spring constant; and $f(t)$ is the externally applied force.

Equation (2.115) can be written as a system of two first-order ODEs, as discussed via the example of equations (1.3) to (1.9), but for our purpose, this is not necessary, i.e., equation (2.115) can be analyzed directly; also, initial conditions for equation (2.115) are not necessary for the present discussion.

If a solution to equation (2.115) of the form

$$x = be^{\lambda t} \tag{2.116}$$

is assumed where b and λ are constants to be determined, substitution of this trial solution in equation (2.115) gives (with $f(t) = 0$)

$$m\lambda^2 be^{\lambda t} + c\lambda be^{\lambda t} + k\lambda be^{\lambda t} = 0$$

or

$$m\lambda^2 + c\lambda + k = 0 \tag{2.117}$$

Equation (2.117) is the characteristic equation of equation (2.115); it can be factored by the quadratic formula for the eigenvalues of equation (2.115)

$$\lambda_1, \lambda_2 = \frac{-c \pm \sqrt{c^2 - 4mk}}{2m} \tag{2.118}$$

If c is large so that $c^2 \gg 4mk$ (a stiff mechanical system), then equation (2.118) gives from a Taylor series expansion of the square root around c^2

$$\lambda_1, \lambda_2 = \frac{-c \pm \sqrt{c^2} \pm (1/2)(c^2)^{-1/2}(-4mk)}{2m}$$

or

$$\lambda_1 = -(k/c), \qquad \lambda_2 = -(c/m) + (k/c) \sim -(c/m) \tag{2.119}$$

If, for example, $c = 1$, $m = k = 10^{-3}$, $\lambda_1 = -10^{-3}$, $\lambda_2 = -10^3$ so that the eigenvalues are in the ratio of $10^3/10^{-3} = 10^6$, i.e., the stiffness ratio is 10^6.

It can be shown [Schiesser (1991)] that condition (2.113) must be satisfied for each eigenvalue of a system of linear, first-order constant coefficient ODEs. Thus for $\lambda_2 = -10^3$,

$$|-10^3 h| < 2$$

so the maximum integration step for a stable numerical integration of equation (2.115) by a first- or second-order Runge-Kutta method is $h_{\max} = 0.002$. However, the problem time scale for equation (2.115) is determined by λ_1. This follows from the general solution to equation (2.115) (again, with $f(t) = 0$) which is a superposition of two solutions of the form of equation (2.116) for the two eigenvalues λ_1 and λ_2

$$x = b_1 e^{\lambda_1 t} + b_2 e^{\lambda_2 t} \tag{2.120}$$

The first exponential, $e^{\lambda_1 t}$, will decay to insignificance for $t > 10^4$, i.e., $e^{-10^{-3}(10^4)} = e^{-10}$ is small compared with the initial value at $t = 0$, e^0. Thus, the problem time scale is 10^4 (this is, of course, not a precise figure, but rather, is estimated from the magnitude of the exponential e^{-10}).

Now, we must take enough steps in the numerical integration to go from $t = 0$ to $t = 10^4$, and none of these time steps can be greater than 0.002 to maintain stability. Thus, we must take $10^4/0.002 = 5 \times 10^6$ time steps. This large number results from a combination of stability criterion (2.113) for the explicit first- and second-order Runge-Kutta methods, and the wide separation in the eigenvalues (a stiffness ratio of 10^6). In order words, the limited stability interval of the explicit Runge-Kutta methods requires a large number of steps to maintain stability during the numerical integration of a system of stiff ODEs. Note also that this conclusion will not change substantially by using a higher order Runge-Kutta method, e.g., the classical fourth-order Runge-Kutta of equations (2.71) to (2.75), since the stability interval increases above two but is still less than three (recall the results in Table 2.10). Thus, the use of a higher order Runge-Kutta method, or a higher order explicit method in general, will not lead to a significant reduction in the number of integration steps required to maintain stability; a higher order method will only improve the accuracy of the numerical solution.

We could then ask the question: How can the solution to a stiff ODE system be computed efficiently? The answer in general is to use an *implicit integrator*. We illustrate this with the implicit Euler method (with greater elaboration in the next section when we consider higher order implicit methods)

$$k_1 = hf(y_{i+1}, t_{i+1}) \tag{2.121}$$

$$y_{i+1} = y_i + k_1 \tag{2.122}$$

Equations (2.121) and (2.122) for the implicit Euler method differ from the equations (2.96) and (2.97) for the explicit Euler method only in the definition of k_1; for the implicit Euler method, the derivative function $f(y, t)$ is evaluated at point $i+1$ (rather than point i for the explicit Euler method). This difference may seem inconsequential, but it has two important effects:

1. The implicit Euler method does not have the stability limit of equation (2.113). This can easily be demonstrated for the model problem of equation (2.68). Thus, if we substitute equation (2.68) in equation (2.122)

   $$y_{i+1} = y_i + h\lambda y_{i+1}$$

   or

   $$y_{i+1}/y_i = 1/(1 - \lambda h) \tag{2.123}$$

   $1/(1 - \lambda h)$ of equation (2.123) is the amplification factor for the implicit Euler method, and clearly, for $\lambda < 0$ (and $h > 0$), $1/(1-\lambda h) < 1$ for all h, i.e., the implicit Euler method is unconditionally stable, and therefore the integration step is not limited by stability. Thus the problem of having to take many steps in integrating a system of stiff ODEs as found previously for the explicit Runge-Kutta methods is avoided.

2. The improvement in the stability of the implicit integration comes at a price, however, that is, the need to solve systems of nonlinear equations if the original ODEs are nonlinear. This requirement is clear from combining equations (2.121) and (2.122) so that y_{i+1} appears on both sides of the resulting equation

   $$y_{i+1} = y_i + hf(y_{i+1}, t_{t+1}) \tag{2.124}$$

If $f(y,t)$ is nonlinear in y (in general, if the ODEs are nonlinear), equation (2.124) requires the solution of a nonlinear equation for y_{i+1}, which is typically done with a variant of Newton's method. However, this additional computation is generally worthwhile for a system of stiff ODEs because of the much larger integration steps that can usually be taken (again, the implicit Euler method at least for test ODE (2.68) has no limit on the integration step to maintain stability).

We should also note that the conclusion about the absolute stability of the implicit Euler method followed from a stability analysis based on the model ODE (2.68). However, this is a single, linear ODE, that is, a highly simplified test problem. We cannot conclude in general that the implicit Euler method will be unconditionally or even highly stable based on the preceding analysis for equation (2.68). We can only hope that the implicit Euler method has good stability for systems of nonlinear ODEs; fortunately, experience has indicated that this is generally the case. Thus, the principal limitation of the implicit Euler method is one of accuracy, but not stability (as with the explicit Euler method, it is only first-order correct). We will consider in the next section integration formulas that are implicit and therefore have good stability characteristics and which also have good accuracy.

Finally, we should note that the eigenvalues of an ODE system can be imaginary and complex as well as real. Consider for example λ_1 and λ_2 given by equation (2.118) if $c^2 < 4km$. We might then ask about the stability limit of the explicit methods for complex and imaginary eigenvalues. The answer is straightforward in the case of the first-order Runge-Kutta method since equation (2.113) becomes

$$|1 + \lambda h| < 1 \tag{2.125}$$

which corresponds to a circle in the complex plane ($Im(\lambda h)$ plotted vs. $\text{Re}(\lambda h)$) of unit radius and centered at $-1 + 0j$, that is, the inside of this circle is the stable region for the first-order Runge-Kutta method. Also, the absolute value in condition (2.125) is now interpreted in the sense of the absolute value of a complex variable. However, the implicit Euler method is stable over the entire left plane, i.e., it has absolute stability or A-stability which follows from equation (2.123).

The stability region for second and higher order explicit Runge-Kutta methods is somewhat more complicated than a circle [Gear (1971)], but the important point is that the regions are not substantially larger for the higher order methods, so again, no advantage is gained by using a higher order explicit method on a stiff ODE problem. A Fortran program for the calculation of the stability region of explicit Runge-Kutta methods is given by Schiesser (1991).

2.4 *Multistep Methods*

Thus far we have considered only single-step methods, which are also called Runge-Kutta methods, for the numerical integration of systems of ODEs. Single-step methods have the advantages of being self-starting and having good accuracy through the use of higher order methods; also, for stiff ODEs, implicit single-step methods have good stability characteristics. However, the higher accuracy is achieved by more derivative evaluations (e.g., calls to subroutines `DERV`) for each step along the solution (compare, for example, the number of derivative evaluations required by the Euler or modified Euler methods with the number required by the higher order `RKF45` formulas). For small ODE problems like those considered in the preceding examples, the number of derivative evaluations is not a major consideration; however, for large ODEs problems, e.g., several hundred or thousand ODEs, each derivative evaluation is a significant computation, particularly if the derivatives are complicated functions, and therefore methods for reducing the total number of derivative evaluations are worth pursuing. To this end, we now consider the multistep methods.

It would seem intuitively reasonable that if the solution to an ODE problem system is smooth over an interval of the independent variable containing several computed solution points, then the use of all of these points would provide a better indication of where the next point along the solution will be than using just the preceding point as is done in the single-step methods. This is the basic idea of the multistep methods; in other words, we might consider a polynomial or Taylor series through several preceding points as the mathematical approximation for computing the next solution point. Of course, this means that we must somehow get the required past points or "history" of the solution to the current point along the solution, and for this reason, multistep methods are not self-starting since at the beginning of the solution, we have only the one initial condition point to start the numerical integration. We could, however, use a single-step (Runge-Kutta) method to compute the first several required solution points, then switch to a multistep method to continue the solution; in fact, this is the usual starting procedure for multistep methods.

As in the case of single-step methods, explicit multistep methods are available for nonstiff ODEs, and implicit multistep methods are available for stiff ODEs. We consider the explicit methods first, specifically the Adams-Bashforth methods [Gear (1971)]. We can observe that within the multistep methods, the earlier solution points may be used, i.e., past values of the dependent variable, and/or past values of the derivative, i.e., $f(y,t)$ in the ODE $dy/dt = f(y,t)$. For example, the second-order Adams-Bashforth method is [Gear (1971)]

$$y_{i+1} = y_i + h\left(\frac{3}{2}f_i - \frac{1}{2}f_{i-1}\right) \tag{2.126}$$

where again h is the integration step. The following points should be noted about equation (2.126):

1. The calculation of y_{i+1} requires y_i (the numerical solution at the preceding point), and f_i and f_{i-1} (the derivatives at the two preceding points).

2. Because f_{i-1} is required, equation (2.126) is not self-starting from the initial condition $y(t_0) = y_0$; in other words, equation (2.126) is a two-step method. Therefore, a one-step (Runge-Kutta) method must be used at the beginning of the calculation to compute one solution point past the initial condition, i.e., to compute y_1 from y_0. Once y_1 (at $t_1 = t_0 + h$) is known, $f_1 = f(y_1, t_1)$ can be calculated from the problem ODE, $dy/dt = f(y,t)$. Then, f_0 and f_1 are known for use in equation (2.126) to calculate y_2, etc.

3. Equation (2.126) requires only one derivative evaluation per step. This is a very important and rather remarkable property of all of the Adams-Bashforth methods and contrasts with the Runge-Kutta methods, for example, which require a minimum of p derivative evaluations per step for a pth order method, e.g., the classical fourth-order Runge-Kutta method requires four derivative evaluations per step. It is this one derivative per step property of the Adams-Bashforth methods that accounts for their computational efficiency.

To complete this introductory discussion of the Adams-Bashforth methods, we list in Table 2.11 the weighting coefficients for the methods up to sixth order [Gear (1971), Table 7.3]. Note that the coefficients in equation (2.126) (i.e., $3/2$, $-1/2$) appear in the

Table 2.11 Weighting Coefficients for the Adams-Bashforth Method of Orders 1 to 6.

j	1	2	3	4	5	6
β_{1j}	1					
$2\beta_{2j}$	3	-1				
$12\beta_{3j}$	23	-16	5			
$24\beta_{4j}$	55	-59	37	-9		
$720\beta_{5j}$	1901	-2774	2616	-1274	251	
$1440\beta_{6j}$	4277	-7923	9982	-7298	2877	-475

second row of this table. In general, the weighting coefficients in the table are used in the stepping formula for a pth order Adams-Bashforth method

$$y_{i+1} = y_i + h\sum_{j=1}^{p} \beta_{pj}\, f_{i+1-j} \tag{2.127}$$

Note for $p = 1$, equation (2.127) becomes the explicit Euler method or first-order Runge-Kutta. Also, equations (2.126) and (2.127) apply to systems of first-order ODEs, i.e., f and y in $dy/dt = f(y,t)$ can be considered as N-vectors, where N is the number of first order ODEs. The order of the Adams-Bashforth method, p, defines an approximating pth order polynomial through $p+1$ points of the numerical solution.

The second-order Adams-Bashforth (SOAB) method of equation (2.126) is implemented in main program ADAMS in Appendix A for a fixed integration step, with starting values computed either by the first-order Runge-Kutta (FORK) or a second-order Runge-Kutta (SORK). Note in particular the similarity of the coding in main program `ADAMS` and main program `ORBIT2` of Program 2.2a. Of particular importance are the calls by main program `ADAMS` to subroutines `INITAL`, `DERV`, and `PRINT`, which, as before, define the initial conditions, the derivatives, and the output of the solution produced for an ODE problem by the `SOAB`, `FORK`, and `SORK` numerical integration algorithms programmed in `ADAMS`.

We now consider the use of the `SOAB` integrator code in terms of a series of examples. We start with a second-order ODE for which an analytical solution is easily derived that can be used as a standard to evaluate the numerical solution. The ODE problem is [Lake (1985)]

$$\frac{d^2a}{dt^2} = \frac{8\pi G\rho}{3}a \tag{2.128}$$

which is to be integrated numerically subject to the initial conditions

$$a(0) = 1, \qquad \frac{da(0)}{dt} = 1 \tag{2.129)(2.130}$$

In equation (2.128), a is Einstein's scale factor for the universe; t is time; G is Newton's gravitational constant; and ρ is the vacuum density. Note that equation (2.128) is linear in a, so that it can easily be integrated analytically. The resulting exact solution can then be used to evaluate the numerical solution. In particular, the exact global error can be computed as a function of the integration step size h.

In order to program equation (2.128) for solution by the `SOAB`, we must write it as two first-order ODEs, as explained with equations (1.3) to (1.9). If we define two state variables $A1$ and $A2$ as

$$A1 = a, \qquad A2 = \frac{da}{dt} = \frac{dA1}{dt} \tag{2.131)(2.132}$$

then equations (2.128) to (2.130) can be written as

$$\frac{dA1}{dt} = A2, \qquad \frac{dA2}{dt} = \frac{8\pi G\rho}{3}A1 \qquad (2.133)(2.134)$$

$$A1(0) = 1, \qquad A2(0) = 1 \qquad (2.135)(2.136)$$

Initial conditions (2.135) and (2.136) are programmed in subroutine `INITAL` of Program 2.5a (again, `INITAL` is called by main program `ADAMS` in Appendix A). Note as before that the dependent variables `A1` and `A2` are in `COMMON/Y/`, and their derivatives `A1T` and `A2T` are in `COMMON/F/`. Also, we use $C = (8\pi G\rho)/3 = 1$.

Equations (2.133) and (2.134) are programmed in subroutine `DERV` in Program 2.5b. Note as before that the principal requirement of `DERV` is to compute the derivatives in `COMMON/F/`, in this case, `A1T` and `A2T`.

Finally, a series of calculations relating to the numerical solution is programmed in subroutine `PRINT`. In particular, the analytical solution to equations (2.133) to (2.136) is computed (the analytical solution is derived in subroutine `INITAL`), and the difference between the numerical and analytical solutions is then computed for printing and plotting. Also, the error in the numerical solution is plotted vs. the integration step size by a call to the point plotting routine `SPLOTS`. By plotting the error on a log-log basis, the slope of the resulting line indicates the apparent order of the numerical integration algorithm, as discussed subsequently.

We now consider a series of runs of Progam 2.5:

1. Integration by the Euler method: Five runs are programmed for the FORK method, for which the data file is listed in Program 2.5d. This data file is read by main program `ADAMS` (in Appendix A). For each run of `ADAMS`, three lines of data are read:

   - Line 1: A documentation line read with a 20A4 format. This line of data is then printed at the beginning of the output for the run.
   - Line 2: The initial, final, and print interval values of the independent variable, read with a 3E10.0 format.
   - Line 3: The number of ODEs, the ratio of the integration step to print interval, the starting integration algorithm (either `FORK=0` or `SORK=1`), and the continuing integration algorithm (`FORK=0`, `SORK=1`, or `SOAB=2`), read with a 4I5 format.

   Therefore, in the case of Program 2.5d, the initial, final, and print interval values of t in equations (2.133) and (2.134) are 0, 1, and 0.1, respectively. The third line of data specifies two ODEs (equations (2.133) and (2.134)), a ratio of the integration step to print interval of 1, 5, 10, 50, and 100 (so that $h = 0.1, 0.02, 0.01, 0.002$, and 0.001 in equation (2.126)), `FORK` as the starting integration algorithm, and `FORK` as the continuing integration algorithm; thus, the first-order Runge-Kutta (Euler method) is used throughout the calculations.

The output from Program 2.5a to d is listed in Table 2.12. The following points should be noted about the output in Table 2.12:

1. The errors in the numerical solution decrease with decreasing integration step, h, as expected (compare the errors in runs 1 to 5 at a given value of t, e.g., $t = 0.5$).

2. The log-log plot of integration error (at $t = 0.5$) vs. integration step, h, is linear with a value close to unity, indicating that the Euler method, at least for this problem, is first-order correct.

```
      SUBROUTINE INITAL
C...
C...  EINSTEIN EQUATION FOR THE SCALE FACTOR OF THE UNIVERSE
C...
C...  THE FOLLOWING ORDINARY DIFFERENTIAL EQUATION (ODE) WAS PROPOSED
C...  BY EINSTEIN FOR THE SCALE FACTOR, A(T), OF THE EXPANSION OF THE
C...  UNIVERSE (1)
C...
C...      2    2
C...     D A/DT   = 8*PI*G*RHO*A/3                                (1)
C...
C...  WHERE
C...
C...     A(T)          SCALE FACTOR
C...
C...     T             TIME
C...
C...     G             NEWTON*S GRAVITATIONAL CONSTANT
C...
C...     RHO           VACUUM DENSITY
C...
C...  (1)  LAKE, GEORGE, WINDOWS ON A NEW COSMOLOGY, SCIENCE, V. 224,
C...  NO. 4650, PP 675-682, 18 MAY 85
C...
C...  EQUATION (1) IS INTEGRATED NUMERICALLY SUBJECT TO THE FOLLOWING
C...  INITIAL CONDITIONS FOR THE CURRENT EPOCH
C...
C...     A(0) = DA(0)/DT = 1                                      (2)(3)
C...
C...  AN ANALYTICAL SOLUTION TO EQUATIONS (1) TO (3) CAN EASILY BE
C...  DERIVED.  IF A SOLUTION OF THE FORM
C...
C...     A(T) = D*EXP(C*T)
C...
C...  IS ASSUMED, WHERE C = (8*PI*G*RHO/3)**(1/2), SUBSTUTION IN
C...  EQUATION (1) CONFIRMS THE SOLUTION.  THUS, THE GENERAL SOLUTION
C...  TO EQUATION (1) IS
C...
C...     A(T) = D1*EXP(C*T) + D2*EXP(-C*T)
C...
C...  WHERE D1 AND D2 ARE CONSTANTS TO BE DETERMINED FROM EQUATIONS (2)
C...  AND (3).  APPLYING THESE INITIAL CONDITIONS, WE HAVE
C...
C...     A(0) = D1 + D2 = 1
C...
C...     DA(0)/DT = C*D1 - C*D2 = 1
C...
C...  THUS, D2 = 1 - D1 AND
C...
C...     C*D1 - C*(1 - D1) = 1
C...
C...     2*C*D1 = 1 + C
```

Program 2.5a Subroutine INITAL for Initial Conditions (2.135) and (2.136). *Continued next page.*

```
C...
C...      D1 = (1 + C)/(2*C)
C...
C...   AND
C...
C...      D2 = 1 - (1 + C)/(2*C) = (C - 1)/(2*C)
C...
C...   THE SOLUTION IS THEREFORE
C...
C...      A(T) = (1 + C)/(2*C)*EXP(C*T) + (C - 1)/(2*C)*EXP(-C*T)    (4)
C...
C...   THE SOLUTION IS VERIFIED AS
C...
C...       2   2
C...      D A/DT  =  (C/2)*(1 + C)*EXP(C*T) - (C/2)*(1 - C)*EXP(-C*T)
C...
C...    -(C**2)*A = -(C/2)*(1 + C)*EXP(C*T) + (C/2)*(1 - C)*EXP(-C*T)
C...
C...      A(0) = (1 + C)/(2*C) - (1 - C)/(2*C) = 1
C...
C...      DA(0)/DT = (1/2)*(1 + C) + (1/2)*(1 - C) = 1
C...
C...   EQUATION (4) CAN THEN BE USED TO CHECK THE NUMERICAL SOLUTION.
C...
       COMMON/T/          T,      NSTOP,      NORUN
      1      /Y/         A1,         A2
      2      /F/        A1T,        A2T
      3      /C/          C,         IP
      4      /H/          H
C...
C...   SET THE MODEL PARAMETERS
       C=1.
       C=SQRT(C)
C...
C...   INITIAL CONDITIONS
       A1=1.0
       A2=1.0
C...
C...   COMPUTE THE INITIAL DERIVATIVES
       CALL DERV
       IP=0
       RETURN
       END
```

Program 2.5a *Continued.*

3. The computed slope of the line in the plot is either 0.986 if the errors for $h = 0.1$ and 0.001 are used to calculate the slope (in subroutine PRINT) or 1.002 if the errors for $h = 0.01$ and 0.001 are used to calculate the slope. This naturally raises the question of why this difference in the computed slopes occurred. This difference is probably due to small departures of the points in the plot from a straight line. These departures from a straight line in turn are probably due to the fact that for relatively large h, e.g., $h = 0.1$, the higher order terms in the

```
      SUBROUTINE DERV
      COMMON/T/           T,      NSTOP,      NORUN
     1      /Y/          A1,         A2
     2      /F/         A1T,        A2T
     3      /C/           C,         IP
     4      /H/           H
C...
C...  EQUATION (1) IS PROGRAMMED AS TWO FIRST-ORDER ODES
      A1T=A2
      A2T=C*A1
      RETURN
      END
```

Program 2.5b Subroutine DERV for ODEs (2.133) and (2.134).

Taylor series approximated by the Euler solution (beyond the h^2) terms are not negligible; only when h becomes sufficiently small are these higher order terms negligible. This explanation is supported by the results: the slope for $h = 0.01$ and $h = 0.001$ is closer to unity than for $h = 0.1$ and 0.001.

We next consider integration by the modified Euler method. Five runs are programmed for the SORK method, for which the data file is listed in Program 2.5e. Note that Programs 2.5d and e are the same except that for the latter, the modified Euler method (SORK) is specified for the integrator.

The output from Program 2.5e is listed in Table 2.13. The following points should be noted about the output in Table 2.13:

1. The errors in the numerical solution are substantially less than for the solution from the Euler method (compare Tables 2.12 and 2.13). This improved accuracy is due to the second-order accuracy of the modified Euler method. Note, however, that this improved accuracy required two derivative evaluations per step along the solution compared with one derivative evaluation per step for the Euler method (see the coding in main program ADAMS of Appendix A).

2. The log-log plot of the error vs. integration step is not linear. Also, the computed order of the modified Euler method is 1.219 from points 1 and 5 and 0.470 from points 3 and 5 when we would expect this to be 2 (since the modified Euler method is second order). However, these unexpected results are not attributable to a failure of our error analysis of the modified Euler method, but rather, to the small integration errors which are below the threshold of the precision of the computer. For the preceding results in Table 2.13, a computer was used with 32-bit arithmetic in single-precision Fortran, which corresponds to about six significant figures. Since the numerical and analytical solutions agree to five figures, for $h = 0.02$, 0.01, 0.002, and 0.001 (four of the five runs), the calculation of the slope with only six-figure accuracy gives 1.219 and 0.470. Note that the error for $h = 0.001$ is actually greater than for $h = 0.002$ which also supports the conclusion that the loss of significant figures is more severe for $h = 0.001$ than for 0.002. The solution to this problem is, of course, to use double-precision Fortran. However, this is a worthwhile lesson, i.e., the precision of the computer must be consistent with the results that are sought.

```
      SUBROUTINE PRINT(NI,NO)
      COMMON/T/          T,      NSTOP,      NORUN
     1      /Y/         A1,         A2
     2      /F/        A1T,        A2T
     3      /C/          C,         IP
     4      /H/          H
C...
C...  DIMENSION THE ARRAYS FOR PLOTTING
      DIMENSION HP(5), EP(5)
C...
C...  PRINT A HEADING FOR THE OUTPUT
      IP=IP+1
      IF(IP.EQ.1)THEN
         WRITE(NO,2)
2        FORMAT(1H ,//,' NORUN',9X,'T',9X,'H',8X,'A(T)',3X,
     1                    'ANAL A(T)',3X,'0/0 ERROR')
      END IF
C...
C...  COMPUTE THE ANALYTICAL SOLUTION
      AE=(1.0+C)/(2.0*C)*EXP(C*T)-(1.0-C)/(2.0*C)*EXP(-C*T)
      ERR=(A1-AE)/AE*100.0
C...
C...  PRINT THE NUMERICAL AND ANALYTICAL SOLUTIONS
      WRITE(NO,1)NORUN,T,H,A1,AE,ERR
1     FORMAT(I6,2F10.4,3F12.4)
C...
C...  STORE THE ERROR IN THE NUMERICAL SOLUTION AND THE INTEGRATION
C...  STEP SIZE FOR SUBSEQUENT PLOTTING TO DETERMINE THE ORDER OF THE
C...  NUMERICAL METHOD (AT THE SIXTH POINT ALONG THE SOLUTION, WHICH
C...  IS SELECTED ARBITRARILY)
      IF(IP.EQ.6)THEN
         HP(NORUN)=ALOG10(ABS(H))
         EP(NORUN)=ALOG10(ABS(A1-AE))
      END IF
C...
C...  AT THE END OF THE FIFTH RUN, PRINT AND PLOT RESULTS FOR THE
C...  FIVE RUNS
      IF((NORUN.EQ.5).AND.(IP.EQ.11))THEN
C...
C...     PLOT THE ERROR VS INTEGRATION STEP
         CALL SPLOTS(1,NORUN,HP,EP)
         WRITE(NO,11)
11       FORMAT(1H ,//,' LOG(ERROR) VS LOG(H)')
C...
C...     COMPUTE AND PRINT THE ESTIMATED ORDER OF THE NUMERICAL
C...     ALGORITHM USING THE FIRST AND FIFTH POINTS
         ORDER=(EP(5)-EP(1))/(HP(5)-HP(1))
         WRITE(NO,12)ORDER
12       FORMAT(1H1,//,' ORDER OF METHOD = ',F5.3,' PTS 1 AND 5')
C...
C...     COMPUTE AND PRINT THE ESTIMATED ORDER OF THE NUMERICAL
```

Program 2.5c Subroutine PRINT for Equations (2.135) and (2.136). *Continued next page.*

```
C...      ALGORITHM USING THE THIRD AND FIFTH POINTS
          ORDER=(EP(5)-EP(3))/(HP(5)-HP(3))
          WRITE(NO,13)ORDER
13        FORMAT(1H1,//,' ORDER OF METHOD = ',F5.3,' PTS 3 AND 5')
        END IF
        RETURN
        END
```

Program 2.5c *Continued.*

```
SECOND ORDER ODE, NSTART = 0, NCONT = 0
0.        1.0       0.1
    2   1   0   0
SECOND ORDER ODE, NSTART = 0, NCONT = 0
0.        1.0       0.1
    2   5   0   0
SECOND ORDER ODE, NSTART = 0, NCONT = 0
0.        1.0       0.1
    2  10   0   0
SECOND ORDER ODE, NSTART = 0, NCONT = 0
0.        1.0       0.1
    2  50   0   0
SECOND ORDER ODE, NSTART = 0, NCONT = 0
0.        1.0       0.1
    2 100   0   0

END OF RUNS
```

Program 2.5d Data File for Equations (2.133) and (2.134) First-Order Runge-Kutta (Euler Method, FORK) Integration.

Finally, we now consider the execution of the preceding program for the SOAB method. Five runs are programmed for this SOAB method, for which the data file is listed in Program 2.5f.

Note that Programs 2.5d and f are the same except that for the latter, the SOAB method is specified for the integrator (with the SORK as the starting method).

The output from Program 2.5f is listed in Table 2.14. The output in Table 2.14 is qualitatively the same as in Table 2.13, i.e., the numerical and analytical solutions agree to five figures for small h, and the calculation of the order of the SOAB method is inaccurate because of the limited precision of the 32-bit arithmetic.

However, there is an important difference between the SORK and SOAB calculations. For SORK, two derivative evaluations are required for each step along the solution, while for the SOAB only one is required, i.e., a 50% reduction in the computational effort for SOAB. This is the major advantage of multistep methods in general relative to single-step methods. This advantage becomes even more pronounced for higher order methods. For example, the fifth-order Adams-Bashforth method defined by the next to last row of weighting coefficients in Table 2.11 still requires only

Table 2.12 Output of Programs 2.5a to d. *Continued next pages.*

```
RUN NO. -    1  SECOND ORDER ODE, NSTART = 0, NCONT = 0

INITIAL T -  0.000E+00

  FINAL T -  0.100E+01

  PRINT T -  0.100E+00

NUMBER OF DIFFERENTIAL EQUATIONS -   2

PRINT INTERVAL/INTEGRATION INTERVAL -     1

STARTING ALGORITHM -  0
  0 -  FIRST ORDER, RUNGE KUTTA (EULER METHOD)
  1 - SECOND ORDER, RUNGE KUTTA (MODIFIED EULER METHOD)

CONTINUING ALGORITHM -  0
  0 -  FIRST ORDER, RUNGE KUTTA (EULER METHOD)
  1 - SECOND ORDER, RUNGE KUTTA (MODIFIED EULER METHOD)
  2 - SECOND ORDER, ADAMS BASHFORTH


NORUN         T         H          A(T)   ANAL A(T)   0/0 ERROR
    1    0.0000    0.1000      1.0000      1.0000      0.0000
    1    0.1000    0.1000      1.1000      1.1052     -0.4679
    1    0.2000    0.1000      1.2100      1.2214     -0.9336
    1    0.3000    0.1000      1.3310      1.3499     -1.3971
    1    0.4000    0.1000      1.4641      1.4918     -1.8584
    1    0.5000    0.1000      1.6105      1.6487     -2.3176
    1    0.6000    0.1000      1.7716      1.8221     -2.7747
    1    0.7000    0.1000      1.9487      2.0138     -3.2296
    1    0.8000    0.1000      2.1436      2.2255     -3.6823
    1    0.9000    0.1000      2.3579      2.4596     -4.1330
    1    1.0000    0.1000      2.5937      2.7183     -4.5816


RUN NO. -    2  SECOND ORDER ODE, NSTART = 0, NCONT = 0

INITIAL T -  0.000E+00

  FINAL T -  0.100E+01

  PRINT T -  0.100E+00

NUMBER OF DIFFERENTIAL EQUATIONS -   2

PRINT INTERVAL/INTEGRATION INTERVAL -     5

STARTING ALGORITHM -  0
  0 -  FIRST ORDER, RUNGE KUTTA (EULER METHOD)
  1 - SECOND ORDER, RUNGE KUTTA (MODIFIED EULER METHOD)
```

Table 2.12 *Continued.*

```
CONTINUING ALGORITHM -  0
  0 -  FIRST ORDER, RUNGE KUTTA (EULER METHOD)
  1 - SECOND ORDER, RUNGE KUTTA (MODIFIED EULER METHOD)
  2 - SECOND ORDER, ADAMS BASHFORTH
```

NORUN	T	H	A(T)	ANAL A(T)	0/0 ERROR
2	0.0000	0.0200	1.0000	1.0000	0.0000
2	0.1000	0.0200	1.1041	1.1052	-0.0986
2	0.2000	0.0200	1.2190	1.2214	-0.1972
2	0.3000	0.0200	1.3459	1.3499	-0.2956
2	0.4000	0.0200	1.4859	1.4918	-0.3940
2	0.5000	0.0200	1.6406	1.6487	-0.4922
2	0.6000	0.0200	1.8114	1.8221	-0.5904
2	0.7000	0.0200	1.9999	2.0138	-0.6884
2	0.8000	0.0200	2.2080	2.2255	-0.7864
2	0.9000	0.0200	2.4379	2.4596	-0.8842
2	1.0000	0.0200	2.6916	2.7183	-0.9820

```
RUN NO. -   3  SECOND ORDER ODE, NSTART = 0, NCONT = 0

INITIAL T -  0.000E+00

  FINAL T -  0.100E+01

  PRINT T -  0.100E+00


NUMBER OF DIFFERENTIAL EQUATIONS -   2

PRINT INTERVAL/INTEGRATION INTERVAL -    10

STARTING ALGORITHM -  0
  0 -  FIRST ORDER, RUNGE KUTTA (EULER METHOD)
  1 - SECOND ORDER, RUNGE KUTTA (MODIFIED EULER METHOD)

CONTINUING ALGORITHM -  0
  0 -  FIRST ORDER, RUNGE KUTTA (EULER METHOD)
  1 - SECOND ORDER, RUNGE KUTTA (MODIFIED EULER METHOD)
  2 - SECOND ORDER, ADAMS BASHFORTH
```

NORUN	T	H	A(T)	ANAL A(T)	0/0 ERROR
3	0.0000	0.0100	1.0000	1.0000	0.0000
3	0.1000	0.0100	1.1046	1.1052	-0.0497
3	0.2000	0.0100	1.2202	1.2214	-0.0993
3	0.3000	0.0100	1.3478	1.3499	-0.1489
3	0.4000	0.0100	1.4889	1.4918	-0.1985
3	0.5000	0.0100	1.6446	1.6487	-0.2480
3	0.6000	0.0100	1.8167	1.8221	-0.2975
3	0.7000	0.0100	2.0068	2.0138	-0.3470
3	0.8000	0.0100	2.2167	2.2255	-0.3965

Table 2.12 *Continued.*

```
    3     0.9000     0.0100        2.4486        2.4596        -0.4460
    3     1.0000     0.0100        2.7048        2.7183        -0.4954


RUN NO. -    4  SECOND ORDER ODE, NSTART = 0, NCONT = 0

INITIAL T -  0.000E+00

  FINAL T -  0.100E+01

  PRINT T -  0.100E+00

NUMBER OF DIFFERENTIAL EQUATIONS -   2

PRINT INTERVAL/INTEGRATION INTERVAL -     50

STARTING ALGORITHM -  0
  0 -  FIRST ORDER, RUNGE KUTTA (EULER METHOD)
  1 - SECOND ORDER, RUNGE KUTTA (MODIFIED EULER METHOD)

CONTINUING ALGORITHM -  0
  0 -  FIRST ORDER, RUNGE KUTTA (EULER METHOD)
  1 - SECOND ORDER, RUNGE KUTTA (MODIFIED EULER METHOD)
  2 - SECOND ORDER, ADAMS BASHFORTH


NORUN          T          H          A(T)    ANAL A(T)    0/0 ERROR
    4     0.0000     0.0020        1.0000        1.0000       0.0000
    4     0.1000     0.0020        1.1051        1.1052      -0.0100
    4     0.2000     0.0020        1.2212        1.2214      -0.0199
    4     0.3000     0.0020        1.3495        1.3499      -0.0299
    4     0.4000     0.0020        1.4912        1.4918      -0.0399
    4     0.5000     0.0020        1.6479        1.6487      -0.0499
    4     0.6000     0.0020        1.8210        1.8221      -0.0598
    4     0.7000     0.0020        2.0123        2.0137      -0.0696
    4     0.8000     0.0020        2.2238        2.2255      -0.0795
    4     0.9000     0.0020        2.4574        2.4596      -0.0893
    4     1.0000     0.0020        2.7156        2.7183      -0.0992


RUN NO. -    5  SECOND ORDER ODE, NSTART = 0, NCONT = 0

INITIAL T -  0.000E+00

  FINAL T -  0.100E+01

  PRINT T -  0.100E+00

NUMBER OF DIFFERENTIAL EQUATIONS -   2

PRINT INTERVAL/INTEGRATION INTERVAL -    100
```

Table 2.12 *Continued.*

```
STARTING ALGORITHM -  0
  0 -  FIRST ORDER, RUNGE KUTTA (EULER METHOD)
  1 - SECOND ORDER, RUNGE KUTTA (MODIFIED EULER METHOD)

CONTINUING ALGORITHM -  0
  0 -  FIRST ORDER, RUNGE KUTTA (EULER METHOD)
  1 - SECOND ORDER, RUNGE KUTTA (MODIFIED EULER METHOD)
  2 - SECOND ORDER, ADAMS BASHFORTH


NORUN         T         H          A(T)   ANAL A(T)    0/0 ERROR
    5    0.0000    0.0010        1.0000      1.0000       0.0000
    5    0.1000    0.0010        1.1051      1.1052      -0.0049
    5    0.2000    0.0010        1.2213      1.2214      -0.0100
    5    0.3000    0.0010        1.3497      1.3499      -0.0149
    5    0.4000    0.0010        1.4915      1.4918      -0.0198
    5    0.5000    0.0010        1.6483      1.6487      -0.0247
    5    0.6000    0.0010        1.8216      1.8221      -0.0296
    5    0.7000    0.0010        2.0130      2.0137      -0.0345
    5    0.8000    0.0010        2.2246      2.2255      -0.0394
    5    0.9000    0.0010        2.4585      2.4596      -0.0443
    5    1.0000    0.0010        2.7169      2.7183      -0.0492


           ..1....1....1....1....1....1....1....1....1....1....1..
-0.142E+01+                                                     1  +I
          -                                                        -I
          -                                                        -I
          -                                                        -I
          -                                                        -I
-0.208E+01+                                    1                   +I
          -                                                        -I
          -                            1                           -I
          -                                                        -I
          -                                                        -I
-0.273E+01+                                                        +I
          -                                                        -I
          -          1                                             -I
          -                                                        -I
          -                                                        -I
-0.339E+01+  1                                                     +I
           ..1....1....1....1....1....1....1....1....1....1....1..
       -0.300E+01 -0.26E+01 -0.22E+01 -0.18E+01 -0.14E+01 -0.10E+01

LOG(ERROR) VS LOG(H)

ORDER OF METHOD = 0.986 PTS 1 AND 5

ORDER OF METHOD = 1.002 PTS 3 AND 5
```

```
SECOND ORDER ODE, NSTART = 1, NCONT = 1
0.          1.0         0.1
    2    1    1    1
SECOND ORDER ODE, NSTART = 1, NCONT = 1
0.          1.0         0.1
    2    5    1    1
SECOND ORDER ODE, NSTART = 1, NCONT = 1
0.          1.0         0.1
    2   10    1    1
SECOND ORDER ODE, NSTART = 1, NCONT = 1
0.          1.0         0.1
    2   50    1    1
SECOND ORDER ODE, NSTART = 1, NCONT = 1
0.          1.0         0.1
    2  100    1    1
END OF RUNS
```

Program 2.5e Data File for Equations (2.133) and (2.134) Second-Order Runge-Kutta (Modified Euler Method, SORK) Integration.

one derivative evaluation per step to obtain a fifth-order result, while the fifth-order RKF45 formulas require six derivative evaluations per step; thus the Adams-Bashforth method achieves a sixfold reduction in the computational effort. Again, though, the fifth-order Adams-Bashforth method is not self-starting and five starting points would be required; for example, a switch could be made from RKF45 to the fifth-order Adams-Bashforth after five points at the beginning of the solution are computed by RKF45.

We now consider a second ODE application with numerical solution by the SOAB integrator of Appendix A, a system of first-order linear ODEs:

$$\begin{aligned}
dy_1/dt + y_1 &= 1, \;\; y_1(0) = 0 \\
dy_2/dt + 2y_2 &= y_1, \;\; y_2(0) = 0 \\
dy_3/dt + 3y_3 &= y_2, \;\; y_3(0) = 0 \\
&\vdots \qquad\quad \vdots \\
dy_n/dt + ny_n &= y_{n-1}, \;\; y_n(0) = 0
\end{aligned} \tag{2.137}$$

Equations (2.137) have several interesting properties:

1. An exact solution can easily be derived (as indicated below).

2. The Jacobian matrix is bidiagonal and therefore the eigenvalues are available by inspection.

3. A minor modification, described below, can be made to make the system arbitrarily stiff.

These points will now be considered.

Table 2.13 Output of Programs 2.5a to c and 2.5e. *Continued next pages.*

```
RUN NO. -    1  SECOND ORDER ODE, NSTART = 1, NCONT = 1

INITIAL T -  0.000E+00

  FINAL T -  0.100E+01

  PRINT T -  0.100E+00

NUMBER OF DIFFERENTIAL EQUATIONS -   2

PRINT INTERVAL/INTEGRATION INTERVAL -     1

STARTING ALGORITHM -  1
  0 -  FIRST ORDER, RUNGE KUTTA (EULER METHOD)
  1 - SECOND ORDER, RUNGE KUTTA (MODIFIED EULER METHOD)

CONTINUING ALGORITHM -  1
  0 -  FIRST ORDER, RUNGE KUTTA (EULER METHOD)
  1 - SECOND ORDER, RUNGE KUTTA (MODIFIED EULER METHOD)
  2 - SECOND ORDER, ADAMS BASHFORTH


NORUN         T         H          A(T)   ANAL A(T)   0/0 ERROR
    1    0.0000    0.1000        1.0000      1.0000      0.0000
    1    0.1000    0.1000        1.1050      1.1052     -0.0155
    1    0.2000    0.1000        1.2210      1.2214     -0.0309
    1    0.3000    0.1000        1.3492      1.3499     -0.0464
    1    0.4000    0.1000        1.4909      1.4918     -0.0618
    1    0.5000    0.1000        1.6474      1.6487     -0.0773
    1    0.6000    0.1000        1.8204      1.8221     -0.0928
    1    0.7000    0.1000        2.0116      2.0138     -0.1082
    1    0.8000    0.1000        2.2228      2.2255     -0.1237
    1    0.9000    0.1000        2.4562      2.4596     -0.1391
    1    1.0000    0.1000        2.7141      2.7183     -0.1546


RUN NO. -    2  SECOND ORDER ODE, NSTART = 1, NCONT = 1

INITIAL T -  0.000E+00

  FINAL T -  0.100E+01

  PRINT T -  0.100E+00

NUMBER OF DIFFERENTIAL EQUATIONS -   2

PRINT INTERVAL/INTEGRATION INTERVAL -     5

STARTING ALGORITHM -  1
  0 -  FIRST ORDER, RUNGE KUTTA (EULER METHOD)
  1 - SECOND ORDER, RUNGE KUTTA (MODIFIED EULER METHOD)
```

Table 2.13 *Continued.*

```
CONTINUING ALGORITHM -  1
  0 -  FIRST ORDER, RUNGE KUTTA (EULER METHOD)
  1 - SECOND ORDER, RUNGE KUTTA (MODIFIED EULER METHOD)
  2 - SECOND ORDER, ADAMS BASHFORTH


NORUN        T          H           A(T)   ANAL A(T)   0/0 ERROR
    2   0.0000     0.0200         1.0000      1.0000      0.0000
    2   0.1000     0.0200         1.1052      1.1052     -0.0007
    2   0.2000     0.0200         1.2214      1.2214     -0.0013
    2   0.3000     0.0200         1.3498      1.3499     -0.0020
    2   0.4000     0.0200         1.4918      1.4918     -0.0026
    2   0.5000     0.0200         1.6487      1.6487     -0.0033
    2   0.6000     0.0200         1.8220      1.8221     -0.0039
    2   0.7000     0.0200         2.0137      2.0138     -0.0046
    2   0.8000     0.0200         2.2254      2.2255     -0.0052
    2   0.9000     0.0200         2.4595      2.4596     -0.0059
    2   1.0000     0.0200         2.7181      2.7183     -0.0065


RUN NO. -    3  SECOND ORDER ODE, NSTART = 1, NCONT = 1

INITIAL T -   0.000E+00

  FINAL T -   0.100E+01

  PRINT T -   0.100E+00

NUMBER OF DIFFERENTIAL EQUATIONS -    2

PRINT INTERVAL/INTEGRATION INTERVAL -     10

STARTING ALGORITHM -  1
  0 -  FIRST ORDER, RUNGE KUTTA (EULER METHOD)
  1 - SECOND ORDER, RUNGE KUTTA (MODIFIED EULER METHOD)

CONTINUING ALGORITHM -  1
  0 -  FIRST ORDER, RUNGE KUTTA (EULER METHOD)
  1 - SECOND ORDER, RUNGE KUTTA (MODIFIED EULER METHOD)
  2 - SECOND ORDER, ADAMS BASHFORTH

NORUN        T          H           A(T)   ANAL A(T)   0/0 ERROR
    3   0.0000     0.0100         1.0000      1.0000      0.0000
    3   0.1000     0.0100         1.1052      1.1052     -0.0002
    3   0.2000     0.0100         1.2214      1.2214     -0.0003
    3   0.3000     0.0100         1.3499      1.3499     -0.0005
    3   0.4000     0.0100         1.4918      1.4918     -0.0007
    3   0.5000     0.0100         1.6487      1.6487     -0.0008
    3   0.6000     0.0100         1.8221      1.8221     -0.0010
    3   0.7000     0.0100         2.0137      2.0138     -0.0011
```

Table 2.13 *Continued.*

```
     3    0.8000    0.0100       2.2255       2.2255      -0.0013
     3    0.9000    0.0100       2.4596       2.4596      -0.0014
     3    1.0000    0.0100       2.7182       2.7183      -0.0016


RUN NO. -   4  SECOND ORDER ODE, NSTART = 1, NCONT = 1

INITIAL T -  0.000E+00

  FINAL T -  0.100E+01

  PRINT T -  0.100E+00

NUMBER OF DIFFERENTIAL EQUATIONS -   2

PRINT INTERVAL/INTEGRATION INTERVAL -    50

STARTING ALGORITHM -  1
  0 -  FIRST ORDER, RUNGE KUTTA (EULER METHOD)
  1 - SECOND ORDER, RUNGE KUTTA (MODIFIED EULER METHOD)

CONTINUING ALGORITHM -  1
  0 -  FIRST ORDER, RUNGE KUTTA (EULER METHOD)
  1 - SECOND ORDER, RUNGE KUTTA (MODIFIED EULER METHOD)
  2 - SECOND ORDER, ADAMS BASHFORTH


NORUN         T         H         A(T)   ANAL A(T)   0/0 ERROR
     4   0.0000    0.0020       1.0000       1.0000      0.0000
     4   0.1000    0.0020       1.1052       1.1052      0.0000
     4   0.2000    0.0020       1.2214       1.2214      0.0000
     4   0.3000    0.0020       1.3499       1.3499      0.0000
     4   0.4000    0.0020       1.4918       1.4918     -0.0001
     4   0.5000    0.0020       1.6487       1.6487     -0.0001
     4   0.6000    0.0020       1.8221       1.8221      0.0000
     4   0.7000    0.0020       2.0138       2.0137      0.0001
     4   0.8000    0.0020       2.2255       2.2255      0.0002
     4   0.9000    0.0020       2.4596       2.4596      0.0003
     4   1.0000    0.0020       2.7183       2.7183      0.0004


RUN NO. -   5  SECOND ORDER ODE, NSTART = 1, NCONT = 1

INITIAL T -  0.000E+00

  FINAL T -  0.100E+01

  PRINT T -  0.100E+00

NUMBER OF DIFFERENTIAL EQUATIONS -   2

PRINT INTERVAL/INTEGRATION INTERVAL -   100
```

Table 2.13 *Continued.*

```
STARTING ALGORITHM -  1
  0 -  FIRST ORDER, RUNGE KUTTA (EULER METHOD)
  1 - SECOND ORDER, RUNGE KUTTA (MODIFIED EULER METHOD)

CONTINUING ALGORITHM -  1
  0 -  FIRST ORDER, RUNGE KUTTA (EULER METHOD)
  1 - SECOND ORDER, RUNGE KUTTA (MODIFIED EULER METHOD)
  2 - SECOND ORDER, ADAMS BASHFORTH


NORUN         T         H         A(T)   ANAL A(T)   O/O ERROR
    5    0.0000    0.0010       1.0000      1.0000      0.0000
    5    0.1000    0.0010       1.1052      1.1052     -0.0002
    5    0.2000    0.0010       1.2214      1.2214      0.0000
    5    0.3000    0.0010       1.3499      1.3499     -0.0001
    5    0.4000    0.0010       1.4918      1.4918      0.0001
    5    0.5000    0.0010       1.6487      1.6487      0.0003
    5    0.6000    0.0010       1.8221      1.8221      0.0003
    5    0.7000    0.0010       2.0138      2.0137      0.0005
    5    0.8000    0.0010       2.2255      2.2255      0.0006
    5    0.9000    0.0010       2.4596      2.4596      0.0008
    5    1.0000    0.0010       2.7183      2.7183      0.0008


           ..1....1....1....1....1....1....1....1....1....1....1..
-0.289E+01+                                                     1 +I
          -                                                       -I
          -                                                       -I
          -                                                       -I
          -                                                       -I
-0.382E+01+                                                       +I
          -                                                       -I
          -                                   1                   -I
          -                                                       -I
          -                                                       -I
-0.474E+01+                           1                           +I
          -                                                       -I
          -                                                       -I
          -  1                                                    -I
          -                                                       -I
-0.567E+01+          1                                            +I
           ..1....1....1....1....1....1....1....1....1....1....1..
       -0.300E+01 -0.26E+01 -0.22E+01 -0.18E+01 -0.14E+01 -0.10E+01

LOG(ERROR) VS LOG(H)

ORDER OF METHOD = 1.219 PTS 1 AND 5

ORDER OF METHOD = 0.470 PTS 3 AND 5
```

```
SECOND ORDER ODE, NSTART = 1, NCONT = 2
0.         1.0        0.1
    2    1    1    2
SECOND ORDER ODE, NSTART = 1, NCONT = 2
0.         1.0        0.1
    2    5    1    2
SECOND ORDER ODE, NSTART = 1, NCONT = 2
0.         1.0        0.1
    2   10    1    2
SECOND ORDER ODE, NSTART = 1, NCONT = 2
0.         1.0        0.1
    2   50    1    2
SECOND ORDER ODE, NSTART = 1, NCONT = 2
0.         1.0        0.1
    2  100    1    2
END OF RUNS
```

Program 2.5f Data File for Equations (2.133) and (2.134) SOAB Integration.

An analytical solution to equations (2.137) can be derived in various ways. Here we use the Laplace transform, defined for dependent variable $y_i(t)$ as

$$L\{y_i(t)\} = \int_0^\infty y_i(t)e^{-st}\,dt = \bar{y}_i(s) \tag{2.138}$$

We also make use of three basic properties of the Laplace transform

$$L\{dy_i/dt\} = \int_0^\infty dy_i(t)/dt\, e^{-st}\,dt = s\bar{y}_i(s) - y_i(0) \tag{2.139}$$

$$L\{cy_i(t)\} = c\bar{y}_i(s) \tag{2.140}$$

$$L\{1\} = 1/s \tag{2.141}$$

Equations (2.137) then transform to

$$s\bar{y}_1(s) + \bar{y}_1(s) = 1/s$$

$$s\bar{y}_2(s) + 2\bar{y}_2(s) = \bar{y}_1(s),$$

$$s\bar{y}_3(s) + 3\bar{y}_3(s) = \bar{y}_2(s)$$

$$\vdots \quad \vdots$$

$$s\bar{y}_n(s) + n\bar{y}_n(s) = \bar{y}_{n-1}(s) \tag{2.142}$$

Table 2.14 Output of Programs 2.5a to c and f. *Continued next pages.*

```
RUN NO. -    1  SECOND ORDER ODE, NSTART = 1, NCONT = 2

INITIAL T -   0.000E+00

  FINAL T -   0.100E+01

  PRINT T -   0.100E+00

NUMBER OF DIFFERENTIAL EQUATIONS -   2

PRINT INTERVAL/INTEGRATION INTERVAL -     1

STARTING ALGORITHM -  1
  0 -  FIRST ORDER, RUNGE KUTTA (EULER METHOD)
  1 - SECOND ORDER, RUNGE KUTTA (MODIFIED EULER METHOD)

CONTINUING ALGORITHM -  2
  0 -  FIRST ORDER, RUNGE KUTTA (EULER METHOD)
  1 - SECOND ORDER, RUNGE KUTTA (MODIFIED EULER METHOD)
  2 - SECOND ORDER, ADAMS BASHFORTH


NORUN         T         H          A(T)   ANAL A(T)   0/0 ERROR
    1    0.0000    0.1000        1.0000      1.0000      0.0000
    1    0.1000    0.1000        1.1050      1.1052     -0.0155
    1    0.2000    0.1000        1.2200      1.2214     -0.1148
    1    0.3000    0.1000        1.3480      1.3499     -0.1377
    1    0.4000    0.1000        1.4892      1.4918     -0.1759
    1    0.5000    0.1000        1.6452      1.6487     -0.2148
    1    0.6000    0.1000        1.8175      1.8221     -0.2536
    1    0.7000    0.1000        2.0079      2.0138     -0.2925
    1    0.8000    0.1000        2.2182      2.2255     -0.3313
    1    0.9000    0.1000        2.4505      2.4596     -0.3701
    1    1.0000    0.1000        2.7072      2.7183     -0.4089


RUN NO. -    2  SECOND ORDER ODE, NSTART = 1, NCONT = 2

INITIAL T -   0.000E+00

  FINAL T -   0.100E+01

  PRINT T -   0.100E+00

NUMBER OF DIFFERENTIAL EQUATIONS -   2

PRINT INTERVAL/INTEGRATION INTERVAL -     5

STARTING ALGORITHM -  1
  0 -  FIRST ORDER, RUNGE KUTTA (EULER METHOD)
  1 - SECOND ORDER, RUNGE KUTTA (MODIFIED EULER METHOD)
```

Table 2.14 *Continued.*

```
CONTINUING ALGORITHM -  2
  0 -  FIRST ORDER, RUNGE KUTTA (EULER METHOD)
  1 - SECOND ORDER, RUNGE KUTTA (MODIFIED EULER METHOD)
  2 - SECOND ORDER, ADAMS BASHFORTH


NORUN         T         H          A(T)    ANAL A(T)    O/O ERROR
    2    0.0000    0.0200        1.0000       1.0000       0.0000
    2    0.1000    0.0200        1.1052       1.1052      -0.0018
    2    0.2000    0.0200        1.2214       1.2214      -0.0035
    2    0.3000    0.0200        1.3498       1.3499      -0.0051
    2    0.4000    0.0200        1.4917       1.4918      -0.0068
    2    0.5000    0.0200        1.6486       1.6487      -0.0084
    2    0.6000    0.0200        1.8219       1.8221      -0.0101
    2    0.7000    0.0200        2.0135       2.0138      -0.0117
    2    0.8000    0.0200        2.2252       2.2255      -0.0134
    2    0.9000    0.0200        2.4592       2.4596      -0.0150
    2    1.0000    0.0200        2.7178       2.7183      -0.0166


RUN NO. -    3  SECOND ORDER ODE, NSTART = 1, NCONT = 2

INITIAL T -  0.000E+00

  FINAL T -  0.100E+01

  PRINT T -  0.100E+00

NUMBER OF DIFFERENTIAL EQUATIONS -   2

PRINT INTERVAL/INTEGRATION INTERVAL -     10

STARTING ALGORITHM -  1
  0 -  FIRST ORDER, RUNGE KUTTA (EULER METHOD)
  1 - SECOND ORDER, RUNGE KUTTA (MODIFIED EULER METHOD)

CONTINUING ALGORITHM -  2
  0 -  FIRST ORDER, RUNGE KUTTA (EULER METHOD)
  1 - SECOND ORDER, RUNGE KUTTA (MODIFIED EULER METHOD)
  2 - SECOND ORDER, ADAMS BASHFORTH

NORUN         T         H          A(T)    ANAL A(T)    O/O ERROR
    3    0.0000    0.0100        1.0000       1.0000       0.0000
    3    0.1000    0.0100        1.1052       1.1052      -0.0005
    3    0.2000    0.0100        1.2214       1.2214      -0.0009
    3    0.3000    0.0100        1.3498       1.3499      -0.0013
    3    0.4000    0.0100        1.4918       1.4918      -0.0017
    3    0.5000    0.0100        1.6487       1.6487      -0.0021
    3    0.6000    0.0100        1.8221       1.8221      -0.0025
    3    0.7000    0.0100        2.0137       2.0138      -0.0029
    3    0.8000    0.0100        2.2255       2.2255      -0.0033
```

Table 2.14 *Continued.*

```
    3   0.9000    0.0100     2.4595     2.4596     -0.0037
    3   1.0000    0.0100     2.7182     2.7183     -0.0041


RUN NO. -   4  SECOND ORDER ODE, NSTART = 1, NCONT = 2

INITIAL T -  0.000E+00

  FINAL T -  0.100E+01

  PRINT T -  0.100E+00

NUMBER OF DIFFERENTIAL EQUATIONS -   2

PRINT INTERVAL/INTEGRATION INTERVAL -    50

STARTING ALGORITHM -  1
  0 -  FIRST ORDER, RUNGE KUTTA (EULER METHOD)
  1 - SECOND ORDER, RUNGE KUTTA (MODIFIED EULER METHOD)

CONTINUING ALGORITHM -  2
  0 -  FIRST ORDER, RUNGE KUTTA (EULER METHOD)
  1 - SECOND ORDER, RUNGE KUTTA (MODIFIED EULER METHOD)
  2 - SECOND ORDER, ADAMS BASHFORTH


NORUN        T         H         A(T)    ANAL A(T)   0/0 ERROR
    4   0.0000    0.0020     1.0000     1.0000      0.0000
    4   0.1000    0.0020     1.1052     1.1052      0.0000
    4   0.2000    0.0020     1.2214     1.2214      0.0000
    4   0.3000    0.0020     1.3499     1.3499      0.0000
    4   0.4000    0.0020     1.4918     1.4918     -0.0001
    4   0.5000    0.0020     1.6487     1.6487     -0.0001
    4   0.6000    0.0020     1.8221     1.8221      0.0000
    4   0.7000    0.0020     2.0138     2.0137      0.0002
    4   0.8000    0.0020     2.2255     2.2255      0.0002
    4   0.9000    0.0020     2.4596     2.4596      0.0003
    4   1.0000    0.0020     2.7183     2.7183      0.0004


RUN NO. -   5  SECOND ORDER ODE, NSTART = 1, NCONT = 2

INITIAL T -  0.000E+00

  FINAL T -  0.100E+01

  PRINT T -  0.100E+00

NUMBER OF DIFFERENTIAL EQUATIONS -   2

PRINT INTERVAL/INTEGRATION INTERVAL -   100
```

Table 2.14 *Continued.*

```
STARTING ALGORITHM -  1
  0 -   FIRST ORDER, RUNGE KUTTA (EULER METHOD)
  1 - SECOND ORDER, RUNGE KUTTA (MODIFIED EULER METHOD)

CONTINUING ALGORITHM -  2
  0 -   FIRST ORDER, RUNGE KUTTA (EULER METHOD)
  1 - SECOND ORDER, RUNGE KUTTA (MODIFIED EULER METHOD)
  2 - SECOND ORDER, ADAMS BASHFORTH
```

NORUN	T	H	A(T)	ANAL A(T)	0/0 ERROR
5	0.0000	0.0010	1.0000	1.0000	0.0000
5	0.1000	0.0010	1.1052	1.1052	0.0000
5	0.2000	0.0010	1.2214	1.2214	-0.0001
5	0.3000	0.0010	1.3499	1.3499	0.0000
5	0.4000	0.0010	1.4918	1.4918	0.0001
5	0.5000	0.0010	1.6487	1.6487	0.0003
5	0.6000	0.0010	1.8221	1.8221	0.0004
5	0.7000	0.0010	2.0138	2.0137	0.0006
5	0.8000	0.0010	2.2255	2.2255	0.0007
5	0.9000	0.0010	2.4596	2.4596	0.0008
5	1.0000	0.0010	2.7183	2.7183	0.0009

```
           ..1....1....1....1....1....1....1....1....1....1....1..
-0.245E+01+                                                    1  +I
          -                                                       -I
          -                                                       -I
          -                                                       -I
          -                                                       -I
-0.358E+01+                                                       +I
          -                                   1                   -I
          -                                                       -I
          -                           1                           -I
          -                                                       -I
-0.471E+01+                                                       +I
          -                                                       -I
          -  1                                                    -I
          -                                                       -I
          -                                                       -I
-0.584E+01+         1                                             +I
           ..1....1....1....1....1....1....1....1....1....1....1..
       -0.300E+01 -0.26E+01 -0.22E+01 -0.18E+01 -0.14E+01 -0.10E+01

LOG(ERROR) VS LOG(H)

ORDER OF METHOD = 1.415 PTS 1 AND 5

ORDER OF METHOD = 0.816 PTS 3 AND 5
```

Equations (2.142) can then easily be combined to give

$$\bar{y}_1(s) = \left(\frac{1}{s}\right)\left(\frac{1}{s+1}\right)$$

$$\bar{y}_2(s) = \left(\frac{1}{s+2}\right)\bar{y}_1(s) = \left(\frac{1}{s}\right)\left(\frac{1}{s+1}\right)\left(\frac{1}{s+2}\right)$$

$$\bar{y}_3(s) = \left(\frac{1}{s+3}\right)\bar{y}_2(s) = \left(\frac{1}{s}\right)\left(\frac{1}{s+1}\right)\left(\frac{1}{s+2}\right)\left(\frac{1}{s+3}\right)$$

$$\vdots \quad \vdots$$

$$\bar{y}_n(s) = \left(\frac{1}{s+n}\right)\bar{y}_{n-1}(s)$$

$$= \left(\frac{1}{s}\right)\left(\frac{1}{s+1}\right)\left(\frac{1}{s+2}\right)\left(\frac{1}{s+3}\right)\cdots\left(\frac{1}{s+n}\right) \tag{2.143}$$

Equation (2.143) can now be inverted by the usual *partial fractions expansion*

$$\bar{y}_n(s) = \left(\frac{1}{s}\right)\left(\frac{1}{s+1}\right)\left(\frac{1}{s+2}\right)\left(\frac{1}{s+3}\right)\cdots\left(\frac{1}{s+n}\right)$$

$$= \frac{a_0}{s} + \frac{a_1}{(s+1)} + \frac{a_2}{(s+2)} + \frac{a_3}{(s+3)} + \cdots \frac{a_n}{(s+n)}$$

The coefficients a_0 to a_n are then obtained from

$$a_i = \lim_{s \to -i} (s+i)\bar{y}_i(s), \qquad i = 0, 1, 2, 3, \ldots n$$

or

$$a_i = \prod_{\substack{j=0 \\ i \neq j}}^{n} \frac{1}{(j-i)}, \qquad i = 0, 1, 2, 3, \ldots n \tag{2.144}$$

Finally, inversion of equation (2.143) gives

$$y_i(t) = a_0 + a_1 e^{-t} + a_2 e^{-2t} + a_3 e^{-3t} + \cdots a_n e^{-nt} \tag{2.145}$$

where the coefficients $a_0, \ldots, a_n$ are given by equation (2.144).

The Jacobian matrix (coefficient matrix) of equations (2.137) is simply

$$\bar{J} = \bar{A} = \begin{bmatrix} -1 & 0 & 0 & \cdots \\ 1 & -2 & 0 & \cdots \\ 0 & 1 & -3 & \cdots \\ \cdots & 0 & 1 & -n \end{bmatrix} \tag{2.146}$$

for which the corresponding eigenvalue problem is

$$\det(\bar{A} - \lambda\bar{I}) = 0 \tag{2.147}$$

Because of the special (bidiagonal) structure of $\bar{A}$ in equation (2.146), the determinant in equation (2.147) can be evaluated by inspection (using cofactors and keeping in mind the $n-1$ zeros in the first row, the $n-2$ zeros in the second row, etc.). Thus, the characteristic equation is

$$(-1-\lambda)(-2-\lambda)(-3-\lambda)\cdots(-n-\lambda) = 0 \tag{2.148}$$

```
      SUBROUTINE INITAL
C...
C...     DYI/DT + I*YI = YI-1, YI(0) = 0, (Y0 = 1)                (1)(2)
C...
      PARAMETER (N=5)
      COMMON/T/          T,      NSTOP,      NORUN
     1      /Y/       Y(N)
     2      /F/    DYDT(N)
C...
C...  INITIAL CONDITION (2)
      DO 1 I=1,N
      Y(I)=0.
1     CONTINUE
C...
C...  COMPUTE THE INITIAL DERIVATIVES
      CALL DERV
      RETURN
      END
```

Program 2.6a Subroutine `INITAL` for the Homogeneous Initial Conditions of Equations (2.137).

and therefore the eigenvalues from equation (2.148) are $\lambda_1 = -1$, $\lambda_2 = -2$, $\lambda_3 = -3, \ldots, \lambda_n = -n$. This result, of course, agrees with equation (2.145) (the eigenvalues appear in the exponentials).

We can generalize equations (2.137) with only minor changes in the preceding results

$$
\begin{aligned}
dy_1/dt + \lambda_1 y_1 &= 1, \; y_1(0) = 0 \\
dy_2/dt + \lambda_2 y_2 &= y_1, \; y_2(0) = 0 \\
dy_3/dt + \lambda_3 y_3 &= y_2, \; y_3(0) = 0 \\
&\vdots \qquad \vdots \\
dy_n/dt + \lambda_n y_n &= y_{n-1}, \; y_n(0) = 0
\end{aligned}
\tag{2.149}
$$

Then the eigenvalues $\lambda_1, \ldots, \lambda_n$ can be selected to make equations (2.149) arbitrarily stiff (by separating the eigenvalues).

We now develop a SOAB solution to equations (2.149) for the case $\lambda_1 = -1, \ldots, \lambda_n = -n$, with $n = 5$. Subroutine `INITAL` in Program 2.6a defines the homogeneous initial conditions of equations (2.137).

Subroutine `DERV` in Program 2.6b is a straightforward implementation of equations (2.137). Note that for the first ODE the right-hand side is set to one.

Subroutine `PRINT` in Program 2.6c computes the analytical solution, equation (2.145), and the difference between the numerical and analytical solutions (`DIFF`). Also, at $t = 0$ and the final value of t (when `IP=1` and `11`) subroutine `JMAP` is called to map and evaluate the Jacobian matrix of the ODEs. Subroutine `EIGEN` is then called to compute the eigenvalues of the Jacobian matrix.

Finally, the data file read by the main program `ADAMS` in Appendix A is listed in Program 2.6d. This data file indicates: (a) subroutine `PRINT` is called 11 times

```
      SUBROUTINE DERV
      PARAMETER (N=5)
      COMMON/T/           T,      NSTOP,      NORUN
     1      /Y/       Y(N)
     2      /F/    DYDT(N)
C...
C...  EQUATION (1)
      DO 1 I=1,N
      IF(I.EQ.1)THEN
         DYDT(I)=-FLOAT(I)*Y(I)+1.0
      ELSE
         DYDT(I)=-FLOAT(I)*Y(I)+Y(I-1)
      END IF
1     CONTINUE
      RETURN
      END
```

Program 2.6b Subroutine DERV for the Derivatives of Equations (2.137).

($0 \leq t \leq 10$ with a print interval of 1); (b) 5 ODEs; (c) 100 integration steps in each print interval; and (d) FORK/SOAB integration.

The output from this program is listed in Table 2.15. The following points should be noted about this output:

1. Turning first to the numerical and analytical solutions, we see they are generally in good agreement (note the small values of the difference, DIFF). The following special cases should be noted for these solutions:

(1.1) At $t = 0$, the analytical solution, equation (2.145), reduces to

$$y_i(0) = a_0 + a_1 + a_2 + a_3 + \cdots a_n = 0$$

where we have made use of the initial conditions for equations (2.137). In other words, the coefficients a_i in the analytical solution should sum to zero. Note that in the preceding output, they sum to $-0.9313E-09$. This clearly is a good test for detecting any errors in the calculation of the a_i.

(1.2) At steady state ($t \rightarrow \infty$), equations (2.137) reduce to

$$0 + y_1 = 1$$

$$0 + 2y_2 = y_1$$

$$0 + 3y_3 = y_2$$

$$\vdots \quad \vdots$$

$$0 + ny_n = y_{n-1}$$

which can easily be combined to give

$$y_n = \frac{1}{n!}$$

Thus, for $n = 5$, $y_5 = 1/5! = 0.00833\ldots$. From the preceding output, it is clear that y_5 is approaching a steady-state value, e.g., $y_5(10) = 0.8331E-02$.

```
      SUBROUTINE PRINT(NI,NO)
      PARAMETER (N=5)
      COMMON/T/          T,      NSTOP,      NORUN
     1      /Y/      Y(N)
     2      /F/   DYDT(N)
C...
C...  ABSOLUTE DIMENSIONING OF THE ARRAYS REQUIRED BY SUBROUTINE JMAP
C...  (A, X, XOLD, F, FOLD) AND SUBROUTINE EIGEN (WI, WR, Z, RW, IW)
      REAL
     1           A(5,5), WR(5), WI(5), Z(5,5),   RW(5),
     2                                 X(5), XOLD(5),
     3                                 F(5), FOLD(5)
      INTEGER                     IW(5)
C...
C...  ARRAY FOR THE COEFFICIENTS IN THE EXACT SOLUTION
      REAL      AC(0:5)
C...
C...  THE FOLLOWING EQUIVALENCE FACILITATES TRANSFER OF THE DEPENDENT
C...  VARIABLE VECTOR IN /Y/ TO ARRAY Y, AND THE DERIVATIVE VECTOR IN
C...  IN /F/ TO ARRAY F FOR USE IN THE SUBSEQUENT CALL TO SUBROUTINE
C...  JMAP
      EQUIVALENCE (Y(1),X(1)),(DYDT(1),F(1))
C...
C...  INITIALIZE AN INTEGER COUNTER TO CONTROL THE PRINTING
      DATA IP/0/
      IP=IP+1
C...
C...  MONITOR THE PROGRESS OF THE SOLUTION ON THE SCREEN
      WRITE(*,*)' IP = ',IP
C...
C...  MAP THE JACOBIAN MATRIX OF THE ODE SYSTEM DEFINED IN SUBROUTINE
C...  DERV, AND COMPUTE ITS TEMPORAL EIGENVALUES.  SUBROUTINE EIGEN
C...  CALLS A SERIES OF EISPACK ROUTINES TO COMPUTE THE TEMPORAL
C...  EIGENVALUES, AND OPTIONALLY, THE EIGENVECTORS OF THE 5-ODE
C...  SYSTEM JACOBIAN MATRIX (THE CALLS TO JMAP AND EIGEN ARE MADE
C...  ONLY AT THE BEGINNING AND END OF EACH RUN TO REDUCE THE VOLUME
C...  OF OUTPUT)
      IF(IP.EQ.1)THEN
         CALL  JMAP(N,A, X,XOLD,F,FOLD    )
         CALL EIGEN(N,A,WR,  WI,Z,  RW,IW)
         WRITE(NO,5)
5        FORMAT(1H ,//)
C...
C...  PRINT A HEADING FOR THE OUTPUT
         WRITE(NO,2)
2        FORMAT(1H ,//,9X,'T'10X,'Y5(T)',5X,'ANAL Y5(T)',11X,'DIFF')
C...
C...  COMPUTE THE ANALYTICAL SOLUTION AND THE PERCENT DIFFERENCE WITH
C...  THE NUMERICAL SOLUTION.  THE COEFFIENTS IN THE ANALYTICAL SOLU-
C...  TION ARE CALCULATED ONLY ONCE
```

Program 2.6c Subroutine PRINT to Print the Solution and Analyze the Jacobian Matrix of Equations (2.137). *Continued next page.*

```
          DO 3 I=0,N
             AC(I)=1.0
          DO 3 J=0,N
             IF(I.NE.J)THEN
             AC(I)=AC(I)*1.0/FLOAT(J-I)
             END IF
3        CONTINUE
         END IF
C...
C...     SUM THE EXPONENTIALS IN THE EXACT SOLUTION
         YA=0.
         DO 4 I=0,N
         YA=YA+AC(I)*EXP(-FLOAT(I)*T)
4        CONTINUE
C...
C...     PRINT THE NUMERICAL AND ANALYTICAL SOLUTIONS
         IF(IP.EQ.1)THEN
            DIFF=0.
         ELSE
            DIFF=(Y(N)-YA)/YA*100.
         END IF
         WRITE(NO,1)T,Y(N),YA,DIFF
1        FORMAT(F10.3,3E15.4)
C...
C...     MAP THE JACOBIAN MATRIX OF THE ODE SYSTEM DEFINED IN SUBROUTINE
C...     DERV, AND COMPUTE ITS TEMPORAL EIGENVALUES AT THE END OF THE
C...     RUN
         IF(IP.EQ.11)THEN
            CALL  JMAP(N,A, X,XOLD,F,FOLD    )
            CALL EIGEN(N,A,WR,  WI,Z,   RW,IW)
            WRITE(NO,5)
         END IF
C...
C...     RESET THE COUNTER IP AT THE END OF EACH RUN
         IF(IP.EQ.11)IP=0
         RETURN
         END
```

Program 2.6c *Continued.*

```
DYI/DT + I*Y = YI-1, YI(0) = 1, NSTART = 0, NCONT = 2
0.         10.0       1.0
    5  100    0    2
END OF RUNS
```

Program 2.6d Data File for the Solution of Equations (2.137).

2. The map of the ODE Jacobian matrix, produced by the call to `JMAP`, appears first for $t = 0$ (`IP=1` in subroutine `PRINT`). The map for $t = 10$ (`IP=11`) printed at the end is identical to the one for $t = 0$. These maps require a step-by-step explanation:

(2.1) The Jacobian matrix of an nth order ODE system is the $n \times n$ matrix of all first-order partial derivatives, with the general element

$$a_{ij} = \frac{\partial(dy_i/dt)}{\partial y_j}$$

where $i = 1, 2, \ldots, n$ and $j = 1, 2, \ldots, n$. Thus, the general element of the Jacobian matrix, a_{ij}, is the partial derivative of the ith ODE with respect to the jth dependent variable. The index of the dependent variables, j, is listed across the top of the map, i.e., $1 \leq j \leq 5$, and the index of the ODEs, i, is listed down the left margin of the map, i.e., $1 \leq i \leq 5$. The magnitude of the ijth element of the Jacobian matrix is then indicated in terms of a number in the above map. For example, the first ODE (with derivative dy_1/dt) has only one element in the Jacobian matrix corresponding to the first dependent variable (y_1) with a magnitude denoted with 4; thus the first row of the map reflects the structure of the first of equations (2.137), i.e.,

$$dy_1/dt + y_1 = 1$$

The corresponding section of the map is

```
     12345
   1  4
```

Similarly, the second row of the map, for the second differential equation (with derivative dy_2/dt) indicates a dependency on y_1 and y_2, which reflects the structure of the second of equations (2.137),

$$dy_2/dt + 2y_2 = y_1$$

The corresponding section of the map (including the first ODE) is

```
     12345
   1  4
   2  45
```

The term $2y_2$ in the ODE contributes a 5 to the map while the right-hand term y_1 contributes a 4 to the map. In this way, the mathematical structure of the entire ODE system can be visualized through the Jacobian matrix map.

(2.2) The calculation of the elements of the Jacobian matrix is done by finite difference approximations of the partial derivatives

$$a_{ij} = \frac{\partial(dy_i/dt)}{\partial y_j}$$

The coding for the finite difference approximations in subroutine JMAP is

```
A(IROW,JCOL)=(F(IROW)-FOLD(IROW))/(Y(JCOL)-YOLDJ)
```

which gives the a_{ij}th element (=A(IROW,JCOL)) of the Jacobian matrix as the ratio of the change in the right-hand side of the ith ODE (=(F(IROW)-FOLD(IROW))) divided by the corresponding change in the jth dependent variable (=(Y(JCOL)-YOLDJ)). Additional details about this finite difference approximation are available in subroutine JMAP which is thoroughly commented (and is available on the software diskette).

Table 2.15 Output from Program 2.6a to d. *Continued next page.*

```
RUN NO. -    1  DYI/DT + I*Y = YI-1, YI(0) = 1, NSTART = 0, NCONT = 2

INITIAL T -  0.000E+00

  FINAL T -  0.100E+02

  PRINT T -  0.100E+01

NUMBER OF DIFFERENTIAL EQUATIONS -   5

PRINT INTERVAL/INTEGRATION INTERVAL -   100

STARTING ALGORITHM -  0
  0 -  FIRST ORDER, RUNGE KUTTA (EULER METHOD)
  1 - SECOND ORDER, RUNGE KUTTA (MODIFIED EULER METHOD)

CONTINUING ALGORITHM -  2
  0 -  FIRST ORDER, RUNGE KUTTA (EULER METHOD)
  1 - SECOND ORDER, RUNGE KUTTA (MODIFIED EULER METHOD)
  2 - SECOND ORDER, ADAMS BASHFORTH

DEPENDENT VARIABLE COLUMN INDEX J (FOR YJ) IS PRINTED HORIZONTALLY

DERIVATIVE ROW INDEX I (FOR DYI/DT = FI(Y1,Y2,...,YJ,...,YN) IS
PRINTED VERTICALLY

JACOBIAN MATRIX ELEMENT IN THE MAP WITH INDICES I,J IS FOR PFI/PYJ
WHERE P DENOTES A PARTIAL DERIVATIVE


         12345
      1  4
      2  45
      3   45
      4    45
      5     45


   I          REAL            IMAG
   1         -5.000           0.000
   2         -4.000           0.000
   3         -3.000           0.000
   4         -2.000           0.000
   5         -1.000           0.000


      T          Y5(T)      ANAL Y5(T)          DIFF
    0.000     0.0000E+00   -0.9313E-09     0.0000E+00
    1.000     0.8411E-03    0.8410E-03     0.3585E-02
    2.000     0.4028E-02    0.4028E-02     0.5388E-02
    3.000     0.6455E-02    0.6455E-02    -0.2236E-03
    4.000     0.7598E-02    0.7598E-02    -0.7539E-03
    5.000     0.8056E-02    0.8056E-02    -0.5433E-03
```

Table 2.15 *Continued.*

```
    6.000     0.8231E-02     0.8231E-02    -0.2829E-03
    7.000     0.8295E-02     0.8295E-02    -0.1347E-03
    8.000     0.8319E-02     0.8319E-02    -0.5597E-04
    9.000     0.8328E-02     0.8328E-02    -0.3355E-04
   10.000     0.8331E-02     0.8331E-02    -0.3354E-04

DEPENDENT VARIABLE COLUMN INDEX J (FOR YJ) IS PRINTED HORIZONTALLY

DERIVATIVE ROW INDEX I (FOR DYI/DT = FI(Y1,Y2,...,YJ,...,YN) IS
PRINTED VERTICALLY

JACOBIAN MATRIX ELEMENT IN THE MAP WITH INDICES I,J IS FOR PFI/PYJ
WHERE P DENOTES A PARTIAL DERIVATIVE


          12345
       1  4
       2  45
       3   45
       4    45
       5     45


   I         REAL          IMAG
   1        -5.000         0.000
   2        -4.000         0.000
   3        -3.000         0.000
   4        -2.000         0.000
   5        -1.000         0.000
```

(2.3) The numbers in the map are not the actual magnitudes of the partial derivatives (not the magnitudes of the elements of the Jacobian matrix). Rather, they are indications of order of magnitude. The reason the actual values of the Jacobian matrix elements are not printed is simply that there is not enough room (this is a small map; when we are considering hundreds or thousands of ODEs the impracticality of printing the magnitudes of the elements becomes apparent). Thus, an element with a magnitude indicator of 4 is smaller than an element with a magnitude indicator of 5. We can, therefore, quickly see from the map where the large and small elements occur, and thus infer which dependent variables most strongly affect which derivatives. The coding in JMAP which produced the 4s and 5s in the preceding map is

```
C...
C...  ABSOLUTE DIMENSIONS FOR THE ARRAYS USED IN SUBROUTINE JMAP
      DIMENSION SYM(11),SYMBOL(100),N1(100),N10(100),N100(100)
C...
C...  DEFINE THE SYMBOLS USED IN PRINTING THE JACOBIAN MAP
      DATA SYM/1H ,1H0,1H1,1H2,1H3,1H4,1H5,1H6,1H7,1H8,1H9/
            .                               .
            .                               .
            .                               .
```

```
C...
C...  DETERMINE WHICH SYMBOL IS TO BE PRINTED IN EACH POSITION OF THE
C...  CURRENT SECTION OF THE JACOBIAN MAP
      DO 9 IROW=1,N
      DO 10 JCOL=JL,JU
      JS=JCOL-JL+1
      IF(ABS(A(IROW,JCOL)).LE.0.00001*SCALE)SYMBOL(JS)=SYM(1)
      IF(ABS(A(IROW,JCOL)).GT.0.00001*SCALE)SYMBOL(JS)=SYM(2)
      IF(ABS(A(IROW,JCOL)).GT.0.00010*SCALE)SYMBOL(JS)=SYM(3)
      IF(ABS(A(IROW,JCOL)).GT.0.00100*SCALE)SYMBOL(JS)=SYM(4)
      IF(ABS(A(IROW,JCOL)).GT.0.01000*SCALE)SYMBOL(JS)=SYM(5)
      IF(ABS(A(IROW,JCOL)).GT.0.10000*SCALE)SYMBOL(JS)=SYM(6)
      IF(ABS(A(IROW,JCOL)).GT.1.00000*SCALE)SYMBOL(JS)=SYM(7)
      IF(ABS(A(IROW,JCOL)).GT.10.0000*SCALE)SYMBOL(JS)=SYM(8)
      IF(ABS(A(IROW,JCOL)).GT.100.000*SCALE)SYMBOL(JS)=SYM(9)
      IF(ABS(A(IROW,JCOL)).GT.1000.00*SCALE)SYMBOL(JS)=SYM(10)
      IF(ABS(A(IROW,JCOL)).GT.10000.0*SCALE)SYMBOL(JS)=SYM(11)
10    CONTINUE
        .
        .
        .
 9    CONTINUE
```

In this case, SCALE=1. The 4s were stored, and subsequently printed, by the statement

```
IF(ABS(A(IROW,JCOL)).GT.0.10000*SCALE)SYMBOL(JS)=SYM(6)
```

(note that SYM(6) contains the character 4) and the 5s were stored, and subsequently printed, by the statement

```
IF(ABS(A(IROW,JCOL)).GT.1.00000*SCALE)SYMBOL(JS)=SYM(7)
```

which is in agreement with the coefficients in the ODEs, e.g., for the second ODE, $0.1 \leq 1 \leq 1$ in $y_1 = 1y_1$ produced a 4 and $1 \leq 2 \leq 10$ in $2y_2$ produced a 5.

(2.4) Also, as noted previously, subroutine JMAP does compute a numerical Jacobian matrix by finite differences and provides this *numerical Jacobian* as an output, in the present case, in array A (see the call to JMAP in subroutine PRINT). This numerical Jacobian is then available for processing by subroutine EIGEN which computes the temporal eigenvalues of the Jacobian matrix (matrix A now becomes an input to EIGEN). The real and imaginary parts of the eigenvalues are then printed by EIGEN. For the present problem, the five eigenvalues are

I	REAL	IMAG
1	-5.000	0.000
2	-4.000	0.000
3	-3.000	0.000
4	-2.000	0.000
5	-1.000	0.000

which agree to four figures with the exact values of equation (2.148).

(2.5) The printing of the Jacobian matrix map, and the calculation of the eigenvalues, is generally done to determine the structure of the ODEs and their stiffness and time scale characteristics. These concepts are discussed in detail by Schiesser (1991). We note here only that by looking at the ODE eigenvalues, we can establish the stiffness and time scale of the ODEs, and therefore make a more informed choice of an integrator and its implementation.

As a third example of the use of the SOAB integrator, we consider a system of first-order, nonlinear ODEs [Gear (1971), pp. 218–220]. The problem, as stated here because of limited space, is fourth order, but can be of any order; in fact, for the SOAB integration we use a 20th order version. The ODE problem in matrix form (an overbar denotes a vector or matrix) is

$$d\bar{y}/dt = -\overline{B}\bar{y} + \overline{U}\overline{w} \tag{2.150}$$

where U is the unitary matrix

$$\overline{U} = (1/2)\begin{bmatrix} -1 & 1 & 1 & 1 \\ 1 & -1 & 1 & 1 \\ 1 & 1 & -1 & 1 \\ 1 & 1 & 1 & -1 \end{bmatrix} \tag{2.151}$$

and $\overline{B}$ and $\overline{w}$ are

$$\overline{B} = \overline{U}\begin{bmatrix} \beta_1 & 0 & 0 & 0 \\ 0 & \beta_2 & 0 & 0 \\ 0 & 0 & \beta_3 & 0 \\ 0 & 0 & 0 & \beta_4 \end{bmatrix}\overline{U} \tag{2.152}$$

$$\overline{w} = [z_1^2, z_2^2, z_3^2, z_4^2]^T \tag{2.153}$$

$$z_i = \frac{\beta_i}{1 + c_i e^{\beta_i t}}, i = 1,2,3,4 \tag{2.154}$$

where $c_i = -(1+\beta_i)$ if $z_i(0) = -1$. $\beta_1, \beta_2, \beta_3, \beta_4$ are arbitrary constants with useful properties for testing an ODE integrator that will be discussed subsequently.

The solution to equation (2.150) is

$$\bar{y} = \overline{U}\bar{z} \tag{2.155}$$

where

$$\bar{z} = [z_1, z_2, z_3, z_4]^T \tag{2.156}$$

The Jacobian matrix of equation (2.150), $\bar{J}$, with a typical element $a_{ij} = \partial(dy_i/dt)/\partial y_j$, is

$$\bar{J} = \overline{U}\begin{bmatrix} -\beta_1 + 2z_1 & 0 & 0 & 0 \\ 0 & -\beta_2 + 2z_2 & 0 & 0 \\ 0 & 0 & -\beta_3 + 2z_3 & 0 \\ 0 & 0 & 0 & -\beta_4 + 2z_4 \end{bmatrix}\overline{U} \tag{2.157}$$

for which the eigenvalues are $2z_i - \beta_i$, $i = 1,2,3,4$ since $\overline{U}^{-1} = \overline{U}$. With the initial values

$$\bar{y}(0) = [-1, -1, -1, -1]^T = \bar{z}(0) \tag{2.158}$$

we have from equation (2.154)

$$z_i \rightarrow \left\{ \begin{array}{ll} \beta_i & \text{if } \beta_i < 0 \\ 0 & \text{if } \beta_i > 0 \end{array} \right\} \text{ as } t \rightarrow \infty \tag{2.159}$$

In the subsequent use of the integrator, we consider both cases, i.e., $\beta_i < 0$ and $\beta_i > 0$. In either case, as stated by Gear (1971), p. 218, the eigenvalues of equation (2.150)

$\rightarrow -|\beta_i|$ as $t \rightarrow \infty$. Also, for $\beta_i > 0$, $c_i < -1$ and for $\beta_i < 0$, $c_i > -1$, we see from equation (2.154) that z_i is finite and negative. All of the preceding equations can easily be extended to higher order (more ODEs); in fact, we now consider the use of this test problem for 20 ODEs rather than four.

Thus, the following programming of equation (2.150) is for the 20th order system, subject to initial condition (2.158). Again, subroutine `INITAL` in Program 2.7a has the primary function of setting the initial conditions. Also, since this subroutine is called only once for each run of the integrator, it is well suited for setting problem parameters such as β_1 to β_{20}. This is done with a `READ` statement in `INITAL`, and the 20 values of β_i are included in the data file after the usual three lines of data. Note that in the following subroutine `INITAL`, the unit number `NI` for the `READ` statement (to read the β_i) is accessed through `COMMON/IO/`.

The following points should be noted about subroutine `INITAL`:

1. The number of ODEs (`N`) and the values of β_i (`BETA`) are read via `DO` loop 10 from the data file; these data are then printed for confirmation. Note also that the 20 dependent variables appear in `COMMON/Y/` (as array `Y(20)`), and their derivative vector appears in `COMMON/F/` (as array `DYDT(20)`).

2. Initial condition (2.158) is then set in `DO` loop 20.

3. Arrays $\overline{U}$ and $\overline{B}$ in equation (2.150) are defined numerically (they are constant matrices and therefore can be defined in `INITAL`).

4. A heading for the output is printed at the end of `INITAL`.

Now that all 20 dependent variables are initialized, their derivatives can be computed in subroutine `DERV` in Program 2.7b according to equation (2.150). Some intermediate calculations are required, e.g., for matrix $\overline{w}$ in equation (2.150), before the derivative vector $d\overline{y}/dt$ (`DYDT(I)`) is computed in `DO` loop 30. Now everything in the right-hand side of equation (2.150) has been computed for the integrator, which then computes a solution (again, recall that it receives the derivative vector through `COMMON/F/`).

Subroutine `PRINT` in Program 2.7c prints the solution and performs some other operations to be described subsequently. Subroutine `PRINT` evaluates: (a) the exact solution, equation (2.155) (array `YEX(I)`), (b) the percent difference between the numerical and exact solutions for the first and 20th components (`PER1` and `PERN`), and (c) the maximum percent error for all 20 dependent variables (`PERMAX`). It then prints all of these quantities vs. t (`T`).

The data file used with the preceding subroutines `INITAL`, `DERV`, and `PRINT` is listed in Program 2.7d. This data file specifies two runs with negative and positive values of β_i, $i = 1, 2, \ldots, 20$, respectively.

The output from these two runs is presented in several parts to facilitate the discussion. The data summary, numerical solution, and errors in the numerical solution for the first run ($\beta_i < 0$) are considered first.

The maximum percent error in the numerical solution occurs in the 20th component, and has a value of 0.5606% at $t = 0.1$. Therefore, the ratio of the print interval/integration interval of 100 is adequate for most practical purposes (this ratio is set in the third line of data for each run).

The Jacobian map at $t = 0$ produced by the call to `JMAP` in subroutine `PRINT` is listed in Table 2.16b. The Jacobian matrix in this case is nearly full, which is unusual, i.e., for most applications in science and engineering the Jacobian matrix consists of mostly zero elements as we will observe in subsequent examples. Although the explicit ODE integrators we have considered thus far do not require any special knowledge of the

```
      SUBROUTINE INITAL
C...
C...  KROGH'S ODE PROBLEM
C...
C...  KROGH'S PROBLEM IS A NONLINEAR ODE PROBLEM WITH AN EXACT
C...  SOLUTION.  THE NUMBER OF ODES AND THE STIFFNESS OF THESE
C...  EQUATIONS CAN BE VARIED BY THE CHOICE OF PARAMETERS.  IN
C...  THIS CASE, N = 20, AND THE STIFFNESS IS SET BY CONSTANTS
C...  READ FROM A DATA FILE.
C...
C...  THE NONLINEAR ODE PROBLEM PROPOSED BY KROGH (GEAR (1), PP.
C...  218-220), IS AN IDEAL TEST PROBLEM FOR ODE SOFTWARE FOR THE
C...  FOLLOWING REASONS:
C...
C...     (1)  THE ORDER OF THE PROBLEM (NUMBER OF ODES) CAN BE
C...          SELECTED BY THE USER (IN THE PRESENT CASE, N = 20).
C...
C...     (2)  THE STIFFNESS OF THE ODES CAN BE SELECTED BY THE USER.
C...          THIS POINT REQUIRES SOME CLARIFICATION SINCE THE ODES
C...          ARE NONLINEAR, BUT STIFFNESS IS A LINEAR CONCEPT BASED
C...          ON EIGENVALUES (A STIFF SYSTEM IS GENERALLY DEFINED AS:
C...
C...                  (PROBLEM TIME)*(LARGEST EIGENVALUE)
C...
C...                         IS MUCH GREATER THAN 1)
C...
C...          GENERALLY, THE CONCEPT OF STIFFNESS APPLIED TO NONLINEAR
C...          PROBLEMS MEANS APPLIED TO THE LINEARIZED VERSION OF THE
C...          NONLINEAR EQUATIONS, FOR WHICH EIGENVALUES CAN BE COM-
C...          PUTED.  HOWEVER, SINCE LINEARIZATION IS PERFORMED AROUND
C...          A BASE POINT, THE EIGENVALUES OF THE LINEARIZED SYSTEM
C...          WILL CHANGE WITH THE BASE POINT, I.E., THE LOCAL EIGEN-
C...          VALUES OF A NONLINEAR SYSTEM ARE NOT CONSTANT.
C...
C...          THE NONLINEAR ODES IN THE KROGH PROBLEM HAVE IN THEIR
C...          ANALYTICAL SOLUTION EXPONENTIALS OF THE FORM EXP(B(I)*T)
C...          WHERE T IS THE INDEPENDENT VARIABLE AND THE B(I), I =
C...          1, 2,..., N ARE PARAMETERS SELECTED BY THE USER.  THUS
C...          THE B(I) ARE ANALOGOUS TO EIGENVALUES, BUT THE EXPONENT-
C...          IALS (EXP(B(I)*T)) APPEAR NONLINEARLY IN THE ANALYTICAL
C...          SOLUTION (RATHER THAN AS A LINEAR COMBINATION IN THE
C...          ANALYTICAL SOLUTION OF A SYSTEM OF CONSTANT COEFFICIENT
C...          LINEAR ODES).
C...
C...          IN THIS SENSE, THE STIFFNESS OF THE KROGH TEST PROBLEM
C...          CAN BE SPECIFIED BY THE USER BY SELECTING THE B(I).
C...
C...     (3)  AS THE PRECEDING DISCUSSION INDICATES, THE KROGH TEST
C...          HAS AN ANALYTICAL (EXACT) SOLUTION, WHICH CAN BE USED
C...          AS A STANDARD IN EVALUATING A NUMERICAL INTEGRATOR,
C...          I.E., THE ERRORS IN NUMERICAL SOLUTIONS CAN BE COMPUTED.
```

Program 2.7a Subroutine INITIAL for Initial Condition (2.158). *Continued next pages.*

```
C...
C...  ALL OF THE FEATURES OF THE KROGH TEST PROBLEM ARE ILLUSTRATED IN
C...  THE FOLLOWING CODING FOR TWO NONSTIFF CASES:
C...
C...      (1)  WE TAKE B(I) = -I, I = 1, 2,..., 20
C...
C...          THESE CONSTANTS ARE READ FROM A DATA FILE IN SUBROUTINE
C...          INITAL AT THE BEGINNING OF THE NUMERICAL SOLUTION, AND
C...          THE INTEGRATION IS PERFORMED BY BY THE SECOND-ORDER
C...          ADAMS-BASHFORTH METHOD
C...
C...      (2)  WE TAKE B(I) = I, I = 1, 2,..., 20 (NOTE THE ONLY
C...           CHANGE IS IN THE SIGN OF B(I)).
C...
C...  THE MATHEMATICAL DETAILS OF THE PROBLEM ARE GIVEN IN REFERENCE
C...  (1) CITED PREVIOUSLY.  THEY CAN ALSO BE INFERRED FROM THE
C...  FOLLOWING CODING.
C...
      COMMON /T/         T
      COMMON /Y/     Y(20)
      COMMON /F/  DYDT(20)
      COMMON /C/ BETA(20), U(20,20), B(20,20), N
      COMMON /IO/  NI, NO
C...
C...  READ THE NUMBER OF ODES AND THE CONSTANT ARRAY, BETA, WHICH
C...  DEFINES THE PROBLEM TIME SCALE AND STIFFNESS
      READ(NI,100)N
      DO 10 J=1,N
         READ(NI,101)BETA(J)
   10 CONTINUE
      WRITE(NO,102)N,(BETA(J),J=1,N)
C...
C...  INITIAL CONDITIONS
      DO 20 J=1,N
         Y(J) = -1.0
   20 CONTINUE
C...
C...  SET THE CONSTANT ARRAY U
      TERM = 2.0/FLOAT(N)
      DO 30 I=1,N
      DO 40 J=1,N
         U(I,J) = TERM
   40 CONTINUE
         U(I,I) = TERM - 1.0
   30 CONTINUE
C...
C...  SET THE CONSTANT ARRAY B
      DO 50 J=1,N
      DO 60 I=1,N
         SUM = 0.0
```

Program 2.7a *Continued.*

```
      DO 70 K=1,N
         SUM = SUM + U(I,K)*BETA(K)*U(K,J)
   70 CONTINUE
         B(I,J) = SUM
   60 CONTINUE
   50 CONTINUE
C...
C...  FORMATS ARE LISTED HERE
  100 FORMAT(I5)
  101 FORMAT(F20.5)
  102 FORMAT(/,11X,'N = ',I3,10X,'BETA',20(/,17X,F15.3))
      RETURN
      END
```

Program 2.7a *Continued.*

```
      SUBROUTINE DERV
      COMMON /T/          T
      COMMON /Y/     Y(20)
      COMMON /F/ DYDT(20)
      COMMON /C/ BETA(20), U(20,20), B(20,20), N
      DIMENSION W(20)
C...
C...  CALCULATE THE ARRAY W
      DO 10 I=1,N
         SUM = 0.0
      DO 20 J=1,N
         SUM = SUM + U(I,J)*Y(J)
   20 CONTINUE
         W(I) = SUM
   10 CONTINUE
C...
C...  CALCULATE THE N ODE DERIVATIVES
      DO 30 I=1,N
         SUM = 0.0
      DO 40 J=1,N
         SUM = SUM + U(I,J)*W(J)*W(J) - B(I,J)*Y(J)
   40 CONTINUE
         DYDT(I) = SUM
   30 CONTINUE
      RETURN
      END
```

Program 2.7b Subroutine DERV for Equation (2.150).

structure of the ODE Jacobian matrix, the implicit integrators we will subsequently consider for stiff ODEs do use the Jacobian matrix, and the structure of the Jacobian matrix, e.g., full, banded, and sparse, can therefore be a major consideration in the selection of an implicit integrator.

```
      SUBROUTINE PRINT(NI,NO)
      COMMON /T/        T
      COMMON /Y/     Y(20)
      COMMON /F/ DYDT(20)
      COMMON /C/ BETA(20), U(20,20), B(20,20), N
      DIMENSION W(20), YEX(20)
C...
C... ABSOLUTE DIMENSIONING OF THE ARRAYS REQUIRED BY SUBROUTINE JMAP
C... (A, X, XOLD, F, FOLD) AND SUBROUTINE EIGEN (WI, WR, Z, RW, IW)
      REAL
     1           A(20,20), WR(20), WI(20), Z(20,20), RW(20),
     2                              X(20), XOLD(20),
     3                              F(20), FOLD(20)
      INTEGER                      IW(20)
C...
C... THE FOLLOWING EQUIVALENCE FACILITATES TRANSFER OF THE DEPENDENT
C... VARIABLE VECTOR IN /Y/ TO ARRAY Y, AND THE DERIVATIVE VECTOR IN
C... IN /F/ TO ARRAY F FOR USE IN THE SUBSEQUENT CALL TO SUBROUTINE
C... JMAP
      EQUIVALENCE (Y(1),X(1)),(DYDT(1),F(1))
C...
C... INITIALIZE A COUNTER FOR THE CALLS TO SUBROUTINES JMAP AND EIGEN
      DATA IP/0/
      IP=IP+1
C...
C... MONITOR THE PROGRESS OF THE SOLUTION ON THE SCREEN
      WRITE(*,*)' IP = ',IP
C...
C... MAP THE JACOBIAN MATRIX OF THE ODE SYSTEM DEFINED IN SUBROUTINE
C... DERV, AND COMPUTE ITS TEMPORAL EIGENVALUES.  SUBROUTINE EIGEN
C... CALLS A SERIES OF EISPACK ROUTINES TO COMPUTE THE TEMPORAL
C... EIGENVALUES, AND OPTIONALLY, THE EIGENVECTORS OF THE 20-ODE
C... SYSTEM JACOBIAN MATRIX (THE CALLS TO JMAP AND EIGEN ARE MADE
C... ONLY AT THE BEGINNING AND END OF EACH RUN TO REDUCE THE VOLUME
C... OF OUTPUT)
      IF(IP.EQ.1)THEN
         CALL  JMAP(N,A, X,XOLD,F,FOLD   )
         CALL EIGEN(N,A,WR,  WI,Z,  RW,IW)
C...
C...     PRINT A HEADING FOR THE SOLUTION
         WRITE(NO,2)
2        FORMAT(//,3X,'TIME',5X,'Y(1)',3X,'YEX(1)',5X,'PER1',
     1                      4X,'Y(20)',2X,'YEX(20)',4X,'PER20',
     2                      3X,'PERMAX')
      END IF
C...
C... CALCULATE THE N EXACT SOLUTIONS
      DO 10 J=1,N
         BETAT = BETA(J)*T
      IF (BETAT.GT.200.) THEN
         W(J) = 0.0
```

Program 2.7c Subroutine PRINT to Print the Numerical and Analytical Solutions to Equation (2.150). *Continued next page.*

```
      ELSE IF (BETAT.LT.-200.) THEN
         W(J) = BETA(J)
      ELSE
         W(J) = BETA(J)/(1.0 - (1.0 + BETA(J))*EXP(BETAT))
      END IF
   10 CONTINUE
      DO 20 I=1,N
         SUM = 0.0
      DO 30 J=1,N
         SUM = SUM + U(I,J)*W(J)
   30 CONTINUE
         YEX(I) = SUM
   20 CONTINUE
C...
C...  CALCULATE THE MAXIMUM ERROR
      ERRMAX = ABS((Y(1) - YEX(1))/YEX(1))*100.
      DO 40 J=2,N
         ERR = ABS((Y(J) - YEX(J))/YEX(J))*100.
         IF (ERR.GT.ERRMAX) ERRMAX = ERR
   40 CONTINUE
C...
C...  PRINT THE NUMERICAL AND EXACT SOLUTIONS FOR I = 1 AND N, AND
C...  THE CORRESPONDING ERRORS, AND THE MAXIMUM ERROR
      PER1=ABS((Y(1)-YEX(1))/YEX(1))*100.
      PERN=ABS((Y(N)-YEX(N))/YEX(N))*100.
      PERMAX=ERRMAX
      WRITE(NO,3)T,Y(1),YEX(1),PER1,Y(N),YEX(N),PERN,PERMAX
3     FORMAT(F7.2,7F9.4)
C...
C...  MAP THE JACOBIAN MATRIX OF THE ODE SYSTEM DEFINED IN SUBROUTINE
C...  DERV, AND COMPUTE ITS TEMPORAL EIGENVALUES AT THE END OF THE RUN
      IF(IP.EQ.11)THEN
         CALL  JMAP(N,A, X,XOLD,F,FOLD    )
         CALL EIGEN(N,A,WR,  WI,Z,  RW,IW)
      END IF
C...
C...  RESET IP AT THE END OF EACH RUN
      IF(IP.EQ.11)IP=0
      RETURN
      END
```

Program 2.7c *Continued.*

The eigenvalues computed by the call to EIGEN at $t = 0$ for the first run are listed in Table 2.16c. The following points should be noted about the eigenvalues:

1. They are entirely real, i.e., there is no oscillatory component to the solution.

2. Some of the real parts are negative, corresponding to stable components of the solution to equation (2.150), and some are positive corresponding to unstable components. Thus, at the beginning, $t = 0$, the solution is actually unstable.

Also, we should mention that the concepts of eigenvalues and eigen functions apply strictly to only linear ODE systems, while equation (2.150) is highly nonlinear. There-

```
KROGH PROBLEM FOR N = 20, NONSTIFF CASE, NEGATIVE BETA
0.         1.0       0.1
   20  100    0    2
   20
-1.
-2.
-3.
-4.
-5.
-6.
-7.
-8.
-9.
-10.
-11.
-12.
-13.
-14.
-15.
-16.
-17.
-18.
-19.
-20.
KROGH PROBLEM FOR N = 20, NONSTIFF CASE, POSITIVE BETA
0.         4.0       0.4
   20  100    0    2
   20
1.
2.
3.
4.
5.
6.
7.
8.
9.
10.
11.
12.
13.
14.
15.
16.
17.
18.
19.
20.
END OF RUNS
```

Program 2.7d Data File for the Solution of Equation (2.150).

Table 2.16 (a) Data Summary, Numerical, and Analytical Solutions for Run No. 1, Programs 2.7a to d. *Continued next page.*

```
RUN NO. -    1  KROGH PROBLEM FOR N = 20, NONSTIFF CASE, NEGATIVE BETA

INITIAL T -  0.000E+00

  FINAL T -  0.100E+01

  PRINT T -  0.100E+00

NUMBER OF DIFFERENTIAL EQUATIONS -  20

PRINT INTERVAL/INTEGRATION INTERVAL -   100

STARTING ALGORITHM -  0
  0 -  FIRST ORDER, RUNGE KUTTA (EULER METHOD)
  1 - SECOND ORDER, RUNGE KUTTA (MODIFIED EULER METHOD)

CONTINUING ALGORITHM -  2
  0 -  FIRST ORDER, RUNGE KUTTA (EULER METHOD)
  1 - SECOND ORDER, RUNGE KUTTA (MODIFIED EULER METHOD)
  2 - SECOND ORDER, ADAMS BASHFORTH

          N =  20          BETA
                          -1.000
                          -2.000
                          -3.000
                          -4.000
                          -5.000
                          -6.000
                          -7.000
                          -8.000
                          -9.000
                         -10.000
                         -11.000
                         -12.000
                         -13.000
                         -14.000
                         -15.000
                         -16.000
                         -17.000
                         -18.000
                         -19.000
                         -20.000

 TIME      Y(1)    YEX(1)      PER1     Y(20)   YEX(20)     PER20    PERMAX
 0.00   -1.0000   -1.0000    0.0000   -1.0000   -1.0000    0.0000    0.0000
 0.10   -4.4928   -4.4933    0.0119    0.1062    0.1068    0.5606    0.5606
 0.20  -11.0707  -11.0711    0.0034    2.7655    2.7657    0.0081    0.0494
 0.30  -15.9548  -15.9549    0.0008    2.1455    2.1455    0.0042    0.0220
 0.40  -18.1078  -18.1078    0.0003    0.7655    0.7655    0.0021    0.0074
```

Table 2.16 (a) *Continued.*

0.50	-19.0184	-19.0184	0.0001	-0.0357	-0.0357	0.0062	0.0062
0.60	-19.4424	-19.4424	0.0000	-0.4448	-0.4448	0.0008	0.0008
0.70	-19.6607	-19.6607	0.0001	-0.6610	-0.6610	0.0008	0.0008
0.80	-19.7826	-19.7826	0.0001	-0.7827	-0.7827	0.0007	0.0007
0.90	-19.8551	-19.8551	0.0001	-0.8551	-0.8551	0.0003	0.0003
1.00	-19.9004	-19.9004	0.0000	-0.9004	-0.9004	0.0003	0.0003

Table 2.16 (b) Jacobian Map at $t = 0$ of Run No. 1, Programs 2.7a to d.

```
DEPENDENT VARIABLE COLUMN INDEX J (FOR YJ) IS PRINTED HORIZONTALLY

DERIVATIVE ROW INDEX I (FOR DYI/DT = FI(Y1,Y2,...,YJ,...,YN) IS
PRINTED VERTICALLY

JACOBIAN MATRIX ELEMENT IN THE MAP WITH INDICES I,J IS FOR PFI/PYJ
WHERE P DENOTES A PARTIAL DERIVATIVE


                   11111111112
          12345678901234567890
     1    45555555554444444442
     2    55555555544444444423
     3    55555555444444444234
     4    55555554444444442344
     5    55555544444444423444
     6    55555544444444144444
     7    55554454444441344444
     8    55544445444413444444
     9    55444444544 34444444
    10    5444444445 444444444
    11    444444444 5444444444
    12    44444444134544444445
    13    44444431444464444455
    14    444443 4444446445555
    15    44443144444444655555
    16    4443 344444444565555
    17    443 4444444445556555
    18    44 44444444445555655
    19    4 344444444455555565
    20    23444444444555555556
```

fore, what we are considering in this example are the eigenvalues and eigen functions of a locally linearized version of the problem. In fact, the finite difference method used in subroutine JMAP is actually a local linearization of the nonlinear ODEs. We will not get into a detailed discussion of this point but merely conclude by saying that because of the local linearization of the nonlinear problem, the eigenvalues can change as the solution evolves (and therefore the base point of the local linearization changes). This is a basic characteristic of a nonlinear system, i.e., nonconstant eigenvalues.

Table 2.16 (c) Eigenvalues at $t = 0$ of Run No. 1, Programs 2.7a to d.

I	REAL	IMAG
1	18.003	0.000
2	-1.003	0.000
3	17.006	0.000
4	-0.002	0.000
5	16.005	0.000
6	0.999	0.000
7	15.006	0.000
8	1.998	0.000
9	14.006	0.000
10	2.998	0.000
11	13.006	0.000
12	4.004	0.000
13	12.005	0.000
14	5.003	0.000
15	11.005	0.000
16	8.005	0.000
17	7.004	0.000
18	9.003	0.000
19	10.003	0.000
20	6.003	0.000

As the solution to equation (2.150) evolves through time, the Jacobian map and associated eigenvalues change, again because equation (2.150) is nonlinear. This is evident in considering the Jacobian map and eigenvalues at $t = 1$ in Table 2.16d. The following points should be noted about the output in Table 2.16d:

1. The map has changed substantially from the one at $t = 0$, again, because the elements of the Jacobian matrix are not constant for a nonlinear system, but rather, are functions of the dependent variables.

2. The eigenvalues are still entirely real, but they now all have negative real parts, indicating that by $t = 1$, the ODE solution is now completely stable. Also, the eigenvalues appear to be approaching the values indicated by the statement "the eigenvalues of equation (2.150) $\rightarrow -|\beta_i|$ as $t \rightarrow \infty$" just after equation (2.159) (the eigenvalues would be closer to $-|\beta_i|$ for larger t).

3. We can estimate the time scale of the solution by noticing that the smallest nonzero eigenvalue is -0.986 corresponding to a component in the solution with the magnitude $\exp(-0.986t)$. This exponential will decay to insignificance in a time in the interval $1 \leq t \leq 10$, and that is what we observe in the numerical solution (the solution has not reached a steady state at $t = 1$).

If we now consider the output from the second run (with positive β_i) in Table 2.17a, we observe that the solution has decayed to a small value at $t = 4$ (note in the second set of data the final value of t is 4 rather than 1). Also, the agreement between the numerical and exact solutions is better than 0.5% (see the maximum error at $t = 0.4$).

The Jacobian map and eigenvalues at $t = 0$ are tabulated in Table 2.17b. Note that the map and eigenvalues have changed substantially from the first run, e.g., all of the eigenvalues have negative real parts so the system is stable initially.

Table 2.16 (d) Eigenvalues and Jacobian Map at $t = 1$ of Run No. 1, Programs 2.7a to d.

```
DEPENDENT VARIABLE COLUMN INDEX J (FOR YJ) IS PRINTED HORIZONTALLY

DERIVATIVE ROW INDEX I (FOR DYI/DT = FI(Y1,Y2,...,YJ,...,YN) IS
PRINTED VERTICALLY

JACOBIAN MATRIX ELEMENT IN THE MAP WITH INDICES I,J IS FOR PFI/PYJ
WHERE P DENOTES A PARTIAL DERIVATIVE

                   11111111112
          12345678901234567890
       1  55555555544444444433
       2  55555555544444444433
       3  55555555444444444334
       4  55555554444444443344
       5  55555544444444423444
       6  55555544444443144444
       7  55554454444433444444
       8  55544445444334444444
       9  55444444543344444444
      10  44444444463444444444
      11  44444444336444444444
      12  44444443344644444455
      13  44444433444464444555
      14  44444334444446445555
      15  44444244444444655555
      16  44442444444444565555
      17  44433444444445556555
      18  44334444444455555655
      19  33344444444555555565
      20  33444444445555555556

   I          REAL           IMAG
   1        -0.986          0.000
   2        -1.513          0.000
   3       -19.913          0.000
   4        -2.448          0.000
   5       -18.951          0.000
   6        -3.575          0.000
   7       -17.994          0.000
   8        -4.726          0.000
   9       -17.009          0.000
  10        -5.842          0.000
  11       -15.988          0.000
  12        -6.915          0.000
  13       -15.001          0.000
  14        -7.949          0.000
  15       -13.990          0.000
  16        -8.970          0.000
  17       -11.994          0.000
  18       -10.987          0.000
  19       -12.993          0.000
  20        -9.981          0.000
```

Table 2.17 (a) Data Summary, Numerical and Analytical Solutions for Run No. 2, Programs 2.7a to d. *Continued next page.*

```
RUN NO. -   2  KROGH PROBLEM FOR N = 20, NONSTIFF CASE, POSITIVE BETA

INITIAL T -  0.000E+00

  FINAL T -  0.400E+01

  PRINT T -  0.400E+00

NUMBER OF DIFFERENTIAL EQUATIONS -  20

PRINT INTERVAL/INTEGRATION INTERVAL -    100

STARTING ALGORITHM -  0
  0 -  FIRST ORDER, RUNGE KUTTA (EULER METHOD)
  1 - SECOND ORDER, RUNGE KUTTA (MODIFIED EULER METHOD)

CONTINUING ALGORITHM -  2
  0 -  FIRST ORDER, RUNGE KUTTA (EULER METHOD)
  1 - SECOND ORDER, RUNGE KUTTA (MODIFIED EULER METHOD)
  2 - SECOND ORDER, ADAMS BASHFORTH

          N =  20          BETA
                          1.000
                          2.000
                          3.000
                          4.000
                          5.000
                          6.000
                          7.000
                          8.000
                          9.000
                         10.000
                         11.000
                         12.000
                         13.000
                         14.000
                         15.000
                         16.000
                         17.000
                         18.000
                         19.000
                         20.000
```

TIME	Y(1)	YEX(1)	PER1	Y(20)	YEX(20)	PER20	PERMAX
0.00	-1.0000	-1.0000	0.0000	-1.0000	-1.0000	0.0000	0.0000
0.40	0.3412	0.3413	0.0127	-0.1626	-0.1625	0.0221	0.4094
0.80	0.2333	0.2333	0.0010	-0.0565	-0.0565	0.0112	0.0275
1.20	0.1503	0.1503	0.0008	-0.0270	-0.0270	0.0072	0.0388
1.60	0.0975	0.0975	0.0012	-0.0148	-0.0148	0.0050	0.0134
2.00	0.0639	0.0639	0.0012	-0.0087	-0.0087	0.0038	0.0255
2.40	0.0422	0.0422	0.0011	-0.0054	-0.0054	0.0029	0.3378
2.80	0.0280	0.0280	0.0010	-0.0034	-0.0034	0.0022	0.0251

Table 2.17 (a) *Continued.*

3.20	0.0186	0.0186	0.0008	-0.0022	-0.0022	0.0017	0.0107
3.60	0.0124	0.0124	0.0005	-0.0014	-0.0014	0.0012	0.0061
4.00	0.0083	0.0083	0.0003	-0.0009	-0.0009	0.0008	0.0040

The map and eigenvalues at $t = 4$ are listed in Table 2.17c. Note that the eigenvalues remain entirely real. Also, we observe that the eigenvalues $\rightarrow -|\beta_i|$ in agreement with the statement immediately after equation (2.159). As noted previously, the same appears to be happening to the eigenvalues of the first run although the final time is only $t = 1$, thus the condition for $t \rightarrow \infty$ is not as clear. However, we again have the confirmation of the time scale of the problem as $1 \leq t \leq 10$ since we now have a solution closer to $t = 10$ ($t = 1$ for the first run, $t = 4$ for the second run).

Generally we can conclude that the combination of the FORK or SORK plus SOAB produces a solution to equation (2.150) of acceptable accuracy (as indicated by the errors in the numerical solution given in Tables 2.16a and 2.17a). However, these explicit integrators could be used because the selection of β_i in Program 2.7d corresponds to a nonstiff ODE problem. If widely different values of β_i are used the ODE problem will be stiff, and we must use an implicit integrator, which we illustrate again through equation (2.150); the integrator in this case is `LSODE` [Byrne and Hindmarsh (1986)].

We will not consider `LSODE` in detail. Rather, we briefly mention the underlying multistep integrator based on backward differentiation formulas (BDFs), and then illustrate the use of `LSODE`. Additional discussion of `LSODE` is given by Byrne and Hindmarsh (1986), and Schiesser (1991). The BDF are implicit, multistep integration formulas with the general format

$$y_{i+1} = \sum_{k=0}^{q-1} \alpha_k y_{i-k} + h\beta_0 dy_{i+1}/dt \tag{2.160}$$

where q is the order of the method, and α_k and β_0 are constants for a particular order. Note that for $q = 1$, with $\alpha_0 = \beta_0 = 1$, $y_{i-k} = y_{i-0} = y_i$, equation (2.160) reduces to

$$y_{i+1} = y_i + hdy_{i+1}/dt$$

which is the implicit Euler method (see equations (2.121) and (2.122)).

Equation (2.160) is explicit in the solution y_{i-k} since it uses only the past values $y_i, y_{i-1}, y_{i-2}, \ldots, y_{i-(q-1)}$, but is implicit in the one derivative, dy_{i+1}/dt. Equation (2.160) can be considered a formula for dy_{i+1}/dt in terms of $y_{i+1}, y_i, y_{i-1}, \ldots, y_{i-(q-1)}$ and is therefore termed a BDF (actually, a series of formulas for different orders); the implicit derivative term $h\beta_0 dy_{i+1}/dt$ gives the BDF its good stability properties. Also, since the BDFs use more than one past value of the solution, it is a multistep method (in constrast with the `RK` methods which require only the base point value, y_i).

The requirement, now, in using equation (2.160) is to select the constants α_k and β_0 so that the resulting integration algorithm has both good accuracy and stability properties. This has been done by C. W. Gear (1971) who assigned the "normal form" values shown in Table 2.18 to these constants for orders one to six ($1 \leq q \leq 6$).

Note that the coefficients in equation (2.160) are for orders one to six. Thus, in using the $q = 6$ coefficients, the integration from y_i to y_{i+1} is approximated by a sixth-order polynomial, i.e., the first six terms of the Taylor series. This relatively high

Table 2.17 (b) Jacobian Map and Eigenvalues at t = 0 of Run No. 2, Programs 2.7a to d.

```
DEPENDENT VARIABLE COLUMN INDEX J (FOR YJ) IS PRINTED HORIZONTALLY

DERIVATIVE ROW INDEX I (FOR DYI/DT = FI(Y1,Y2,...,YJ,...,YN) IS
PRINTED VERTICALLY

JACOBIAN MATRIX ELEMENT IN THE MAP WITH INDICES I,J IS FOR PFI/PYJ
WHERE P DENOTES A PARTIAL DERIVATIVE

                11111111112
       12345678901234567890
   1   55555555544444444432
   2   55555555444444444324
   3   55555554444444443244
   4   55555544444444432444
   5   55555444444444324444
   6   55554544444443244444
   7   55544454444432444444
   8   554444464443 3444444
   9   55444444644 34444444
  10   5444444446 444444444
  11   44444444416444444444
  12   44444444 44644444445
  13   44444432444464444555
  14   44444314444446445555
  15   44444 44444444655555
  16   4444 444444444565555
  17   444 4444444445556555
  18   44 44444444455555655
  19   4 444444444555555565
  20   23444444444555555556

  I          REAL          IMAG
  1       -21.992         0.000
  2        -3.002         0.000
  3       -20.994         0.000
  4        -4.003         0.000
  5       -19.993         0.000
  6        -5.001         0.000
  7       -18.993         0.000
  8        -6.000         0.000
  9        -7.001         0.000
 10       -17.992         0.000
 11       -16.992         0.000
 12        -8.003         0.000
 13       -15.993         0.000
 14        -9.000         0.000
 15       -14.994         0.000
 16       -11.994         0.000
 17       -10.994         0.000
 18       -12.994         0.000
 19        -9.997         0.000
 20       -13.994         0.000
```

Table 2.17 (c) Jacobian Map and Eigenvalues at $t = 4$ of Run No. 2, Programs 2.7a to d.

```
DEPENDENT VARIABLE COLUMN INDEX J (FOR YJ) IS PRINTED HORIZONTALLY

DERIVATIVE ROW INDEX I (FOR DYI/DT = FI(Y1,Y2,...,YJ,...,YN) IS
PRINTED VERTICALLY

JACOBIAN MATRIX ELEMENT IN THE MAP WITH INDICES I,J IS FOR PFI/PYJ
WHERE P DENOTES A PARTIAL DERIVATIVE

                     11111111112
            12345678901234567890
        1   55555555544444444432
        2   55555555544444444403
        3   55555555444444444034
        4   55555554444444440344
        5   55555544444444403444
        6   55555544444444034444
        7   55554454444440344444
        8   55544445444403444444
        9   55444444544034444444
       10   44444444460344444444
       11   44444444416444444444
       12   44444444134644444445
       13   44444441344464444455
       14   44444413444446444555
       15   44444134444444645555
       16   44441344444444465555
       17   44413444444444556555
       18   44034444444445555655
       19   30344444444455555565
       20   23444444444455555556

     I          REAL           IMAG
     1       -20.001          0.000
     2        -1.019          0.000
     3       -19.001          0.000
     4        -2.001          0.000
     5       -18.001          0.000
     6        -3.001          0.000
     7       -17.001          0.000
     8        -4.001          0.000
     9       -16.001          0.000
    10        -5.001          0.000
    11       -15.001          0.000
    12        -6.001          0.000
    13       -14.001          0.000
    14        -7.001          0.000
    15       -13.001          0.000
    16       -10.001          0.000
    17       -11.001          0.000
    18        -9.001          0.000
    19        -8.001          0.000
    20       -12.001          0.000
```

Table 2.18 Normal Form Coefficients for the BDFs.

q	β_0	α_0	α_1	α_2	α_3	α_4	α_5
1	1	1					
2	$\frac{2}{3}$	1	$\frac{1}{3}$				
3	$\frac{6}{11}$	1	$\frac{6}{11}$	$\frac{1}{11}$			
4	$\frac{24}{50}$	1	$\frac{35}{50}$	$\frac{1}{5}$	$\frac{1}{50}$		
5	$\frac{120}{274}$	1	$\frac{225}{274}$	$\frac{85}{274}$	$\frac{15}{274}$	$\frac{1}{274}$	
6	$\frac{720}{1764}$	1	$\frac{1624}{1764}$	$\frac{735}{1764}$	$\frac{175}{1764}$	$\frac{21}{1764}$	$\frac{1}{1764}$

accuracy is achieved by using $y_i, y_{i-1}, \ldots, y_{i-5}$ as well as dy_{i+1}/dt. However, at the beginning of the solution, only the initial condition, y_0, is available, so the calculation must start with the $q = 1$ formula (the implicit Euler method). Then, as additional points are computed along the solution, the higher order BDFs can be used. In other words, the computer implementation of the BDFs must not only handle the implicit term dy_{i+1}/dt and adjust the step size to achieve the required accuracy, but it must also vary the order of the method starting with $q = 1$ and eventually working up to the higher order formulas. This approach is therefore a *variable step, variable order*, implicit (stiff) method.

Additionally, the past values of the solution must be stored for use in the multistep formulas, and they must be available at the current integration step size. This requires interpolation of past values to obtain the necessary values at the current integration step size. Therefore, the computer implementation of the BDF method is relatively complicated; fortunately, all of this has been done, and the BDF method is available in well-established computer codes like subroutine `LSODE` which we will consider next.

The coefficients in Table 2.18 were selected to achieve a compromise between accuracy and stability. This compromise is required because for methods higher than second order, absolute stability is not possible, i.e., the entire left half of the complex plane cannot be a stable region for linear multistep methods above second order (recall again that the implicit Euler method, which is a first-order method, is stable for λh over the entire left half plane, i.e., Re(λh) plotted vs. Im(λh), as discussed previously just after equation (2.125)). The coefficients in Table 2.18 were selected so that the negative real axis remains in the stable region (real, negative eigenvalues of the ODEs will be handled with stability in the numerical integration, no matter what their separation or stiffness). Fortunately, for many applications, the ODE eigenvalues are real, and therefore the stability properties of the BDFs with the coefficients from Table 2.18 will be good. For highly oscillatory systems with eigenvalues along the imaginary axis, the coefficients of Table 2.18 will not give good stability at higher orders, and therefore restricting the maximum order to two (for which stability is achieved over the entire left half plane) may be necessary.

The good stability properties of the BDF are achieved at a price, however. Equation (2.160) is implicit in y_{i+1}, and therefore a system of nonlinear equations must usually be solved for y_{i+1} (considering the case of n simultaneous nonlinear ODEs so that y_{i+1} is a vector of length n). The solution of the nonlinear equations is usually done by a variant of Newton's method; thus implementation of the BDFs of equation (2.160) requires the use of the ODE Jacobian matrix, and the iterative solution of a linear system until converegence to the solution of the nonlinear equations (i.e., solution for y_{i+1}) is achieved. An alternative approach that has been receiving attention recently is to use an indirect (iterative) method for the solution of the linear algebraic system in Newton's method (rather than a direct method such as Gauss row reduction, or equivalently, LU factorization). This iterative method is usually based on a *Krylov subspace* as implemented in the code `VODPK` written by Byrne and Hindmarsh (1987). A major advantage of the iterative method is the reduced storage requirement and consequent reduction in the number of arithmetic operations since the Jacobian matrix is not required at each point along the solution in contrast with a direct method applied to the linear algebra (recall again that the full Jacobian matrix for a system of n ODEs is of size n^2, and generally requires n^3 operations for solution of the linear equations; these numbers become very large with increasing n; thus, for large n, the structure of the Jacobian matrix must be exploited to reduce the storage requirement and number of arithmetic operations when using a direct method).

We now return to the ODE system of equation (2.150), but with the β_i selected to produce a stiff ODE system. Rather than read the values of β_i from the data file, the values are set in subroutine `INITAL`.

The following points should be noted about Program 2.8a:

1. The coding is in double precision which is recommended for stiff problems on short word length computers, e.g., 32-bit computers for which the precision is about seven figures in single precision, and about 14 figures in double precision. As a rule, if the stiffness ratio of the ODE problem system is computed as

$$\text{largest real part}/\text{smallest real part} = 10^m$$

   where "real part" refers to the real part of an ODE eigenvalue, then computer precision substantially greater than m significant figures should be used. For example, if the ODE system has two eigenvalues $\lambda_{\max} = -10000$ and $\lambda_{\min} = 0.0001$, then $-10000/-0.0001 = 10^8$, and the computer precision should be substantially greater than eight figures, which would be the case for double-precision Fortran (about 14 figures), but not single-precision Fortran (about seven figures).

2. The values of β_i are set in `DO` loop 10 of subroutine `INITAL` for two cases (two solutions): (a) the nonstiff values are the same as for the first case of Program 2.7d, i.e., `B(I)=-I`, $I = 1, 2, \ldots, 20$ and (b) the stiff values `B(I)=-1000, I LE 10` and `-0.001, I GT 10`. The latter require an implicit integrator which is the reason for the use of `LSODE`.

3. The initial conditions are set using the analytical solution in `DO` loops 70, 80, and 90 of subroutine `INITAL` since for the second (stiff) case, the initial value of time is $t = 0.000001$ (rather than $t = 0$) as indicated in the data file which follows.

4. The calls to subroutine `JMAP` and `EIGEN` in subroutine `PRINT` are commented out so that these calls are not executed in order to reduce the volume of output.

```
      SUBROUTINE INITAL
C...
C...  KROGH'S ODE PROBLEM - SOLUTION OF A STIFF CASE BY LSODE
C...
C...  KROGH'S PROBLEM (1) IS A NONLINEAR ODE PROBLEM WITH AN EXACT
C...  SOLUTION.  THE NUMBER OF ODES AND THE STIFFNESS OF THESE
C...  EQUATIONS CAN BE VARIED BY THE CHOICE OF PARAMETERS.  IN
C...  THIS CASE, N = 20, AND THE STIFFNESS IS SET BY CONSTANTS
C...  READ FROM A DATA FILE.
C...
C...  (1)  GEAR, C. W., NUMERICAL INITIAL VALUE PROBLEMS IN ORDI-
C...       NARY DIFFERENTIAL EQUATIONS, PRENTICE-HALL, ENGLEWOOD
C...       CLIFFS, NJ, 1971, PP 218-220
C...
C...  THE NONLINEAR ODE PROBLEM PROPOSED BY KROGH (1) IS AN IDEAL
C...  TEST PROBLEM FOR ODE SOFTWARE FOR THE FOLLOWING REASONS:
C...
C...     (1)  THE ORDER OF THE PROBLEM (NUMBER OF ODES) CAN BE
C...          SELECTED BY THE USER (IN THE PRESENT CASE, N = 20).
C...
C...     (2)  THE STIFFNESS OF THE ODES CAN BE SELECTED BY THE USER.
C...          THIS POINT REQUIRES SOME CLARIFICATION SINCE THE ODES
C...          ARE NONLINEAR, BUT STIFFNESS IS A LINEAR CONCEPT BASED
C...          ON EIGENVALUES (A STIFF SYSTEM IS GENERALLY DEFINED AS:
C...
C...                 (PROBLEM TIME)*(LARGEST EIGENVALUE)
C...
C...                       IS MUCH GREATER THAN 1)
C...
C...          GENERALLY, THE CONCEPT OF STIFFNESS APPLIED TO NONLINEAR
C...          PROBLEMS MEANS APPLIED TO THE LINEARIZED VERSION OF THE
C...          NONLINEAR EQUATIONS, FOR WHICH EIGENVALUES CAN BE COM-
C...          PUTED.  HOWEVER, SINCE LINEARIZATION IS PERFORMED AROUND
C...          A BASE POINT, THE EIGENVALUES OF THE LINEARIZED SYSTEM
C...          WILL CHANGE WITH THE BASE POINT, I.E., THE LOCAL EIGEN-
C...          VALUES OF A NONLINEAR SYSTEM ARE NOT CONSTANT.
C...
C...          THE NONLINEAR ODES IN THE KROGH PROBLEM HAVE IN THEIR
C...          ANALYTICAL SOLUTION EXPONENTIALS OF THE FORM EXP(B(I)*T)
C...          WHERE T IS THE INDEPENDENT VARIABLE AND THE B(I), I =
C...          1, 2,..., N ARE PARAMETERS SELECTED BY THE USER.  THUS
C...          THE B(I) ARE ANALOGOUS TO EIGENVALUES, BUT THE EXPONENT-
C...          IALS (EXP(B(I)*T)) APPEAR NONLINEARLY IN THE ANALYTICAL
C...          SOLUTION (RATHER THAN AS A LINEAR COMBINATION IN THE
C...          ANALYTICAL SOLUTION OF A SYSTEM OF CONSTANT COEFFICIENT
C...          LINEAR ODES).
C...
C...          IN THIS SENSE, THE STIFFNESS OF THE KROGH TEST PROBLEM
C...          CAN BE SPECIFIED BY THE USER BY SELECTING THE B(I).
C...
```

Program 2.8a Subroutines INITAL, DERV, and PRINT for the LSODE Solution of Equation (2.150). *Continued next pages.*

```
C...      (3)  AS THE PRECEDING DISCUSSION INDICATES, THE KROGH TEST
C...           HAS AN ANALYTICAL (EXACT) SOLUTION, WHICH CAN BE USED
C...           AS A STANDARD IN EVALUATING A NUMERICAL INTEGRATOR,
C...           I.E., THE ERRORS IN NUMERICAL SOLUTIONS CAN BE COMPUTED.
C...
C...  ALL OF THE FEATURES OF THE KROGH TEST PROBLEM ARE ILLUSTRATED IN
C...  THE FOLLOWING CODING FOR TWO CASES:
C...
C...      (1)  NONSTIFF CASE
C...
C...           HERE WE TAKE B(I) = -I, I = 1, 2,..., 20.  THESE CON-
C...           STANTS ARE SET IN DO LOOP 10 IN SUBROUTINE INITAL AT
C...           THE BEGINNING OF THE NUMERICAL SOLUTION, AND INTEGRATION
C...           IS BY A NONSTIFF OPTION OF LSODE.
C...
C...      (2)  STIFF CASE
C...
C...           HERE WE TAKE B(I), I = 1, 2,..., 20 TO HAVE THE VALUE
C...           -1000 FOR I LE 10 AND -0.001 FOR I GT 10.  THEY ARE SET
C...           IN DO LOOP 10 IN SUBROUTINE INITAL AT THE BEGINNING OF
C...           THE NUMERICAL SOLUTION, AND THE INTEGRATION IS PERFORMED
C...           BY A STIFF OPTION OF LSODE FOR A FULL NUMERICAL JACOBIAN
C...           MATRIX. SINCE THESE VALUES OF B(I) SPAN SIX ORDERS OF
C...           MAGNITUDE, THIS PROBLEM CAN BE CONSIDERED STIFF (AND
C...           THEREFORE SHOULD NOT BE DONE BY A NONSTIFF (EXPLICIT)
C...           INTEGRATOR).
C...
C...  THE MATHEMATICAL DETAILS OF THE PROBLEM ARE GIVEN IN REFERENCE
C...  (1) CITED PREVIOUSLY.  THEY CAN ALSO BE INFERRED FROM THE
C...  FOLLOWING CODING.
C...
      IMPLICIT DOUBLE PRECISION (A-H,O-Z)
      COMMON /T/          T,     NSTOP,     NORUN
      COMMON /Y/     Y(20)
      COMMON /F/  DYDT(20)
      COMMON /C/ BETA(20), U(20,20), B(20,20), W(20), N
      COMMON /IO/  NI, NO
C...
C...  READ THE NUMBER OF ODES.  THEN SET THE CONSTANT ARRAY BETA
C...  WHICH DEFINES THE PROBLEM TIME SCALE AND STIFFNESS
      READ(NI,100)N
      DO 10 J=1,N
C...
C...  NONSTIFF CASE, B(I) = -I, I = 1, 2,..., N
      IF(NORUN.EQ.1)THEN
         IF(J.EQ.1)THEN
            BETA(J)=-1.0D0
         ELSE
            BETA(J)=BETA(J-1)-1.0D0
         END IF
C...
```

Program 2.8a *Continued.*

```
C...  STIFF CASE, B(I) = -1000, I LE N/2, B(I) = -0.001, I GT N/2
      ELSE IF(NORUN.EQ.2)THEN
         IF(J.LE.N/2)THEN
            BETA(J)=-1.0D+03
         ELSE
            BETA(J)=-1.0D-03
         END IF
      END IF
   10 CONTINUE
      WRITE(NO,102)N,(BETA(J),J=1,N)
C...
C...  SET THE CONSTANT ARRAY U
      TERM = 2.0D0/DFLOAT(N)
      DO 20 I=1,N
      DO 30 J=1,N
         U(I,J) = TERM
   30 CONTINUE
         U(I,I) = TERM - 1.0D0
   20 CONTINUE
C...
C...  SET THE CONSTANT ARRAY B
      DO 40 J=1,N
      DO 50 I=1,N
         SUM = 0.0D0
      DO 60 K=1,N
         SUM = SUM + U(I,K)*BETA(K)*U(K,J)
   60 CONTINUE
         B(I,J) = SUM
   50 CONTINUE
   40 CONTINUE
C...
C...  INITIAL CONDITIONS FROM THE EXACT SOLUTION
      DO 70 J=1,N
         BETAT = BETA(J)*T
C...
C...  AVOID AN OVERFLOW OF THE EXPONENTIAL FUNCTION, EXP(BETAT), IN
C...  THE EXACT SOLUTION
      IF (BETAT.GT.50.0D0) THEN
         W(J) = 0.0D0
C...
C...  AVOID AN UNDERFLOW OF THE EXPONENTIAL FUNCTION, EXP(BETAT), IN
C...  THE EXACT SOLUTION
      ELSE IF (BETAT.LT.-50.0D0) THEN
         W(J) = BETA(J)
      ELSE
         W(J) = BETA(J)/(1.0D0 - (1.0D0 + BETA(J))*DEXP(BETAT))
      END IF
   70 CONTINUE
      DO 80 I=1,N
         SUM = 0.0D0
```

Program 2.8a *Continued.*

```
      DO 90 J=1,N
         SUM = SUM + U(I,J)*W(J)
   90 CONTINUE
         Y(I) = SUM
   80 CONTINUE
C...
C...  PRINT THE OUTPUT HEADING
      WRITE(NO,103)
      WRITE(NO,104)
C...
C...  FORMATS ARE LISTED HERE
  100 FORMAT(I5)
  101 FORMAT(F20.5)
  102 FORMAT(1H ,10X,'N = ',I3,10X,'BETA',20(/,17X,F15.3),//)
  103 FORMAT(1H1)
  104 FORMAT(1H ,6X,'TIME',8X, 'Y(1)',6X, 'YEX(1)',8X, 'PER1',/,
     1                    18X,'Y(20)',5X,'YEX(20)',7X,'PER20',/,
     2                    29X,'PERMAX',/)
      RETURN
      END

      SUBROUTINE DERV
      IMPLICIT DOUBLE PRECISION (A-H,O-Z)
      COMMON /T/        T,    NSTOP,    NORUN
      COMMON /Y/    Y(20)
      COMMON /F/ DYDT(20)
      COMMON /C/ BETA(20), U(20,20), B(20,20), W(20), N
C...
C...  CALCULATE THE ARRAY W
      DO 10 I=1,N
         SUM = 0.0D0
      DO 20 J=1,N
         SUM = SUM + U(I,J)*Y(J)
   20 CONTINUE
         W(I) = SUM
   10 CONTINUE
C...
C...  CALCULATE THE N ODE DERIVATIVES
      DO 30 I=1,N
         SUM = 0.0D0
      DO 40 J=1,N
         SUM = SUM + U(I,J)*W(J)*W(J) - B(I,J)*Y(J)
   40 CONTINUE
         DYDT(I) = SUM
   30 CONTINUE
      RETURN
      END

      SUBROUTINE PRINT(NI,NO)
      IMPLICIT DOUBLE PRECISION (A-H,O-Z)
```

Program 2.8a *Continued.*

```
      COMMON /T/        T,     NSTOP,     NORUN
      COMMON /Y/     Y(20)
      COMMON /F/ DYDT(20)
      COMMON /C/ BETA(20), U(20,20), B(20,20), W(20), N
      DIMENSION YEX(20)
C...
C...  ABSOLUTE DIMENSIONING OF THE ARRAYS REQUIRED BY SUBROUTINE JMAP
C...  (A, X, XOLD, F, FOLD) AND SUBROUTINE EIGEN (WI, WR, Z, RW, IW).
C...  NOTE THAT ARRAYS A, WR, WI, Z AND RW ARE SINGLE PRECISION IN
C...  ORDER TO CONFORM TO THE CODING IN SUBROUTINES JMAP AND EIGEN
      REAL       A(20,20),   WR(20), WI(20),  Z(20,20), RW(20)
      DIMENSION     X(20), XOLD(20),  F(20),  FOLD(20),PER(20)
      INTEGER      IW(20)
C...
C...  THE FOLLOWING EQUIVALENCE FACILITATES TRANSFER OF THE DEPENDENT
C...  VARIABLE VECTOR IN /Y/ TO ARRAY Y, AND THE DERIVATIVE VECTOR IN
C...  IN /F/ TO ARRAY F FOR USE IN THE SUBSEQUENT CALL TO SUBROUTINE
C...  JMAP
      EQUIVALENCE (Y(1),X(1)),(DYDT(1),F(1))
C...
C...  CALCULATE THE N EXACT SOLUTIONS
      DO 10 J=1,N
         BETAT = BETA(J)*T
C...
C...  AVOID AN OVERFLOW OF THE EXPONENTIAL FUNCTION, EXP(BETAT), IN
C...  THE EXACT SOLUTION
      IF (BETAT.GT.50.0D0) THEN
         W(J) = 0.0D0
C...
C...  AVOID AN UNDERFLOW OF THE EXPONENTIAL FUNCTION, EXP(BETAT), IN
C...  THE EXACT ASOLUTION
      ELSE IF (BETAT.LT.-50.0D0) THEN
         W(J) = BETA(J)
      ELSE
         W(J) = BETA(J)/(1.0D0 - (1.0D0 + BETA(J))*DEXP(BETAT))
      END IF
   10 CONTINUE
      DO 20 I=1,N
         SUM = 0.0D0
      DO 30 J=1,N
         SUM = SUM + U(I,J)*W(J)
   30 CONTINUE
         YEX(I) = SUM
   20 CONTINUE
C...
C...  CALCULATE THE MAXIMUM ERROR
      ERRMAX = DABS((Y(1) - YEX(1))/YEX(1))*100.0D0
      DO 40 J=2,N
         ERR = DABS((Y(J) - YEX(J))/YEX(J))*100.0D0
         IF (ERR.GT.ERRMAX) ERRMAX = ERR
   40 CONTINUE
```

Program 2.8a *Continued.*

```
C...
C...  PRINT THE NUMERICAL AND EXACT SOLUTIONS FOR I = 1 AND N, AND
C...  THE CORRESPONDING ERRORS, AND THE MAXIMUM ERROR
      PER1=DABS((Y(1)-YEX(1))/YEX(1))*100.0D0
      PERN=DABS((Y(N)-YEX(N))/YEX(N))*100.0D0
      PERMAX=ERRMAX
      WRITE(NO,101)T,Y(1),YEX(1),PER1,Y(N),YEX(N),PERN,PERMAX
 101  FORMAT(F11.6,2E12.4,F12.3,/,
     +          11X,2E12.4,F12.3,/,
     +          23X,       F12.3,/)
C...
C...  MAP THE JACOBIAN MATRIX OF THE ODE SYSTEM DEFINED IN SUBROUTINE
C...  DERV, AND COMPUTE ITS TEMPORAL EIGENVALUES
C...  CALL JMAP(N,A,X,XOLD,F,FOLD)
C...
C...  SUBROUTINE EIGEN CALLS A SERIES OF EISPACK ROUTINES TO
C...  COMPUTE THE TEMPORAL EIGENVALUES, AND OPTIONALLY, THE
C...  EIGENVECTORS OF THE 20-ODE SYSTEM JACOBIAN MATRIX
C...  CALL EIGEN(N,A,WR,WI,Z,RW,IW)
      RETURN
      END
```

Program 2.8a *Continued.*

The main program to call subroutine LSODE is listed in Program 2.8b. The following points should be noted about Program 2.8b:

1. The absolute dimensioning is for 250 ODEs as indicated by the following COMMON block:

```
      COMMON/T/          T,       NSTOP,       NORUN
     1       /Y/     Y(250)
     2       /F/     F(250)
```

   Again, the ODE dependent variable is in COMMON/Y/ and the temporal derivative vector is in COMMON/F/. LSODE also requires a real work array and an integer work array with sizes defined by the number of ODEs; in Program 2.8b, the absolute dimensioning for the real work array is 6600 and for the integer work array is 275.

```
      DIMENSION YV(250), RWORK(6600), IWORK(275)
```

   The sizes of these arrays are defined by formulas given in the documentation of LSODE (an extensive set of comments at the beginning of the code). An example calculation of the size of the real work array is given in the comments (in main program LSODTE in Program 2.8b).

2. Two external subroutines are call by LSODE, as defined by the EXTERNAL statement:

```
C...
C...  EXTERNAL THE DERIVATIVE AND ODE JACOBIAN MATRIX
C...  ROUTINES CALLED BY LSODE
      EXTERNAL FCN, JAC
```

```
      PROGRAM LSODET
C...
C...  THE FOLLOWING PROGRAM CALLS AN ODE APPLICATION, DEFINED AS THE
C...  USER-SUPPLIED SUBROUTINES INITAL, DERV AND PRINT, PLUS DATA.
C...  THE SYSTEM OF ODES PROGRAMMED IN SUBROUTINE DERV IS INTEGRATED
C...  BY LSODE (THE BASIC SOLVER IN ODEPACK, A LIBRARY OF INTEGRATORS
C...  DEVELOPED BY ALAN C. HINDMARSH, LAWRENCE LIVERMORE NATIONAL LAB-
C...  ORATORY).
C...
C...  THE MODEL INITIAL CONDITIONS ARE SET IN SUBROUTINE INITAL, AND
C...  THE MODEL DERIVATIVES ARE PROGRAMMED IN SUBROUTINE DERV.  THE
C...  NUMERICAL SOLUTION IS PRINTED AND PLOTTED IN SUBROUTINE PRINT.
C...
C...  SUBROUTINES INITAL, DERV AND PRINT.  THE FOLLOWING CODING IS FOR
C...  250 ORDINARY DIFFERENTIAL EQUATIONS (ODES).  IF MORE ODES ARE TO
C...  BE INTEGRATED, ALL OF THE 250*S SHOULD BE CHANGED TO THE
C...  REQUIRED NUMBER.  DOUBLE PRECISION IS USED FOR THE INTEGRATION
C...  OF STIFF ODES
      IMPLICIT DOUBLE PRECISION (A-H,O-Z)
      COMMON/T/          T,      NSTOP,      NORUN
     1      /Y/    Y(250)
     2      /F/    F(250)
C...
C...  COMMON AREA TO PROVIDE THE INPUT/OUTPUT UNIT NUMBERS TO OTHER
C...  SUBROUTINES
      COMMON/IO/       NI,        NO
C...
C...  ABSOLUTE DIMENSIONING OF THE ARRAYS REQUIRED BY LSODE.  ARRAY
C...  WORK IS SIZED BY THE FOLLOWING FORMULA FOR BANDED PROCESSING
C...  OF THE ODE JACOBIAN MATRIX VIA FINITE DIFFERENCES: 22+17*NEQN+
C...  (2*ML+MU)*NEQN.  FOR NEQN = 250, ML = MU = 3, 22+17*250+(2*3+3)*
C...  250 = 6600 (APPROXIMATELY).  IF FEWER THAN 250 ODES ARE TO BE
C...  INTEGRATED, THEN ML AND MU CAN BE LARGER FOR THE GIVEN SIZE OF
C...  WORK.  IF NEQN GT 250 AND/OR ML AND MU GT 3, WORK MUST BE DIMEN-
C...  SIONED TO A LARGER SIZE
      DIMENSION YV(250), RWORK(6600), IWORK(275)
C...
C...  EXTERNAL THE DERIVATIVE AND ODE JACOBIAN MATRIX ROUTINES CALLED
C...  BY LSODE
      EXTERNAL FCN, JAC
C...
C...  ARRAY FOR THE TITLE (FIRST LINE OF DATA), CHARACTERS  END OF
C...  RUNS
      CHARACTER TITLE(20)*4, ENDRUN(3)*4
C...
C...  VARIABLE FOR THE TYPE OF ERROR CRITERION
      CHARACTER*3 ABSREL
C...
C...  DEFINE THE CHARACTERS  END OF RUNS
      DATA ENDRUN/'END ','OF R','UNS '/
C...
```

Program 2.8b Main Program LSODET and Subordinate Routines to Call Subroutine LSODE. *Continued next pages.*

```
C...  DEFINE THE INPUT/OUTPUT UNIT NUMBERS
      NI=5
      NO=6
C...
C...  OPEN THE INPUT/OUTPUT FILES
      OPEN(NI,FILE='DATA',   STATUS='OLD')
      OPEN(NO,FILE='OUTPUT',STATUS='NEW')
C...
C...  INITIALIZE THE RUN COUNTER
      NORUN=0
C...
C...  BEGIN A RUN
1     NORUN=NORUN+1
C...
C...  INITIALIZE THE RUN TERMINATION VARIABLE
      NSTOP=0
C...
C...  READ THE FIRST LINE OF DATA
      READ(NI,1000,END=999)(TITLE(I),I=1,20)
C...
C...  TEST FOR  END OF RUNS  IN THE DATA
      DO 2 I=1,3
      IF(TITLE(I).NE.ENDRUN(I))GO TO 3
2     CONTINUE
C...
C...  AN END OF RUNS HAS BEEN READ, SO TERMINATE EXECUTION
999   STOP
C...
C...  READ THE SECOND LINE OF DATA
3     READ(NI,1001,END=999)T0,TF,TP
C...
C...  READ THE THIRD LINE OF DATA
      READ(NI,1002,END=999)NEQN,ERROR
C...
C...  PRINT A DATA SUMMARY
      WRITE(NO,1003)NORUN,(TITLE(I),I=1,20),T0,TF,TP,NEQN,ERROR
C...
C...  INITIALIZE TIME
      T=T0
C...
C...  SET THE INITIAL CONDITIONS
      CALL INITAL
C...
C...  PRINT THE INITIAL CONDITIONS
      CALL PRINT(NI,NO)
C...
C...  SET THE INITIAL CONDITIONS FOR SUBROUTINE LSODE
      DO 5 I=1,NEQN
      YV(I)=Y(I)
5     CONTINUE
C...
```

Program 2.8b *Continued.*

```
C...  SET THE PARAMETERS FOR SUBROUTINE LSODE
      TV=T0
      ITOL=1
      RTOL=ERROR*1.0D-02
      ATOL=ERROR*1.0D-02
      LRW=6600
      LIW=275
      IOPT=0
      ITASK=1
      ISTATE=1
C...
C...  TWO LSODE OPTIONS, AS SELECTED BY THE RUN NUMBER NORUN, ARE
C...  PROGRAMMED HERE. THE TERMINOLOGY IS
C...
C...     IMPLICIT ADAMS METHOD FOR NONSTIFF PROBLEMS
C...
C...     METHOD BASED ON BACKWARD DIFFERENTIATION FORMULAS (BDF) FOR
C...     STIFF PROBLEMS
C...
C...  EACH OF THESE METHODS (NONSTIFF AND STIFF) GENERATES NONLINEAR
C...  ALGEBRAIC EQUATIONS DURING THE NUMERICAL INTEGRATION WHICH ARE
C...  SOLVED BY ONE OF THE FOLLOWING METHODS
C...
C...     FUNCTIONAL ITERATION (A PICARD ITERATION)
C...
C...     CHORD ITERATION, ESSENTIALLY NEWTON*S METHOD, WITH THE ODE
C...     JACOBIAN MATRIX PARTIAL DERIVATIVES APPROXIMATED BY FINITE
C...     DIFFERENCES.  A FULL, BANDED OR DIAGONAL MATRIX MAY BE USED.
C...     THE BANDED AND DIAGONAL FORMS TYPICALLY ARE APPROXIMATIONS TO
C...     THE FULL MATRIX, I.E., CONTAIN A SUBSET OF THE ELEMENTS OF THE
C...     FULL MATRIX.
C...
C...     (1)  NORUN = 1
C...
C...          IMPLICIT ADAMS METHOD, FUNCTIONAL ITERATION
              IF(NORUN.EQ.1)THEN
              METH=1
              MITER=0
C...
C...     (2)  NORUN = 2
C...
C...          IMPLICIT BDF METHOD, CHORD ITERATION WITH AN INTERNALLY
C...          GENERATED (DIFFERENCE QUOTIENT) FULL JACOBIAN
              ELSE IF(NORUN.EQ.2)THEN
              METH=2
              MITER=2
C...
      END IF
C...
C...  THE METHOD FLAG, MF, IS COMPUTED FROM METH AND MITER
      MF=10*METH+MITER
C...
```

Program 2.8b *Continued.*

```
C...  FOR THE NONSTIFF CASE (NORUN = 1), THE INDEPENDENT VARIABLE IS
C...  ADVANCED BY AN ADDITIVE INCREMENT
4        IF(NORUN.EQ.1)THEN
            TOUT=TV+TP
C...
C...  FOR THE STIFF CASE (NORUN = 2), THE INDEPENDENT VARIABLE IS
C...  ADVANCED BY A MULIPLICATIVE FACTOR
         ELSE IF(NORUN.EQ.2)THEN
            TOUT=TV*10.0D0
         END IF
C...
C...  CALL SUBROUTINE LSODE TO COVER ONE PRINT INTERVAL
      CALL LSODE(FCN,NEQN,YV,TV,TOUT,ITOL,RTOL,ATOL,ITASK,ISTATE,
     1           IOPT,RWORK,LRW,IWORK,LIW,JAC,MF)
C...
C...  PRINT THE SOLUTION
      T=TV
      DO 6 I=1,NEQN
      Y(I)=YV(I)
6     CONTINUE
      CALL PRINT(NI,NO)
C...
C...  TEST FOR AN ERROR CONDITION
      IF(ISTATE.LT.0)THEN
C...
C...     PRINT A MESSAGE INDICATING AN ERROR CONDITION
         WRITE(NO,1004)ISTATE
C...
C...     GO ON TO THE NEXT RUN
         GO TO 1
      END IF
C...
C...  CHECK FOR A RUN TERMINATION
      IF(NSTOP.NE.0)GO TO 1
C...
C...  MONITOR THE PROGRESS OF THE CALCULATION
      WRITE(*,*)TV,TF
C...
C...  CHECK FOR THE END OF THE RUN
      IF(TV.LT.(TF-0.5D0*TP))GO TO 4
C...
C...  THE CURRENT RUN IS COMPLETE, SO PRINT THE COMPUTATIONAL STAT-
C...  ISTICS FOR LSODE AND GO ON TO THE NEXT RUN
      WRITE(NO,8)RWORK(11),IWORK(14),IWORK(11),IWORK(12),IWORK(13)
8     FORMAT(1H ,//,' LSODE COMPUTATIONAL STATISTICS'             ,//,
     1 ' LAST STEP SIZE                       ',           D10.3,//,
     2 ' LAST ORDER OF THE METHOD             ',             I10,//,
     3 ' TOTAL NUMBER OF STEPS TAKEN          ',             I10,//,
     4 ' NUMBER OF FUNCTION EVALUATIONS       ',             I10,//,
     5 ' NUMBER OF JACOBIAN EVALUATIONS       ',             I10,/)
```

Program 2.8b *Continued.*

```
      GO TO 1
C...
C...  ****************************************************************
C...
C...  FORMATS
C...
1000  FORMAT(20A4)
1001  FORMAT(3D10.0)
1002  FORMAT(I5,20X,D10.0)
1003  FORMAT(1H1,
     1 ' RUN NO. - ',I3,2X,20A4,//,
     2 ' INITIAL T - ',D10.3,//,
     3 '   FINAL T - ',D10.3,//,
     4 '   PRINT T - ',D10.3,//,
     5 ' NUMBER OF DIFFERENTIAL EQUATIONS - ',I3,//,
     6 ' MAXIMUM INTEGRATION ERROR - ',D10.3,//,
     7 1H1)
1004  FORMAT(1H ,//,' ISTATE = ',I3,//,
     1 ' INDICATING AN INTEGRATION ERROR, SO THE CURRENT RUN'     ,/,
     2 ' IS TERMINATED.  PLEASE REFER TO THE DOCUMENTATION FOR'   ,/,
     3 ' SUBROUTINE',//,25X,'LSODE',//,
     4 ' FOR AN EXPLANATION OF THESE ERROR INDICATORS'            )
      END

      SUBROUTINE FCN(NEQN,TV,YV,YDOT)
C...
C...  SUBROUTINE FCN IS AN INTERFACE ROUTINE BETWEEN SUBROUTINES LSODE
C...  AND DERV
C...
C...  COMMON AREA
      IMPLICIT DOUBLE PRECISION (A-H,O-Z)
      COMMON/T/          T,      NSTOP,      NORUN
     1      /Y/      Y(1)
     2      /F/      F(1)
C...
C...  VARIABLE DIMENSION THE DEPENDENT AND DERIVATIVE ARRAYS
      DIMENSION YV(NEQN), YDOT(NEQN)
C...
C...  TRANSFER THE INDEPENDENT VARIABLE, DEPENDENT VARIABLE VECTOR
C...  FOR USE IN SUBROUTINE DERV
      T=TV
      DO 1 I=1,NEQN
      Y(I)=YV(I)
1     CONTINUE
C...
C...  EVALUATE THE DERIVATIVE VECTOR
      CALL DERV
C...
```

Program 2.8b *Continued.*

```
C...  TRANSFER THE DERIVATIVE VECTOR FOR USE BY SUBROUTINE LSODE
      DO 2 I=1,NEQN
      YDOT(I)=F(I)
2     CONTINUE
      RETURN
      END

      SUBROUTINE JAC(NEQ,T,Y,ML,MU,PD,NROWPD)
C...
C...  SUBROUTINE JAC IS CALLED ONLY IF AN OPTION OF LSODE IS SELECTED
C...  FOR WHICH THE USER MUST SUPPLY THE ANALYTICAL JACOBIAN MATRIX OF
C...  THE ODE SYSTEM.  THE PROGRAMMING OF THE ANALYTICAL JACOBIAN IN
C...  SUBROUTINE JAC IS DESCRIBED IN DETAIL IN THE DOCUMENTATION
C...  COMMENTS AT THE BEGINNING OF SUBROUTINE LSODE.
C...
      RETURN
      END
```

Program 2.8b *Continued.*

FCN defines the ODE derivative vector as in the case of RKF45 (see Program 2.2a). JAC defines the ODE Jacobian matrix if an option of LSODE is selected which requires the user to define the Jacobian matrix. Generally, this option is not selected because for large ODE problems, programming the Jacobian matrix is impractical; rather, an option is selected to have LSODE compute a numerical Jacobian by finite differences. However, a dummy subroutine JAC is provided to satisfy the loader (this requirement varies with the requirements of the loader for different computers).

3. A set of parameters is defined before the first call to LSODE:

```
C...
C...  SET THE PARAMETERS FOR SUBROUTINE LSODE
      TV=T0
      ITOL=1
      RTOL=ERROR*1.0D-02
      ATOL=ERROR*1.0D-02
      LRW=6600
      LIW=275
      IOPT=0
      ITASK=1
      ISTATE=1
```

Briefly these parameters are (a) the independent variable is set to its initial value, TV=T0, (b) one error tolerance is applied to each of the ODE dependent variables, ITOL=1 (the alternative is to apply a different error tolerance to each dependent variable so that ITOL=2), (c) a relative error tolerance (RTOL) and an absolute error tolerance (ATOL) are applied to the dependent variables of magnitude ERROR*1.0D-02 where ERROR is read from the data file as 1.0D-05; the additional factor of 1.0D-02 was added because of the stiff case which appeared to require a tighter error tolerance, (d) the lengths of the real and integer work arrays are defined, LRW=6600, LIW=275 (these lengths may be overspecified, and the only disadvantage is the use of additional memory), (e) no additional options will be set when LSODE is called, i.e., only default options will be used, IOPT=0, (f) ITASK=1 for "normal computation" of the solution at the specified output value of the independent variable (this is a technical detail that we will not consider here, but has to do with the use of interpolation of the solution at the output points if ITASK=1), and (g) LSODE is initialized before the first call by ISTATE=1. LSODE has many options, and the preceding parameters basically select the defaults. The available options are described in detail in a set of comments at the beginning of LSODE; sample problems are also included in the comments, including representative output so that the operation of LSODE can be confirmed on the user's local computer(s).

4. LSODE has an option for nonstiff problems, the implicit Adams method, selected by setting METH=1, and an option for stiff problems, the BDF method, selected by setting METH=2. Within each of these two methods, several options can be selected with variable MITER. For the nonstiff problem of Program 2.8a (with NORUN=1), MITER=0 corresponding to functional iteration to solve the linear algebra of the implicit Adams methods. For the stiff problem of Program 2.8a (with NORUN=2), MITER=2 corresponding to an internally generated Jacobian matrix (a numerical Jacobian from finite differences). With these choices of METH and MITER, a method flag, MF, is computed as MF=10*METH+MITER (10 for NORUN=1 and 22 for NORUN=2), which becomes an input to LSODE:

```
C...
C...     (1)  NORUN = 1
C...
C...          IMPLICIT ADAMS METHOD, FUNCTIONAL
C...          ITERATION
              IF(NORUN.EQ.1)THEN
              METH=1
              MITER=0
C...
C...     (2)  NORUN = 2
C...
C...          IMPLICIT BDF METHOD, CHORD ITERATION WITH AN
C...          INTERNALLY GENERATED (DIFFERENCE QUOTIENT)
C...          FULL JACOBIAN
              ELSE IF(NORUN.EQ.2)THEN
              METH=2
              MITER=2
C...
      END IF
C...
C...  THE METHOD FLAG, MF, IS COMPUTED FROM METH AND MITER
      MF=10*METH+MITER
```

5. The independent variable in equation (2.160), t, is advanced either by an additive increment, TP (read from the data file in Program 2.8a) for the nonstiff case, or by a multiplicative factor of 10 in the stiff case. The latter demonstrates an important property of an implicit integrator to accelerate over orders of magnitude

in the independent variable, in this case from 0.000001 to 100 (as specifed in the subsequent data file of Program 2.8c):

```
C...
C...  FOR THE NONSTIFF CASE (NORUN = 1), THE INDEPENDENT
C...  VARIABLE IS ADVANCED BY AN ADDITIVE INCREMENT
4        IF(NORUN.EQ.1)THEN
            TOUT=TV+TP
C...
C...  FOR THE STIFF CASE (NORUN = 2), THE INDEPENDENT
C...  VARIABLE IS ADVANCED BY A MULTIPLICATIVE FACTOR
         ELSE IF(NORUN.EQ.2)THEN
            TOUT=TV*10.0D0
         END IF
```

6. The call to LSODE moves the solution forward from TV to TOUT:

```
C...
C...  CALL SUBROUTINE LSODE TO COVER ONE PRINT INTERVAL
      CALL LSODE(FCN,NEQN,YV,TV,TOUT,ITOL,RTOL,ATOL,ITASK,ISTATE,
     1           IOPT,RWORK,LRW,IWORK,LIW,JAC,MF)
```

Thus, before the call to LSODE, the solution at $t = TV$ is in array YV, and on return from LSODE, the solution at $t = TOUT$ is in array YV. Note again that FCN and JAC are the external subroutines described previously.

7. After printing the solution, checks are made for any error conditions (ISTATE<0) and the end of the run ($t = TF$, where TF is read from the data file of Program 2.8c). Also, the run is terminated if NSTOP is set to a nonzero value in any of the subroutines where it appears, e.g., INITAL, DERV or PRINT; this stopping variable is provided so that a run can be terminated if the solution has any obvious errors (e.g., physically unrealistic values):

```
C...
C...  TEST FOR AN ERROR CONDITION
      IF(ISTATE.LT.0)THEN
C...
C...     PRINT A MESSAGE INDICATING AN ERROR CONDITION
         WRITE(NO,1004)ISTATE
C...
C...     GO ON TO THE NEXT RUN
         GO TO 1
      END IF
C...
C...  CHECK FOR A RUN TERMINATION
      IF(NSTOP.NE.0)GO TO 1
C...
C...  CHECK FOR THE END OF THE RUN
      IF(TV.LT.(TF-0.5D0*TP))GO TO 4
```

8. At the end of each run, the LSODE computational statistics are printed. These are (a) the last integration step size (which in the case of a stiff problem such

```
KROGH PROBLEM FOR N = 20, NONSTIFF CASE
0.          1.0        0.1
   20                       0.00001
   20
KROGH PROBLEM FOR N = 20, STIFF CASE
0.000001  100.0      0.000001
   20                       0.00001
   20
END OF RUNS
```

Program 2.8c Data File Read by Program 2.8b.

as for NORUN=2 is usually much larger than would be possible with an explicit integrator because of the superior stability properties of the implicit (BDF) integrator), (b) the last order of the BDF, i.e., the last value of q in equation (2.160), (c) the total number of steps taken (which again will usually be much smaller than would be possible with an explicit integrator), (d) the total number of function evaluations, i.e., the number of times $f(y,t)$ is evaluated in the ODE system $dy/dt = f(y,t)$, and (e) the number of Jacobian evaluations required to solve the nonlinear equations resulting from the use of equation (2.160) as discussed previously:

```
C...
C...  THE CURRENT RUN IS COMPLETE, SO PRINT THE COMPUTATIONAL STAT-
C...  ISTICS FOR LSODE AND GO ON TO THE NEXT RUN
      WRITE(NO,8)RWORK(11),IWORK(14),IWORK(11),IWORK(12),IWORK(13)
8     FORMAT(1H ,//,' LSODE COMPUTATIONAL STATISTICS'            ,//,
     1 ' LAST STEP SIZE                         ',         D10.3,//,
     2 ' LAST ORDER OF THE METHOD               ',           I10,//,
     3 ' TOTAL NUMBER OF STEPS TAKEN            ',           I10,//,
     4 ' NUMBER OF FUNCTION EVALUATIONS         ',           I10,//,
     5 ' NUMBER OF JACOBIAN EVALUATIONS         ',           I10,/)
```

The data file read by main program LSODET in Program 2.8b is listed in Program 2.8c. Note for the second case that time starts at 0.000001 and ends at 100 and thus it covers eight orders of magnitude; the ability to accelerate over orders of magnitude in the independent variable is a characteristic of a quality implicit integrator such as LSODE.

The output from Program 2.8a to 2.8c is given in Table 2.19. The following points can be noted about the output of Table 2.19:

1. The solution for both cases (NORUN=1, 2) is in close agreement with the analytical solution. Also, the solution for the first case is in close agreement with the solution of Table 2.16 as expected since the values of β_i are the same.

2. The number of Jacobian evaluations for the nonstiff case is zero since the Jacobian matix is not used in the nonstiff option of LSODE (for METH=1 or MF=10 in Program 2.8b). Also, the order of the BDF used last (at the end of the solution) is three ($q = 3$ in equation (2.160)) and the number of derivative evaluations is 170, a modest number to compute the entire solution which is a result of the relatively high accuracy of the BDF methods; by contrast the SOAB method would use approximately 100 calls/print interval $\times$ 10 print intervals (for the entire

Table 2.19 Output from Programs 2.8a to c. *Continued next pages.*

```
RUN NO. -    1  KROGH PROBLEM FOR N = 20, NONSTIFF CASE

INITIAL T -  0.000D+00

  FINAL T -  0.100D+01

  PRINT T -  0.100D+00

NUMBER OF DIFFERENTIAL EQUATIONS -  20

MAXIMUM INTEGRATION ERROR -  0.100D-04


          N =  20          BETA
                         -1.000
                         -2.000
                         -3.000
                         -4.000
                         -5.000
                         -6.000
                         -7.000
                         -8.000
                         -9.000
                        -10.000
                        -11.000
                        -12.000
                        -13.000
                        -14.000
                        -15.000
                        -16.000
                        -17.000
                        -18.000
                        -19.000
                        -20.000

       TIME         Y(1)       YEX(1)          PER1
                   Y(20)      YEX(20)         PER20
                               PERMAX

    0.000000 -0.1000E+01 -0.1000E+01          0.000
             -0.1000E+01 -0.1000E+01          0.000
                                0.000

    0.100000 -0.4493E+01 -0.4493E+01          0.000
              0.1068E+00  0.1068E+00          0.003
                                0.003

    0.200000 -0.1107E+02 -0.1107E+02          0.000
              0.2766E+01  0.2766E+01          0.000
                                0.001
```

Table 2.19 *Continued.*

```
  0.300000 -0.1595E+02 -0.1595E+02        0.000
            0.2146E+01  0.2146E+01        0.000
                             0.000

  0.400000 -0.1811E+02 -0.1811E+02        0.000
            0.7655E+00  0.7655E+00        0.000
                             0.001

  0.500000 -0.1902E+02 -0.1902E+02        0.000
           -0.3569E-01 -0.3569E-01        0.000
                             0.000

  0.600000 -0.1944E+02 -0.1944E+02        0.000
           -0.4448E+00 -0.4448E+00        0.000
                             0.000

  0.700000 -0.1966E+02 -0.1966E+02        0.000
           -0.6610E+00 -0.6610E+00        0.000
                             0.000

  0.800000 -0.1978E+02 -0.1978E+02        0.000
           -0.7827E+00 -0.7827E+00        0.000
                             0.000

  0.900000 -0.1986E+02 -0.1986E+02        0.000
           -0.8551E+00 -0.8551E+00        0.000
                             0.000

  1.000000 -0.1990E+02 -0.1990E+02        0.000
           -0.9004E+00 -0.9004E+00        0.000
                             0.000

LSODE COMPUTATIONAL STATISTICS

LAST STEP SIZE                          0.204D-01

LAST ORDER OF THE METHOD                        3

TOTAL NUMBER OF STEPS TAKEN                   148

NUMBER OF FUNCTION EVALUATIONS                170

NUMBER OF JACOBIAN EVALUATIONS                  0


RUN NO. -   2  KROGH PROBLEM FOR N = 20, STIFF CASE

INITIAL T -  0.100D-05

  FINAL T -  0.100D+03

  PRINT T -  0.100D-05
```

Table 2.19 *Continued.*

```
NUMBER OF DIFFERENTIAL EQUATIONS -  20

MAXIMUM INTEGRATION ERROR -  0.100D-04

          N =  20            BETA
                        -1000.000
                        -1000.000
                        -1000.000
                        -1000.000
                        -1000.000
                        -1000.000
                        -1000.000
                        -1000.000
                        -1000.000
                        -1000.000
                           -0.001
                           -0.001
                           -0.001
                           -0.001
                           -0.001
                           -0.001
                           -0.001
                           -0.001
                           -0.001
                           -0.001
      TIME         Y(1)       YEX(1)          PER1
                  Y(20)      YEX(20)         PER20
                              PERMAX

  0.000001 -0.1000E+01 -0.1000E+01           0.000
           -0.1001E+01 -0.1001E+01           0.000
                               0.000

  0.000010 -0.1000E+01 -0.1000E+01           0.000
           -0.1010E+01 -0.1010E+01           0.000
                               0.000

  0.000100 -0.9999E+00 -0.9999E+00           0.000
           -0.1105E+01 -0.1105E+01           0.000
                               0.000

  0.001000 -0.9990E+00 -0.9990E+00           0.000
           -0.2714E+01 -0.2714E+01           0.000
                               0.000

  0.010000 -0.9901E+00 -0.9901E+00           0.000
           -0.9566E+03 -0.9566E+03           0.000
                               0.000

  0.100000 -0.9092E+00 -0.9092E+00           0.000
           -0.1000E+04 -0.1000E+04           0.000
                               0.000
```

Table 2.19 *Continued.*

```
  1.000000 -0.5004E+00 -0.5004E+00          0.000
           -0.1000E+04 -0.1000E+04          0.000
                             0.000

 10.000000 -0.9141E-01 -0.9141E-01          0.000
           -0.1000E+04 -0.1000E+04          0.000
                             0.000

100.000000 -0.1041E-01 -0.1041E-01          0.001
           -0.1000E+04 -0.1000E+04          0.000
                             0.001

LSODE COMPUTATIONAL STATISTICS

LAST STEP SIZE                           0.291D+01

LAST ORDER OF THE METHOD                         3

TOTAL NUMBER OF STEPS TAKEN                    386

NUMBER OF FUNCTION EVALUATIONS                1441

NUMBER OF JACOBIAN EVALUATIONS                  47
```

solution) = 1000 derivative evaluations for a solution of comparable accuracy (compare percent errors in the last columns of Tables 2.16 and 2.19).

3. The final integration step of 0.291D+01 indicates how `LSODE` accelerated toward the end of the solution (the initial integration step must have been of the order of `0.100D-05` which is the initial output time).

4. For the stiff case (second run), the number of Jacobian evaluations, 47, and the number of steps taken, 316, indicate on average that the Jacobian matrix was updated only 316/47 $\sim$ 7 time steps. This result indicates an important feature of `LSODE` in achieving computational efficiency: the Jacobian matrix is updated only as often as required to converge the Newton iterative solution of equation (2.160) for y_{i+1}. The use of a Jacobian matrix which is not completely up to date does not introduce any error in the numerical solution as long as the Newton iteration to compute the solution, y_{i+1}, converges in accordance with the user-specified error tolerance. This feature of using infrequent Jacobian updates is an important factor in determining the overall computational efficieny of a BDF integrator since the formation of the numerical Jacobian is a computationally intensive part of the BDF integration.

In summary, the use of an implicit integrator can result in substantial reductions in the number of steps taken in the solution of stiff ODEs since the stability constraints of explicit integrators such as condition (2.113) are circumvented. In the case of very stiff ODE systems, the reduction in computational effort can be by many orders of magnitude, which is the justification for the additional computational complexity of implicit integrators, e.g., the Newton iteration and associated linear algebra.

2.5 *Some Unconventional Uses of ODE Integrators*

We conclude this chapter with a series of examples of the use of ODE integrators for unconventional applications, in particular, one-dimensional integrals, multidimensional integrals, solution of integro-differential equations, solution of boundary value ODEs, and solution of systems of nonlinear algebraic and transcendental equations. We start with the integration of the *normal or Gaussian probability distribution* defined by the integral

$$I(t) = 1/2 + \frac{1}{\sqrt{2\pi}} \int_0^t e^{-(1/2)t^2}\, dt \tag{2.161}$$

We can evaluate one-dimensional integrals such as $I(t)$ of equation (2.161) by integrating an equivalent ODE of the form

$$\frac{dI}{dt} = f(t) \tag{2.162}$$

subject to the initial condition

$$I(0) = I_0 \tag{2.163}$$

Note that equation (2.162) is a special case of the general first-order ODE $dy/dt = f(y,t)$ in which the derivative function depends only on the independent variable, t. The solution to equation (2.162), subject to initial condition (2.163), is

$$I(t) = I_0 + \int_0^t f(t)\, dt \tag{2.164}$$

which includes equation (2.161) as a special case. Thus, to use this idea for integrating one-dimensional integrals, we program the integrand as the derivative function of equation (2.162).

Program 2.9a with subroutines `INITAL`, `DERV`, and `PRINT` illustrates this idea applied to equation (2.161). The integration is done by subroutine `RKF45`.

The following points can be noted about Program 2.9a:

1. In subroutine `INITAL`, π in the constant $1/\sqrt{2\pi}$ is calculated as $\pi = 4\tan^{-1}(1)$. Initial condition (2.163) for equation (2.161) is then defined as $I(0) = 0.5$.

2. The integrand $(1/\sqrt{2\pi})e^{-(1/2)t^2}$ of equation (2.161) is programmed in subroutine `DERV`.

3. The integral is printed in subroutine `PRINT` at $t = 0.5, 1, 1.5, \ldots, 4$.

The main program which calls `RKF45` is essentially identical to several main programs listed previously, e.g., Program 2.2a, and therefore it is not listed here. The data file for subroutines `INITAL`, `DERV`, and `PRINT` is listed in Program 2.9b.

The output of Programs 2.9a and b is listed in Table 2.20. The computed integrals are in agreement with the tabulated values in subroutine `INITAL` to about four figures, which is consistent with the error tolerance of Program 2.9b (0.00001). The advantage of using numerical integration to evaluate integrals, generally called *numerical quadrature,* is the ease with which complicated integrals can be evaluated. This is demonstrated in the present case with the programming of the integrand in subroutine `DERV`. In principle, any integand which can be expressed in Fortran can be integrated.

```
      SUBROUTINE INITAL
C...
C...  NUMERICAL INTEGRATION OF THE NORMAL PROBABILITY DISTRIBUTION
C...  FUNCTION
C...
C...  THE INTEGRAL TO BE EVALUATED IS
C...
C...                             TUP
C...        I = 1/2 + 1/SQRT(2*PI) INT EXP(-(1/2)*(T**2))DT          (1)
C...                              0
C...
C...  THE INTEGRAND IS THE NORMAL PROBABILITY DISTRIBUTION FUNCTION.
C...  THE UPPER LIMIT OF THE INTEGRAL TUP DEFINES THE VALUE OF THE
C...  INTEGRAL.  TABULATED VALUES ARE GIVEN BELOW FOR COMPARISON WITH
C...  THE NUMERICAL SOLUTION (REFERENCE - HANDBOOK OF MATHEMATICAL
C...  SCIENCES, CRC PRESS, INC., WEST PALM BEACH, FL, 5TH EDITION, 1975,
C...  PP 747-754)
C...
C...                       TUP           I
C...                       0.5         0.6915
C...                       1.0         0.8413
C...                       1.5         0.9332
C...                       2.0         0.9772
C...                       2.5         0.9938
C...                       3.0         0.9987
C...                       3.5         0.9998
C...                       4.0         1.0000
C...
      COMMON/T/T/Y/I/F/DIDT/CONST/A
      REAL I
C...
C...  INITIAL CONDITION AND MULTIPLYING CONSTANT
      A=1.0/SQRT(2.0*4.0*ATAN(1.))
      I=0.5
      RETURN
      END

      SUBROUTINE DERV
      COMMON/T/T/Y/I/F/DIDT/CONST/A
      REAL I
C...
C...  EVALUATE THE INTEGRAND
      DIDT=A*EXP(-0.5*(T**2))
      RETURN
      END
      SUBROUTINE PRINT(NI,NO)
      COMMON/T/T/Y/I/F/DIDT/CONST/A
      REAL I
      WRITE(NO,1)T,I
1     FORMAT(5H T = ,F3.1,3X,4HI = ,F8.6)
      RETURN
      END
2
```

Program 2.9a Subroutines `INITAL`, `DERV`, and `PRINT` for the Normal Probability Distribution Function.

```
INTEGRATION OF (1/(2*PI)**0.5)*EXP(-(1/2)*(T**2))
0.        4.0        0.5
   1                      0.00001
END OF RUNS
```

Program 2.9b Data File for Subroutines `INITAL`, `DERV` and `PRINT` of Program 2.9a.

Table 2.20 Output of Programs 2.9a and b.

```
RUN NO. -    1  INTEGRATION OF (1/(2*PI)**0.5)*EXP(-(1/2)*(T**2))

INITIAL T -  0.000E+00

  FINAL T -  0.400E+01

  PRINT T -  0.500E+00

NUMBER OF DIFFERENTIAL EQUATIONS -   1

INTEGRATION ALGORITHM - RKF45

MAXIMUM INTEGRATION ERROR -  0.100E-04

T = 0.0   I = 0.500000

T = 0.5   I = 0.691463

T = 1.0   I = 0.841346

T = 1.5   I = 0.933194

T = 2.0   I = 0.977251

T = 2.5   I = 0.993791

T = 3.0   I = 0.998650

T = 3.5   I = 0.999768

T = 4.0   I = 0.999969
```

Further, by using a quality ODE integrator, the accuracy of the integral can be assured through the user-specified error tolerance; this contrasts with a less sophisticated fixed interval numerical quadrature algorithm, such as Simpson's rule, for which the error can be controlled only by comparing numerical integrals with differing numbers of integration steps, with possibly an extrapolation to zero interval width, a so-called *Richardson extrapolation*.

As a second example of numerical quadrature using an ODE integrator, we consider four *elliptic integrals*

$$E_{1k} = \int_0^{\pi/2} \frac{dt}{\sqrt{1 - k^2 \sin^2(t)}} \tag{2.165}$$

$$E_{2k} = \int_0^{\pi/2} \sqrt{1 - k^2 \sin^2(t)}\, dt \tag{2.166}$$

$$E_{3k} = \int_0^{\pi/2} \frac{dt}{\sqrt{1 - \left\{1 - k^2 \sin^2(t)\right\}}} \tag{2.167}$$

$$E_{4k} = \int_0^{\pi/2} \sqrt{1 - \left\{1 - k^2 \sin^2(t)\right\}}\, dt \tag{2.168}$$

We can now apply equations (2.162) and (2.163) to each of these integrals, as illustrated in subroutines `INITAL`, `DERV`, and `PRINT` in Program 2.10a.

We can note the following points about Program 2.10a:

1. The four integrals of equations (2.165) to (2.168) are contained in a single array, `E(4)`, which appears in `COMMON/Y/` as expected. The initial conditions for the four integrals are zero since $I_0 = 0$ in equation (2.163).

2. The four integrands of equations (2.165) to (2.168) are in array `DEDT(4)`, which appears in `COMMON/F/` as expected. These four integrands are evaluated in subroutine `DERV`.

3. Subroutine `PRINT` prints the four integrals when the upper limit $\pi/2$ has been reached. Also, *Legendre's relation* between the four integrals

   $$E_{1k}E_{4k} + E_{2k}E_{3k} - E_{1k}E_{3k} = \pi/2 \tag{2.169}$$

   is printed as a residual in subroutine `PRINT`.

The data file for Program 2.10a is listed in Program 2.10b (again, the integration was by `RKF45` called by a main program which is not listed here).

The output from Programs 2.10a and b is listed in Table 2.21. The agreement between the numerical integrals and the tabulated integrals in subroutine `INITAL` is to about five figures, which is consistent with the error tolerance of 0.00001 in Program 2.10b.

As a final example of numerical quadrature, we consider a two-dimensional integral in which the integration in one variable is done with the SOAB algorithm considered previously, and the integration in the second variable is done with `RKF45`. The integral is

$$I = \int_{y_l}^{y_u} \int_{x_l}^{x_u} f(x,y)\, dx\, dy \tag{2.170}$$

To illustrate how integral I in equation (2.170) can be computed using two ODE integrators, we choose $f(x,y) = 1$ and $x_l = 0$, $x_u = (1 - y^2)^{1/2}, y_l = 0$, $y_u = 1$. These limits

```
      SUBROUTINE INITAL
C...
C...  NUMERICAL INTEGRATION OF THE COMPLETE ELLIPTIC INTEGRALS
C...
C...  THE INTEGRALS TO BE EVALUATED ARE
C...
C...            PI/2            DT
C...      E1 = INT -----------------------------
C...             0  (1 - (K**2)*(SIN(T)**2))**0.5
C...
C...            PI/2
C...      E2 = INT (1 - (K**2)*(SIN(T)**2))**0.5 DT
C...             0
C...
C...            PI/2                DT
C...      E3 = INT --------------------------------
C...             0  (1 - (1 - K**2)*(SIN(T)**2))**0.5
C...
C...            PI/2
C..       E4 = INT (1 - (1 - K**2)*(SIN(T)**2))**0.5 DT
C...             0
C...
C...  THESE INTEGRALS ARE TABULATED BELOW FOR COMPARISON WITH THE
C...  NUMERICAL VALUES (REFERENCE - HANDBOOK OF MATHEMATICAL SCIENCES,
C...  CRC PRESS, INC., WEST PALM BEACH, FL, 5TH EDITION, 1975, PP 535
C...  -547).  IN ALL CASES, WE TAKE K = 0.5.
C...
C...                                 E1 = 1.6858
C...
C...                                 E2 = 1.4675
C...
C...                                 E3 = 2.1565
C...
C...                                 E4 = 1.2111
C...
      PARAMETER (N=4)
      COMMON/T/       T,  NSTOP,  NORUN
     +       /Y/   E(N)
     +       /F/DEDT(N)
     +       /C/       K
      REAL             K
C...
C...  CONSTANT K
      K=0.5
C...
C...  INITIAL CONDITION
      DO 1 I=1,N
        E(I)=0.
1     CONTINUE
      RETURN
      END
```

Program 2.10a Subroutines `INITAL`, `DERV`, and `PRINT` for the Evaluation of Four Elliptic Integrals. *Continued next page.*

```
      SUBROUTINE DERV
      PARAMETER (N=4)
      COMMON/T/       T,  NSTOP,  NORUN
     +      /Y/   E(N)
     +      /F/DEDT(N)
     +      /C/       K
      REAL            K
C...
C...  EVALUATE THE INTEGRANDS
         ARG1=SQRT(1.0-(K**2)*(SIN(T)**2))
         ARG2=SQRT(1.0-(1.0-K**2)*(SIN(T)**2))
         DEDT(1)=1.0/ARG1
         DEDT(2)=    ARG1
         DEDT(3)=1.0/ARG2
         DEDT(4)=    ARG2
      RETURN
      END

      SUBROUTINE PRINT(NI,NO)
      PARAMETER (N=4)
      COMMON/T/       T,  NSTOP,  NORUN
     +      /Y/   E(N)
     +      /F/DEDT(N)
     +      /C/       K
      REAL            K
C...
C...  WRITE THE VALUES OF THE FOUR ELLIPTIC INTEGRALS
      IF(T.GT.1.0)THEN
         WRITE(NO,1)T,(E(I),I=1,N)
1        FORMAT(//,5F10.5)
C...
C...     LEGENDRE*S RELATION
         R=E(1)*E(4)+E(2)*E(3)-E(1)*E(3)-1.5707693
C...
C...     WRITE THE RESIDUAL FOR LEGENDRE*S RELATION
         WRITE(NO,2)R
2        FORMAT(/,' RESIDUAL = ',F8.5)
      END IF
      RETURN
      END
```

Program 2.10a *Continued.*

```
ELLIPTIC INTEGRALS E1, E2, E3, E4
0.        1.5707963 1.5707963
    4                   0.00001
END OF RUNS
```

Program 2.10b Data File for Subroutines INITAL, DERV, PRINT of Program 2.10a.

Table 2.21 Output of Programs 2.10a and b.

```
RUN NO. -    1  ELLIPTIC INTEGRALS E1, E2, E3, E4

INITIAL T -   0.000E+00

  FINAL T -   0.157E+01

  PRINT T -   0.157E+01

NUMBER OF DIFFERENTIAL EQUATIONS -    4

INTEGRATION ALGORITHM - RKF45

MAXIMUM INTEGRATION ERROR -  0.100E-04


  1.57080   1.68575   1.46746   2.15635   1.21107

RESIDUAL =  0.00010
```

correspond to integration over one quarter of a circle with unit radius for which the area is $\pi/4$. This can easily be demonstrated analytically

$$I = \int_0^1 \int_0^{(1-y^2)^{1/2}} 1\,dx\,dy = \int_0^1 x\,\Big|_0^{(1-y^2)^{1/2}} dy = \int_0^1 (1-y^2)^{1/2}\,dy$$

$$= (1/2)\left\{y(1-y^2)^{1/2} + \sin^{-1}(y)\right\}\Big|_0^1 = \pi/4 \tag{2.171}$$

The preceding integration was stated in detail to give some insight into the numerical procedure implemented in the following subroutines INITAL, DERV, and PRINT of Program 2.11a.

The following points can be noted about Program 2.11a:

1. The integration in y is performed through COMMON/T/, /Y/, and /F/ (using subroutine RKF45). Thus, an initial condition for the outer integral in equation (2.170) is required which is zero (since the upper limit of the y integration initially is y_l). The coding in subroutine INITAL for this initial condition is

```
      COMMON/T/Y
     +       /Y/I
     +       /F/DIDT
      REAL I
C...
C...  INITIAL CONDITION
      I=0.
```

2. Function FXY to compute the integrand of equation (2.170) is declared EXTERNAL in subroutine DERV. Then the lower and upper limits for the x integration (inner

```
      SUBROUTINE INITAL
C...
C...  TWO DIMENSIONAL INTEGRATION
C...
C...  THE FOLLOWING CODING ILLUSTRATES THE EVALUATION OF AN INTEGRAL
C...  OF THE FORM
C...
C...            YU    XU
C...     I = INT ( INT F(X,Y) DX) DY
C...            YL    XL
C...
C...  WHERE
C...
C...     I        INTEGRAL TO BE EVALUATED
C...
C...     F(X,Y)   GIVEN INTEGRAND
C...
C...     YU,YL    UPPER AND LOWER LIMITS FOR Y
C...
C...     XU,XL    UPPER AND LOWER LIMITS FOR X
C...
C...  XU AND XL ARE GENERALLY OF THE FORM XU = F(Y), XL = G(Y).
C...
C...  IN THE FOLLOWING CODING, THE INTEGRATION IN Y IS DONE BY THE USUAL
C...  ODE INTEGRATION THROUGH COMMON/T/, /Y/ AND /F/.  THE INTEGRATION IN
C...  X IS DONE BY SUBROUTINE INTX CALLED IN DERV.  INTX IMPLEMENTS THE
C...  SECOND ORDER ADAMS BASHFORTH METHOD WITH STARTING BY THE MODIFIED
C...  EULER METHOD.
C...
C...  F(X,Y) IS DEFINED IN A USER-SUPPLIED FUNCTION CALL BY INTX. TO
C...  TEST THE METHOD, A SAMPLE PROBLEM IS CODED FOR F(X,Y) = 1 OVER
C...  A QUARTER OF THE CIRCLE WITH UNIT RADIUS, FOR WHICH THE CORRECT
C...  ANSWER (THE AREA OF A QUARTER CIRCLE) IS PI/4.  THIS RESULT CAN
C...  BE DEMONSTRATED IN THE FOLLOWING WAY, WHICH GIVES SOME INSIGHT
C...  INTO THE OPERATION OF THE FOLLOWING CODING (XU = (1 - Y**2)**0.5
C...  FOR A CIRCLE WITH THE EQUATION X**2 + Y**2 = 1)
C...
C...       1     XU             1                1
C...  I = INT ( INT 1 DX) DY = INT (XU - 0) DY = INT (1 - Y**2)**0.5 DY
C...       0     0              0                0
C...
C...                                             1
C...     = (1/2)(Y*(1 - Y**2)**0.5 + ARCSIN(Y))   = PI/4
C...                                             0
      COMMON/T/Y
     +      /Y/I
     +      /F/DIDT
      REAL I
C...
C...  INITIAL CONDITION
      I=0.
      RETURN
```

Program 2.11a Subroutines `INITAL`, `DERV`, and `PRINT` for the Evaluation of Two-Dimensional Integral (2.170). *Continued next pages.*

```
      END
      SUBROUTINE DERV
      COMMON/T/Y
     +      /Y/I
     +      /F/DIDT
      REAL I
C...
C...  EXTERNAL THE FUNCTION CALLED BY INTX TO EVALUATE THE INTEGRAND
      EXTERNAL FXY
C...
C...  EVALUATE THE INTEGRAND
C...     LIMITS FOR X (BECAUSE OF ROUNDING, Y CAN BECOME
C...     SLIGHTLY GREATER THAN ONE WHICH CAUSES PROBLEMS WITH
C...     THE SQUARE ROOT UNLESS THE ABS IS USED)
            XL=0.
            XU=SQRT(ABS(1.0-Y**2))
C...
C...     NUMBER OF INTEGRATION INTERVALS IN X
            NX=100
C...
C...     INTEGRATION IN X
            CALL INTX(FXY,XL,XU,NX,Y,SUMX)
C...
C...     INTEGRAL IN X BECOMES THE INTEGRAND FOR THE Y
C...     INTEGRATION
            DIDT=SUMX
      RETURN
      END

      SUBROUTINE PRINT(NI,NO)
      COMMON/T/Y
     +      /Y/I
     +      /F/DIDT
      REAL I
C...
C...  WRITE THE SOLUTION (AFTER MULTIPLICATION BY 4 SO THE
C...  ANSWER IS PI, THE AREA OF A CIRCLE WITH UNIT RADIUS)
      WRITE(NO,1)Y,4.0*I
1     FORMAT(//,5H Y = ,F3.1,3X,4HI = ,F8.6)
      RETURN
      END

      FUNCTION FXY(X,Y)
C...
C...  FUNCTION FXY(X,Y) DEFINES THE INTEGRAND AS A FUNCTION OF X AND
C...  Y (IN THE FOLLOWING EXAMPLE, THE INTEGRAND IS CONSTANT, BUT IN
C...  GENERAL, IT WILL BE A FUNCTION OF THE INPUTS X AND Y)
C...
      FXY=1.0
      RETURN
      END
```

Program 2.11a *Continued.*

```
      SUBROUTINE INTX(FXY,XL,XU,NX,Y,SUMX)
C...
C...  SUBROUTINE INTX DOES A ONE DIMENSIONAL INTEGRATION OVER NX IN-
C...  TERVALS IN X AT A GIVEN VALUE OF Y.  THE COMPUTED INTEGRAL IS
C...  RETURNED AS SUMX.
C...
C...  EXTERNAL THE ROUTINE CALLED BY INTX TO EVALUATE THE INTEGRAND
      EXTERNAL FXY
C...
C...  SET THE INTEGRATION INTERVAL
      DX=(XU-XL)/FLOAT(NX)
C...
C...  BEGIN THE INTEGRATION OVER NX STEPS
C...
C...  NSTART = 1 - SECOND ORDER, RUNGE KUTTA (MODIFIED EULER) TO START
C...  THE INTEGRATION
      NSTART=1
      SUMX=0.
      DO 1 I=1,NX
C...
C...  TAKE A SECOND ORDER (NSTART = 1) RUNGE KUTTA STEP AT THE BEGINNING
C...  SINCE THE ADAMS BASHFORTH METHOD IS NOT SELF-STARTING
      IF(NSTART.EQ.1)THEN
C...
C...     FIRST ORDER, RUNGE KUTTA (EULER) STEP FROM N-1 = 1-1 = 0 TO
C...     N = 1
         FNM2=FXY(XL,Y)
         SUMX=SUMX+FNM2*DX
         X=XL+DX
C...
C...     THE DERIVATIVE AT N = 1 BECOMES THE DERIVATIVE AT N-1 FOR
C...     THE FIRST STEP OF THE ADAMS BASHFORTH INTEGRATION
         FNM1=FXY(X,Y)
C...
C...     SECOND ORDER, RUNGE KUTTA STEP
         SUMX=SUMX+0.5*(FNM1-FNM2)*DX
      END IF
C...
C...  TAKE A SECOND ORDER, ADAMS BASHFORTH STEP FROM N-1 TO N
      IF(NSTART.EQ.2)THEN
         SUMX=SUMX+DX*(1.5*FNM1-0.5*FNM2)
         X=X+DX
C...
C...     DERIVATIVES AT N-1 AND N-2 FOR NEXT STEP
         FNM2=FNM1
         FNM1=FXY(X,Y)
      END IF
      NSTART=2
C...
C...  THE FOLLOWING WRITE WAS USED TO CHECK THE OPERATION OF SUBROUTINE
C...  INTX.  WHEN ACTIVATED, IT ALSO GIVES A COMPLETE PICTURE OF HOW THE
```

Program 2.11a *Continued.*

```
C...  TWO DIMENSIONAL INTEGRATION WORKS.  HOWEVER, A WORD OF CAUTION -
C...  THIS WRITE WILL BE EXECUTED MANY TIMES UNLESS THE NUMBER OF INCRE-
C...  MENTS IN X AND Y IS SMALL, E.G., 10 IN EACH DIRECTION RESULTS IN
C...  100 CALLS TO INTX
C...  WRITE(*,*)' I = ',I,' Y = ',Y,' X = ',X,' SUMX = ',SUMX
C...
C...  TAKE THE NEXT STEP IN X
1     CONTINUE
C...
C...  INTEGRATION IS COMPLETE AT Y
      RETURN
      END
```

Program 2.11a *Continued.*

integral) of equation (2.170) are set, using in this case the equation for the circle of unit radius

```
C...
C...  EXTERNAL THE FUNCTION CALLED BY INTX TO EVALUATE THE
C...  INTEGRAND
      EXTERNAL FXY
C...
C...  EVALUATE THE INTEGRAND
C...     LIMITS FOR X (BECAUSE OF ROUNDING, Y CAN BECOME
C...     SLIGHTLY GREATER THAN ONE WHICH CAUSES PROBLEMS
C...     WITH THE SQUARE ROOT UNLESS THE ABS IS USED)
            XL=0.
            XU=SQRT(ABS(1.0-Y**2))
```

3. The inner integral in equation (2.170) is then evaluated over 100 intervals in x by a call to subroutine INTX which implements the SOAB for the x integration. This integral is returned through `SUMX` and becomes the integrand for the outer integral in y through the statement `DIDT=SUMX`:

```
C...
C...     NUMBER OF INTEGRATION INTERVALS IN X
            NX=100
C...
C...     INTEGRATION IN X
            CALL INTX(FXY,XL,XU,NX,Y,SUMX)
C...
C...     INTEGRAL IN X BECOMES THE INTEGRAND FOR THE Y
C...     INTEGRATION
            DIDT=SUMX
```

4. Subroutine `PRINT` receives the two-dimensional integral through `COMMON/Y/`, multiplies it by 4 (to bring the exact answer to π), and prints the result

```
      COMMON/T/Y
     +      /Y/I
```

```
INTEGRATION OF F(X,Y) = 1 OVER A QUARTER OF THE CIRCLE WITH UNIT RADIUS
0.        1.0        1.0
   1                    0.00001
END OF RUNS
```

Program 2.11b Data File for Program 2.11a.

Table 2.22 Output of Programs 2.11a and b.

```
RUN NO. -   1  INTEGRATION OF F(X,Y) = 1 OVER A QUARTER OF THE CIRCLE WITH UNIT
               RADIUS

 INITIAL T -  0.000E+00

   FINAL T -  0.100E+01

   PRINT T -  0.100E+01

 NUMBER OF DIFFERENTIAL EQUATIONS -   1

 INTEGRATION ALGORITHM - RKF45

 MAXIMUM INTEGRATION ERROR -  0.100E-04


 Y = 0.0   I = 0.000000

 Y = 1.0   I = 3.141679
```

```
     +        /F/DIDT
      REAL I
C...
C...  WRITE THE SOLUTION (AFTER MULTIPLICATION BY 4 SO THE
C...  ANSWER IS PI, THE AREA OF A CIRCLE WITH UNIT RADIUS)
      WRITE(NO,1)Y,4.0*I
1     FORMAT(//,5H Y = ,F3.1,3X,4HI = ,F8.6)
```

5. The operation of subroutine INTX follows directly from the previous discussion for the SOAB. The initial integration is done by the second-order Runge-Kutta (modified Euler method) to start the SOAB.

The data file for Program 2.11a is listed in Program 2.11b. As expected, one ODE is declared, and the integration proceeds from zero to one.

The output from Progams 2.11a and b is listed in Table 2.22. As expected, the computed two-dimensional integral is close to π. Although this example is for a simple integrand for which the integral can easily be evaluated analytically, the preceding Programs 2.11a and b are general purpose, and can be used to evaluate any two-dimensional integral with finite limits for which the integrand $f(x, y)$ can be programmed in subroutine INTXY as a function of x and y. However, the complexity of the preceding coding also demonstrates the impracticality of applying this approach to higher dimensional integrals. There is also another limiting factor—the number of grid points required in the various dimensions, that is, the so-called *curse of dimensionality*.

As the next example of a rather unconventional use of an ODE integrator, we consider an *integro-differential equation*

$$a_1\frac{dy}{dt} + a_0 y + a_{-1}\int_0^t y\,dt = 1 \tag{2.172}$$

$$y(0) = 0 \tag{2.173}$$

We first derive an analytical solution to equations (2.172) and (2.173), then consider the programming for a numerical solution.

If the Laplace transform of $y(t)$ is defined as

$$L\{y(t)\} = \bar{y}(s) = \int_0^t y(t)e^{-st}\,dt \tag{2.174}$$

then equations (2.172) and (2.173) transform to

$$a_1 s\bar{y}(s) + a_0\bar{y}(s) + a_{-1}\frac{\bar{y}(s)}{s} = \frac{1}{s} \tag{2.175}$$

or

$$a_1 s^2\bar{y}(s) + a_0 s\bar{y}(s) + a_{-1}\bar{y}(s) = 1 \tag{2.176}$$

We should note parenthetically that equation (2.175) is far from obvious by starting with the definition of $y(s)$ in equation (2.174). Actually, we have made use of a table of Laplace transforms [Strang (1986), pp. 513–523] in transforming equation (2.172) into equation (2.175); and, of course, the advantage of using integral transforms such as in equation (2.174) is to convert a relatively difficult problem, such as integro-differential equation (2.172) into a relatively easy equation to manipulate mathematically, such as algebraic equation (2.175).

If equation (2.176) is solved for $\bar{y}(s)$

$$\bar{y}(s) = \frac{1}{a_1 s^2 + a_0 s + a_{-1}} \tag{2.177}$$

or in factored form

$$\bar{y}(s) = \frac{1/a_1}{(s-r_1)(s-r_2)} \tag{2.178}$$

where

$$r_{1,2} = \frac{-a_0 \pm \left\{a_0^2 - 4a_1 a_{-1}\right\}^{1/2}}{2a_1} \tag{2.179}$$

$\bar{y}(s)$ in equation (2.178) can be inverted by partial fractions

$$\bar{y}(s) = \frac{1/a_1}{(s-r_1)(s-r_2)} = (1/a_1)\left\{\frac{b_1}{s-r_1} + \frac{b_2}{s-r_2}\right\} \tag{2.180}$$

where

$$b_1 = \left\{\frac{1}{s-r_2}\right\}_{s=r_1} = \frac{1}{r_1-r_2}, \qquad b_2 = \left\{\frac{1}{s-r_1}\right\}_{s=r_2} = \frac{1}{r_2-r_1} \tag{2.181}$$

Then, inversion of equation (2.180) gives the solution to equations (2.172) and (2.173)

$$y(t) = (1/a_1)\left\{\frac{1}{r_1 - r_2}e^{r_1 t} + \frac{1}{r_2 - r_1}e^{r_2 t}\right\} \tag{2.182}$$

Initial condition (2.173) is easily verified

$$y(0) = (1/a_1)\left\{\frac{1}{r_1 - r_2} + \frac{1}{r_2 - r_1}\right\} = 0 \tag{2.183}$$

Substitution of equation (2.182) in equation (2.172) gives

$$\begin{aligned}
a_1\frac{dy}{dt} &= b_1 r_1 e^{r_1 t} + b_2 r_2 e^{r_2 t} \\
a_0 y &= (a_0/a_1)\left\{b_1 e^{r_1 t} + b_2 e^{r_2 t}\right\} \\
a_{-1}\int_0^t y\,dt &= (a_{-1}/a_1)\left\{(b_1/r_1)e^{r_1 t} + (b_2/r_2)e^{r_2 t} - (b_1/r_1) - (b_2/r_2)\right\} \\
-1 &= -1.
\end{aligned}$$

The left-hand sides of these four equations sum to zero (from equation (2.172)). The question then is whether the right-hand sides also sum to zero.

Considering first the constant terms (which do not depend on t)

$$\begin{aligned}
(a_{-1}/a_1)\left\{-(b_1/r_1) - (b_2/r_2)\right\} &= \frac{(a_{-1}/a_1)}{r_2 - r_1\left\{\frac{1}{r_1} - \frac{1}{r_2}\right\}} \\
&= (a_{-1}/a_1)\left\{\frac{1}{r_1 r_2}\right\} = (a_{-1}/a_1)(a_1/a_{-1}) = 1
\end{aligned}$$

as required (in the last step we made use of the relationship $r_1 r_2 = a_{-1}/a_1$ which follows from comparing equations (2.177) and (2.178)).

Now, considering the terms with $e^{r_1 t}$

$$\begin{aligned}
&b_1 r_1 e^{r_1 t} + (a_0/a_1)b_1 e^{r_1 t} + (a_{-1}/a_1)(b_1/r_1)e^{r_1 t} \\
&\quad = \left\{r_1 + (a_0/a_1) + (a_{-1}/a_1)(1/r_1)\right\} b_1 e^{r_1 t} \\
&\quad = \left\{a_1 r_1^2 + a_0 r_1 + a_{-1}\right\}\left(\frac{1}{a_1 r_1}\right) b_1 e^{r_1 t} = \{0\}\left(\frac{1}{a_1 r_1}\right) b_1 e^{r_1 t} = 0
\end{aligned}$$

The term in brackets is zero since r_1 is a root of the characteristic equation

$$a_1 s^2 + a_0 s + a_{-1} = 0 \tag{2.184}$$

A similar analysis leads to the same conclusion for the terms involving $e^{r_2 t}$. Thus equation (2.182) is a solution to equations (2.172) and (2.173). Incidentally, the roots of the characteristic equation (2.184), which are eigenvalues for equations (2.172), can be real, complex, or repeated. If they are repeated ($r_1 = r_2$), equation (2.182) assumes the indeterminant form $y(t) = 0/0$, and therefore in order to use it, a minor revision of the Laplace transform derivation is required or l'Hospital's rule must be applied to equation (2.182).

We now consider the programming of equations (2.172) and (2.173), with the analytical solution, equation (2.182), programmed in subroutine `PRINT` for comparison with the numerical solution. Subroutines `INITAL`, `DERV`, and `PRINT` are listed in Program 2.12a. The dependent variables are

$$y(t) = \text{Y}(1), \qquad \int_0^t y(t)\,dt = \text{Y}(2) \tag{2.185)(2.186}$$

Thus, in order to compare `Y(2)` in the Fortran with the analytical solution we require

$$\int_0^t y(t)\,dt = \int_0^t (1/a_1)\left\{\frac{1}{r_1 - r_2}e^{r_1 t} + \frac{1}{r_2 - r_1}e^{r_2 t}\right\}dt$$

$$= (1/a_1)\left\{\frac{1}{r_1(r_1 - r_2)}(e^{r_1 t} - 1) + \frac{1}{r_2(r_2 - r_1)}(e^{r_2 t} - 1)\right\} \tag{2.187}$$

Note that equation (2.186) also indicates the initial condition for `Y(2)` is zero (the initial condition for `Y(1)` is given by equation (2.173)).

The following points can be noted about Program 2.12a:

1. The three coefficients in equation (2.172) are set in subroutine `INITAL` as $a_{-1} = 1$, $a_0 = 3$, $a_1 = 1$. The eigenvalues r_1 and r_2 from equation (2.179) are then real and distinct. Initial conditions (2.173) and (2.187) with $t = 0$ are then programmed.

2. Subroutine `DERV` contains straightforward programming of equations (2.172) and (2.186).

3. Analytical solutions (2.182) and (2.187) are programmed in subroutine `PRINT` for comparison with the numerical solutions (the programming of the analytical solutions is unnecessarily repetitive and therefore inefficient, such as the repeated calculation of exponential terms, but this coding was used to facilitate comparison with the analytical solutions, equations (2.182) and (2.187)).

The integration was done by subroutine `RKF45`. The main program is not listed since it is very similar to preceding main programs that call `RKF45`, e.g., Program 2.2a. The output of Table 2.23 indicates the numerical and analytical solutions agree to five figures.

As the next example, we consider the solution of a boundary value ODE, which is an unconventional use of an initial value ODE integrator, `RKF45` in this case. As discussed in Sections 1.1.4 to 1.1.6, boundary value ODEs have the feature of having the dependent variables specified at more than one value of the independent variable. We start by considering two special cases of an elementary example

$$\frac{d^2y}{dt^2} + \alpha\left(1 - \left(\frac{dy}{dt}\right)^2\right)y = 0 \tag{2.188}$$

1. For a given value of α, the boundary conditions are specified as

$$y(0.5) = 0.3, \qquad y(1) = 0 \tag{2.189)(2.190}$$

Thus, we require a solution of equation (2.188) which has the values 0.3 and 0 at $t = 0.5$ and 1, respectively.

```
      SUBROUTINE INITAL
      IMPLICIT DOUBLE PRECISION (A-H,O-Z)
      COMMON/T/       T,  NSTOP,  NORUN
     +      /Y/   Y(2)
     +      /F/  YT(2)
     +      /C/A(-1:1)
C...
C...  COEFFICIENTS (REAL EIGENVALUES)
      A(-1)=1.0D0
      A( 0)=3.0D0
      A( 1)=1.0D0
C...
C...  INITIAL CONDITIONS
      Y(1)=0.0D0
      Y(2)=0.0D0
C...
C...  INITIAL DERIVATIVES
      CALL DERV
      RETURN
      END

      SUBROUTINE DERV
      IMPLICIT DOUBLE PRECISION (A-H,O-Z)
      COMMON/T/       T,  NSTOP,  NORUN
     +      /Y/   Y(2)
     +      /F/  YT(2)
     +      /C/A(-1:1)
C...
C...  INTEGRAL TERM
      YT(2)=Y(1)
C...
C...  DERIVATIVE TERM
      YT(1)=(1.0D0/A(1))*(1.0D0-A(0)*Y(1)-A(-1)*Y(2))
      RETURN
      END

      SUBROUTINE PRINT(NI,NO)
      IMPLICIT DOUBLE PRECISION (A-H,O-Z)
      COMMON/T/       T,  NSTOP,  NORUN
     +      /Y/   Y(2)
     +      /F/  YT(2)
     +      /C/A(-1:1)
C...
C...  MONITOR CALCULATION
      WRITE(*,*)T
C...
C...  HEADING FOR THE SOLUTION
      IF(T.LT.0.01D0)THEN
         WRITE(NO,3)
3        FORMAT(6X,'T',8X,'Y1',7X,'Y1A',3X,'DIFF Y1',
     +                    8X,'Y2',7X,'Y2A',3X,'DIFF Y2',/)
      END IF
```

Program 2.12a Subroutine `INITAL`, `DERV`, and `PRINT` for the Solution of Equations (2.172) and (2.173). *Continued next page.*

```
C...
C...  ANALYTICAL SOLUTION
      DIS=A(0)**2-4.0D0*A(1)*A(-1)
      IF(DIS.GT.0.0D0)THEN
C...
C...     REAL EIGENVALUES
         R1=(-A(0)+DSQRT(DIS))/(2.0D0*A(1))
         R2=(-A(0)-DSQRT(DIS))/(2.0D0*A(1))
         Y1A=(1.0D0/A(1))*((1.0D0/(R1-R2))*DEXP(R1*T)
     +                    +(1.0D0/(R2-R1))*DEXP(R2*T))
         Y2A=(1.0D0/A(1))*(( 1.0D0/(R1*(R1-R2)))*(DEXP(R1*T)-1.0D0)
     +                   + ( 1.0D0/(R2*(R2-R1)))*(DEXP(R2*T)-1.0D0))
      ELSE
C...
C...     REPEATED, COMPLEX EIGENVALUES NOT PROGRAMMED
         WRITE(NO,2)DIS
2        FORMAT(' DIS = ',E12.4,' (NOT POSITIVE)')
         STOP
      END IF
C...
C...
C...  DIFFERENCE BETWEEN NUMERICAL AND ANALYTICAL SOLUTIONS
      DIFF1=Y(1)-Y1A
      DIFF2=Y(2)-Y2A
C...
C...  PRINT NUMERICAL, ANALYTICAL SOLUTIONS, AND THE DIFFERENCE
      WRITE(NO,1)T,Y(1),Y1A,DIFF1,
     +             Y(2),Y2A,DIFF2
1     FORMAT(F7.3,6F10.5)
      RETURN
      END
```

Program 2.12a *Continued.*

```
LINEAR INTEGRO-DIFFERENTIAL EQUATION
0.        10.0      1.0
    2                     0.00001
END OF RUNS
```

Program 2.12b Data File for Program 2.12a.

2. We require the value of α for which the solution to equation (2.188) has the values

$$y(0.5) = 0.3, \quad \frac{dy(0.5)}{dt} = 0, \quad y(1) = 0 \qquad (2.191)\text{–}(2.193)$$

Note that in case (1), as expected, since equation (2.188) is second-order, two boundary conditions are specified. Also, in case (2), three conditions are specified for a second-order equation, but it may be possible to satisfy all three conditions by choosing α correctly. These problems suggest that boundary value problems may be more difficult to solve from a theoretical point of view in the sense that solutions which satisfy all of the specified boundary conditions may not exist.

Table 2.23 Output from Programs 2.12a and b.

```
RUN NO. -    1  LINEAR INTEGRO-DIFFERENTIAL EQUATION

INITIAL T -   0.000D+00

  FINAL T -   0.100D+02

  PRINT T -   0.100D+01

NUMBER OF DIFFERENTIAL EQUATIONS -    2

MAXIMUM INTEGRATION ERROR -   0.100D-04

      T          Y1         Y1A    DIFF Y1          Y2         Y2A    DIFF Y2

  0.000   0.00000   0.00000   0.00000   0.00000   0.00000   0.00000
  1.000   0.27261   0.27261   0.00000   0.21335   0.21335   0.00000
  2.000   0.20595   0.20595   0.00000   0.45550   0.45550   0.00000
  3.000   0.14201   0.14201   0.00000   0.62782   0.62782   0.00000
  4.000   0.09703   0.09703   0.00000   0.74594   0.74594   0.00000
  5.000   0.06623   0.06623   0.00000   0.82660   0.82660   0.00000
  6.000   0.04521   0.04521   0.00000   0.88165   0.88165   0.00000
  7.000   0.03085   0.03085   0.00000   0.91922   0.91922   0.00000
  8.000   0.02106   0.02106   0.00000   0.94487   0.94487   0.00000
  9.000   0.01437   0.01437   0.00000   0.96237   0.96237   0.00000
 10.000   0.00981   0.00981   0.00000   0.97432   0.97432   0.00000
```

This contrasts with initial value problems for which solutions generally do exist for a given set of initial conditions.

We now consider case (2) (the method to be developed can also be applied to case (1)). Basically, we first assume a value for α, then start the integration at $t = 0.5$ using boundary conditions (2.191) and (2.192) and integrate to $t = 1$. We then observe if boundary condition (2.193) has been satisfied; initially it most likely will not, thus we use the departure from $y(1) = 0$ to adjust α and repeat the calculation. If the adjustment of α was in the right direction with an appropriate magnitude, $y(1)$ in the second calculation should be closer to zero. But again, it most likely will not be zero, so again we adjust α and repeat the calculation from $t = 0.5$ for a third time. If the process converges, we should eventually arrive at a value of α for which equations (2.191)–(2.193) are satisfied. We now consider the implementation of this algorithm in subroutines `INITAL`, `DERV`, and `PRINT` of Program 2.13a.

We can note the following points about Program 2.13a:

1. In subroutine `INITAL`, α is initiated with a value of one and an increment of two (these values are subsequently refined through iterative execution of the program; `FLAGM` and `FLAGP` are variables which control the iteration). Initial conditions (2.191) and (2.192) are then set. Also, the independent variable t in equation (2.188) is the variable `S` in the programming (note that it is the first element in `COMMON/T/`).

2. Equation (2.188) is programmed in subroutine `DERV` as two first-order ODEs.

3. The programming in subroutine `PRINT` adjusts α according to the departure of

```
      SUBROUTINE INITAL
C...
C...  BOUNDARY VALUE ODE INTEGRATED AS AN INITIAL VALUE ODE
C...
      IMPLICIT DOUBLE PRECISION (A-H,O-Z)
      COMMON/T/S,NFIN,NORUN/Y/Y1,Y2/F/DY1DS,DY2DS/PARM/A,DA,FLAGM,FLAGP
C...
C...  INITIAL PARAMETER A, INCREMENT DA, CONTROL VARIABLES FOR
C...  ITERATION ON A
      IF(NORUN.EQ.1)THEN
         A=1.0D0
         DA=2.0D0
         FLAGM=-1.0D0
         FLAGP=-1.0D0
C...
C...  INITIAL CONDITIONS
      END IF
      Y1=0.3D0
      Y2=0.0D0
      RETURN
      END

      SUBROUTINE DERV
      IMPLICIT DOUBLE PRECISION (A-H,O-Z)
      COMMON/T/S,NFIN,NORUN/Y/Y1,Y2/F/DY1DS,DY2DS/PARM/A,DA,FLAGM,FLAGP
C...
C...  ODES
      DY2DS=-A*((1.0D0-Y2**2)**0.5D0)*Y1
      DY1DS=Y2
      RETURN
      END

      SUBROUTINE PRINT(NI,NO)
      IMPLICIT DOUBLE PRECISION (A-H,O-Z)
      COMMON/T/S,NFIN,NORUN/Y/Y1,Y2/F/DY1DS,DY2DS/PARM/A,DA,FLAGM,FLAGP
C...
C...  TEST FOR THE END OF THE RUN
      IF(S.LT.0.95D0)RETURN
C...
C...  MONITOR THE RUN
      WRITE(*,*)NORUN,A,Y1
C...
C...  AT THE END OF THE RUN, PRINT Y1 TO DETERMINE HOW CLOSELY THE
C...  BOUNDARY CONDITION IS SATISFIED
      WRITE(NO,1)NORUN,S,A,Y1
1     FORMAT(10X,' NORUN = ',I5,/,
     +       10X,'     S = ',F10.2,/,
     +       10X,'     A = ',F10.6,/,
     +       10X,'    Y1 = ',F10.6,/)
C...
C...  AT THE END OF THE RUN, INCREMENT A DEPENDING ON THE SIGN DEVIATION
```

Program 2.13a Subroutines INITAL, DERV, and PRINT for the Solution of Equations (2.188), (2.191)–(2.193). *Continued next page.*

```
C...  FROM THE PRESCRIBED HOMOGENEOUS BOUNDARY CONDITION, Y1(1) = 0
C...
C...        IF Y1(1) IS POSITIVE, INCREASE A
            IF(Y1.LT.0.0D0)GO TO 2
C...
C...            IF A SMALLER INCREMENT IN A IS REQUIRED, HALVE DA
                IF(FLAGM.LT.0.0D0)GO TO 3
                DA=DA/2.0D0
C...
C...            INCREMENT A POSITIVELY FOR THE NEXT RUN
3               A=A+DA
                FLAGP=1.0D0
                RETURN
C...
C...        IF Y1(1) IS NEGATIVE, DECREASE A
C...
C...            IF A SMALLER INCREMENT IN A IS REQUIRED, HALVE DA
2               IF(FLAGP.LT.0.0D0)GO TO 4
                DA=DA/2.0D0
C...
C...            INCREMENT A NEGATIVELY FOR THE NEXT RUN
4               A=A-DA
                FLAGM=1.0D0
                RETURN
      END
```

Program 2.13a *Continued.*

```
CARNAHAN ET AL, APPLIED NUMERICAL METHODS, PROB. 6.26, P. 420
0.5       1.0       0.5
    2                   0.00001
END OF RUNS
```

Program 2.13b Data File for Program 2.13a.

the solution from $y(1) = 0$. The logic of this programming is easily understood so we will not discuss it in detail. Generally, the increment in α, `DA`, is reduced with successive executions of the program.

The data file read by the main program is listed in Program 2.13b. Again, the main program that calls `RKF45` is similar to preceding main programs, e.g., Program 2.2a, so it is not listed here. Within this main program, a `DO` loop is executed for 22 runs. This number was selected to provide convergence of α to a final value with about five-figure accuracy to satisfy the boundary condition $y(1) = 0$.

Abbreviated output from Programs 2.13a and b is listed in Table 2.24 for the runs 1, 2, 21, and 22. Note that for the 22nd run, the value of α to about five significant figures (judging from successive values) is $A = 11.417542$ and the value of $y(1)$ is `Y(1)` $= -0.000001$.

This same procedure could have been applied to the first case discussed previously with equations (2.189) and (2.190) as boundary conditions. For this case, $dy(0.5)/dt$ is used in the same way as α, i.e., a value of this initial derivative is assumed in

Table 2.24 Output from Programs 2.13a and b. *Continued next page.*

```
RUN NO. -   1  CARNAHAN ET AL, APPLIED NUMERICAL METHODS, PROB. 6.26, P. 420

 INITIAL T -  0.500D+00

   FINAL T -  0.100D+01

   PRINT T -  0.500D+00

 NUMBER OF DIFFERENTIAL EQUATIONS -   2

 MAXIMUM INTEGRATION ERROR -  0.100D-04

          NORUN =      1
              S =        1.00
              A =   1.000000
             Y1 =   0.263339

RUN NO. -   2  CARNAHAN ET AL, APPLIED NUMERICAL METHODS, PROB. 6.26, P. 420

 INITIAL T -  0.500D+00

   FINAL T -  0.100D+01

   PRINT T -  0.500D+00

 NUMBER OF DIFFERENTIAL EQUATIONS -   2

 MAXIMUM INTEGRATION ERROR -  0.100D-04

          NORUN =      2
              S =        1.00
              A =   3.000000
             Y1 =   0.195789
                    .
                    .
                    .
          NORUN =     21
              S =        1.00
              A =  11.417603
             Y1 =  -0.000002

RUN NO. -  22  CARNAHAN ET AL, APPLIED NUMERICAL METHODS, PROB. 6.26, P. 420

 INITIAL T -  0.500D+00

   FINAL T -  0.100D+01

   PRINT T -  0.500D+00

 NUMBER OF DIFFERENTIAL EQUATIONS -   2

 MAXIMUM INTEGRATION ERROR -  0.100D-04
```

Table 2.24 *Continued.*

```
NORUN =     22
    S =          1.00
    A =  11.417542
   Y1 =  -0.000001
```

subroutine INITAL (as variable Y(2)), then adjusted in subroutine PRINT for the next run depending on the deviation of Y(1) from the boundary condition $y(1) = 0$. This is the *shooting method* for two-point boundary value ODEs.

As the final point for this example, we mention the important characteristic of ODE integration algorithms and that generally they are valid for either positive or negative integration steps, that is, they work correctly in either direction of the independent variable (increasing as we have considered, or decreasing). To illustrate this property, we now consider the preceding problem, using integration from $t = 1$ to $t = 0.5$ with $y(1) = 0$, $dy(1)/dt = -0.873849$ (the value of Y(2) at $S = 1$ from the 22nd run of Programs 2.13a and b) and $A = 11.417542$ (the value of α from the 22nd run). These values are set in subroutine INITAL, listed in Program 2.14a, and the data file is changed to reflect the reverse integration, Program 2.14b (note the negative print interval in the data file which sets a negative integration step in subroutine RKF45).

Essentially everything else in the preceding Program 2.13a remains the same; subroutine PRINT now tests on a final value of $S = 0.5$ rather than $S = 1$. Also the main program is executed only once since the correct value of α is used initially to reproduce boundary conditions (2.191) and (2.192) at $t = 0.5$. The stopping criterion in the main program was changed from

```
IF(TV.LT.(TF-0.5D0*TP))GO TO 4
```

in the case of Programs 2.13a and b to

```
IF(TV.GT.(TF-0.5D0*TP))GO TO 4
```

for Programs 2.14a and b.

The output from Programs 2.14a and b is listed in Table 2.25. Note the values Y1 = 0.300003 and Y2 = 0.000025 which are close approximations to boundary conditions (2.191) and (2.192), $y(0.5) = 0.3$, $dy(0.5)/dt = 0$.

Thus, to reiterate, ODE integration algorithms based on the Taylor series are valid for positive or negative integration steps (h in the preceding integration algorithms such as the Euler and modified Euler methods, and the RKF45 and the BDFs)

Finally, we conclude with a method for applying an ODE numerical integrator to the solution of nonlinear algebraic and transcendental equations of the form

$$\begin{aligned} f_1(x_1, x_2, \ldots, x_n) &= 0 \\ f_2(x_1, x_2, \ldots, x_n) &= 0 \\ f_3(x_1, x_2, \ldots, x_n) &= 0 \\ &\vdots \\ f_n(x_1, x_2, \ldots, x_n) &= 0 \end{aligned} \tag{2.194}$$

We seek the solution vector $[x_1, x_2, \ldots, x_n]^T$ to this $n \times n$ system by *Davidenko's method* (1953), which is a differential form of Newton's method. We therefore start with

```
      SUBROUTINE INITAL
C...
C...  BOUNDARY VALUE ODE INTEGRATED AS AN INITIAL VALUE ODE
C...
      IMPLICIT DOUBLE PRECISION (A-H,O-Z)
      COMMON/T/S,NFIN,NORUN/Y/Y1,Y2/F/DY1DS,DY2DS/PARM/A
C...
C...  PARAMETER A
      A=11.417542
C...
C...  INITIAL CONDITIONS
      Y1= 0.0D0
      Y2=-0.873849D0
      RETURN
      END

      SUBROUTINE DERV
      IMPLICIT DOUBLE PRECISION (A-H,O-Z)
      COMMON/T/S,NFIN,NORUN/Y/Y1,Y2/F/DY1DS,DY2DS/PARM/A
C...
C...  ODES
      DY2DS=-A*((1.0D0-Y2**2)**0.5D0)*Y1
      DY1DS=Y2
      RETURN
      END

      SUBROUTINE PRINT(NI,NO)
      IMPLICIT DOUBLE PRECISION (A-H,O-Z)
      COMMON/T/S,NFIN,NORUN/Y/Y1,Y2/F/DY1DS,DY2DS/PARM/A
C...
C...  TEST FOR THE END OF THE RUN
      IF(S.GT.0.55D0)RETURN
C...
C...  MONITOR THE RUN
      WRITE(*,*)NORUN,A,Y1,Y2
C...
C...  AT THE END OF THE RUN, PRINT Y1 TO DETERMINE HOW CLOSELY THE
C...  BOUNDARY CONDITION IS SATISFIED
      WRITE(NO,1)NORUN,S,A,Y1,Y2
1     FORMAT(10X,' NORUN = ',I5,/,
     +       10X,'     S = ',F10.2,/,
     +       10X,'     A = ',F10.6,/,
     +       10X,'    Y1 = ',F10.6,/,
     +       10X,'    Y2 = ',F10.6,/)
      RETURN
      END
```

Program 2.14a Subroutines INITAL, DERV, and PRINT for the Reverse Time Integration of Equation (2.188).

```
CARNAHAN ET AL, APPLIED NUMERICAL METHODS, PROB. 6.26, P. 420
1.0         0.5         -0.5
   2                         0.00001
END OF RUNS
```

Program 2.14b The Data File for Program 2.14a.

Table 2.25 Output from Programs 2.14a and b.

```
 RUN NO. -   1  CARNAHAN ET AL, APPLIED NUMERICAL METHODS, PROB. 6.26,
                P. 420

 INITIAL T -  0.100D+01

   FINAL T -  0.500D+00

   PRINT T - -0.500D+00

 NUMBER OF DIFFERENTIAL EQUATIONS -   2

 MAXIMUM INTEGRATION ERROR -  0.100D-04

           NORUN =      1
               S =        0.50
               A =  11.417542
              Y1 =   0.300003
              Y2 =   0.000025
```

Newton's method in n dimensions

$$\bar{J}\Delta\bar{x} = -\bar{f} \tag{2.195}$$

or in the alternate form

$$\Delta\bar{x} = -\bar{J}^{-1}\bar{f} \tag{2.196}$$

where f_{ij} is the ith row and jth column component of the Jacobian matrix

$$f_{ij} = \frac{\partial f_i\left(x_1, x_2, \ldots, x_n\right)}{\partial x_j} \tag{2.197}$$

$$\bar{J} = \begin{bmatrix} f_{11} & f_{12} & \cdot & f_{1n} \\ f_{21} & f_{22} & \cdot & f_{2n} \\ \cdot & \cdot & \cdot & \cdot \\ f_{n1} & f_{n2} & \cdot & f_{nn} \end{bmatrix} \tag{2.198}$$

$$\bar{f} = \begin{bmatrix} f_1(x_1, x_2, \ldots, x_n) \\ f_2(x_1, x_2, \ldots, x_n) \\ \cdot \\ f_n(x_1, x_2, \ldots, x_n) \end{bmatrix} \tag{2.199}$$

Equation (2.195) (or (2.196)) can be used to calculate a *vector of Newton corrections*, $\Delta\bar{x}$, which are finite changes in the dependent variables.

$$\Delta\bar{x} = \begin{bmatrix} \Delta x_1 \\ \Delta x_2 \\ \cdot \\ \Delta x_n \end{bmatrix} \tag{2.200}$$

The calculation can diverge if the corrections are too large, and they may therefore be "damped" by multiplying them by a factor between zero and one so that the full Newton step is not taken.

We consider now an alternative approach to the calculation of changes in the dependent variables to move them to a solution of equation (2.194) which is based on differential changes rather than finite changes. The idea follows immediately from equations (2.195) and (2.196) by converting them to differential equations

$$\bar{J}d\bar{x}/dt = -\bar{f} \tag{2.201}$$

or

$$d\bar{x}/dt = -\bar{J}^{-1}\bar{f} \tag{2.202}$$

where t is a *continuation parameter* introduced into the original problem, equation (2.194). We now integrate equation (2.201) or (2.202) until $d\bar{x}/dt = \bar{0}$ which implies $\bar{f} = \bar{0}$ from equation (2.201) or (2.202), i.e., the solution vector x satisfies equation (2.194). Of course, as with Newton's method, we still have the problem that $\bar{J}$ may become singular as the solution proceeds, and we therefore include in our implementation of equations (2.201) and (2.202) the continuous calculation of the condition number of the Jacobian matrix $\bar{J}$ of equation (2.198).

To start the integration of equation (2.201) or (2.202), we choose an initial condition, $(x_1(t_0), x_2(t_0), \ldots, x_n(t_0))^T$, arbitrarily, which is analogous to the choice of a trial solution in Newton's method (and, of course, the closer the choice of the initial condition to the final solution, the better the chance of arriving at the solution, but experience has indicated this is not a critical requirement as long as $\bar{J}$ does not become ill conditioned as the solution proceeds). In fact, one of the major advantages of this differential approach, Davidenko's method, over Newton's method is a substantially larger domain of attraction or convergence, i.e., the starting condition can be further away from the solution, which can be important when the choice of a trial solution is not obvious (the usual case).

In practice, for large n in equation (2.194), we integrate equation (2.201) rather than (2.202) for the same reason we use equation (2.195) rather than equation (2.196), that is, we can exploit the structure of $\bar{J}$ such as bandedness or sparseness, while this structure is generally lost if we work with $\bar{J}^{-1}$. Thus, we are required to integrate a system of semi-implicit ODEs, as characterized by equations (1.12) and (1.13), or equations (1.14) and (1.15) in matrix form. We can approach this integration in one of two ways, depending on whether we use equation (2.201) or (2.202). Equation (2.202) does have the disadvantage associated with $\bar{J}^{-1}$, but it has the advantage that a library integrator for explicit ODEs can be used, for example, `RKF45` (since equation (2.202) is a system of explicit ODEs). Alternatively, equation (2.201) requires an integrator designed specifically for semi-implicit ODEs; quality integrators for this ODE structure are available such as `LSODI` [Byrne and Hindmarsh (1987)] and `DASSL` [Brenan et al. (1989)].

We will now use equations (2.201) and (2.202) to illustrate the application of

Davidenko's method to the 2×2 nonlinear system [Biegler (1991)]

$$\begin{aligned} x_1^2 + x_2^2 &= 17 \\ 2x_1^{1/3} + x_2^{1/2} &= 4 \end{aligned} \tag{2.203}$$

which has the the two solutions $x_1 = 1$, $x_2 = 4$, and $x_1 = 4.07150$, $x_2 = 0.65027$. In terms of the preceding notation, we have

$$\begin{aligned} f_1(x_1, x_2) &= x_1^2 + x_2^2 - 17 \\ f_2(x_1, x_2) &= 2x_1^{1/3} + x_2^{1/2} - 4 \end{aligned} \tag{2.204}$$

or

$$\bar{f} = \begin{bmatrix} x_1^2 + x_2^2 - 17 \\ 2x_1^{1/3} + x_2^{1/2} - 4 \end{bmatrix} \tag{2.205}$$

$$\bar{J} = \begin{bmatrix} 2x_1 & 2x_2 \\ (2/3)x_1^{-2/3} & (1/2)x_2^{-1/2} \end{bmatrix} \tag{2.206}$$

If we now substitute equations (2.205) and (2.206) in equation (2.201)

$$\begin{bmatrix} 2x_1 & 2x_2 \\ (2/3)x_1^{-1/3} & (1/2)x_2^{-1/2} \end{bmatrix} \cdot \begin{bmatrix} dx_1/dt \\ dx_2/dt \end{bmatrix} = - \begin{bmatrix} x_1^2 + x_2^2 - 17 \\ 2x_1^{1/3} + x_2^{1/2} - 4 \end{bmatrix} \tag{2.207}$$

Equation (2.207) can be written in the alternate forms

$$\begin{aligned} (2x_1)dx_1/dt + (2x_2)dx_2/dt &= -(x_1^2 + x_2^2 - 17) \\ ((2/3)x_1^{-1/3})dx_1/dt + ((1/2)x_2^{-1/2})dx_2/dt &= -(2x_1^{1/3} + x_2^{1/2} - 4) \end{aligned} \tag{2.208}$$

or

$$\begin{bmatrix} dx_1/dt \\ dx_2/dt \end{bmatrix} = - \begin{bmatrix} 2x_1 & 2x_2 \\ (2/3)x_1^{-1/3} & (1/2)x_2^{-1/2} \end{bmatrix}^{-1} \cdot \begin{bmatrix} x_1^2 + x_2^2 - 17 \\ 2x_1^{1/3} + x_2^{1/2} - 4 \end{bmatrix} \tag{2.209}$$

Equation (2.209) is, of course, just an example of equation (2.202).

We now consider the programming of equation (2.208) using RKF45. Since this is an integrator for explicit ODEs, we must first compute dx_1/dt and dx_2/dt explicitly, which, in Program 2.15a that follows, is accomplished by calls to the two linear equation solvers DECOMP and SOLVE [Forsythe et al. (1977)] (note that equations (2.208) are linear in dx_1/dt and dx_2/dt). Subroutines INITAL, DERV, and PRINT for the integration of equation (2.208) are listed in Program 2.15a.

We can note the following points about Program 2.15a:

1. Subroutine INITAL defines the initial conditions for equations (2.208) for two runs to compute the two sets of roots $x_1 = 1$, $x_2 = 4$, and $x_1 = 4.07150$, $x_2 = 0.65027$:

```
C...
C...  INITIAL ESTIMATE OF THE SOLUTION
      IF(NORUN.EQ.1)THEN
         X(1)=2.0D0
         X(2)=1.0D0
      ELSE IF(NORUN.EQ.2)THEN
         X(1)=1.0D0
         X(2)=2.0D0
      END IF
```

```
      SUBROUTINE INITAL
C...
C...  THE DAVIDENKO DIFFERENTIAL EQUATION CAN BE WRITTEN IN MATRIX
C...  FORM AS
C...
C...     - -          - -       -
C...     J*DX/DT = -F, X(T0) = X0                                (1)(2)
C...
C...  WHERE
C...
C...     -
C...     J              JACOBIAN MATRIX FOR THE NONLINEAR EQUATIONS
C...                    (NLE)
C...     -
C...     X              SOLUTION VECTOR FOR THE DAVIDENKO ODE
C...
C...     T              INDEPENDENT VARIABLE IN THE DAVIDENKO METHOD
C...
C...     T0             INITIAL VALUE OF T
C...
C...     -
C...     X0             INITIAL CONDITION VECTOR FOR EQUATION (1)
C...
C...                                  -                        -
C...  IN THE CODING IN SUBROUTINE DERV, J IS STORED IN ARRAY A, AND F IS
C...  STORED IN ARRAY B.  THE UNCOUPLING OF THE DERIVATIVES IN EQUATION
C...  (1) IS DONE IN SUBROUTINE DERV BY A CALL TO SUBROUTINES DECOMP AND
C...  SOLVE (LISTED AND DISCUSSED IN DETAIL IN FORSYTHE, G. E., M. A.
C...  MALCOLM AND C. B. MOLER, COMPUTER METHODS FOR MATHEMATICAL COMPU-
C...  TATIONS, CHAPTER 3, PRENTICE-HALL, ENGLEWOOD CLIFFS, NJ, 1977).
C...  ANY OTHER ROUTINE(S) FOR LINEAR ALGEBRAIC EQUATIONS CAN BE USED IN
C...                                                         --   -
C...  PLACE OF DECOMP AND SOLVE (THE GENERAL PROBLEM IS AX = B).
C...
C...                                                         -  -
C...  THE PROGRAMMING OF EQUATION (1) IN TERMS OF VECTORS X, DX/DT AND
C...  -          -
C...  F, AND MATRIX J, IS INTENDED TO ILLUSTRATE THE GENERAL FEATURES OF
C...  DAVIDENKO*S METHOD, I.E., THESE VECTORS AND MATRIX CAN BE EXPANDED
C...  TO ACCOMMODATE A PROBLEM OF ANY SIZE.  HERE WE DIMENSION FOR
C...  TWO NLE PROVIDED BY L. T. BIEGLER:
C...
C...     X1**2 + X2**2 = 17
C...
C...     2*X1**(1/3) + X2**(1/2) = 4
C...
C...  DOUBLE PRECISION IS USED TO POSSIBLY ENHANCE THE PERFORMANCE
C...  OF DAVIDENKO*S METHOD WHEN THE JACOBIAN MATRIX IS POORLY
C...  CONDITIONED
      IMPLICIT DOUBLE PRECISION (A-H,O-Z)
      PARAMETER(N=2)
      COMMON/T/          T,      NSTOP,      NORUN
```

Program 2.15a Subroutines INITAL, DERV, and PRINT for the Davidenko Solution of System (2.208). *Continued next pages.*

```
     1       /Y/      X(N)
     2       /F/   DXDT(N)
     3      /CV/    A(N,N),       B(N),    WORK(N),    IPVT(N),
     4                 COND
C...
C...  INITIAL ESTIMATE OF THE SOLUTION
      IF(NORUN.EQ.1)THEN
         X(1)=2.0D0
         X(2)=1.0D0
      ELSE IF(NORUN.EQ.2)THEN
         X(1)=1.0D0
         X(2)=2.0D0
      END IF
C...
C...  INITIALIZE ALL OF THE MODEL CALCULATIONS
      CALL DERV
      RETURN
      END

      SUBROUTINE DERV
      IMPLICIT DOUBLE PRECISION (A-H,O-Z)
      PARAMETER(N=2)
      COMMON/T/          T,      NSTOP,      NORUN
     1       /Y/      X(N)
     2       /F/   DXDT(N)
     3      /CV/    A(N,N),       B(N),    WORK(N),    IPVT(N),
     4                 COND
C...
C...  JACOBIAN MATRIX
      A(1,1)=2.0D0*X(1)
      A(1,2)=2.0D0*X(2)
      A(2,1)=2.0D0*(1.0D0/3.0D0)*X(1)**(-2.0D0/3.0D0)
      A(2,2)=      (1.0D0/2.0D0)*X(2)**(-1.0D0/2.0D0)
C...
C...  COMPUTE THE RIGHT HAND SIDE VECTOR FOR THE DAVIDENKO DIFFERENTIAL
C...  EQUATION.  NOTE THAT THESE STATEMENTS CONSTITUTE A DEFINITION OF
C...  THE ELEMENTS OF B (OR VECTOR F IN EQUATION (1)).  THEY FOLLOW
C...  DIRECTLY FROM THE STATEMENT OF THE NONLINEAR EQUATIONS
      B(1)=-(      X(1)**2             +X(2)**2             -17.0D0)
      B(2)=-(2.0D0*X(1)**(1.0D0/3.0D0)+X(2)**(1.0D0/2.0D0)- 4.0D0)
C...
C...  SOLVE FOR THE VECTOR OF DERIVATIVES, DX/DT, IN EQUATION (1) (IN
C...  EFFECT, EQUATION (1) IS PREMULTIPLIED BY THE INVERSE OF JACOBIAN
C...  MATRIX J)
      CALL DECOMP(N,N,A,COND,IPVT,WORK)
C...
C...  CHECK THE CONDITION OF MATRIX A BEFORE COMPUTING THE DERIVATIVE
C...  VECTOR, DX/DT.  IF A IS NEAR SINGULAR, TERMINATE THE CURRENT RUN
      IF(COND.LT.1.0D+20)GO TO 3
      WRITE(NO,4)COND
4     FORMAT(1H ,//,
```

Program 2.15a *Continued.*

```
     1 30H THE CONDITION NUMBER OF A IS ,E10.3,33H SO THE CURRENT RUN IS
     1 TERMINATED)
      NSTOP=1
C...
C...  COMPUTE THE DERIVATIVE VECTOR DX/DT (INDIRECTLY VIA ARRAY B)
3     CALL SOLVE(N,N,A,B,IPVT)
C...
C...  TRANSFER ARRAY B TO ARRAY DXDT SO THE DERIVATIVE VECTOR, DX/DT,
C...  IS AVAILABLE IN COMMON/F/
      DO 2 I=1,N
      DXDT(I)=B(I)
2     CONTINUE
      RETURN
      END

      SUBROUTINE PRINT(NI,NO)
      IMPLICIT DOUBLE PRECISION (A-H,O-Z)
      PARAMETER(N=2)
      COMMON/T/          T,      NSTOP,      NORUN
     1      /Y/      X(N)
     2      /F/   DXDT(N)
     3     /CV/    A(N,N),      B(N),    WORK(N),    IPVT(N),
     4                COND
C...
C...  DIMENSION THE ARRAYS FOR PLOTTING THE EQUATION RESIDUALS AND
C...  CONDITION NUMBER
      DIMENSION F1R(11), F2R(11), CONP(11), TP(11)
C...
C...  INITIALIZE OUTPUT
      IF(T.LT.0.001D0)THEN
C...
C...     PRINT HEADING FOR THE SOLUTION
         WRITE(NO,1)
1        FORMAT(9X,   'T',8X,'X1',    8X,'X2',/,
     +           10X,       4X,'DX1/DT',4X,'DX2/DT',/,
     +           10X,       6X,'RES1',  6X,'RES2',/,
     +           10X,       5X,'CONDITION NUMBER',/)
C...
C...     INITIALIZE COUNTER FOR PLOTTED SOLUTION
         IP=0
      END IF
C...
C...  PRINT THE NUMERICAL SOLUTION.  THE ELEMENTS OF B MUST FIRST BE
C...  RECOMPUTED SINCE B CONTAINS THE VECTOR DX/DT AFTER THE CALL TO
C...  SOLVE IN SUBROUTINE DERV
3     B(1)=-(       X(1)**2              +X(2)**2              -17.0D0)
      B(2)=-(2.0D0*X(1)**(1.0D0/3.0D0)+X(2)**(1.0D0/2.0D0)- 4.0D0)
C...
C...  PRINT THE SOLUTION, DERIVATIVES OF EQUATION (1) AND THE RESI-
C...  DUALS OF THE NONLINEAR EQUATIONS AT THE BEGINNING AND END OF
C...  EACH RUN
```

Program 2.15a *Continued.*

```
      IP=IP+1
      IF((IP.EQ.1).OR.(IP.EQ.11))THEN
         WRITE(NO,2)T,(X(I),I=1,N),(DXDT(I),I=1,N),(B(I),I=1,N),COND
2        FORMAT(F10.1,2F10.5,/,10X,2F10.5,/,10X,2F10.5,/,16X,E11.3,/)
      END IF
C...
C...  STORE THE RESIDUALS AND CONDITION NUMBER FOR PLOTTING
       F1R(IP)=DLOG10(DABS(B(1)))
       F2R(IP)=DLOG10(DABS(B(2)))
      CONP(IP)=DLOG10(DABS(COND))
      TP(IP)=T
C...
C...  PLOT THE LOG OF THE RESIDUALS OF THE NLE TO INVESTIGATE THE
C...  EXPONENTIAL CONVERGENCE OF DAVIDENKO*S ALGORITHM
      IF(IP.LT.11)RETURN
      CALL SPLOTS(1,IP,TP,F1R)
      WRITE(NO,11)
      CALL SPLOTS(1,IP,TP,F2R)
      WRITE(NO,12)
      CALL SPLOTS(1,IP,TP,CONP)
      WRITE(NO,13)
11    FORMAT(1H ,//,16X,'LOG10 OF RESIDUAL OF EQUATION 1 VS T')
12    FORMAT(1H ,//,16X,'LOG10 OF RESIDUAL OF EQUATION 2 VS T')
13    FORMAT(1H ,//,16X,'LOG10 OF CONDITION NUMBER VS T')
      RETURN
      END
```

Program 2.15a *Continued.*

Array X containing the dependent variable vector of equations (2.208), x_1 and x_2, is in COMMON/Y/. Note that the initial estimates of the roots are not close to the actual roots; rather, these initial estimates merely have to be on the two sides of a line in the x_1–x_2 plane along which the Jacobian matrix of equation (2.206) is singular [Biegler (1991)].

2. In subroutine DERV, the Jacobian matrix is first computed according to equation (2.206):

```
C...
C...  JACOBIAN MATRIX
      A(1,1)=2.0D0*X(1)
      A(1,2)=2.0D0*X(2)
      A(2,1)=2.0D0*(1.0D0/3.0D0)*X(1)**(-2.0D0/3.0D0)
      A(2,2)=       (1.0D0/2.0D0)*X(2)**(-1.0D0/2.0D0)
```

3. Then the right-hand side vector is computed according to equation (2.205):

```
C...
C...  COMPUTE THE RIGHT HAND SIDE VECTOR FOR THE DAVIDENKO DIFFERENTIAL
C...  EQUATION.  NOTE THAT THESE STATEMENTS CONSTITUTE A DEFINITION OF
C...  THE ELEMENTS OF B (OR VECTOR F IN EQUATION (1)).  THEY FOLLOW
C...  DIRECTLY FROM THE STATEMENT OF THE NONLINEAR EQUATIONS
      B(1)=-(       X(1)**2             +X(2)**2              -17.0D0)
      B(2)=-(2.0D0*X(1)**(1.0D0/3.0D0)+X(2)**(1.0D0/2.0D0)- 4.0D0)
```

4. Equation (2.208) can then be solved for dx_1/dt and dx_2/dt by calls to the linear solvers DECOMP and SOLVE. This is a two-step process. First DECOMP is called to decompose the Jacobian matrix (in array A) and compute its condition number, COND. Then a test is made on the magnitude of the condition number. If it is greater than 10^{20}, the algebraic system for dx_1/dt and dx_2/dt of equation (2.208) is considered ill conditioned (numerically singular) and program execution is stopped (via NSTOP=1). If the system is not ill conditioned, subroutine SOLVE is then called to compute dx_1/dt and dx_2/dt according to equations (2.208) (these derivatives are returned by SOLVE in array B):

```
C...
C...  SOLVE FOR THE VECTOR OF DERIVATIVES, DX/DT, IN EQUATION (1) (IN
C...  EFFECT, EQUATION (1) IS PREMULTIPLIED BY THE INVERSE OF JACOBIAN
C...  MATRIX J)
      CALL DECOMP(N,N,A,COND,IPVT,WORK)
C...
C...  CHECK THE CONDITION OF MATRIX A BEFORE COMPUTING THE DERIVATIVE
C...  VECTOR, DX/DT.  IF A IS NEAR SINGULAR, TERMINATE THE CURRENT RUN
      IF(COND.LT.1.0D+20)GO TO 3
      WRITE(NO,4)COND
4     FORMAT(1H ,//,
     1 30H THE CONDITION NUMBER OF A IS ,E10.3,33H SO THE CURRENT RUN IS
     1 TERMINATED)
      NSTOP=1
C...
C...  COMPUTE THE DERIVATIVE VECTOR DX/DT (INDIRECTLY VIA ARRAY B)
3     CALL SOLVE(N,N,A,B,IPVT)
```

5. Finally, in DERV the derivatives dx_1/dt and dx_2/dt computed by DECOMP and SOLVE (in array B) are stored in the derivative array DXDT:

```
      DO 2 I=1,N
      DXDT(I)=B(I)
2     CONTINUE
```

Note that DXDT is in COMMON/F/ so that it is then integrated by RKF45 which returns x_1 and x_2 in array X through COMMON/Y/.

6. In subroutine PRINT, a heading for the numerical solution is first printed:

```
C...
C...  INITIALIZE OUTPUT
      IF(T.LT.0.001D0)THEN
C...
C...     PRINT HEADING FOR THE SOLUTION
         WRITE(NO,1)
1        FORMAT(9X,     'T',8X,'X1',     8X,'X2',/,
     +          10X,        4X,'DX1/DT',4X,'DX2/DT',/,
     +          10X,        6X,'RES1',  6X,'RES2',/,
     +          10X,        5X,'CONDITION NUMBER',/)
C...
C...     INITIALIZE COUNTER FOR PLOTTED SOLUTION
         IP=0
      END IF
```

7. The residuals of the nonlinear equations (2.204) (i.e., the residuals of the functions in equation (2.205)) are then computed:

```
C...
C...  PRINT THE NUMERICAL SOLUTION.  THE ELEMENTS OF B MUST FIRST BE
C...  RECOMPUTED SINCE B CONTAINS THE VECTOR DX/DT AFTER THE CALL TO
C...  SOLVE IN SUBROUTINE DERV
3     B(1)=-(        X(1)**2              +X(2)**2              -17.0D0)
      B(2)=-(2.0D0*X(1)**(1.0D0/3.0D0)+X(2)**(1.0D0/2.0D0)- 4.0D0)
```

8. Next the numerical solution and the equation residuals are printed at the beginning and end of the solution (when $IP = 1$ and $IP = 11$):

```
C...
C...  PRINT THE SOLUTION, DERIVATIVES OF EQUATION (1) AND THE RESI-
C...  DUALS OF THE NONLINEAR EQUATIONS AT THE BEGINNING AND END OF
C...  EACH RUN
      IP=IP+1
      IF((IP.EQ.1).OR.(IP.EQ.11))THEN
         WRITE(NO,2)T,(X(I),I=1,N),(DXDT(I),I=1,N),(B(I),I=1,N),COND
2        FORMAT(F10.1,2F10.5,/,10X,2F10.5,/,10X,2F10.5,/,16X,D11.3,/)
      END IF
```

9. At each point along the solution, $\log_{10}$ of the equation residuals and the condition number of the Jacobian matrix are stored along with the independent variable t for plotting at the end of the solution.

```
C...
C...  STORE THE RESIDUALS AND CONDITION NUMBER FOR PLOTTING
       F1R(IP)=DLOG10(DABS(B(1)))
       F2R(IP)=DLOG10(DABS(B(2)))
      CONP(IP)=DLOG10(DABS(COND))
      TP(IP)=T
```

10. Finally, at the end of the solution, three plots are produced by calls to the point plotting routine SPLOTS:

```
C...
C...  PLOT THE LOG OF THE RESIDUALS OF THE NLE TO INVESTIGATE THE
C...  EXPONENTIAL CONVERGENCE OF DAVIDENKO*S ALGORITHM
      IF(IP.LT.11)RETURN
      CALL SPLOTS(1,IP,TP,F1R)
      WRITE(NO,11)
      CALL SPLOTS(1,IP,TP,F2R)
      WRITE(NO,12)
      CALL SPLOTS(1,IP,TP,CONP)
      WRITE(NO,13)
11    FORMAT(1H ,//,16X,'LOG10 OF RESIDUAL OF EQUATION 1 VS T')
12    FORMAT(1H ,//,16X,'LOG10 OF RESIDUAL OF EQUATION 2 VS T')
13    FORMAT(1H ,//,16X,'LOG10 OF CONDITION NUMBER VS T')
```

The three plots produced by the calls to SPLOTS: (a) $\log_{10}(f_1(x_1, x_2))$ vs. t, (b) $\log_{10}(f_2(x_1, x_2))$ vs. t, and (c) $\log_{10}(\text{cond})$ vs. t, are semi-log plots of the residuals of equations (2.204) and the condition number of the Jacobian matrix of equation (2.206). In the case of the residual plots, we can observe the progress of the solution toward

```
x1**2 + x2**2 = 17, 2*x1**(1/3) + x2**(1/2) = 4
0.          10.0       1.0
   2                              1.0D-08
x1**2 + x2**2 = 17, 2*x1**(1/3) + x2**(1/2) = 4
0.          10.0       1.0
   2                              1.0D-08
END OF RUNS
```

Program 2.15b Data File for the Davidenko Solution of System (2.208).

a solution of equations (2.203). There is also an additional reason for constructing the residual plots. To explain this, consider the scalar case $n = 1$, for which equation (2.201) becomes

$$(df/dx)(dx/dt) = -f \tag{2.210}$$

Equation (2.210) can be rearranged to

$$df/f = -dt \tag{2.211}$$

which, with the initial condition $f_0(x_0)$, $t = t_0$, integrates to

$$\ln(f/f_0) = -(t - t_0)$$

or

$$f/f_0 = e^{-(t-t_0)} \tag{2.212}$$

Thus for the scalar case, $n = 1$, the equation residual varies semi-logarithmically with the continuation parameter t. In subroutine PRINT of Program 2.15a, we investigate whether this same semi-logarithmic relationship applies to the multivariable case ($n = 2$ in the case of equations (2.203)).

The data file read by the main program which calls RKF45 is listed in Program 2.15b. Program 2.15b specifies two runs with integration of two ODEs from $t = 0$ to $t = 10$ and an error tolerance of 10^{-8}. The main program which calls RKF45 is essentially identical to several main programs listed previously, e.g., Program 2.2a, and therefore it is not listed here.

The output of Programs 2.25a and b is listed below in Table 2.26. We can note the following points about the output in Table 2.26:

1. The solution for the first run starts at $x_1 = 2$, $x_2 = 1$ with residuals 12 and 0.48016, respectively, and proceeds to a solution $x_1 = 4.07144$, $x_2 = 0.65026$ with residuals 0.00054 and 0.00002, respectively. Thus, the computed roots are close to the values $x_1 = 4.07150$, $x_2 = 0.65027$ which are correct to six figures.

```
      T          X1          X2
              DX1/DT      DX2/DT
                RES1        RES2
               CONDITION NUMBER

    0.0     2.00000     1.00000
            4.34436    -2.68872
           12.00000     0.48016
                 0.227D+02
```

```
10.0    4.07144    0.65026
        0.00007    0.00001
        0.00054    0.00002
             0.168D+02
```

2. The continuation parameter t varied over the interval $0 \leq t \leq 10$. This interval was selected as a compromise to achieve reasonable convergence to the solution (which set the lower limit on the final value of t) without excessive computation (which set the upper limit on the final value of t).

3. The initial and final condition numbers of the Jacobian matrix of equation (2.206), 22.7 and 16.8, respectively, indicate that the ODE system of equations (2.208) is well conditioned. In other words, only if the condition number approached the reciprocal of the machine epsilon or unit roundoff, which in the present case for double-precision Fortran is approximately $\epsilon = 10^{-14}$ (so $1/\epsilon = 10^{14}$), would the ODE system of equation (2.208) be numerically singular.

4. The semi-log plots of the condition number indicate that the condition number remained small as the solution proceeded, i.e., ODEs (2.208) did not approach a region in the x_1–x_2 plane where the Jacobian matrix of equation (2.206) becomes ill conditioned. For the second run, the condition number at $t = 0$, 12.8, increased moderately to 15.9 at $t = 10$.

5. For the second run, the solution converged to the root $x_1 = 0.99998$, $x_2 = 3.99994$,

Table 2.26 Output from Programs 2.15a and b. *Continued next pages.*

```
  RUN NO. -    1  x1**2 + x2**2 = 17, 2*x1**(1/3) + x2**(1/2) = 4

INITIAL T -  0.000D+00

  FINAL T -  0.100D+02

  PRINT T -  0.100D+01

NUMBER OF DIFFERENTIAL EQUATIONS -   2

MAXIMUM INTEGRATION ERROR -  0.100D-07

        T          X1          X2
                DX1/DT      DX2/DT
                  RES1        RES2
                 CONDITION NUMBER

      0.0    2.00000     1.00000
             4.34436    -2.68872
            12.00000     0.48016
                  0.227D+02

     10.0    4.07144     0.65026
             0.00007     0.00001
             0.00054     0.00002
                  0.168D+02
```

Table 2.26 *Continued.*

```
            ..1....1....1....1....1....1....1....1....1....1....1..
  0.108D+01+  1                                                  +I
           -       1                                             -I
           -            1                                        -I
           -                                                     -I
           -                 1                                   -I
 -0.368D+00+                      1                              +I
           -                           1                         -I
           -                                1                    -I
           -                                                     -I
           -                                                     -I
 -0.182D+01+                                     1               +I
           -                                          1          -I
           -                                                     -I
           -                                               1     -I
           -                                                     -I
 -0.326D+01+                                                   1 +I
            ..1....1....1....1....1....1....1....1....1....1....1..
        0.000D+00  0.20D+01  0.40D+01  0.60D+01  0.80D+01  0.10D+02


              LOG10 OF RESIDUAL OF EQUATION 1 VS T

            ..1....1....1....1....1....1....1....1....1....1....1..
 -0.319D+00+  1                                                  +I
           -       1                                             -I
           -            1                                        -I
           -                                                     -I
           -                 1                                   -I
 -0.177D+01+                      1                              +I
           -                                                     -I
           -                           1                         -I
           -                                1                    -I
           -                                                     -I
 -0.321D+01+                                     1               +I
           -                                          1          -I
           -                                                     -I
           -                                               1     -I
           -                                                     -I
 -0.466D+01+                                                   1 +I
            ..1....1....1....1....1....1....1....1....1....1....1..
        0.000D+00  0.20D+01  0.40D+01  0.60D+01  0.80D+01  0.10D+02
```

with the small residuals 0.00054 and 0.00003; this solution is an accurate approximation of the exact roots $x_1 = 1$, $x_2 = 4$. However, in general, we should avoid drawing any conclusions about the accuracy of roots from the corresponding values of the equation residuals (e.g., equation residuals can appear "small" when in fact the corresponding computed "roots" are not very accurate).

6. The semi-logarithmic relationship of equation (2.212) appears to apply to the

Table 2.26 *Continued.*

```
              LOG10 OF RESIDUAL OF EQUATION 2 VS T

            ..1....1....1....1....1....1....1....1....1....1....1..
  0.136D+01+  1                                                   +I
            -                                                     -I
            -                                                     -I
            -                                                     -I
            -                                                     -I
  0.129D+01+                                                      +I
            -                                                     -I
            -                                                     -I
            -                                                     -I
            -                                                     -I
  0.123D+01+                      1    1    1    1    1    1    1 +I
            -                1                                    -I
            -           1                                         -I
            -                                                     -I
            -                                                     -I
  0.117D+01+     1                                                +I
            ..1....1....1....1....1....1....1....1....1....1....1..
        0.000D+00  0.20D+01  0.40D+01  0.60D+01  0.80D+01  0.10D+02


              LOG10 OF CONDITION NUMBER VS T

 RUN NO. -   2  x1**2 + x2**2 = 17, 2*x1**(1/3) + x2**(1/2) = 4

INITIAL T -  0.000D+00

  FINAL T -  0.100D+02

  PRINT T -  0.100D+01

NUMBER OF DIFFERENTIAL EQUATIONS -   2

MAXIMUM INTEGRATION ERROR -  0.100D-07

        T         X1         X2
               DX1/DT     DX2/DT
                 RES1       RES2
                CONDITION NUMBER

      0.0    1.00000    2.00000
            -0.96935    3.48467
            12.00000    0.58579
                  0.128D+02
```

multidimensional case, at least for this problem, since the plots of $\log_{10}$ (equation residual) vs. t are nearly linear in all four cases (at least to within the resolution of the point plots). This convergence property is important since it gives us

Table 2.26 *Continued.*

```
      10.0     0.99998    3.99994
               0.00002    0.00006
               0.00054    0.00003
                     0.159D+02

                ..1....1....1....1....1....1....1....1....1....1....1..
    0.108D+01+   1                                                    +I
             -         1                                              -I
             -              1                                         -I
             -                                                        -I
             -                   1                                    -I
   -0.368D+00+                        1                               +I
             -                                                        -I
             -                             1                          -I
             -                                  1                     -I
             -                                                        -I
   -0.182D+01+                                       1                +I
             -                                            1           -I
             -                                                        -I
             -                                                 1      -I
             -                                                        -I
   -0.326D+01+                                                      1 +I
                ..1....1....1....1....1....1....1....1....1....1....1..
             0.000D+00  0.20D+01  0.40D+01  0.60D+01  0.80D+01  0.10D+02

                  LOG10 OF RESIDUAL OF EQUATION 1 VS T

                ..1....1....1....1....1....1....1....1....1....1....1..
   -0.232D+00+   1                                                    +I
             -         1                                              -I
             -              1                                         -I
             -                                                        -I
             -                   1                                    -I
   -0.168D+01+                        1                               +I
             -                                                        -I
             -                             1                          -I
             -                                  1                     -I
             -                                                        -I
   -0.313D+01+                                       1                +I
             -                                            1           -I
             -                                                        -I
             -                                                 1      -I
             -                                                        -I
   -0.458D+01+                                                      1 +I
                ..1....1....1....1....1....1....1....1....1....1....1..
             0.000D+00  0.20D+01  0.40D+01  0.60D+01  0.80D+01  0.10D+02
```

some assurance that if the numerical integration of equation (2.201) or (2.202) is carried far enough, the corresponding dependent variable vector $\bar{x}$ will converge to the solution of system (2.194).

Table 2.26 *Continued.*

```
              LOG10 OF RESIDUAL OF EQUATION 2 VS T

             ..1....1....1....1....1....1....1....1....1....1....1..
   0.120D+01+                        1    1    1    1    1    1    1  +I
            -                                                         -I
            -                   1                                     -I
            -                                                         -I
            -                                                         -I
   0.117D+01+              1                                          +I
            -                                                         -I
            -                                                         -I
            -                                                         -I
            -                                                         -I
   0.114D+01+                                                         +I
            -                                                         -I
            -                                                         -I
            -                                                         -I
            -  1                                                      -I
   0.111D+01+       1                                                 +I
             ..1....1....1....1....1....1....1....1....1....1....1..
          0.000D+00  0.20D+01  0.40D+01  0.60D+01  0.80D+01  0.10D+02


              LOG10 OF CONDITION NUMBER VS T
```

To conclude this example, we note the following advantages of the Davidenko method for the solution of systems of nonlinear equations:

1. The domain of convergence is usually substantially larger than for Newton's method. The same conclusion also applies to modified Newton methods, e.g., *Powell's dogleg method* and various forms of finite stepping *homotopy continuation* (homotopy continuation is discussed by Rheinboldt (1980) and Watson (1986, 1987), and will subsequently be described briefly).

2. Implementation can be accomplished with readily available library software, for example `RKF45`, `DECOMP`, and `SOLVE` as in Programs 15.2a and b, or by using integrators designed for semi-implicit ODEs, for example, `LSODI` and `DASSL`.

3. The Jacobian matrix can be computed numerically by finite differences rather than analytically as was done in the preceding example. In fact, we have used the numerical Jacobian computed by subroutine `JMAP` for this purpose. And, of course, for relatively large problems, the structure of the Jacobian matrix can be exploited.

4. In this discussion, we have presented just the basic ideas of the Davidenko method. It is really an open-ended algorithm in which the analyst can put together combinations of software in accordance with the properties and requirements of the particular nonlinear problem system. We should also mention that the resulting ODEs might be stiff, thereby requiring an implicit integrator; this issue has been addressed by Boggs (1971) and Hachtel et al. (1974).

Table 2.27 Example Calculation in Homotopy Continuation.

t	$h(x,t)=0$	x
0	$g(x)=0$	x_0
0.1	$0.1f(x)+0.9g(x)=0$	$x_{0.1}$
0.2	$0.2f(x)+0.8g(x)=0$	$x_{0.2}$
$\vdots$	$\vdots$	$\vdots$
0.8	$0.8f(x)+0.2g(x)=0$	$x_{0.8}$
0.9	$0.9f(x)+0.1g(x)=0$	$x_{0.9}$
1.0	$f(x)=0$	x_r

Thus, based on our experience, we can recommend Davidenko's method, particularly when Newton's method fails, assuming that the Jacobian matrix is well behaved. If the latter condition is not satisfied, we recommend the use of *steepest descent* [Edelen, (1976a,b), Byrne et al. (1973)].

To conclude this discussion, we consider Davidenko's method as a special case of homotopy continuation. This name comes from the following features of this approach to the solution of nonlinear systems:

1. A vector-valued homotopy function $\overline{H}(\bar{x},t)$ is defined, which as the arguments indicate, is a function of the dependent variable vector, $\bar{x}$, of the nonlinear system, i.e., equation (2.194) and the continuation parameter, t. The dimension of $\overline{H}(\bar{x},t)$ is n.

2. An initial condition vector, $\bar{x}(t_0)$, is assumed to start the calculation at a prescribed initial value of t, t_0. In general, $\bar{x}(t_0)$ is a solution to another problem which is known, say the system of equations $\bar{g}(\bar{x})=\bar{0}$, that is, $\bar{g}(\bar{x}(t_0))=\bar{0}$. We then gradually change t from its initial value, t_0, to its final value, t_f. Along this homotopy path, we require $\overline{H}(\bar{x},t)=\bar{0}$. Also, we formulate $\overline{H}(\bar{x},t)$ so that $\overline{H}(\bar{x}(t_0),t_0)=\bar{g}(\bar{x}(t_0))=0$ and $\overline{H}(\bar{x}(t_f),t_f)=\bar{f}(\bar{x}(t_f))=\bar{0}$, that is, at the end of the calculation when $t=t_f$, we have solved the problem of interest, $\bar{f}(\bar{x}(t_f))=\bar{0}$. In other words, we continue the solution of the known problem, $\bar{g}(\bar{x}(t_0))=\bar{0}$, to the solution of the problem we wish to solve, $\bar{f}(\bar{x}(t_f))=\bar{0}$, along the homotopy path, $\overline{H}(\bar{x},t)=\bar{0}$.

This description may give the impression that the method of homotopy continuation is complicated. In fact, this is not the case. We illustrate its use now with a few examples for the scalar case, $n=1$. The first requirement is to formulate the homotopy function (which we now write as a scalar, but the ideas carry over to the n dimensional case).

Here is a possibility

$$h(x,t)=tf(x)+(1-t)g(x) \tag{2.213}$$

where the problem of interest is $f(x)=0$ and we choose $g(x)$ so that $g(x_0)=0$. Note that $h(x,t)$ has the required properties $h(x,0)=g(x)$ and $h(x,1)=f(x)$ (i.e., $t_0=0$ and $t_f=1$). We can now visualize the use of the homotopy function of equation (2.213) through the use of a table in which we step through a series of values of t, $0, 0.1, 0.2, \ldots, 1$.

At $t=0$, we have simply $g(x_0)=0$, that is, for our selection of $g(x)$, we know x_0 (by definition of $g(x)$). The choice of $g(x)$ is not critical, but the performance of the method will generally improve if $g(x)$ is selected to be similar to $f(x)$. However, even a linear function can be used, that is, $g(x) = x - x_0$.

We now advance t to $t = 0.1$ and the requirement then is to find the root of $0.1f(x) + 0.9g(x) = 0$, that is, the value of x which satisfies this equation. Typically, x is computed using Newton's method. The important point, however, is that the root, call it $x_{0.1}$, will not be far from x_0, or in other words, x_0 is a good initial guess for the solution of $0.1f(x) + 0.9g(x) = 0$. This follows since this equation is not far from $g(x) = 0$, i.e, we are using 90% of $g(x)$ and 10% of $f(x)$ in $0.1f(x) + 0.9g(x) = 0$.

Now, having computed $x_{0.1}$, we advance t to $t = 0.2$, and solve the equation $0.2f(x)+0.8g(x) = 0$. Again, the solution, $x_{0.2}$, will not be far from $x_{0.1}$, that is, $x_{0.1}$ will be a good initial guess of the solution for whatever root-finding method we use. This procedure is then continued through the values of $t = 0.3, 0.4, \ldots, 1$. When $t = 1$, we are solving the problem of interest, $f(x) = 0$, for the root x_r, using as a starting value $x_{0.9}$. Thus, the initial problem $g(x) = 0$, with a known solution, is continued to the problem of interest, $f(x) = 0$, by the variation of the continuation parameter t from its initial value, $t_0 = 0$, to its final value, $t_f = 1$.

If this procedure fails along the way, the changes in t can be reduced so that t might, for example, be stepped through the values $0, 0.01, 0.02, \ldots, 1$. Now the initial estimate of the solution of $H(x,t) = 0$ at each step along the way will be even closer to the required root. Other complications can develop in the use of homotopy continuation, e.g., the root of $H(x,t) = 0$ at each value of t along the homotopy path might not be a single-valued function of t; that is, the homotopy function may have turning points at which $dx/dt = \infty$. If this condition develops, the usual procedure is to move along the arc length of the x–t path. The details of these special cases are beyond the scope of this discussion. Additional details are provided by Rheinboldt (1980) and Watson (1986, 1987).

To conclude, we now consider some other homotopy functions, including one which leads to Davidenko's differential equation. For example, if t again covers a finite interval, we could consider

$$h(x,t) = f(x) - (1-t)f(x_0) = 0 \tag{2.214}$$

$$0 \le t \le 1$$

$$t = 0, \qquad x = x_0$$

$$t = 1, \qquad f(x_r) = 0$$

Note that for equation (2.214), $g(x) = f(x)$, and we need only select x_0, that is, the solution to equation (2.214) at $t = 0$ is clearly $x = x_0$. We then construct a table as before, letting t step through the values $t = 0, 0.1, 0.2, \ldots, 1$ to arrive at the final solution $x = x_r$.

Another possibility is to construct a homotopy function in which the continuation parameter varies over a semi-infinite interval, for example

$$h(x,t) = f(x) - e^{-t}f(x_0) = 0 \tag{2.215}$$

$$0 \le t \le \infty$$

$$t = 0, \qquad x = x_0$$

$$t \to \infty, \qquad f(x_r) = 0$$

We now let $t \to \infty$, and from equation (2.215) $f(x_r) = 0$.

We can analyze this last case a bit further. Since along the homotopy path $h(x,t) = 0$, the differential of $h(x,t)$ is also zero, that is

$$dh = \frac{\partial h}{\partial x}dx + \frac{\partial h}{\partial t}dt = 0 \tag{2.216}$$

If $h(x,t)$ from equation (2.215) is substituted into equation (2.216),

$$dh = \frac{df(x)}{dx}dx + (e^{-t}f(x_0))dt = 0$$

or

$$\frac{dx}{dt} = \frac{e^{-t}f(x_0)}{df(x)/dx} \tag{2.217}$$

Now we can substitute equation (2.215) in equation (2.217) to arrive at the *scalar Davidenko differential equation*

$$\frac{dx}{dt} = -\frac{f(x)}{df(x)/dx} \tag{2.218}$$

which is also equation (2.202) for $n = 1$. Then, the extension to the multidimensional case follows as discussed previously, that is, equations (2.201) and (2.202).

Thus, Davidenko's method can be considered as a special case of homotopy continuation. Interestingly, other forms of homotopy continuation do not have as large a domain of convergence as Davidenko's method for the 2×2 problem of equation (2.204) [Biegler (1991)]. This superior performance is probably due to the differential steps taken during the integration of equations (2.201) and (2.202) rather than the finite steps of Newton's method and various forms of finite stepping homotopy continuation (e.g., as suggested by the calculations of Table 2.17).

In this chapter we have considered ODE integrators in some detail. Clearly these integrators are useful in their own right, but they also serve as the basis for PDE solvers through the numerical method of lines. In the remaining chapters we therefore consider the numerical integration of PDEs.

References

Biegler, L. T. (1991). Private communication.

Boggs, G. T. (1971). "The Solution of Nonlinear Systems of Equations by A-stable Integration Techniques," *SIAM J. Numer. Anal.*, **8**, 767–785.

Brenan, K. E., S. L. Campbell and L. R. Petzold (1989). *Numerical Solution of Initial-Value Problems in Differential-Algebraic Equations*. North-Holland, New York.

Byrne, G. D. and C. A. Hall (1973). *Numerical Solution of Systems of Nonlinear Algebraic Equations*. Academic Press, New York.

Byrne, G. D. and A. C. Hindmarsh (1987). "Stiff ODE Solvers: A Review of Current and Coming Attractions," *J. Comput. Phys.*, **70**, 1–62.

Davidenko, D. F. (1953). "On a New Method for Numerical Solution of Systems of Nonlinear Equations" (Russian), *Dokaldy Akad. Nauk. SSSR*, **88**, 601–602 (translated as *Soviet Math. Dokl.*); also, *Ukr. Mat. Z.5*.

Edelen, D. G. B. (1976a). "On the Construction of Differential Systems for the Solution of Nonlinear Algebraic and Transcendental Systems of Equations," in *Numerical Methods for Differential Systems*, L. Lapidus and W. E. Schiesser (eds.), pp. 67–84. Academic Press, New York.

Edelen, D. G. B. (1976b). "Differential Procedures for Systems of Implicit Relations and Implicitly Coupled Nonlinear Boundary-Value Problems," *Numerical Methods for Differential Systems*, L. Lapidus and W. E. Schiesser, (eds.), pp. 85–95. Academic Press, New York.

Fehlberg, E. (1970). *Low Order Classical Runge Kutta Formulas with Stepsize Control*. NASA Technical Report R-315.

Feynman, R. P., R. B. Leighton and M. Sands (1963). *The Feynman Lectures on Physics*. Addison-Wesley, Reading, MA.

Forsythe, G. E., M. A. Malcolm and C. B. Moler (1977). *Computer Methods for Mathematical Computations*. Prentice-Hall, Englewood Cliffs, NJ.

Gear, C. W. (1971). *Numerical Initial Value Problems in Ordinary Differential Equations*. Prentice-Hall, Englewood Cliffs, NJ.

Greenspan, Donald (1990). "A Counterexample of the Use of Energy as a Measure of Computational Accuracy," *J. Computational Phys.*, **91**, 490–494.

Hachtel, G., and M. Mack (1974). "A Pseudo Dynamic Method for Solving Nonlinear Algebraic Equations," in *Stiff Differential Systems*, R. A. Willoughby (ed.), pp. 135–150. Plenum Press, New York.

Kahaner, D., C. Moler and S. Nash (1989). *Numerical Methods and Software*. Prentice-Hall, Englewood Cliffs, NJ.

Lake, George (1985). "Windows on a New Cosmology," *Science*, **224**(4650), 675–682.

Rheinboldt, W. C. (1980). "Solution Fields on Nonlinear Equations and Continuation Methods," *SIAM J. Numer. Anal.*, **17**(2), 221–237.

Schiesser, W. E. (1991). *The Numerical Method of Lines Integration of Partial Differential Equations*. Academic Press, San Diego.

Silebi, C. A. and W. E. Schiesser (1992). *Dynamic Modeling of Transport Process Systems*. Academic Press, San Diego.

Stengle, G. (1991). Private communication.

Strang, G. (1986). *Introduction to Applied Mathematics*. Wellesley-Cambridge Press, Wellesley, MA.

Watson, L. T. (1986). "Numerical Linear Algebra Aspects of Globally Convergent Homotopy Methods," *SIAM Review*, **28**(4).

Watson, L. T., S. C. Billups and A. P. Morgan (1987). "Algorithm 652: HOMPACK: A Suite of Codes for Globally Convergent Homotopy Algorithms," *ACM Trans. Math. Software*, **13**(3), 281–310.

chapter three

Partial Differential Equations First Order in Time

In Chapter 2 we considered differential equations with only time or a time-like (initial value) independent variable. Thus we considered ODEs. In analyzing many physical systems, however, we must consider how the state variables or dependent variable(s) vary with both time and one or more spatial variables. Thus, we have at least two dependent variables to consider which naturally leads to PDEs. In this chapter, we consider PDEs in which time (or the initial value variable) appears only in first-order derivatives. To illustrate how such PDEs occur in applications, we consider a series of examples, starting with a one dimensional heat transfer system.

3.1 *PDEs with Zeroth-Order and First-Order Spatial Derivatives*

The heat transfer system of Figure 3.1 is rather typical in many applications; heat is exchanged between a flowing fluid with temperature $u_1(x,t)$ and a solid with temperature $u_2(x,t)$. This might occur, for example, when a coolant is passed through a computer to remove the heat generated in the electronic circuitry. The problem can be transient in which case time, t, must be included in the analysis. Also, the temperatures $u_1(x,t)$ and $u_2(x,t)$ will vary with spatial position, x, along the direction of flow of the fluid, and therefore this spatial variable must also be included in the analysis. We therefore have two independent variables, x and t, that lead to a mathematical model based on PDEs.

In order to develop the PDEs for $u_1(x,t)$ and $u_2(x,t)$, we write energy balances for the fluid and for the solid for a section of the fluid-solid system of Figure 3.1 of length Δx. For the fluid flowing with constant linear velocity, v, through a channel with cross-sectional area A_c, the energy balance is

$$A_c \Delta x \rho C_p \frac{\partial u_1}{\partial t} = v A_c \rho C_p u_1|_x - v A_c \rho C_p u_1|_{x+\Delta x} + A_h \Delta x U(u_2 - u_1) \tag{3.1}$$

where u_1 = fluid temperature (K); u_2 = solid temperature (K); t = time (s); x = position along the solid (m); u_1 = fluid linear velocity (m/s); A_c = fluid cross-sectional area (m^2); ρ = fluid density (kg/m^3); C_p = fluid heat capacity $(j/kg\text{-}K)$; A_h = heat transfer area per unit length along the solid (m); U = heat transfer coefficient $(j/s\text{-}m^2\text{-}K = w/m^2\text{-}K)$. We have assumed fluid density, heat capacity, and linear velocity do vary with temperature (which is probably a good assumption for liquids, but not for gases or vapors). Also, variations in the fluid and solid temperatures transverse to the direction of flow are neglected; thus x is the only significant spatial variable. The units for each

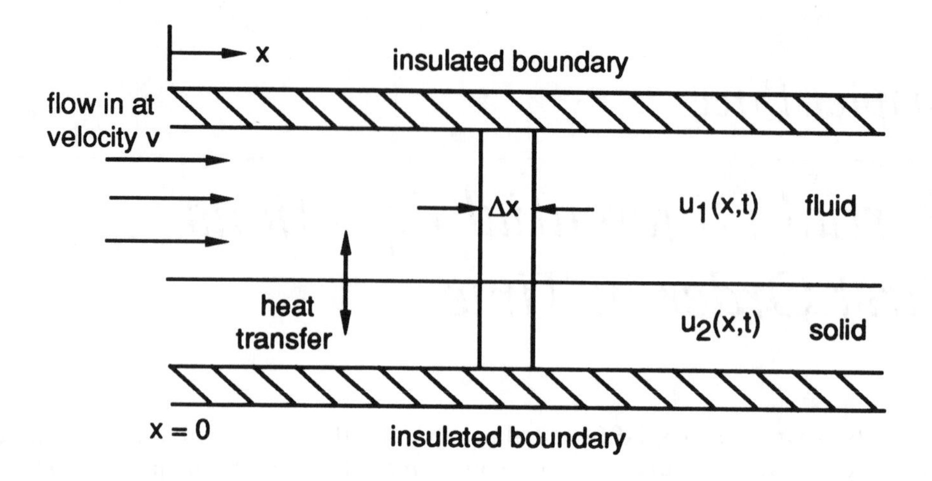

Figure 3.1 One-Dimensional Heat Transfer System.

term in a new equation should be checked for consistency. Thus, equation (3.1) has the units

$A_c \Delta x \rho C_p \dfrac{\partial u_1}{\partial t}$ $(m^2)(m)(kg/m^3)(j/kg\text{-}K)(K/s) = j/s$
(the j/s accumulating in the fluid)

$vA_c\rho C_p u_1|_x$ $(m/s)(m^2)(kg/m^3)(j/kg\text{-}K)(K) = j/s$
(the j/s flowing into the incremental volume of length Δx at x, or out of
$vA_c\rho C_p u_1|_{x+\Delta x}$ the volume at $x + \Delta x$)

$A_h \Delta x U(u_2 - u_1)(m^2/m)(m)(j/s\text{-}m^2\text{-}K)(K) = j/s$
(the j/s transferred between the fluid and solid due to the temperature difference $(u_2 - u_1)$)

Division of equation (3.1) by the coefficient of the time derivative gives

$$\frac{\partial u_1}{\partial t} = -v\frac{\{u_1|_{x+\Delta x} - u_1|_x\}}{\Delta x} + \left(\frac{A_h U}{A_c \rho C_p}\right)(u_2 - u_1) \tag{3.2}$$

In the limit as $\Delta x \to 0$, the first right-hand term becomes a partial derivative in x, thus equation (3.2) becomes

$$\frac{\partial u_1}{\partial t} = -v\frac{\partial u_1}{\partial x} + \left(\frac{A_h U}{A_c \rho C_p}\right)(u_2 - u_1) \tag{3.3}$$

Equation (3.3) has a first-order derivative in t (an initial value independent variable) and x (a spatial or boundary value independent variable), thus it is classified as a *first-order hyperbolic* PDE, as discussed in Section 1.2.4. Since it is first order in t, it requires one initial condition

$$u_1(x, 0) = f(x) \tag{3.4}$$

It also requires one boundary condition since it is first order in x

$$u_1(0, t) = g(t) \tag{3.5}$$

Note that equation (3.5) specifies the way $u_1(x, t)$ varies at $x = 0$. The position $x = 0$ can be considered a physical boundary of the system of Figure 3.1, hence the name "boundary condition."

Equation (3.3) has two dependent variables, u_1 and u_2, and therefore a second equation is required (so that we have two equations in the two unknowns u_1 and u_2). We obtain a PDE for u_2 by writing an energy balance for the solid, which will be similar to the energy balance for the fluid except there will be no convective (flow) terms (and we neglect conduction in the solid in the x direction)

$$A_s \Delta x \rho_s C_{ps} \frac{\partial u_2}{\partial t} = A_h \Delta x U (u_1 - u_2) \tag{3.6}$$

where A_s = solid cross-sectional area (m^2); ρ_s = solid density (kg/m^3); C_{ps} = solid heat capacity (j/kg-K).

Division of equation (3.6) by the coefficient of the time derivative gives

$$\frac{\partial u_2}{\partial t} = \left(\frac{A_h U}{A_s \rho_s C_{ps}} \right) (u_1 - u_2) \tag{3.7}$$

Equation (3.7) has a first-order derivative in t, and therefore it requires one initial condition

$$u_2(x, 0) = h(x) \tag{3.8}$$

However, equation (3.7) does not have a derivative in x (it can be considered to be zeroth order in x). This does not mean, however, that since there is only an explicit derivative in t, this derivative should be written as an ordinary or total derivative (why?).

Equations (3.3)–(3.5), (3.7) and (3.8) constitute the mathematical model to compute $u_1(x, t)$ and $u_2(x, t)$. In order to compute a numerical solution to this model, we must have some integration algorithms for PDEs available, and we will develop these subsequently. However, we now proceed with a discussion of a numerical method of lines program for the integration of this system of PDEs. After we have considered the details of this program, we can discuss the numerical methods that were used in the formulation of the program.

Also, we will evaluate the numerical solution of equations (3.3) to (3.5) and (3.7) and (3.8) with an analytical solution to these equations, which we derive next. First, we take

$$c_1 = \frac{A_h U}{A_c \rho C_p}, \qquad c_2 = \frac{A_h U}{A_s \rho_s C_{ps}} \qquad (3.9)(3.10)$$

in equations (3.3) and (3.7). Therefore, the equations to be integrated analytically are

$$\frac{\partial u_1}{\partial t} + v \frac{\partial u_1}{\partial x} = c_1 (u_2 - u_1) \tag{3.11}$$

$$\frac{\partial u_2}{\partial t} = c_2 (u_1 - u_2) \tag{3.12}$$

$$u_1(x, 0) = f(x), \quad u_1(0, t) = g(t), \quad u_2(x, 0) = h(x) \qquad (3.13)\text{–}(3.15)$$

These equations were originally presented by Bateman (1923, pp. 123–128).

If Laplace transforms of $u_1(x,t)$ and $u_2(x,t)$ with respect t are defined as

$$L\{u_1(x,t)\} = \int_0^\infty u_1(x,t)e^{-st}\,dt = \bar{u}_1(x,s) \tag{3.16}$$

$$L\{u_2(x,t)\} = \int_0^\infty u_2(x,t)e^{-st}\,dt = \bar{u}_2(x,s) \tag{3.17}$$

Application of these transforms to equations (3.11) to (3.15) gives

$$s\bar{u}_1(x,s) - u_1(x,0) + v\frac{d\bar{u}_1(x,s)}{dx} = c_1(\bar{u}_2(x,s) - \bar{u}_1(x,s)) \tag{3.18}$$

$$s\bar{u}_2(x,s) - u_2(x,0) = -c_2(\bar{u}_2(x,s) - \bar{u}_1(x,s)) \tag{3.19}$$

Equation (3.19) can be solved for $\bar{u}_2(x,s)$

$$\bar{u}_2(x,s) = \frac{c_2}{s+c_2}\bar{u}_1(x,s) + \frac{1}{s+c_2}u_2(x,0) \tag{3.20}$$

Substitution of equation (3.20) in equation (3.18) then gives

$$v\frac{d\bar{u}_1(x,s)}{dx} + (s+c_1)\bar{u}_1(x,s) = c_1\bar{u}_2(x,s) + u_1(x,0)$$

or

$$v\frac{d\bar{u}_1(x,s)}{dx} + (s+c_1)\bar{u}_1(x,s) = c_1\left\{\frac{c_2}{s+c_2}\bar{u}_1(x,s) + \frac{1}{s+c_2}u_2(x,0)\right\} + u_1(x,0) \tag{3.21}$$

Equation (3.21) is easily rearranged to

$$v\frac{d\bar{u}_1(x,s)}{dx} + \left(s + c_1 - \frac{c_1c_2}{s+c_2}\right)\bar{u}_1(x,s) = \frac{c_1}{s+c_2}u_2(x,0) + u_1(x,0) \tag{3.22}$$

or if $f(s)$ is defined as

$$f(s) = \left(s + c_1 - \frac{c_1c_2}{s+c_2}\right)/v \tag{3.23}$$

equation (3.22) becomes

$$\frac{d\bar{u}_1(x,s)}{dx} + f(s)\bar{u}_1(x,s) = \frac{c_1/v}{s+c_2}u_2(x,0) + (1/v)u_1(x,0) \tag{3.24}$$

Equation (3.24) is a first-order, ordinary differential equation defining $\bar{u}_1(x,s)$ as a function of x. If both sides of equation (3.24) are multiplied by $e^{f(s)x}$, which introduces an integrating factor [Strang (1986), pp. 472–473] then

$$e^{f(s)x}\left(\frac{d\bar{u}_1(x,s)}{dx} + f(s)\bar{u}_1(x,s)\right) = e^{f(s)x}\left(\frac{c_1/v}{s+c_2}u_2(x,0) + (1/v)u_1(x,0)\right) \tag{3.25}$$

integration of both sides of equation (3.25) with respect to x gives

$$e^{f(s)x}\bar{u}_1(x,s) - u_1(0,s) = \int_0^x e^{f(s)\lambda}\left(\frac{c_1/v}{s+c_2}u_2(\lambda,0) + (1/v)u_1(\lambda,0)\right)d\lambda \tag{3.26}$$

or

$$\bar{u}_1(x,s) = u_1(0,s)e^{-f(s)x} + \int_0^x e^{f(s)(\lambda-x)}\left(\frac{c_1/v}{s+c_2}u_2(\lambda,0) + (1/v)u_1(\lambda,0)\right)d\lambda \tag{3.27}$$

$u_1(x,t)$ can now be obtained by an inverse Laplace transform

$$\begin{aligned} u_1(x,t) &= L^{-1}\left\{\bar{u}_1(x,s)\right\} \\ &= \frac{1}{2\pi i}\int_{\gamma-i\infty}^{\gamma+i\infty}\Big\{u_1(0,s)e^{-f(s)x} \\ &\quad + \int_0^x e^{f(s)(\lambda-x)}\left(\frac{c_1/v}{s+c_2}u_2(\lambda,0)+(1/v)u_1(\lambda,0)\right)d\lambda\Big\}e^{st}\,ds \end{aligned} \tag{3.28}$$

Equation (3.28) is the final solution we seek. However, in order to produce a solution which can be evaluated numerically with reasonable effort, we take the initial conditions, equations (3.13) and (3.15), to be homogeneous

$$u_1(x,0) = u_2(x,0) = 0 \tag{3.29)(3.30}$$

and equation (3.28) reduces to

$$u_1(x,t) = \frac{1}{2\pi i}\int_{\gamma-i\infty}^{\gamma+i\infty}\left\{u_1(0,s)e^{-\left(\left(s+c_1-\frac{c_1c_2}{s+c_2}\right)/v\right)x}\right\}e^{st}\,ds$$

which rearranges to

$$u_1(x,t) = \frac{1}{2\pi i}\int_{\gamma-i\infty}^{\gamma+i\infty} e^{-(c_1/v)x}\left\{u_1(0,s)e^{-(s/v)x}e^{\left(\frac{c_1c_2/v}{s+c_2}\right)x}\right\}e^{st}\,ds \tag{3.31}$$

At this point, we can consider four types of boundary conditions (specifications of $u_1(0,t)$):

Case (1): A boundary condition which is consistent with the initial condition, and thereby avoids a discontinuity. For example

$$u_1(0,t) = 1 - e^{-c_2 t} \tag{3.32}$$

has the property $u_1(0,0) = 0$ and is therefore consistent with the homogeneous initial condition, equation (3.29), $u_1(x,0) = 0$. The transform of equation (3.32), which will be used in equation (3.31), is

$$\bar{u}_1(0,s) = \frac{1}{s}\frac{c_2}{s+c_2} \tag{3.33}$$

Case (2): A boundary condition which is not consistent with the initial condition, and therefore introduces a discontinuity. For example

$$u_1(0,t) = e^{-c_2 t} \tag{3.34}$$

has the property $u_1(0,0) = 1$ and therefore is not consistent with the homogeneous initial condition, equation (3.29), $u_1(0,0) = 0$ (i.e., there is a unit jump at $x = 0$ and $t = 0$). The transform of equation (3.34), which will be used in equation (3.31), is

$$\bar{u}_1(0,s) = \frac{1}{s+c_2} \tag{3.35}$$

Case (3): A boundary condition which can be analyzed in terms of two or more other boundary conditions. For example, the Heaviside unit step function, $h(t)$, defined as

$$u_1(0,t) = h(t) = \left\{ \begin{array}{ll} 0, & t < 0 \\ 1, & t > 0 \end{array} \right\} \tag{3.36}$$

can be analyzed by considering a combination of (3.33) and (3.35). This is easily demonstrated; the Laplace transform of equation (3.36) is

$$\bar{u}_1(0,s) = 1/s \tag{3.37}$$

But the right-hand sides of equations (3.33) and (3.35) when added give

$$\frac{1}{s}\frac{c_2}{s+c_2} + \frac{1}{s+c_2} = \frac{1}{s} \tag{3.38}$$

as expected from adding the right-hand sides of equations (3.32) and (3.34). Thus we can get the solution for Case 3 merely by adding the solutions for Cases 1 and 2 (owing to the linearity of equations (3.11) and (3.12)).

Case (4): A boundary condition which is the integral of another boundary condition. For example, the ramp function, $r(t)$

$$u_1(0,t) = r(t) = \left\{ \begin{array}{ll} t, & 0 \le t \le 1 \\ 1, & t > 1 \end{array} \right\} \tag{3.39}$$

is given by

$$r(t) = \int_0^t \{h(t) - h(t-1)\}\, dt \tag{3.40}$$

where $h(t)$ is given by equation (3.36). Thus we can get the solution for $r(t)$ merely by integrating the solution for the boundary condition $u_1(0,t) = h(t) - h(t-1)$.

The solutions for all four cases will now be derived analytically, for comparison with the numerical solutions.

We first note the following transform pairs

$$L^{-1}\left\{\tfrac{1}{s}e^{k/s}\right\} = I_o(2\sqrt{kt}) \tag{3.41}$$

[Carslaw and Jaeger (1959), p. 495]

$$L^{-1}\left\{\bar{f}(s+a)\right\} = e^{-at}f(t) \tag{3.42}$$

$$L^{-1}\left\{e^{-as}\bar{f}(s)\right\} = h(t-a)f(t-a) \tag{3.43}$$

$$L^{-1}\left\{\tfrac{1}{s}\bar{f}(s)\right\} = \int_0^t f(\lambda)\,d\lambda \tag{3.44}$$

[Beyer (1978), p. 600]

$u_1(x,t)$ for the Case 1 boundary condition (3.32) is therefore (from substituting equation (3.33) in equation (3.31))

$$u_1(x,t) = e^{-(c_1/v)x}L^{-1}\left\{\frac{1}{s}\frac{c_2}{s+c_2}e^{-(s/v)x}e^{\left(\frac{c_1c_2/v}{s+c_2}\right)x}\right\} \tag{3.45}$$

or

$$u_1(x,t) = e^{-(c_1/v)x}c_2\int_0^t h(\lambda - x/v)e^{-c_2(\lambda-x/v)}I_o\left\{2\sqrt{\frac{c_1c_2}{v}x(\lambda - x/v)}\right\}d\lambda \tag{3.46}$$

A special case of equation (3.46) can be checked by applying the *final value theorem* of the Laplace transform to equation (3.45)

$$\lim_{t\to\infty} u_1(x,t) = \lim_{s\to 0} s\bar{u}_1(x,s) = \lim_{s\to 0} e^{-(c_1/v)x}\left\{\frac{c_2}{s+c_2}e^{-(s/v)x}e^{\left(\frac{c_1c_2/v}{s+c_2}\right)x}\right\}$$

or

$$e^{-(c_1/v)x}e^{(c_1/v)x} = 1$$

as expected (i.e., $u_1(0,t) = 1 - e^{-c_2t}$ at $x = 0$ gives $u_1(x,\infty) = 1$ since $c_2 > 0$).

For boundary condition (3.34) we have (from substituting equation (3.35) in equation (3.31))

$$u_1(x,t) = e^{-(c_1/v)x}L^{-1}\left\{\frac{1}{s+c_2}e^{-(s/v)x}e^{\left(\frac{c_1c_2/v}{s+c_2}\right)x}\right\} \tag{3.47}$$

or

$$u_1(x,t) = e^{-(c_1/v)x}h(\lambda - x/v)e^{-c_2(\lambda-x/v)}I_o\left\{2\sqrt{\frac{c_1c_2}{v}x(\lambda - x/v)}\right\} \tag{3.48}$$

The final value theorem of the Laplace transform applied to equation (3.47) gives

$$\lim_{t\to\infty} u_1(x,t) = \lim_{s\to 0} s\bar{u}_1(x,s) = \lim_{s\to 0} e^{-(c_1/v)x}s\left\{\frac{1}{s+c_2}e^{-(s/v)x}e^{\left(\frac{c_1c_2/v}{s+c_2}\right)x}\right\} = 0$$

as expected (i.e., $u_1(0,t) = e^{-c_2t}$ gives $u_1(x,\infty) = 0$).

For boundary condition (3.36) we have (from substituting equation (3.37) in equation (3.31))

$$u_1(x,t) = e^{-(c_1/v)x} L^{-1}\left\{\frac{1}{s} e^{-(s/v)x} e^{\left(\frac{c_1c_2/v}{s+c_2}\right)x}\right\} \tag{3.49}$$

or

$$u_1(x,t) = e^{-(c_1/v)x} L^{-1}\left\{\left(\frac{1}{s+c_2} + \frac{1}{s}\frac{c_2}{s+c_2}\right) e^{-(s/v)x} e^{\left(\frac{c_1c_2/v}{s+c_2}\right)x}\right\} \tag{3.50}$$

Equation (3.50) is just a superposition of equations (3.45) and (3.47). Thus, from equation (3.50),

$$\begin{aligned} u_1(x,t) = {} & e^{-(c_1/v)x} c_2 \int_0^t h(\lambda - x/v) e^{-c_2(\lambda - x/v)} I_o\left\{2\sqrt{\frac{c_1c_2}{v} x(\lambda - x/v)}\right\} d\lambda \\ & + \left\{e^{-(c_1/v)x} h(\lambda - x/v) e^{-c_2(\lambda - x/v)} I_o\left\{2\sqrt{\frac{c_1c_2}{v} x(\lambda - x/v)}\right\}\right\} \end{aligned} \tag{3.51}$$

The final value theorem of the Laplace transform applied to equation (3.50) gives

$$\lim_{t\to\infty} u_1(x,t) = \lim_{s\to 0} s\bar{u}_1(x,s) = 1$$

as expected (i.e., $u_1(0,t) = h(t)$ gives $u_1(x,\infty) = 1$).

Finally, for boundary condition (3.39) we have (from substituting equation (3.40) in equation (3.31)), or more directly, from equation (3.51) and superposition,

$$\begin{aligned} u_1(x,t) = {} & e^{-(c_1/v)x} c_2 \int_0^t h(\lambda - x/v) e^{-c_2(\lambda - x/v)} I_o\left\{2\sqrt{\frac{c_1c_2}{v} x(\lambda - x/v)}\right\} d\lambda \\ & + \left\{e^{-(c_1/v)x} h(\lambda - x/v) e^{-c_2(\lambda - x/v)} I_o\left\{2\sqrt{\frac{c_1c_2}{v} x(\lambda - x/v)}\right\}\right\} \\ & - e^{-(c_1/v)x} c_2 \int_0^t h(\lambda - x/v - 1) e^{-c_2(\lambda - x/v - 1)} I_o\left\{2\sqrt{\frac{c_1c_2}{v} x(\lambda - x/v - 1)}\right\} d\lambda \\ & - \left\{e^{-(c_1/v)x} h(\lambda - x/v - 1) e^{-c_2(\lambda - x/v - 1)} I_o\left\{2\sqrt{\frac{c_1c_2}{v} x(\lambda - x/v - 1)}\right\}\right\} \end{aligned} \tag{3.52}$$

Equation (3.52) results from inversion of

$$\bar{u}_1(x,s) = e^{-(c_1/v)x}\left\{\frac{1}{s^2} e^{-(s/v)x} e^{\left(\frac{c_1c_2/v}{s+c_2}\right)x}\right\} - e^{-(c_1/v)x}\left\{\frac{e^{-s}}{s^2} e^{-(s/v)x} e^{\left(\frac{c_1c_2/v}{s+c_2}\right)x}\right\} \tag{3.53}$$

Equation (3.53) follows from equations (3.40) and (3.49), plus the superposition of solutions. Application of the final value theorem to equation (3.53) gives

$$\begin{aligned} \lim_{t\to\infty} u_1(x,t) &= \lim_{s\to 0} s\bar{u}_1(x,s) \\ &= \lim_{s\to 0} s\left\{e^{-(c_1/v)x} e^{-(s/v)x} e^{\left(\frac{c_1c_2/v}{s+c_2}\right)x}\left\{\frac{1}{s^2} - \frac{e^{-s}}{s^2}\right\}\right\} \\ &= \lim_{s\to 0}\left\{e^{-(c_1/v)x} e^{-(s/v)x} e^{\left(\frac{c_1c_2/v}{s+c_2}\right)x}\left\{\frac{1}{s} - \frac{e^{-s}}{s}\right\}\right\} = 1 \end{aligned}$$

as expected (i.e., $u_1(0,t) = r(t)$ gives $u_1(x,\infty) = 1$).

```
      SUBROUTINE INITAL
      IMPLICIT DOUBLE PRECISION (A-H,O-Z)
      COMMON/T/          T,      NSTOP,      NORUN
C...  V(0,T) = 1 - EXP(-C2*T)
      IF((NORUN-1)*(NORUN-2)*(NORUN-3).EQ.0)CALL INIT1
C...
C...  V(0,T) = EXP(-C2*T)
      IF((NORUN-4)*(NORUN-5)*(NORUN-6).EQ.0)CALL INIT2
C...
C...  V(0,T) = 1 (UNIT STEP)
      IF((NORUN-7)*(NORUN-8)*(NORUN-9).EQ.0)CALL INIT3
C...
C...  V(0,T) = T (UNIT RAMP)
      IF((NORUN-10)*(NORUN-11)*(NORUN-12).EQ.0)CALL INIT4
      RETURN
      END
```

Program 3.1a Subroutine `INITAL` for Initial Conditions (3.13) and (3.15).

The four analytical solutions, equations (3.46), (3.48), (3.51), and (3.52) are included in the following programming for comparison with the four numerical solutions. These four cases provide a spectrum of test problems; for example, by varying c_2, the rate of change of $u_1(0,t)$ can be varied for both the continuous and discontinuous cases.

In developing a computer program for the four preceding problems, we follow the same procedure as in Chapter 2 of writing three subroutines `INITAL`, `DERV`, and `PRINT`. This suggests the first important characteristic of method of lines solutions—we will be integrating a system of ODEs which approximates the PDEs of interest—equations (3.11) and (3.12) in this case. We start by approximating the derivative $\partial u_1/\partial x$ in equation (3.11) over a 21-point grid in x, with grid spacing Δx (not the same Δx as was used in derving equations (3.11) and (3.12), but rather, a finite spacing that will be used in various finite difference approximations of $\partial u_1/\partial x$). Thus, we will have 21 values of u_1 and 21 values of u_2 along this grid, and at each grid point we will write an ODE for u_1 and u_2. There will therefore be a total of $2 \times 21 = 42$ ODEs. This system of ODEs is indicated in the `/Y/` and `/F/` sections of `COMMON` in subroutines `INITAL`, `DERV`, and `PRINT`. We start with subroutine `INITAL` in Program 3.1a for initial conditions (3.13) and (3.15) with $f(x) = h(x) = 0$ (since we assumed homogeneous initial conditions (3.29) and (3.30) in deriving the four analytical solutions that will be compared with the numerical solutions).

Twelve cases are programmed for `NORUN=1` to `NORUN=12`. For cases 1 to 3 (`NORUN=1` to 3), subroutine `INIT1` is called to set initial conditions (3.29) and (3.30) (and boundary condition (3.32) is used in the first derivative subroutine, `DERV1`); for Cases 4 to 6, subroutine `INIT2` is called to again set initial conditions (3.29) and (3.30) (but with boundary condition (3.34) used in the second derivative subroutine, `DERV2`), etc. Thus in all four initialization routines, `INIT1` to `INIT4`, the same initial conditions (3.29) and (3.30) are used, but with different boundary conditions in each of the four associated derivative subroutines `DERV1` to `DERV4`, that is, boundary conditions (3.32), (3.34), (3.36), and (3.40), which are generally stated as boundary condition (3.14).

The four initialization subroutines, INIT1 to INIT4, are listed in Program 3.1b.

We can note the following points about Program 3.1b:

1. The four initialization routines, INIT1 to INIT4, are similar since they all implement initial conditions (3.29) and (3.30).

2. In each of these routines, u_1 and u_2 in equations (3.11) and (3.12) are each defined on 21 points by arrays U1(NX) and U2(NX) (with NX=21). These two arrays appear in COMMON/Y/ since they are dependent variables computed by the integration of $2 \times 21 = 42$ ODEs, as we will observe when considering the derivatives subroutines DERV1 to DERV4. Also, in subroutines INIT1 to INIT3, a 43rd dependent variable, U1A, is in COMMON/Y/. This additional dependent variable is required to evaluate the integrals in the analytical solutions of equations (3.46) and (3.51) (analytical solution (3.48) does not have an integral, thus the 43rd variable, U1A, in subroutine INIT2 is not actually used in the analytical solution). The integrals are computed by integrating an associated ODE, as described in Section 2.4 (see equations (2.162) and (2.163)). Therefore, we are not strictly evaluating an analytical solution, but rather a "semi-analytical" solution in the sense that the integrals in equations (3.46) and (3.51) are computed numerically; these integrals would be difficult to evaluate analytically, and thus this approach demonstrates the advantage of combining analytical and numerical methods, i.e., the analytical portion of the solution is of reasonable complexity, and the numerical method is applied when continuing analytical analysis would become quite complicated.

3. Subroutine INIT4 contains three additional dependent variables in array U2A(NT) (with NT=3) since analytical solution (3.52) actually requires three integrals (the two which appear in the right-hand side of equation (3.52) plus a third integral to compute the ramp function solution, as indicated by $r(t)$ expressed as the integral of the difference in two step functions, $h(t) - h(t-1)$, of equation (3.40)).

We can note the following points about DERV1 to DERV4 in Program 3.1c:

1. Boundary conditions (3.32), (3.34), (3.36), and (3.40) are implemented at the beginning of subroutines DERV1, DERV2, DERV3, and DERV4, respectively. Note that the temporal derivatives of the boundary values are set to zero so that the ODE integrator does not move the boundary value of the dependent variable away from its prescribed value. For example, in DERV1 we have the following programming of the boundary condition (3.32) (at $x = 0$ or grid point $i = 1$):

```
C...
C...  BOUNDARY CONDITION
       U1(1)=1.0D+00-DEXP(-C2*T)
      U1T(1)=0.0D+00
```

Another possibility is to program the temporal derivative of the boundary value dependent variable in DERV1; for example

```
C...
C...  BOUNDARY CONDITION
       U1(1)=1.0D+00-DEXP(-C2*T)
      U1T(1)=C2*DEXP(-C2*T)
```

A third possibility is to program the initial value of the boundary dependent

```
      SUBROUTINE INIT1
      IMPLICIT DOUBLE PRECISION (A-H,O-Z)
      PARAMETER (NX=21)
      COMMON/T/          T,      NSTOP,      NORUN
     +      /Y/    U1(NX),    U2(NX),        U1A
     +      /F/   U1T(NX),   U2T(NX),       U1AT
     +      /S/   U1X(NX)
     +      /C/         C1,         C2,          V,          X,
     +                EXP1,        ARG
C...
C...  SET THE PROBLEM PARAMETERS
         C1=0.560D+00
         C2=0.100D+00
          V=2.031D+00
          X=1.0D+00
C...
C...  PRECOMPUTE SOME QUANTITIES USED IN THE ANALYTICAL SOLUTION
      EXP1=DEXP(-C1/V*X)
C...
C...  INITIAL CONDITIONS
      DO 1 I=1,NX
         U1(I)=0.0D+00
         U2(I)=0.0D+00
1     CONTINUE
      U1A=0.0D+00
C...
C...  INITIAL DERIVATIVES
      CALL DERV1
      RETURN
      END

      SUBROUTINE INIT2
      IMPLICIT DOUBLE PRECISION (A-H,O-Z)
      PARAMETER (NX=21)
      COMMON/T/          T,      NSTOP,      NORUN
     +      /Y/    U1(NX),    U2(NX),        U1A
     +      /F/   U1T(NX),   U2T(NX),       U1AT
     +      /S/   U1X(NX)
     +      /C/         C1,         C2,          V,          X,
     +                EXP1,        ARG
C...
C...  SET THE PROBLEM PARAMETERS
         C1=0.560D+00
         C2=0.100D+00
          V=2.031D+00
          X=1.0D+00
C...
C...  PRECOMPUTE SOME QUANTITIES USED IN THE ANALYTICAL SOLUTION
      EXP1=DEXP(-C1/V*X)
C...
```

Program 3.1b Subroutines INIT1, INIT2, INIT3, and INIT4 for Initial Conditions (3.29) and (3.30). *Continued next pages.*

```
C...  INITIAL CONDITIONS
      DO 1 I=1,NX
         U1(I)=0.0D+00
         U2(I)=0.0D+00
1     CONTINUE
      U1A=0.0D+00
C...
C...  INITIAL DERIVATIVES
      CALL DERV2
      RETURN
      END

      SUBROUTINE INIT3
      IMPLICIT DOUBLE PRECISION (A-H,O-Z)
      PARAMETER (NX=21)
      COMMON/T/          T,     NSTOP,     NORUN
     +      /Y/    U1(NX),    U2(NX),        U1A
     +      /F/   U1T(NX),   U2T(NX),       U1AT
     +      /S/   U1X(NX)
     +      /C/         C1,        C2,          V,          X,
     +                EXP1,       ARG
C...
C...  SET THE PROBLEM PARAMETERS
         C1=0.560D+00
         C2=0.100D+00
          V=2.031D+00
          X=1.0D+00
C...
C...  PRECOMPUTE SOME QUANTITIES USED IN THE ANALYTICAL SOLUTION
      EXP1=DEXP(-C1/V*X)
C...
C...  INITIAL CONDITIONS
      DO 1 I=1,NX
         U1(I)=0.0D+00
         U2(I)=0.0D+00
1     CONTINUE
      U1A=0.0D+00
C...
C...  INITIAL DERIVATIVES
      CALL DERV3
      RETURN
      END

      SUBROUTINE INIT4
      IMPLICIT DOUBLE PRECISION (A-H,O-Z)
      PARAMETER (NX=21,NT=3)
      COMMON/T/          T,     NSTOP,     NORUN
     +      /Y/    U1(NX),    U2(NX),    U1A(NT)
     +      /F/   U1T(NX),   U2T(NX),   U1AT(NT)
     +      /S/   U1X(NX)
     +      /C/         C1,        C2,          V,          X,
     +                EXP1,       ARG
```

Program 3.1b *Continued.*

```
C...
C...  SET THE PROBLEM PARAMETERS
         C1=0.560D+00
         C2=0.100D+00
          V=2.031D+00
          X=1.0D+00
C...
C...  PRECOMPUTE SOME QUANTITIES USED IN THE ANALYTICAL SOLUTION
      EXP1=DEXP(-C1/V*X)
C...
C...  INITIAL CONDITIONS
      DO 1 I=1,NX
         U1(I)=0.0D+00
         U2(I)=0.0D+00
1     CONTINUE
      DO 2 I=1,NT
         U1A(I)=0.0D+00
2     CONTINUE
C...
C...  INITIAL DERIVATIVES
      CALL DERV4
      RETURN
      END
```

Program 3.1b *Continued.*

variable in INIT1, and program only its derivative in DERV1; for example in INIT1 (in which $T = 0$)

```
C...
C...  BOUNDARY CONDITION
      U1(1)=1.0D+00-DEXP(-C2*T)
```

Then in DERV1 we program only the temporal derivative

```
C...
C...  BOUNDARY CONDITION
      U1T(1)=C2*DEXP(-C2*T)
```

The three methods are equivalent mathematically, and should therefore give the same numerical solution. The third approach, in which only the temporal derivative is set in DERV1, may be required when using some implicit ODE integrators (e.g., LSODE), for which setting a dependent variable (e.g., U1(1)) in the derivative subroutine might cause a numerical error in the ODE integration.

2. The spatial derivative $u_{1x} = \partial u_1/\partial x$, which is in array U1X, is computed from u_1 in array U1 by one of three spatial differentiation routines, DSS012, DSS018, or DSS020, in DERV1 to DERV4. For example, in DERV1 which is executed for the first, second, and third runs (NORUN= 1 to 3)

```
C...
C...  U1X
```

```
      SUBROUTINE DERV1
      IMPLICIT DOUBLE PRECISION (A-H,O-Z)
      PARAMETER (NX=21)
      COMMON/T/          T,      NSTOP,      NORUN
     +      /Y/   U1(NX),    U2(NX),       U1A
     +      /F/  U1T(NX),   U2T(NX),      U1AT
     +      /S/  U1X(NX)
     +      /C/         C1,         C2,          V,          X,
     +                EXP1,    ARG
C...
C...  BOUNDARY CONDITION
       U1(1)=1.0D+00-DEXP(-C2*T)
      U1T(1)=0.0D+00
C...
C...  U1X
C...
C...     TWO POINT UPWIND
         IF(NORUN.EQ.1)CALL DSS012(0.D+0,X,NX,U1,U1X,1.0D+00)
C...
C...     FOUR POINT BIASED UPWIND
         IF(NORUN.EQ.2)CALL DSS018(0.D+0,X,NX,U1,U1X,1.0D+00)
C...
C...     FIVE POINT BIASED UPWIND
         IF(NORUN.EQ.3)CALL DSS020(0.D+0,X,NX,U1,U1X,1.0D+00)
C...
C...  PDES
      DO 1 I=1,NX
         IF(I.NE.1)THEN
            U1T(I)=C1*(U2(I)-U1(I))-V*U1X(I)
         END IF
            U2T(I)=C2*(U1(I)-U2(I))
1     CONTINUE
C...
C...  TEST FOR THE ELAPSED TIME
C...  T - X/V LT 0
      IF((T-X/V).LT.0.0D+00)THEN
       U1AT=0.0D+00
C...
C...  T - X/V GE 0
      ELSE
         ARG=2.0D+00*DSQRT((C1*C2*X/V)*(T-X/V))
        U1AT=EXP1*C2*DEXP(-C2*(T-X/V))*BESSI0(ARG)
      END IF
      RETURN
      END
```

Program 3.1c Subroutines DERV1, DERV2, DERV3, and DERV4 for Equations (3.11) and (3.12) with Boundary Conditions (3.32), (3.34), (3.36), and (3.40). *Continued next pages.*

```
      SUBROUTINE DERV2
      IMPLICIT DOUBLE PRECISION (A-H,O-Z)
      PARAMETER (NX=21)
      COMMON/T/          T,      NSTOP,     NORUN
     +      /Y/   U1(NX),    U2(NX),       U1A
     +      /F/  U1T(NX),   U2T(NX),      U1AT
     +      /S/  U1X(NX)
     +      /C/         C1,          C2,          V,          X,
     +                EXP1,    ARG
C...
C... BOUNDARY CONDITION
       U1(1)=DEXP(-C2*T)
      U1T(1)=0.0D+00
C...
C... U1X
C...
C...   TWO POINT UPWIND
        IF(NORUN.EQ.4)CALL DSS012(0.0D+00,X,NX,U1,U1X,1.0D+00)
C...
C...   FOUR POINT BIASED UPWIND
        IF(NORUN.EQ.5)CALL DSS018(0.0D+00,X,NX,U1,U1X,1.0D+00)
C...
C...   FIVE POINT BIASED UPWIND
        IF(NORUN.EQ.6)CALL DSS020(0.0D+00,X,NX,U1,U1X,1.0D+00)
C...
C... PDES
      DO 1 I=1,NX
        IF(I.NE.1)THEN
          U1T(I)=C1*(U2(I)-U1(I))-V*U1X(I)
        END IF
          U2T(I)=C2*(U1(I)-U2(I))
1     CONTINUE
C...
C... TEST FOR THE ELAPSED TIME
C... T - X/V LT 0
      IF((T-X/V).LT.0.0D+00)THEN
       U1AT=0.0D+00
C...
C... T - X/V GE 0
      ELSE
        ARG=2.0D+00*DSQRT((C1*C2*X/V)*(T-X/V))
       U1AT=EXP1*DEXP(-C2*(T-X/V))*BESSI0(ARG)
      END IF
      RETURN
      END

      SUBROUTINE DERV3
      IMPLICIT DOUBLE PRECISION (A-H,O-Z)
      PARAMETER (NX=21)
      COMMON/T/          T,      NSTOP,     NORUN
     +      /Y/   U1(NX),    U2(NX),       U1A
     +      /F/  U1T(NX),   U2T(NX),      U1AT
```

Program 3.1c *Continued.*

```
     +      /S/   U1X(NX)
     +      /C/        C1,          C2,           V,          X,
     +                 EXP1,    ARG
C...
C...  BOUNDARY CONDITION
       U1(1)=1.0D+00
      U1T(1)=0.0D+00
C...
C...  U1X
C...
C...     TWO POINT UPWIND
         IF(NORUN.EQ.7)CALL DSS012(0.0D+00,X,NX,U1,U1X,1.0D+00)
C...
C...     FOUR POINT BIASED UPWIND
         IF(NORUN.EQ.8)CALL DSS018(0.0D+00,X,NX,U1,U1X,1.0D+00)
C...
C...     FIVE POINT BIASED UPWIND
         IF(NORUN.EQ.9)CALL DSS020(0.0D+00,X,NX,U1,U1X,1.0D+00)
C...
C...  PDES
      DO 1 I=1,NX
         IF(I.NE.1)THEN
            U1T(I)=C1*(U2(I)-U1(I))-V*U1X(I)
         END IF
            U2T(I)=C2*(U1(I)-U2(I))
1     CONTINUE
C...
C...  TEST FOR THE ELAPSED TIME
C...  T - X/V LT 0
      IF((T-X/V).LT.0.0D+00)THEN
       U1AT=0.0D+00
C...
C...  T - X/V GE 0
      ELSE
        ARG=2.0D+00*DSQRT((C1*C2*X/V)*(T-X/V))
       U1AT=EXP1*C2*DEXP(-C2*(T-X/V))*BESSI0(ARG)
      END IF
      RETURN
      END

      SUBROUTINE DERV4
      IMPLICIT DOUBLE PRECISION (A-H,O-Z)
      PARAMETER (NX=21,NT=3)
      COMMON/T/          T,      NSTOP,      NORUN
     +      /Y/    U1(NX),    U2(NX),    U1A(NT)
     +      /F/   U1T(NX),   U2T(NX),   U1AT(NT)
     +      /S/   U1X(NX)
     +      /C/        C1,          C2,           V,          X,
     +                 EXP1,    ARG
C...
C...  BOUNDARY CONDITION
```

Program 3.1c *Continued.*

```
      IF(T.LE.1.0D+00)THEN
         U1(1)=T
        U1T(1)=1.0D+00
      ELSE IF(T.GT.1.0D+00)THEN
         U1(1)=1.0D+00
        U1T(1)=0.0D+00
      END IF
C...
C...  U1X
C...
C...     TWO POINT UPWIND
         IF(NORUN.EQ.10)CALL DSS012(0.0D+00,X,NX,U1,U1X,1.0D+00)
C...
C...     FOUR POINT BIASED UPWIND
         IF(NORUN.EQ.11)CALL DSS018(0.0D+00,X,NX,U1,U1X,1.0D+00)
C...
C...     FIVE POINT BIASED UPWIND
         IF(NORUN.EQ.12)CALL DSS020(0.0D+00,X,NX,U1,U1X,1.0D+00)
C...
C...  PDES
      DO 1 I=1,NX
         IF(I.NE.1)THEN
            U1T(I)=C1*(U2(I)-U1(I))-V*U1X(I)
         END IF
            U2T(I)=C2*(U1(I)-U2(I))
1     CONTINUE
C...
C...  TEST FOR THE ELAPSED TIME FOR RESPONSE TO STEP AT T = 0
C...  T - X/V LT 0
      IF((T-X/V).LT.0.0D+00)THEN
       U1AT(1)=0.0D+00
C...
C...  T - X/V GE 0
      ELSE
         ARG=2.0D+00*DSQRT((C1*C2*X/V)*(T-X/V))
       U1AT(1)=EXP1*C2*DEXP(-C2*(T-X/V))*BESSI0(ARG)
      END IF
C...
C...  TEST FOR THE ELAPSED TIME FOR RESPONSE TO STEP AT T = 1
C...  T - X/V - 1 LT 0
      IF(((T-X/V)-1.0D+00).LT.0.0D+00)THEN
       U1AT(2)=0.0D+00
C...
C...  T - X/V - 1 GE 0
      ELSE
          ARG=2.0D+00*DSQRT((C1*C2*X/V)*(T-X/V-1.0D+00))
      U1AT(2)=-EXP1*C2*DEXP(-C2*(T-X/V-1.0D+00))*BESSI0(ARG)
      END IF
C...
C...  SUPERIMPOSE TWO STEP RESPONSES
      U1AT(3)=U1AT(1)/C2+U1A(1)+U1AT(2)/C2+U1A(2)
      RETURN
      END
```

Program 3.1c *Continued.*

```
C...
C...      TWO POINT UPWIND
          IF(NORUN.EQ.1)CALL DSS012(0.D+0,X,NX,U1,U1X,1.0D+00)
C...
C...      FOUR POINT BIASED UPWIND
          IF(NORUN.EQ.2)CALL DSS018(0.D+0,X,NX,U1,U1X,1.0D+00)
C...
C...      FIVE POINT BIASED UPWIND
          IF(NORUN.EQ.3)CALL DSS020(0.D+0,X,NX,U1,U1X,1.0D+00)
```

The other arguments of the three spatial differentiation routines (in addition to U1 and U1X) are (a) the left boundary value of x ($x = 0$), the right boundary value of x (x or $X = 1$), the number of points in the spatial grid ($NX = 21$), and (c) for the last arguments an indication of whether the flow of the fluid in Figure 3.1 is left to right (a positive velocity so $1.0D + 00$ is used) or right to left (a negative velocity in which case a negative value would be used for the last argument). Each of the three spatial differentiation routines computes $u_{1x} = \partial u_1/\partial x$ from u_1, but by different finite difference approximations which will be described subsequently. In summary, the first to fourth and sixth arguments are inputs to these differentiation routines, and the fifth argument, the numerical spatial derivative, $u_{1x} = \partial u_1/\partial x$ (in array U1X), is the output (returned by the routines).

3. With the spatial derivative computed, the PDEs, equations (3.11) and (3.12) can now be programmed

```
C...
C...  PDES
      DO 1 I=1,NX
         IF(I.NE.1)THEN
            U1T(I)=C1*(U2(I)-U1(I))-V*U1X(I)
         END IF
            U2T(I)=C2*(U1(I)-U2(I))
1     CONTINUE
```

Note the similarity of the programming of the PDEs to the PDEs themselves (equations (3.11) and (3.12)) which is one of the major advantages of this approach (i.e., the method of lines).

4. At the end of each of the derivative subroutines, the integrands of the integrals in the various analytical solutions, equations (3.46), (3.51), and (3.52), are programmed. For example, in subroutine DERV1

```
C...
C...  TEST FOR THE ELAPSED TIME
C...  T - X/V LT 0
      IF((T-X/V).LT.0.0D+00)THEN
       U1AT=0.0D+00
C...
C...  T - X/V GE 0
      ELSE
         ARG=2.0D+00*DSQRT((C1*C2*X/V)*(T-X/V))
       U1AT=EXP1*C2*DEXP(-C2*(T-X/V))*BESSI0(ARG)
      END IF
```

is the programming for the integrand in equation (3.46)

$$\text{U1AT} = e^{-(c_1/v)x} c_2 h(\lambda - x/v) e^{-c_2(\lambda - x/v)} I_o \left\{ 2\sqrt{\frac{c_1 c_2}{v} x(\lambda - x/v)} \right\}$$

`BESSI0` is a Fortran function to compute the modified Bessel function I_o to be discussed subsequently. `U1AT` is then integrated by putting it in `COMMON/F/`. The integrated result is `U1A` returned by the ODE integrator as the 43rd dependent variable in `COMMON/Y/`, i.e., $u_1(x,t) =$ `U1A` from equation (3.46) with $x = 1$

$$\text{U1A} = e^{-(c_1/v)x} c_2 \int_0^t h(\lambda - x/v) e^{-c_2(\lambda - x/v)} I_o \left\{ 2\sqrt{\frac{c_1 c_2}{v} x(\lambda - x/v)} \right\} d\lambda$$

The unit step function, $h(\lambda - x/v)$, is included as `U1AT=0.0D+00` if $\lambda < x/v$, that is by the `IF` statement `IF((T-X/V).LT.0.0D+00)THEN`, etc. Similar programming is used at the end of subroutines `DERV2`, `DERV3`, and `DERV3` to implement the other three analytical solutions, equations (3.46), (3.51), and (3.52).

Finally, subroutines `PRINT1` to `PRINT4` print the solutions and map the ODE Jacobian matrix produced by calls to subroutine `MAP` for each of the four problems. We can note the following points for subroutines `PRINT1` to `PRINT4` in Program 3.1d:

1. In each case the analytical solution available through `COMMON/Y/` and `/F/` from the corresponding `DERV1` to `DERV4` is printed along with the numerical solution. Thus, in `PRINT1`, `U1A` is printed along with `U1(21)` (the analytical and numerical solutions at $x = 1$).

2. In subroutine `PRINT2`, `U1AT` is printed as the analytical solution rather than `U1A` since equation (3.48) does not have an integral.

3. In subroutine `PRINT3`, a linear combination of `U1AT` and `U1A` is printed as the analytical solution, in accordance with equation (3.51) which has the sum of two terms on the right-hand side (with and without an integral)

```
WRITE(NO,2)T,U1(NX),U1AT/C2+U1A
```

4. In `PRINT4`, `U1A(3)` is printed as the analytical solution, for which the derivative `U1AT(3)` is computed in `DERV4` as

```
U1AT(3)=U1AT(1)/C2+U1A(1)+U1AT(2)/C2+U1A(2)
```

   `U1AT(3)` is then integrated by putting it in `COMMON/F/` to produce `U1A(3)` through `COMMON/Y/`. `U1A(3)` and its derivative `U1AT(3)` are a linear combination of four terms in accordance with equation (3.52).

5. Subroutine `MAP` and function `BESSI0` are listed in Program 3.1e. Subroutine `MAP` essentially just calls subroutine `JMAP` for 42 ODEs at the beginning and end of each run. Function `BESSI0` is taken from Press et al. (1986).

The main program that calls `RKF45`, `INITAL`, and `PRINT`, and indirectly `DERV` through `RKF45`, is similar to the main programs in Chapter 2 that call `RKF45`, e.g., Program 2.2a, and therefore it is not listed here. The data file read by the main program is listed in Program 3.1f.

```
      SUBROUTINE PRINT1(NI,NO)
      IMPLICIT DOUBLE PRECISION (A-H,O-Z)
      PARAMETER (NX=21)
      COMMON/T/          T,      NSTOP,      NORUN
     +      /Y/    U1(NX),    U2(NX),        U1A
     +      /F/   U1T(NX),   U2T(NX),       U1AT
     +      /S/   U1X(NX)
     +      /C/         C1,         C2,          V,          X,
     +                EXP1,      ARG
C...
C...  INITIALIZE A COUNTER FOR THE PRINTING AND PLOTTING
      DATA IP/0/
      IP=IP+1
      IF(IP.EQ.1)THEN
C...
C...  MAP THE ODE JACOBIAN MATRIX AT THE BEGINNING OF THE SOLUTION
      CALL MAP
C...
C...  PRINT A HEADING
      WRITE(NO,1)
1     FORMAT(//,14X,'T',8X,'U1(X,T)',3X,'U1(X,T) ANAL')
      END IF
C...
C...  PRINT THE SOLUTION
      WRITE(NO,2)T,U1(NX),U1A
2     FORMAT(F15.2,2F15.4)
C...
C...  WRITE NORUN, IP AND T ON THE SCREEN TO MONITOR THE PROGRESS OF
C...  THE CALCULATIONS
      WRITE(*,*)' NORUN = ',NORUN,' IP = ',IP,' T = ',T
C...
C...  MAP THE ODE JACOBIAN MATRIX AT THE END OF THE SOLUTION
      IF(IP.LT.11)RETURN
      CALL MAP
C...
C...  RESET THE INTEGER COUNTER FOR THE NEXT RUN
      IP=0
      RETURN
      END

      SUBROUTINE PRINT2(NI,NO)
      IMPLICIT DOUBLE PRECISION (A-H,O-Z)
      PARAMETER (NX=21)
      COMMON/T/          T,      NSTOP,      NORUN
     +      /Y/    U1(NX),    U2(NX),        U1A
     +      /F/   U1T(NX),   U2T(NX),       U1AT
     +      /S/   U1X(NX)
     +      /C/         C1,         C2,          V,          X,
     +                EXP1,         ARG
C...
```

Program 3.1d Subroutines PRINT1, PRINT2, PRINT3, and PRINT4 for Equations (3.11) and (3.12) with Boundary Conditions (3.32), (3.34), (3.36), and (3.40). *Continued next pages.*

```
C...  INITIALIZE A COUNTER FOR THE PRINTING AND PLOTTING
      DATA IP/0/
      IP=IP+1
      IF(IP.EQ.1)THEN
C...
C...  MAP THE ODE JACOBIAN MATRIX AT THE BEGINNING OF THE SOLUTION
C...  CALL MAP
C...
C...  PRINT A HEADING
      WRITE(NO,1)
1     FORMAT(//,14X,'T',8X,'U1(X,T)',3X,'U1(X,T) ANAL')
      END IF
C...
C...  PRINT THE SOLUTION
      WRITE(NO,2)T,U1(NX),U1AT
2     FORMAT(F15.2,2F15.4)
C...
C...  WRITE NORUN, IP AND T ON THE SCREEN TO MONITOR THE PROGRESS OF
C...  THE CALCULATIONS
      WRITE(*,*)' NORUN = ',NORUN,' IP = ',IP,' T = ',T
C...
C...  MAP THE ODE JACOBIAN MATRIX AT THE END OF THE SOLUTION
      IF(IP.LT.11)RETURN
C...  CALL MAP
C...
C...  RESET THE INTEGER COUNTER FOR THE NEXT RUN
      IP=0
      RETURN
      END

      SUBROUTINE PRINT3(NI,NO)
      IMPLICIT DOUBLE PRECISION (A-H,O-Z)
      PARAMETER (NX=21)
      COMMON/T/          T,      NSTOP,      NORUN
     +      /Y/    U1(NX),     U2(NX),        U1A
     +      /F/   U1T(NX),    U2T(NX),       U1AT
     +      /S/   U1X(NX)
     +      /C/        C1,         C2,          V,          X,
     +               EXP1,        ARG
C...
C...  INITIALIZE A COUNTER FOR THE PRINTING AND PLOTTING
      DATA IP/0/
      IP=IP+1
      IF(IP.EQ.1)THEN
C...
C...  MAP THE ODE JACOBIAN MATRIX AT THE BEGINNING OF THE SOLUTION
C...  CALL MAP
C...
C...  PRINT A HEADING
      WRITE(NO,1)
```

Program 3.1d *Continued.*

```
1     FORMAT(//,14X,'T',8X,'U1(X,T)',3X,'U1(X,T) ANAL')
      END IF
C...
C...  PRINT THE SOLUTION
      WRITE(NO,2)T,U1(NX),U1AT/C2+U1A
2     FORMAT(F15.2,2F15.4)
C...
C...  WRITE NORUN, IP AND T ON THE SCREEN TO MONITOR THE PROGRESS OF
C...  THE CALCULATIONS
      WRITE(*,*)' NORUN = ',NORUN,' IP = ',IP,' T = ',T
C...
C...  MAP THE ODE JACOBIAN MATRIX AT THE END OF THE SOLUTION
      IF(IP.LT.11)RETURN
C...  CALL MAP
C...
C...  RESET THE INTEGER COUNTER FOR THE NEXT RUN
      IP=0
      RETURN
      END

      SUBROUTINE PRINT4(NI,NO)
      IMPLICIT DOUBLE PRECISION (A-H,O-Z)
C...
C...  DECLARE SELECTED VARIABLES AS DOUBLE PRECISION
      PARAMETER (NX=21,NT=3)
      COMMON/T/          T,     NSTOP,     NORUN
     +      /Y/    U1(NX),   U2(NX),   U1A(NT)
     +      /F/   U1T(NX),  U2T(NX),  U1AT(NT)
     +      /S/   U1X(NX)
     +      /C/        C1,        C2,         V,         X,
     +               EXP1,       ARG
C...
C...  INITIALIZE A COUNTER FOR THE PRINTING AND PLOTTING
      DATA IP/0/
      IP=IP+1
      IF(IP.EQ.1)THEN
C...
C...  MAP THE ODE JACOBIAN MATRIX AT THE BEGINNING OF THE SOLUTION
C...  CALL MAP
C...
C...  PRINT A HEADING
      WRITE(NO,1)
1     FORMAT(//,14X,'T',8X,'U1(X,T)',3X,'U1(X,T) ANAL')
      END IF
C...
C...  PRINT THE SOLUTION
      WRITE(NO,2)T,U1(NX),U1A(3)
2     FORMAT(F15.2,2F15.4)
C...
C...  WRITE NORUN, IP AND T ON THE SCREEN TO MONITOR THE PROGRESS OF
C...  THE CALCULATIONS
```

Program 3.1d *Continued.*

```
      WRITE(*,*)' NORUN = ',NORUN,' IP = ',IP,' T = ',T
C...
C...  MAP THE ODE JACOBIAN MATRIX AT THE END OF THE SOLUTION
      IF(IP.LT.11)RETURN
C...  CALL MAP
C...
C...  RESET THE INTEGER COUNTER FOR THE NEXT RUN
      IP=0
      RETURN
      END
```

Program 3.1d *Continued.*

The following points can be noted about this data file:

1. In all cases, the independent variable, t, in equations (3.11) and (3.12) covers the interval $0 \leq t \leq 10$, and the integration is performed with an error tolerance of 0.000001.

2. For runs one to nine, 43 ODEs are specified, 42 for the 21-point approximation of equations (3.11) and (3.12) ($2 \times 21 = 42$), plus one ODE for the analytical solutions, equations (3.46), (3.48), and (3.51). For runs 10 to 12, 45 ODEs are specified, 42 for the 21-point approximation of equations (3.11) and (3.12), plus three ODEs for the analytical solution, equation (3.52).

The output from Programs 3.1a to f will be considered for each of the four cases, equations (3.32), (3.34), (3.36), and (3.40), in Tables 3.1a to d, respectively.

The following points can be noted about the output in Table 3.1a:

1. Run No. 1:

   (1.1) The Jacobian maps at $t = 0$ and $t = 10$ give a clear indication of the structure of the 43 ODEs. For example, ODEs 2 to 21, which are the method of lines approximation of equation (3.11), are bidiagonal along the main diagonal, e.g., for ODE no. 10

   ```
   10          66                    4
   ```

   the 66 along the main diagonal is due to the two-point upwind approximation of the derivative $\partial u_1/\partial x$ in equation (3.11) from subroutine `DSS012` (the finite difference approximations in `DSS012`, `DSS018`, and `DSS020` will subsequently be discussed in some detail). Note that the 66 is for `U1(9)` and `U1(10)` as expected for ODE no. 10 since `U1(9)` is *upwind* of `U1(10)`. The 4 in the map of ODE no. 10 reflects the coupling between equations (3.11) and (3.12) through the heat transfer term $c_1(u_2 - u_1)$, which for ODE no. 10 is $c_1($`U2(10)` $-$ `U1(10)`$)$ (note that `U2(10)` is dependent variable $21 + 10 = 31$ in `COMMON/Y/`, thus the 4 corresponds to the 31st dependent variable). This type of banded structure with a concentration of points along the main diagonal and outlying diagonals due to coupling between simultaneous PDEs is typical. It also indicates why a banded ODE solver may not be particularly effective for a simultaneous PDE problem, i.e., to include the outlying diagonals a wide bandwidth must be specifed in using a banded ODE solver, thereby losing much of its computational efficiency relative to a full matrix ODE solver.

```
      SUBROUTINE MAP
      PARAMETER (N=42)
      COMMON/T/       T
     1      /Y/   Y(N)
     2      /F/   F(N)
C...
C...  SUBROUTINE MAP CALLS SUBROUTINE JMAP TO MAP THE JACOBIAN MATRIX
C...  OF AN NTH-ORDER ODE SYSTEM.
C...
C...  DECLARE SELECTED VARIABLES DOUBLE PRECISION
      DOUBLE PRECISION T, Y, F, YOLD(N), FOLD(N)
C...
C...  DEFINE SINGLE PRECISION ARRAY REQUIRED BY SUBROUTINE JMAP (A)
      REAL A(N,N)
C...
C...  MAP THE JACOBIAN MATRIX OF THE ODE SYSTEM DEFINED IN SUBROUTINE
C...  DERV
      CALL JMAP(N,A,Y,YOLD,F,FOLD)
      RETURN
      END

      DOUBLE PRECISION FUNCTION BESSI0(X)
C...  FUNCTION BESSI0(X)
C...
C...  FUNCTION BESSI0 IS A MINOR REVISION BY W. E. SCHIESSER OF THE
C...  ORIGINAL NUMERICAL RECIPES FUNCTION TO BRING IT INTO CLOSER
C...  CONFORMITY WITH STANDARD FORTRAN DOUBLE PRECISION PROGRAMMING.
C...  THE ORIGINAL STATEMENTS THAT HAVE BEEN MODIFIED ARE COMMENTED
C...  OUT
      DOUBLE PRECISION      X,     AX,
     *        Y,     P1,     P2,     P3,     P4,    P5,     P6,     P7,
     *       Q1,     Q2,     Q3,     Q4,     Q5,    Q6,     Q7,     Q8,
     *       Q9
C...  REAL*8 Y,     P1,     P2,     P3,     P4,    P5,     P6,     P7,
C... *       Q1,     Q2,     Q3,     Q4,     Q5,    Q6,     Q7,     Q8,
C... *       Q9
      DATA           P1,            P2,            P3,            P4,
     *               P5,            P6,            P7/
     *           1.0D0,  3.5156229D0, 3.0899424D0,  1.2067492D0,
     *    0.2659732D0,  0.360768D-1,  0.45813D-2/
      DATA           Q1,            Q2,            Q3,            Q4,
     *               Q5,            Q6,            Q7,            Q8,
     *               Q9/
     *   0.39894228D0, 0.1328592D-1, 0.225319D-2, -0.157565D-2,
     *    0.916281D-2,-0.2057706D-1,0.2635537D-1,-0.1647633D-1,
     *    0.392377D-2/
      IF (DABS(X).LT.3.75D0) THEN
C...  IF ( ABS(X).LT.3.75   ) THEN
        Y=(X/3.75D0)**2
```

Program 3.1e Subroutine MAP and Function BESSI0 to Map the ODE Jacobian Matrix and Compute the Modified Bessel Function I_o in Equations (3.46), (3.48), (3.51), and (3.52). *Continued next page.*

```
C...      Y=(X/3.75  )**2
          BESSI0=P1+Y*(P2+Y*(P3+Y*(P4+Y*(P5+Y*(P6+Y*P7)))))
        ELSE
          AX=DABS(X)
C...      AX= ABS(X)
          Y=3.75D0/AX
C...      Y=3.75  /AX
          BESSI0=(DEXP(AX)/DSQRT(AX))*(Q1+Y*(Q2+Y*(Q3+Y*(Q4
     *        +Y*(Q5+Y*(Q6+Y*(Q7+Y*(Q8+Y*Q9))))))))
C...      BESSI0=( EXP(AX)/ SQRT(AX))*(Q1+Y*(Q2+Y*(Q3+Y*(Q4
C... *        +Y*(Q5+Y*(Q6+Y*(Q7+Y*(Q8+Y*Q9))))))))
        END IF
        RETURN
        END
```

Program 3.1e *Continued.*

(1.2) ODE no. 1 has no entries in the map due to boundary condition (3.32) which is reflected in the coding `U1T(1)=0.0D+00` in subroutine `DERV1`; that is, the first derivative, `U1T(1)`, in `COMMON/F/` is constant and does not depend on any of the dependent variables.

(1.3) ODEs 22 to 42 are the method of lines approximation of equation (3.12). They have just two monodiagonals due to the heat transfer term $c_2(u_1 - u_2)$ in equation (3.12).

(1.4) The map is for ODES 1 to 42 only, even though there is a 43rd ODE for the analytical solution, equation (3.46). This limit of 42 is set in subroutine `MAP` of Program 3.1e and was chosen so that only the ODEs for the method of lines approximations of equations (3.11) and (3.12) would be mapped.

(1.5) The numerical and analytical solutions at $x = 1$ agree to 3+ figures (for `NORUN=1`)

T	U1(X,T)	U1(X,T) ANAL
0.00	0.0000	0.0000
1.00	0.0383	0.0379
2.00	0.1089	0.1085
3.00	0.1745	0.1741
4.00	0.2355	0.2350
5.00	0.2921	0.2916
6.00	0.3447	0.3442
7.00	0.3934	0.3931
8.00	0.4387	0.4384
9.00	0.4807	0.4804
10.00	0.5197	0.5194

The lack of agreement in the fourth figure is due principally to the two-point upwind approximations in subroutine `DSS012`, which are only first-order correct in the spatial variable x in equation (3.11), that is, the truncation error is proportional to Δx^1 as will be explained subsequently. This first-order characteristic is designated as $O(\Delta x)$, where "O" (capital oh) means "of order" so $O(\Delta x)$ means of order Δx.

(1.6) The map at $t = 10$ is identical to the one at $t = 0$ because equations (3.11) and (3.12) are linear; thus the ODE Jacobian matrix is a constant matrix.

Table 3.1 (a) Abbreviated Output of Programs 3.1a to f for Boundary Condition (3.32). *Continued next pages.*

```
RUN NO. -    1  BATEMAN TEST PROBLEM - DSS012 - V(0,T) = 1 - EXP(-(K/U)*T)

INITIAL T -   0.000D+00

  FINAL T -   0.100D+02

  PRINT T -   0.100D+01

NUMBER OF DIFFERENTIAL EQUATIONS -  43

MAXIMUM INTEGRATION ERROR -  0.100D-05


DEPENDENT VARIABLE COLUMN INDEX J (FOR YJ) IS PRINTED HORIZONTALLY

DERIVATIVE ROW INDEX I (FOR DYI/DT = FI(Y1,Y2,...,YJ,...,YN) IS PRINTED
VERTICALLY

JACOBIAN MATRIX ELEMENT IN THE MAP WITH INDICES I,J IS FOR PFI/PYJ
WHERE P DENOTES A PARTIAL DERIVATIVE


                 111111111122222222223333333333444
        123456789012345678901234567890123456789012
    1
    2    6                    4
    3    66                    4
    4     66                    4
    5      66                    4
    6       66                    4
    7        66                    4
    8         66                    4
    9          66                    4
   10           66                    4
   11            66                    4
   12             66                    4
   13              66                    4
   14               66                    4
   15                66                    4
   16                 66                    4
   17                  66                    4
   18                   66                    4
   19                    66                    4
   20                     66                    4
   21                      66                    4
   22                            3
   23    3                    3
   24     3                    3
   25      3                    3
   26       3                    3
   27        3                    3
```

Table 3.1 (a) *Continued.*

```
  28         3                    3
  29          3                    3
  30           3                    3
  31            3                    3
  32             3                    3
  33              3                    3
  34               3                    3
  35                3                    3
  36                 3                    3
  37                  3                    3
  38                   3                    3
  39                    3                    3
  40                     3                    3
  41                      3                    3
  42                       3                    3


        T        U1(X,T)   U1(X,T) ANAL
      0.00        0.0000         0.0000
      1.00        0.0383         0.0379
      2.00        0.1089         0.1085
      3.00        0.1745         0.1741
      4.00        0.2355         0.2350
      5.00        0.2921         0.2916
      6.00        0.3447         0.3442
      7.00        0.3934         0.3931
      8.00        0.4387         0.4384
      9.00        0.4807         0.4804
     10.00        0.5197         0.5194


DEPENDENT VARIABLE COLUMN INDEX J (FOR YJ) IS PRINTED HORIZONTALLY

DERIVATIVE ROW INDEX I (FOR DYI/DT = FI(Y1,Y2,...,YJ,...,YN) IS PRINTED
VERTICALLY

JACOBIAN MATRIX ELEMENT IN THE MAP WITH INDICES I,J IS FOR PFI/PYJ
WHERE P DENOTES A PARTIAL DERIVATIVE


                1111111111222222222233333333334444
       123456789012345678901234567890123456789012
   1
   2    6                    4
   3    66                    4
   4     66                    4
   5      66                    4
   6       66                    4
   7        66                    4
   8         66                    4
   9          66                    4
  10           66                    4
  11            66                    4
  12             66                    4
```

Table 3.1 (a) *Continued.*

```
  13              66                   4
  14               66                   4
  15                66                   4
  16                 66                   4
  17                  66                   4
  18                   66                   4
  19                    66                   4
  20                     66                   4
  21                      66                   4
  22                        3
  23     3                   3
  24      3                   3
  25       3                   3
  26        3                   3
  27         3                   3
  28          3                   3
  29           3                   3
  30            3                   3
  31             3                   3
  32              3                   3
  33               3                   3
  34                3                   3
  35                 3                   3
  36                  3                   3
  37                   3                   3
  38                    3                   3
  39                     3                   3
  40                      3                   3
  41                       3                   3
  42                        3                   3


RUN NO. -   2  BATEMAN TEST PROBLEM - DSS018 - V(0,T) = 1 - EXP(-(K/U)*T)

INITIAL T -  0.000D+00

  FINAL T -  0.100D+02

  PRINT T -  0.100D+01

NUMBER OF DIFFERENTIAL EQUATIONS -  43

MAXIMUM INTEGRATION ERROR -  0.100D-05


DEPENDENT VARIABLE COLUMN INDEX J (FOR YJ) IS PRINTED HORIZONTALLY

DERIVATIVE ROW INDEX I (FOR DYI/DT = FI(Y1,Y2,...,YJ,...,YN) IS PRINTED
VERTICALLY

JACOBIAN MATRIX ELEMENT IN THE MAP WITH INDICES I,J IS FOR PFI/PYJ
WHERE P DENOTES A PARTIAL DERIVATIVE
```

Table 3.1 (a) *Continued.*

```
              111111111122222222223333333333444
     123456789012345678901234567890123456789012
 1
 2    665                  4
 3    666                   4
 4    5666                   4
 5     5666                   4
 6      5666                   4
 7       5666                   4
 8        5666                   4
 9         5666                   4
10          5666                   4
11           5666                   4
12            5666                   4
13             5666                   4
14              5666                   4
15               5666                   4
16                5666                   4
17                 5666                   4
18                  5666                   4
19                   5666                   4
20                    5666                   4
21                    6676                    4
22                        3
23    3                    3
24     3                    3
25      3                    3
26       3                    3
27        3                    3
28         3                    3
29          3                    3
30           3                    3
31            3                    3
32             3                    3
33              3                    3
34               3                    3
35                3                    3
36                 3                    3
37                  3                    3
38                   3                    3
39                    3                    3
40                     3                    3
41                      3                    3
42                       3                    3
```

T	U1(X,T)	U1(X,T) ANAL
0.00	0.0000	0.0000
1.00	0.0378	0.0378
2.00	0.1084	0.1084
3.00	0.1740	0.1739
4.00	0.2349	0.2349
5.00	0.2916	0.2915

Table 3.1 (a) *Continued.*

```
   6.00        0.3442        0.3441
   7.00        0.3930        0.3929
   8.00        0.4383        0.4382
   9.00        0.4803        0.4803
  10.00        0.5193        0.5193
                 .
                 .
                 .

RUN NO. -   3  BATEMAN TEST PROBLEM - DSS020 - V(0,T) = 1 - EXP(-(K/U)*T)

INITIAL T -  0.000D+00

  FINAL T -  0.100D+02

  PRINT T -  0.100D+01

NUMBER OF DIFFERENTIAL EQUATIONS -  43

MAXIMUM INTEGRATION ERROR -  0.100D-05


DEPENDENT VARIABLE COLUMN INDEX J (FOR YJ) IS PRINTED HORIZONTALLY

DERIVATIVE ROW INDEX I (FOR DYI/DT = FI(Y1,Y2,...,YJ,...,YN) IS PRINTED
VERTICALLY

JACOBIAN MATRIX ELEMENT IN THE MAP WITH INDICES I,J IS FOR PFI/PYJ
WHERE P DENOTES A PARTIAL DERIVATIVE


               111111111122222222223333333333444
      123456789012345678901234567890123456789012
   1
   2  6665                  4
   3  6465                   4
   4  6666                    4
   5  56666                    4
   6   56666                    4
   7    56666                    4
   8     56666                    4
   9      56666                    4
  10       56666                    4
  11        56666                    4
  12         56666                    4
  13          56666                    4
  14           56666                    4
  15            56666                    4
  16             56666                    4
  17              56666                    4
  18               56666                    4
  19                56666                    4
  20                 56666                    4
  21                 66776                     4
```

Table 3.1 (a) *Continued.*

```
22                        3
23    3                    3
24     3                    3
25      3                    3
26       3                    3
27        3                    3
28         3                    3
29          3                    3
30           3                    3
31            3                    3
32             3                    3
33              3                    3
34               3                    3
35                3                    3
36                 3                    3
37                  3                    3
38                   3                    3
39                    3                    3
40                     3                    3
41                      3                    3
42                       3                    3
```

T	U1(X,T)	U1(X,T) ANAL
0.00	0.0000	0.0000
1.00	0.0378	0.0378
2.00	0.1084	0.1084
3.00	0.1740	0.1740
4.00	0.2349	0.2349
5.00	0.2916	0.2916
6.00	0.3442	0.3442
7.00	0.3930	0.3930
8.00	0.4383	0.4383
9.00	0.4803	0.4803
10.00	0.5193	0.5193

.
.
.

Therefore, in the subsequent discussion, we consider only the Jacobian map at $t = 0$.

2. Run No. 2:

(2.1) ODEs no. 1 to 21 which are the method of lines approximation of equation (3.11), have quaddiagonal structure along the main diagonal with an outlying monodiagonal. For example, for ODE no. 10

```
10          5666                   4
```

The 5666 structure corresponds to `U1(8)`, `U1(9)`, `U1(10)`, and `U1(11)` reflecting the four-point biased upwind approximations in subroutine `DSS018`.

```
BATEMAN TEST PROBLEM - DSS012 - U1(0,T) = 1 - EXP(-C2*T)
0.          10.0       1.0
  43                         0.000001
BATEMAN TEST PROBLEM - DSS018 - U1(0,T) = 1 - EXP(-C2*T)
0.          10.0       1.0
  43                         0.000001
BATEMAN TEST PROBLEM - DSS020 - U1(0,T) = 1 - EXP(-C2*T)
0.          10.0       1.0
  43                         0.000001
BATEMAN TEST PROBLEM - DSS012 - U1(0,T) = EXP(-C2)*T)
0.          10.0       1.0
  43                         0.000001
BATEMAN TEST PROBLEM - DSS018 - U1(0,T) = EXP(-C2)*T)
0.          10.0       1.0
  43                         0.000001
BATEMAN TEST PROBLEM - DSS020 - U1(0,T) = EXP(-C2)*T)
0.          10.0       1.0
  43                         0.000001
BATEMAN TEST PROBLEM - DSS012 - U1(0,T) = 1
0.          10.0       1.0
  43                         0.000001
BATEMAN TEST PROBLEM - DSS018 - U1(0,T) = 1
0.          10.0       1.0
  43                         0.000001
BATEMAN TEST PROBLEM - DSS020 - U1(0,T) = 1
0.          10.0       1.0
  43                         0.000001
BATEMAN TEST PROBLEM - DSS012 - UNIT RAMP
0.          10.0       1.0
  45                         0.000001
BATEMAN TEST PROBLEM - DSS018 - UNIT RAMP
0.          10.0       1.0
  45                         0.000001
BATEMAN TEST PROBLEM - DSS020 - UNIT RAMP
0.          10.0       1.0
  45                         0.000001
END OF RUNS
```

Program 3.1f Data File Read by the Main Program that Calls `RKF45` for the Solution of Equations (3.11) and (3.12).

That is, for ODE no. 10, two points are upwind, `U1(8)` and `U1(9)`, and one point is downwind, `U(11)`. Thus, the approximation uses more upwind than downwind points so it is biased upwind; physically this seems like a reasonable approach since in a convective (flowing) system, what happens upwind of a particular point will have greater influence than what happens downwind.

(2.2) ODEs no. 22 to 42 again are the method of lines approximation of equation (3.12), which has only monodiagonals since there is no partial derivative with respect to x in equation (3.12).

(2.3) The numerical and analytical solutions at $x = 1$ are in agreement to approximately four figures (NORUN=2)

T	U1(X,T)	U1(X,T) ANAL
0.00	0.0000	0.0000
1.00	0.0378	0.0378
2.00	0.1084	0.1084
3.00	0.1740	0.1739
4.00	0.2349	0.2349
5.00	0.2916	0.2915
6.00	0.3442	0.3441
7.00	0.3930	0.3929
8.00	0.4383	0.4382
9.00	0.4803	0.4803
10.00	0.5193	0.5193

This improved accuracy relative to run no. 1 is due to the higher accuracy of the approximations in subroutine DSS018, which are third-order correct in the spatial variable x, i.e., they are $O(\Delta x^3)$. This improved accuracy is achieved through the additional calculation required in DSS018 (four points are used in the calculation of $\partial u_1/\partial x$ of equation (3.11) rather than the two points of DSS012).

3. Run No. 3:

(3.1) ODEs no. 1 to 21 which are the method of lines approximation of equation (3.11), have pentadiagonal structure along the main diagonal with an outlying monodiagonal. For example, for ODE no. 10

```
10          56666                    4
```

The 56666 structure corresponds to U1(7), U1(8), U1(9), U1(10), and U1(11) reflecting the five-point biased upwind approximations in subroutine DSS020. That is, for ODE no. 10, three points are upwind, U1(7), U1(8), and U1(9), and one point is downwind, U(11).

(3.2) ODEs no. 22 to 42 again are the method of lines approximation of equation (3.12), which has only monodiagonals since there is no partial derivative with respect to x in equation (3.12).

(3.3) The numerical and analytical solutions at $x = 1$ are in agreement to four figures (NORUN=3)

T	U1(X,T)	U1(X,T) ANAL
0.00	0.0000	0.0000
1.00	0.0378	0.0378
2.00	0.1084	0.1084
3.00	0.1740	0.1740
4.00	0.2349	0.2349
5.00	0.2916	0.2916
6.00	0.3442	0.3442
7.00	0.3930	0.3930
8.00	0.4383	0.4383
9.00	0.4803	0.4803
10.00	0.5193	0.5193

This improved accuracy relative to runs no. 1 and 2 is due to the higher accuracy of the approximations in subroutine DSS020, which are fourth-order correct in the spatial variable x, i.e., they are $O(\Delta x^4)$. This improved accuracy is achieved through the additional calculation required in DSS020 (five points are used in the calculation of $\partial u_1/\partial x$ of equation (3.11) rather than the two points of DSS012 or the four points of DSS018).

(3.4) Again, the ODE Jacobian matrix at $t = 10$ is the same as at $t = 0$ because the ODEs in the method of lines approximation of equations (3.11) and (3.12) are linear (equations (3.11) and (3.12) are linear in u_1 and u_2).

In conclusion, we have observed for the first problem based on boundary condition (3.32) that using higher order approximations for the PDE spatial derivatives, e.g., the five-point biased upwind approximations of DSS020, produces solutions of greater accuracy with little additional computational effort and no increase in programming requirements (the calls to DSS012, DSS018, and DSS020 are identical).

All of the preceding discussion for runs no. 1, 2, and 3 is based on the results for boundary condition (3.32) which is consistent with a homogeneous initial condition, i.e., $u_1(0,0) = 0$ from boundary condition (3.32) is consistent with the initial condition $u_1(x,0) = 0$. Intuitively this seems like a relatively easy condition to accommodate since a discontinuity is not introduced at $x = 0$ and $t = 0$. We next consider boundary condition (3.34) which does introduce a discontinuity at $x = 0$ and $t = 0$ since from boundary condition (3.34), $u_1(0,0) = 1$ while again, the initial condition is $u_1(x,0) = 0$, that is, a finite unit jump at $x = 0$ and $t = 0$ occurs.

In considering the output of Programs 3.1a to f for runs no. 4, 5, and 6, produced by subroutines INIT2, DERV2, and PRINT2, we consider only the numerical and analytical solutions since the ODE Jacobian maps will be the same as for runs no. 1 to 3 (again, because PDEs (3.11) and (3.12) are linear with constant coefficients). The abbreviated output for runs no. 4 to 6 (NORUN=4 to 6) is given in Table 3.1b.

The following points can be noted about the output in Table 3.1b:

1. Run No. 4:

(4.1) The numerical and analytical solutions again agree to about 3+ figures using the two-point upwind approximations of DSS012. In other words, the boundary condition (3.34) which introduces a discontinuity at $x = 0$ and $t = 0$ does not substantially reduce the accuracy of the numerical solution relative to the smooth case of boundary condition (3.32).

(4.2) This agreement, however, is to some extent a reflection of the characteristics of this particular problem, PDEs (3.11) and (3.12). For example, if we consider the case $c_1 = c_2 = 0$ in equations (3.11) and (3.12), that is, there is no heat transfer between the fluid and solid in Figure 3.1, the solid temperature remains at the initial value since equation (3.12) reduces to

$$\frac{\partial u_2}{\partial t} = 0 \tag{3.54}$$

which indicates that u_2 does not change with time, t. Equation (3.11) reduces to

$$\frac{\partial u_1}{\partial t} + v\frac{\partial u_1}{\partial x} = 0 \tag{3.55}$$

which is the *advection equation*. Equation (3.55) is deceptively simple in appearance; it is actually one of the most difficult PDEs to integrate numerically. To understand this, we can consider the analytical solution to equation (3.55) for the initial and boundary conditions

$$u_1(x,0) = 0, \qquad u_1(0,t) = f(t) \qquad (3.56)(3.57)$$

Table 3.1 (b) Abbreviated Output of Programs 3.1a to f for Boundary Condition (3.34). *Continued next page.*

```
RUN NO. -    4  BATEMAN TEST PROBLEM - DSS012 - V(0,T) = EXP(-(K/U)*T)

INITIAL T -  0.000D+00

  FINAL T -  0.100D+02

  PRINT T -  0.100D+01

NUMBER OF DIFFERENTIAL EQUATIONS -  43

MAXIMUM INTEGRATION ERROR -  0.100D-05

         T        U1(X,T)   U1(X,T) ANAL
       0.00        0.0000         0.0000
       1.00        0.7320         0.7316
       2.00        0.6804         0.6802
       3.00        0.6323         0.6322
       4.00        0.5873         0.5874
       5.00        0.5454         0.5456
       6.00        0.5063         0.5066
       7.00        0.4699         0.4702
       8.00        0.4360         0.4364
       9.00        0.4044         0.4048
      10.00        0.3750         0.3754


RUN NO. -    5  BATEMAN TEST PROBLEM - DSS018 - V(0,T) = EXP(-(K/U)*T)

INITIAL T -  0.000D+00

  FINAL T -  0.100D+02

  PRINT T -  0.100D+01

NUMBER OF DIFFERENTIAL EQUATIONS -  43

MAXIMUM INTEGRATION ERROR -  0.100D-05

         T        U1(X,T)   U1(X,T) ANAL
       0.00        0.0000         0.0000
       1.00        0.7315         0.7316
       2.00        0.6802         0.6802
       3.00        0.6322         0.6322
       4.00        0.5874         0.5874
       5.00        0.5456         0.5456
       6.00        0.5066         0.5066
       7.00        0.4702         0.4702
       8.00        0.4364         0.4364
       9.00        0.4048         0.4048
      10.00        0.3754         0.3754
```

Table 3.1 (b) *Continued.*

```
RUN NO. -    6  BATEMAN TEST PROBLEM - DSS020 - V(0,T) = EXP(-(K/U)*T)

INITIAL T -  0.000D+00

  FINAL T -  0.100D+02

  PRINT T -  0.100D+01

NUMBER OF DIFFERENTIAL EQUATIONS -  43

MAXIMUM INTEGRATION ERROR -  0.100D-05

           T        U1(X,T)   U1(X,T) ANAL
        0.00         0.0000         0.0000
        1.00         0.7311         0.7316
        2.00         0.6802         0.6802
        3.00         0.6322         0.6322
        4.00         0.5874         0.5874
        5.00         0.5456         0.5456
        6.00         0.5066         0.5066
        7.00         0.4702         0.4702
        8.00         0.4364         0.4364
        9.00         0.4048         0.4048
       10.00         0.3754         0.3754
```

The analytical solution to equations (3.55), (3.56), and (3.57) is easily derived as [Schiesser (1991)]

$$u_1(x,t) = f(t - x/v) \tag{3.58}$$

that is, the boundary condition function, $f(t)$, propagates with velocity v. For example, if the boundary condition function is the Heaviside unit step function, $h(t)$, defined by equation (3.36), then this unit step propagates with velocity v, i.e.,

$$u_1(x,t) = h(t - x/v) \tag{3.59}$$

and the numerical method used to integrate equation (3.55) must somehow accommodate a propagating discontinuity. This is a difficult numerical problem, and generally all that can be achieved is an approximation to the exact solution, equation (3.59). For example, the two-point upwind approximations in DSS012 will introduce numerical diffusion into the solution [Schiesser (1991)], while the upwind approximations in subroutines DSS018 and DSS020 will introduce considerably less numerical diffusion, but also some numerical oscillation. These numerical distortions are essentially negligible for the case $c_1 = c_2 = 1$ in equations (3.11) and (3.12) because the heat transfer tends to smooth out the propagating discontinuity (and also, boundary condition (3.34) is somewhat smoother than a unit step). In any case, numerical approximations which work well in one case, such as in runs no. 4, 5, and 6, may not work well for even the same problem depending on the problem parameters such as c_1 and c_2, and the auxiliary conditions such as the boundary condition function, $f(t)$, in equation (3.57). This conclusion is especially true for hyperbolic PDEs which tend to propagate discontinuities; equation (3.55) with analytical solution equation (3.58) illustrates this point. On the other hand, parabolic PDEs, which are first order in time and second order in space, tend to damp out discontinuities; thus the problem of computing numerical solutions with a minimum of distortion is substantially reduced.

2. Run No. 5:

(5.1) The numerical and analytical solutions agree to four figures except at $t = 1$, where the two solutions are

```
1.00          0.7315          0.7316
```

which probably reflects the effect of the discontinuity at $x = 0$; that is, according to equation (3.58), the discontinuity for the advection equation (3.55) would arrive at $x = 1$ at $t = 1$ if $v = 1$ corresponding to $h(t - x/v) = h(1 - 1/1) = h(0)$ for which, according to equation (3.36), a unit jump occurs. Thus, the numerical solution of equations (3.11) and (3.12) is probably most difficult to calculate accurately at $x = 1$ when $t = 1$.

(5.2) Again, these results demonstrate the improved performance of the four-point biased upwind approximations in `DSS018` relative to the two-point upwind approximations in `DSS012`.

3. Run No. 6:

(6.1) The numerical and analytical solutions agree to four figures except at $t = 1$, where the two solutions are

```
1.00          0.7311          0.7316
```

which again probably reflects the effect of the discontinuity at $x = 0$. Note also that this difference between the numerical and analytical solutions is greater than for the four-point biased upwind approximations of `DSS018`. This reflects the greater numerical oscillation that can occur with the five-point approximations which are based on fourth-order polynomials relative to the four-point approximations which are based on third-order polynomials. In other words, higher order polynomials can have more maxima and minima, and can therefore oscillate more than lower order polynomials. However, the differences between the numerical and analytical solutions at $t = 1$ are small in both cases.

(6.2) Again, these results demonstrate the improved performance of the five-point biased upwind approximations in `DSS020` relative to the two-point upwind approximations in `DSS012`.

In summary, the four- and five-point biased upwind approximations in subroutines `DSS018` and `DSS020` work well for equations (3.11) and (3.12) with the discontinuous boundary condition (3.34). In general, these subroutines are recommended for hyperbolic problems, particularly in the case of propagating discontinuities or sharp fronts.

In considering the output of Programs 3.1a to f for runs no. 7, 8, and 9, produced by subroutines `INIT3`, `DERV3`, and `PRINT3`, we again consider only the numerical and analytical solutions since the ODE Jacobian maps will be the same as for runs no. 1 to 3 or 4 to 6. The abbreviated output for runs no. 7 to 9 is given in Table 3.1c.

Table 3.1 (c) Abbreviated Output of Programs 3.1a to f for Boundary Condition (3.36). *Continued next page.*

```
RUN NO. -   7  BATEMAN TEST PROBLEM - DSS012 - V(0,T) = 1

INITIAL T -  0.000D+00

  FINAL T -  0.100D+02

  PRINT T -  0.100D+01

NUMBER OF DIFFERENTIAL EQUATIONS -  43

MAXIMUM INTEGRATION ERROR -  0.100D-05


           T         U1(X,T)   U1(X,T) ANAL
        0.00          0.0000         0.0000
        1.00          0.7703         0.7694
        2.00          0.7894         0.7886
        3.00          0.8068         0.8062
        4.00          0.8228         0.8224
        5.00          0.8375         0.8372
        6.00          0.8510         0.8507
        7.00          0.8633         0.8632
        8.00          0.8747         0.8746
        9.00          0.8851         0.8851
       10.00          0.8946         0.8947


RUN NO. -   8  BATEMAN TEST PROBLEM - DSS018 - V(0,T) = 1

INITIAL T -  0.000D+00

  FINAL T -  0.100D+02

  PRINT T -  0.100D+01

NUMBER OF DIFFERENTIAL EQUATIONS -  43

MAXIMUM INTEGRATION ERROR -  0.100D-05


           T         U1(X,T)   U1(X,T) ANAL
        0.00          0.0000         0.0000
        1.00          0.7693         0.7694
        2.00          0.7886         0.7886
        3.00          0.8062         0.8062
        4.00          0.8224         0.8224
        5.00          0.8372         0.8372
        6.00          0.8508         0.8508
        7.00          0.8632         0.8632
        8.00          0.8746         0.8747
        9.00          0.8851         0.8851
       10.00          0.8947         0.8948
```

Table 3.1 (c) *Continued.*

```
RUN NO. -    9  BATEMAN TEST PROBLEM - DSS020 - V(0,T) = 1

INITIAL T -  0.000D+00

  FINAL T -  0.100D+02

  PRINT T -  0.100D+01

NUMBER OF DIFFERENTIAL EQUATIONS -  43

MAXIMUM INTEGRATION ERROR -  0.100D-05
```

T	U1(X,T)	U1(X,T) ANAL
0.00	0.0000	0.0000
1.00	0.7689	0.7695
2.00	0.7886	0.7887
3.00	0.8062	0.8063
4.00	0.8224	0.8224
5.00	0.8372	0.8372
6.00	0.8508	0.8508
7.00	0.8632	0.8633
8.00	0.8746	0.8747
9.00	0.8851	0.8852
10.00	0.8947	0.8948

The following points can be noted about the output in Table 3.1c:

1. Again, as in the case of runs no. 1 to 6, the two-point upwind approximations in subroutine DSS012 give the poorest agreement between the numerical and analytical solutions, as expected (because of their relatively low order).

2. The unit step function of equation (3.36) produces a discontinuity which has the greatest effect over the interval $0 \leq t \leq 2$ as discussed previously in paragraph (4.2). The four-point biased upwind formulas again give the best results because of the somewhat higher oscillation of the five-point biased upwind formulas. The solutions at $t = 0$, 1, and 2 for these two cases are indicated below.

Four-point biased upwind approximations (run no. 8):

0.00	0.0000	0.0000
1.00	0.7693	0.7694
2.00	0.7886	0.7886

Five-point biased upwind approximations (run no. 9):

0.00	0.0000	0.0000
1.00	0.7689	0.7695
2.00	0.7886	0.7887

These results are quite acceptable considering the unit step discontinuity of boundary condition (3.36) at $x = 0$.

As before, in considering the output of Programs 3.1a to f for runs no. 10, 11 and 12, produced by subroutines `INIT4`, `DERV4`, and `PRINT4`, we again consider only the numerical and analytical solutions since the ODE Jacobian maps will be the same as for runs no. 1 to 3, 4 to 6, and 7 to 9. The abbreviated output for runs no. 10 to 12 is given in Table 3.1d.

The following points can be noted about the output in Table 3.1d:

1. Again, the two-point upwind approximations in `DSS012` perform relatively poorly because of their low order.

2. The numerical solutions of runs no. 11 and 12 for the four- and five-point biased upwind approximations agree with the analytical solution to 3+ figures. Actually, this problem based on the ramp function, $r(t)$, of equation (3.39) is easier numerically than the previous problem based on the unit step, $h(t)$, of equation (3.36) since $r(t)$ is continuous (but is discontinuous in its first derivative).

Also, this last set of runs indicates an advantange of the numerical method, that is, the programming of the numerical solution does not increase in complexity even though the analytical solution may be considerably more complex; for example, the method of lines programming in subroutines `DERV1` to `DERV4` is essentially the same, but the analytical solution for `DERV4`, equation (3.52), is substantially more complicated than in the previous three cases of equations (3.46), (3.48), and (3.51). And, of course, as with ODEs, numerical methods offer a major advantage in the case of nonlinear PDEs for which an analytical solution is generally not available.

Finally, to complete this discussion of PDEs first order in time, t, and zeroth and first order in space, x, as illustrated through equations (3.11) and (3.12), we now briefly consider the differentiation formulas in subroutines `DSS012`, `DSS018`, and `DSS020`. Additional details are available in Schiesser (1991). For the differentiation formulas in `DSS012`, we can consider a two-point finite difference approximation

$$\frac{\partial u}{\partial x}(x,t) \sim \frac{u(x,t) - u(x-\Delta x,t)}{\Delta x} + O(\Delta x) \tag{3.60}$$

This approximation is used for flow from left to right in the positive x direction since it is upwind at x–Δx. Alternatively, we use the approximation

$$\frac{\partial u}{\partial x}(x,t) \sim \frac{u(x+\Delta x,t) - u(x,t)}{\Delta x} + O(\Delta x) \tag{3.61}$$

when the flow is right to left in the negative x direction. Therefore, we must include in the programming of these approximations a method for choosing one or the other depending on the direction of flow; this is the purpose of the last argument in subroutine `DSS012`, which is listed in Program 3.2a.

Note that if sixth argument, V, is positive (and real), the two-point upwind approximation for flow in the positive direction is used (via `DO` loop 1 according to equation (3.60)); conversely, if the sixth argument is negative, a two-point upwind approximation in the other direction is used (via `DO` loop 2) according to equation (3.61). Also, at the first grid point for positive flow, $i = 1$, a downwind approximation is used since an upwind approximation would require a fictitious value, `U`(0); similarly at the first grid point for negative flow, $i = N$, a downwind approximation is used since an upwind approximation would require a fictitious value, `U`$(N+1)$.

Table 3.1 (d) Abbreviated Output of Programs 3.1a to f for Boundary Condition (3.40). *Continued next page.*

```
RUN NO. -  10  BATEMAN TEST PROBLEM - DSS012 - UNIT RAMP

INITIAL T -  0.000D+00

  FINAL T -  0.100D+02

  PRINT T -  0.100D+01

NUMBER OF DIFFERENTIAL EQUATIONS -  45

MAXIMUM INTEGRATION ERROR -  0.100D-05


          T        U1(X,T)   U1(X,T) ANAL
       0.00         0.0000         0.0000
       1.00         0.3937         0.3880
       2.00         0.7800         0.7792
       3.00         0.7982         0.7975
       4.00         0.8149         0.8144
       5.00         0.8303         0.8299
       6.00         0.8443         0.8441
       7.00         0.8572         0.8571
       8.00         0.8691         0.8690
       9.00         0.8800         0.8800
      10.00         0.8899         0.8900


RUN NO. -  11  BATEMAN TEST PROBLEM - DSS018 - UNIT RAMP

INITIAL T -  0.000D+00

  FINAL T -  0.100D+02

  PRINT T -  0.100D+01

NUMBER OF DIFFERENTIAL EQUATIONS -  45

MAXIMUM INTEGRATION ERROR -  0.100D-05


          T        U1(X,T)   U1(X,T) ANAL
       0.00         0.0000         0.0000
       1.00         0.3880         0.3879
       2.00         0.7792         0.7792
       3.00         0.7975         0.7976
       4.00         0.8144         0.8144
       5.00         0.8299         0.8299
       6.00         0.8441         0.8441
       7.00         0.8571         0.8571
       8.00         0.8690         0.8690
       9.00         0.8800         0.8800
      10.00         0.8900         0.8900
```

Table 3.1 (d) *Continued.*

```
RUN NO. -   12  BATEMAN TEST PROBLEM - DSS020 - UNIT RAMP

INITIAL T -   0.000D+00

  FINAL T -   0.100D+02

  PRINT T -   0.100D+01

NUMBER OF DIFFERENTIAL EQUATIONS -   45

MAXIMUM INTEGRATION ERROR -   0.100D-05


           T          U1(X,T)    U1(X,T) ANAL
        0.00           0.0000          0.0000
        1.00           0.3880          0.3879
        2.00           0.7791          0.7791
        3.00           0.7975          0.7975
        4.00           0.8144          0.8144
        5.00           0.8299          0.8299
        6.00           0.8441          0.8440
        7.00           0.8571          0.8571
        8.00           0.8690          0.8690
        9.00           0.8800          0.8799
       10.00           0.8900          0.8900
```

The principal limitation of the two-point approximations in equations (3.60) and (3.61) is their first-order accuracy, that is $O(\Delta x)$, which follows from a Taylor series analysis. In fact, all of the differentiation formulas we have used can be derived from the Taylor series as we will now demonstrate. For example, if we are interested in a four-point biased upwind approximation for use in subroutine DSS018 at a point x_i, we use the dependent variable $u(x)$ at x_{i-2}, x_{i-1}, x_i, and x_{i+1} in the approximation (two points upwind of x_i, that is, x_{i-1} and x_{i-2}, and one point downwind, x_{i+1}). This suggests we write a Taylor series for grid points $i-2$, $i-1$, and $i+1$:

$$u(x_{i-2}) = u(x_i) + (du(x_i)/dx)(-2\Delta x) + (1/2!)(d^2u(x_i)/dx^2)(-2\Delta x)^2 + \cdots \quad (3.62)$$

$$u(x_{i-1}) = u(x_i) + (du(x_i)/dx)(-\Delta x) + (1/2!)(d^2u(x_i)/dx^2)(-\Delta x)^2 + \cdots \quad (3.63)$$

$$u(x_{i+1}) = u(x_i) + (du(x_i)/dx)(\Delta x) + (1/2!)(d^2u(x_i)/dx^2)(\Delta x)^2 + \cdots \quad (3.64)$$

We now take a linear combination of equations (3.62) to (3.64) so as to keep the derivative of interest, $du(x_i)/dx$, and drop as many of the higher derivative terms as possible (to give maximum accuracy in the final differentiation formula).

If we multiply equation (3.62) by a constant a, equation (3.63) by a constant b, and equation (3.64) by a constant c

$$\begin{aligned} au(x_{i-2}) &= au(x_i) + a(du(x_i)/dx)(-2\Delta x \\ &\quad + a(1/2!)(d^2u(x_i)/dx^2)(-2\Delta x)^2 + \cdots \\ bu(x_{i-1}) &= bu(x_i) + b(du(x_i)/dx)(-\Delta x \\ &\quad + b(1/2!)(d^2u(x_i)/dx^2)(-\Delta x)^2 + \cdots \\ cu(x_{i+1}) &= cu(x_i) + c(du(x_i)/dx)(\Delta x) \\ &\quad + c(1/2!)(d^2u(x_i)/dx^2)(\Delta x)^2 + \cdots \end{aligned} \tag{3.65}$$

then sum the resulting equations, we can retain the first derivative term, $du(x_i)/dx$ by imposing the condition

$$-2a - b + c = 1 \tag{3.66}$$

The 1 in the right-hand side of equation (3.66) ensures that when equations (3.65) are added with the numerical values of a, b, and c (still to be determined), the first derivative term will remain. Similarly, in order to eliminate the second derivative term, $d^2u(x_i)/dx^2$, we can impose the condition

$$4a + b + c = 0 \tag{3.67}$$

The 0 in the right-hand side of equation (3.67) ensures that when equations (3.65) are added with the numerical values of a, b, and c, the second derivative term will be eliminated.

In the same way, to eliminate the third derivative terms, we can impose the condition,

$$-8a - b + c = 0 \tag{3.68}$$

We now have three linear algebraic equations ((3.66) to (3.68)) for the three constants a to c. Simultaneous solution gives $a = 1/3!$, $b = -6/3!$, $c = 2/3!$. Then, if equations (3.65) are added with these numerical values for a to c, we obtain for the derivative of interest $du(x_i)/dx$

$$du(x_i)/dx = (1/(3!\Delta x))(u(x_{i-2}) - 6u(x_{i-1}) + 3u(x_i) + 2u(x_{i+1})) + O(\Delta x^3) \tag{3.69}$$

Note that the approximation of equation (3.69) is third-order correct. Also, the weighting coefficients, 1, −6, 3, and 2 sum to zero as required for equation (3.69) to correctly differentiate a constant to zero (this check, which can be made by inspection, is quite useful when deriving a new differentiation formula). Also, of course, the right-hand side of equation (3.69) is a four-point biased upwind approximation as we originally intended.

Equation (3.69) can be applied at the grid points for the method of lines approximation of equations (3.11) and (3.12) for $i = 3, 4, \ldots, 20$ in a 21-point grid. However, for $i = 1$, 2, and 21, additional approximations are required to be used in place of equation (3.69) in order to avoid fictitious points. For example, at $i = 1$, equation (3.69) requires $u(x_{-1})$, $u(x_0)$, $u(x_1)$, and $u(x_2)$, but x_{-1} and x_0 are beyond the left end of the grid. Therefore, the preceding Taylor series analysis is repeated using as starting points $u(x_1)$, $u(x_2)$, $u(x_3)$, and $u(x_4)$. The final result is

$$du(x_1)/dx = (1/(3!\Delta x))(-11u(x_1) + 18u(x_2) - 9u(x_3) + 2u(x_4)) + O(\Delta x^3) \tag{3.70}$$

Similarly, for $i = 2$, we have

$$du(x_2)/dx = (1/(3!\Delta x))(-2u(x_1) - 3u(x_2) + 6u(x_3) - u(x_4)) + O(\Delta x^3) \tag{3.71}$$

```
      SUBROUTINE DSS012(XL,XU,N,U,UX,V)
C...
C...  SUBROUTINE DSS012 IS AN APPLICATION OF FIRST ORDER DIRECTIONAL
C...  DIFFERENCING IN THE NUMERICAL METHOD OF LINES.  IT IS INTENDED
C...  SPECIFICALLY FOR THE ANALYSIS OF CONVECTIVE SYSTEMS MODELED BY
C...  FIRST ORDER HYPERBOLIC PARTIAL DIFFERENTIAL EQUATIONS WITH THE
C...  SIMPLEST FORM
C...
C...                          U  + V*U  = 0                        (1)
C...                           T      X
C...
C...  THE FIRST FIVE PARAMETERS, XL, XU, N, U AND UX, ARE THE SAME
C...  AS FOR SUBROUTINES DSS002 TO DSS010 AS DEFINED IN THOSE ROUTINES.
C...  THE SIXTH PARAMETER, V, MUST BE PROVIDED TO DSS012 SO THAT THE
C...  DIRECTION OF FLOW IN EQUATION (1) CAN BE USED TO SELECT THE
C...  APPROPRIATE FINITE DIFFERENCE APPROXIMATION FOR THE FIRST ORDER
C...  SPATIAL DERIVATIVE IN EQUATION (1), U .  THE CONVENTION FOR THE
C...  SIGN OF V IS                          X
C...
C...     FLOW LEFT TO RIGHT                  V GT 0
C...     (I.E., IN THE DIRECTION             (I.E., THE SIXTH ARGUMENT IS
C...     OF INCREASING X)                    POSITIVE IN CALLING DSS012)
C...
C...     FLOW RIGHT TO LEFT                  V LT 0
C...     (I.E., IN THE DIRECTION             (I.E., THE SIXTH ARGUMENT IS
C...     OF DECREASING X)                    NEGATIVE IN CALLING DSS012)
C...
      REAL U(N),UX(N)
C...
C...  COMPUTE THE SPATIAL INCREMENT, THEN SELECT THE FINITE DIFFERENCE
C...  APPROXIMATION DEPENDING ON THE SIGN OF V IN EQUATION (1).
      DX=(XU-XL)/FLOAT(N-1)
      IF(V.LT.0.)GO TO 10
C...
C...     (1)  FINITE DIFFERENCE APPROXIMATION FOR POSITIVE V
              UX(1)=(U(2)-U(1))/DX
              DO 1 I=2,N
              UX(I)=(U(I)-U(I-1))/DX
1             CONTINUE
              RETURN
C...
C...     (2)  FINITE DIFFERENCE APPROXIMATION FOR NEGATIVE V
10            NM1=N-1
              DO 2 I=1,NM1
              UX(I)=(U(I+1)-U(I))/DX
2             CONTINUE
              UX(N)=(U(N)-U(N-1))/DX
      RETURN
      END
```

Program 3.2a Subroutine DSS012 for the Two-Point Upwind Approximation of First-Order Derivatives.

and for $i = N$ $(= 21)$

$$du(x_N)/dx = (1/(3!\Delta x))(-2u(x_{N-3})+9u(x_{N-2})-18u(x_{N-1})+11u(x_N))+O(\Delta x^3) \quad (3.72)$$

Note that the weighting coefficients of equations (3.70) to (3.72) also sum to zero so that these formulas will differentiate a constant to zero. Equations (3.69) to (3.72) can now be programmed in subroutine DSS018, which is listed in Program 3.2b.

We can note the following points about subroutine DSS018:

1. The arguments are the same as those of subroutine DSS012. In particular, the sixth argument is used to select equations (3.69) to (3.72) for a positive velocity, V, at the beginning of DSS018, or a similar set of approximations for a negative velocity at the end of DSS018.

2. DO loop 1 implements equation (3.69) for $i = 3, 4, \ldots, N-1$. Similarly, DO loop 2 implements equation (3.71) for $i = 2, 3, \ldots, N-2$ (instead of just $i = 2$). Also, note the weighting coefficients in equation (3.71) are reversed in order and opposite in sign to those of equation (3.69). This *antisymmetric* property applies to all of the differentiation formulas for the approximation of odd order derivatives.

3. Equations (3.70) through (3.72) are used for i = 1, 2, and N, respectively, for positive V (at the beginning of DSS018). Similarly, equations (3.70), (3.69), and (3.72) are used for $i = 1$, $N-1$, and N, respectively, for negative V.

This completes all of the details for the four-point biased upwind approximations in subroutine DSS018. A similar analysis can be used to develop the five-point biased upwind approximations in subroutine DSS020. For example, if we start with the points x_{i-3}, x_{i-2}, x_{i-1}, x_i, and x_{i+1}, the following approximation for $du(x_i)/dx$ can be derived

$$du(x_i)/dx = (1/(4!\Delta x))(-2u(x_{i-3}) + 12u(x_{i-2}) - 36u(x_{i-1}) + 20u(x_i) + 6u(x_{i+1})) + O(\Delta x^4) \quad (3.73)$$

Similar formulas for $i = 1, 2, 3$, and N can also be derived in the same way as for the preceding four-point formulas. Then, the four formulas for positive V and four analogous formulas for negative V can be programmed in subroutine DSS020, which is listed in Program 3.2c.

Note that the weighting coefficients in DSS020 for UX(I) and positive V are divided by 12 rather than $4! = 24$; thus they are half those in equation (3.73). The programming in DSS020 is directly analogous to that in DSS018, and therefore the details in DSS020 will not be considered.

In summary, the Taylor series is the starting point for classical finite difference approximations of derivatives. We have used the Taylor series to derive approximations for first-order derivatives, for example, $\partial u_1/\partial x$, in equation (3.11). Similar procedures can be used to obtain finite difference approximations of second and higher order spatial derivatives, as we will see in the subsequent discussion. We now consider PDEs first order in time and second order in space, which we generally classify as parabolic.

3.2 *PDEs with Second-Order Spatial Derivatives*

Parabolic PDEs occur in many applications in science and engineering. Also, they are generally easier to integrate numerically than first-order hyperbolic PDEs such as equation (3.11) or equation (3.55) since they do not propagate discontinuities as

```
      SUBROUTINE DSS018(XL,XU,N,U,UX,V)
C...
C...  SUBROUTINE DSS018 IS AN APPLICATION OF THIRD-ORDER DIRECTIONAL
C...  DIFFERENCING IN THE NUMERICAL METHOD OF LINES.  IT IS INTENDED
C...  SPECIFICALLY FOR THE ANALYSIS OF CONVECTIVE SYSTEMS MODELED BY
C...  FIRST-ORDER HYPERBOLIC PARTIAL DIFFERENTIAL EQUATIONS AS DIS-
C...  CUSSED IN SUBROUTINE DSS012.  THE COEFFICIENTS OF THE FINITE
C...  DIFFERENCE APPROXIMATIONS USED HEREIN ARE TAKEN FROM BICKLEY, W.
C...  G., FORMULAE FOR NUMERICAL DIFFERENTIATION, THE MATHEMATICAL
C...  GAZETTE, PP. 19-27, 1941, N = 3, M = 1, P = 0, 1, 2, 3.  THE
C...  IMPLEMENTATION IS THE **FOUR-POINT BIASED UPWIND FORMULA** OF
C...  M. B. CARVER AND H. W. HINDS, THE METHOD OF LINES AND THE
C...  ADVECTION EQUATION, SIMULATION, VOL. 31, NO. 2, PP. 59-69,
C...  AUGUST, 1978
C...
      DIMENSION U(N),UX(N)
C...
C...  COMPUTE THE COMMON FACTOR FOR EACH FINITE DIFFERENCE APPROXIMATION
C...  CONTAINING THE SPATIAL INCREMENT, THEN SELECT THE FINITE DIFFER-
C...  ENCE APPROXIMATION DEPENDING ON THE SIGN OF V (SIXTH ARGUMENT).
      DX=(XU-XL)/FLOAT(N-1)
      R3FDX=1./(6.*DX)
      IF(V.LT.0.)GO TO 10
C...
C...     (1)  FINITE DIFFERENCE APPROXIMATION FOR POSITIVE V
      UX(  1)=R3FDX*
     1( -11.*U(  1) +18.*U(  2)  -9.*U(  3)  +2.*U(  4))
      UX(  2)=R3FDX*
     1(  -2.*U(  1)  -3.*U(  2)  +6.*U(  3)  -1.*U(  4))
      NM1=N-1
      DO 1 I=3,NM1
      UX(  I)=R3FDX*
     1(  +1.*U(I-2)  -6.*U(I-1)  +3.*U(  I)  +2.*U(I+1))
1     CONTINUE
      UX(  N)=R3FDX*
     1(  -2.*U(N-3)  +9.*U(N-2) -18.*U(N-1) +11.*U(  N))
      RETURN
C...
C...     (2)  FINITE DIFFERENCE APPROXIMATION FOR NEGATIVE V
10    UX(  1)=R3FDX*
     1( -11.*U(  1) +18.*U(  2)  -9.*U(  3)  +2.*U(  4))
      NM2=N-2
      DO 2 I=2,NM2
      UX(  I)=R3FDX*
     1(  -2.*U(I-1)  -3.*U(  I)  +6.*U(I+1)  -1.*U(I+2))
2     CONTINUE
      UX(N-1)=R3FDX*
     1(  +1.*U(N-3)  -6.*U(N-2)  +3.*U(N-1)  +2.*U(  N))
      UX(  N)=R3FDX*
     1(  -2.*U(N-3)  +9.*U(N-2) -18.*U(N-1) +11.*U(  N))
      RETURN
      END
```

Program 3.2b Subroutine DSS018 for the Four-Point Biased Upwind Approximation of First-Order Derivatives.

```
      SUBROUTINE DSS020(XL,XU,N,U,UX,V)
C...
C...  SUBROUTINE DSS020 IS AN APPLICATION OF FOURTH ORDER DIRECTIONAL
C...  DIFFERENCING IN THE NUMERICAL METHOD OF LINES.  IT IS INTENDED
C...  SPECIFICALLY FOR THE ANALYSIS OF CONVECTIVE SYSTEMS MODELED BY
C...  FIRST ORDER HYPERBOLIC PARTIAL DIFFERENTIAL EQUATIONS AS DIS-
C...  CUSSED IN SUBROUTINE DSS012.  THE USE OF FIVE POINT BIASED
C...  UPWIND APPROXIMATIONS WAS PROPOSED AND DESCRIBED IN THE PAPER
C...  M. B. CARVER AND H. W. HINDS, THE METHOD OF LINES AND THE
C...  ADVECTION EQUATION, SIMULATION, VOL. 31, NO. 2, PP. 59-69,
C...  AUGUST, 1978
C...
      REAL U(N),UX(N)
C...
C...  COMPUTE THE COMMON FACTOR FOR EACH FINITE DIFFERENCE APPROXIMATION
C...  CONTAINING THE SPATIAL INCREMENT, THEN SELECT THE FINITE DIFFER-
C...  ENCE APPROXIMATION DEPENDING ON THE SIGN OF V (SIXTH ARGUMENT).
      DX=(XU-XL)/FLOAT(N-1)
      R4FDX=1./(12.*DX)
      IF(V.LT.0.)GO TO 10
C...
C...     (1)  FINITE DIFFERENCE APPROXIMATION FOR POSITIVE V
      UX(  1)=R4FDX*
     1( -25.*U(  1) +48.*U(  2) -36.*U(  3) +16.*U(  4)  -3.*U(  5))
      UX(  2)=R4FDX*
     1(  -3.*U(  1) -10.*U(  2) +18.*U(  3)  -6.*U(  4)  +1.*U(  5))
      UX(  3)=R4FDX*
     1(  +1.*U(  1)  -8.*U(  2)  +0.*U(  3)  +8.*U(  4)  -1.*U(  5))
      NM1=N-1
      DO 1 I=4,NM1
      UX(  I)=R4FDX*
     1(  -1.*U(I-3)  +6.*U(I-2) -18.*U(I-1) +10.*U(  I)  +3.*U(I+1))
1     CONTINUE
      UX(  N)=R4FDX*
     1(   3.*U(N-4) -16.*U(N-3) +36.*U(N-2) -48.*U(N-1) +25.*U(  N))
      RETURN
C...
C...     (2)  FINITE DIFFERENCE APPROXIMATION FOR NEGATIVE V
10    UX(  1)=R4FDX*
     1( -25.*U(  1) +48.*U(  2) -36.*U(  3) +16.*U(  4)  -3.*U(  5))
      NM3=N-3
      DO 2 I=2,NM3
      UX(  I)=R4FDX*
     1(  -3.*U(I-1) -10.*U(  I) +18.*U(I+1)  -6.*U(I+2)  +1.*U(I+3))
2     CONTINUE
      UX(N-2)=R4FDX*
     1(  +1.*U(N-4)  -8.*U(N-3)  +0.*U(N-2)  +8.*U(N-1)  -1.*U(  N))
      UX(N-1)=R4FDX*
     1(  -1.*U(N-4)  +6.*U(N-3) -18.*U(N-2) +10.*U(N-1)  +3.*U(  N))
      UX(  N)=R4FDX*
     1(   3.*U(N-4) -16.*U(N-3) +36.*U(N-2) -48.*U(N-1) +25.*U(  N))
      RETURN
      END
```

Program 3.2c Subroutine DSS020 for the Five-Point Biased Upwind Approximation of First-Order Derivatives.

do the hyperbolic equations; rather they damp out discontinuities which might arise, for example, from inconsistent initial and boundary conditions. To illustrate how parabolic PDEs might arise in applications, we consider as a starting point *Maxwell's equations of electromagnetic field theory*:

$$\nabla \times H = J + J_d = J + \partial D/\partial t \tag{3.74}$$

$$\nabla \times E = -\partial B/\partial t \tag{3.75}$$

$$\nabla \bullet D = \rho \tag{3.76}$$

$$\nabla \bullet B = 0 \tag{3.77}$$

where E = electric field intensity ($volts/m$); H = magnetic field intensity ($amps/m$); D = electric flux density ($coulombs/m^2$); B = magnetic flux density ($webers/m^2$); J = electric current density ($amps/m^2$); and J_d = displacement current density ($amps/m^2$).

Also, we generally require constitutive equations, which for simple matter, are taken as:

$$D = \epsilon E, \qquad B = \mu H, \qquad J = \sigma E \tag{3.78)–(3.80}$$

where ϵ = capacitivity or permittivity ($farads/m$); μ = inductivity or permeability ($henrys/m$); and σ = conductivity ($mohs/m$). Finally, the *equation of continuity* for charge is

$$\partial\rho/\partial t + \nabla \bullet J = 0 \tag{3.81}$$

where ρ = charge density ($columbs/cm^3$) and J = current density ($amps/cm^2$). We now consider the application of the preceding electromagnetic field equations to the calculation of the transient current density, J, in a conductor, as illustrated in Figure 3.2.

The conductor is cylindrical; thus the current density is a function of radial position, r, and time, t, i.e., $J(r, t)$. To analyze this problem system, we start by combining equations (3.75) and (3.80)

$$\nabla \times E = \nabla \times (J/\sigma) = -\partial B/\partial t \tag{3.82}$$

From the identity

$$\nabla \times (\nabla \times A) = \nabla(\nabla \bullet A) - \nabla^2 A \tag{3.83}$$

we can write equation (3.82) as

$$\begin{aligned}
\nabla \times \nabla \times (J/\sigma) &= \nabla(\nabla \bullet (J/\sigma)) - \nabla^2(J/\sigma) \\
&= -\nabla \times (\partial B/\partial t) \\
&= -\partial(\nabla \times B)/\partial t = -\partial(\nabla \times (\mu H))/\partial t \\
&= -\mu\partial J/\partial t
\end{aligned} \tag{3.84}$$

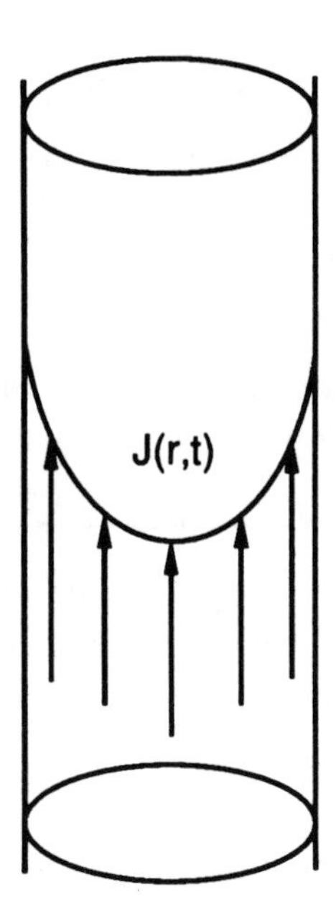

Figure 3.2 Transient Current Density in a Cylindrical Conductor.

where we have used equations (3.79) and (3.74) with $J_d = 0$. Also, we have

$$\nabla(\nabla \bullet (J/\sigma)) - \nabla^2(J/\sigma) = (1/\sigma)\nabla(\nabla \bullet J) - (1/\sigma)\nabla^2 J = -(1/\sigma)\nabla^2 J \qquad (3.85)$$

where we have used equation (3.81) with $\partial\rho/\partial t = 0$, (constant charge density) so $\nabla \bullet J = 0$. Therefore, equation (3.84) becomes

$$(1/\sigma)\nabla^2 J = \mu \partial J/\partial t$$

or

$$\partial J/\partial t = 1/(\sigma\mu)\nabla^2 J \qquad (3.86)$$

Equation (3.86) is generally called *Fourier's second law* or the *diffusion equation.* Note that equation (3.86) is first order in time, t, and second order in space; thus it is a parabolic PDE and it therefore requires one initial condition and two boundary conditions.

Before we state these auxiliary conditions, we first write the right-hand side term in equation (3.86) using the Laplacian operator, $\nabla^2 J$, in cylindrical coordinates

$$\nabla^2 J = \frac{1}{r}\frac{\partial}{\partial r}\left(r\frac{\partial J}{\partial r}\right) + \frac{1}{r^2}\frac{\partial^2 J}{\partial \theta^2} + \frac{\partial^2 J}{\partial z^2}$$

For the conductor of Figure 3.2, the current density, J, will not vary axially or angularly; thus

$$\nabla^2 J = \frac{1}{r}\frac{\partial}{\partial r}\left(r\frac{\partial J}{\partial r}\right) = \frac{\partial^2 J}{\partial r^2} + \frac{1}{r}\frac{\partial J}{\partial r} \qquad (3.87)$$

Substitution of equation (3.87) in equation (3.86) gives the PDE to be integrated

$$\partial J/\partial t = 1/(\sigma\mu)\left\{\frac{\partial^2 J}{\partial r^2} + \frac{1}{r}\frac{\partial J}{\partial r}\right\} \qquad (3.88)$$

The initial condition for equation (3.88) is

$$J(r, 0) = 0 \tag{3.89}$$

One boundary condition for equation (3.88) reflects symmetry in the $J(r, t)$ radial profile at $r = 0$

$$\frac{\partial J(0, t)}{\partial r} = 0 \tag{3.90}$$

The second boundary condition for equation (3.88) requires some additional analysis.

Equation (3.86) can be considered as either a vector or scalar equation. For the present problem it is a scalar equation since $J_r = J_\theta = 0$ (only J_z is nonzero). That is, if

$$J = J_r\vec{r} + J_\theta\vec{\theta} + J_z\vec{z}$$

then $J_r = J_\theta = 0$. Thus, $(\nabla \times J)|_r = (1/r)\partial J_z/\partial\theta = 0$ (from angular symmetry), $(\nabla \times J)|_\theta = -\partial J_z/\partial r$. Also, equation (3.82) must apply at the surface $r = r_1$

$$\nabla \times E = \nabla \times (J/\sigma) = -\partial(\mu H)/\partial t$$

or (θ component equation only)

$$-\partial J_z/\partial r|_{r=r_1} = -(\mu\sigma)\partial H/\partial t|_{r=r_1} \tag{3.91}$$

which is one equation relating J and H.

Also, we have as a second equation relating J and H the integral form of equation (3.74)

$$\int_s (\nabla \times H) \bullet ds = \int_l H \bullet dl = \int_s (J + \partial D/\partial t) \bullet ds$$

where we have applied *Stokes' theorem* to convert a surface integral to a line integral. H has only the component H_θ (which is constant around the circumference of the lead and depends only on t). With

$$\int_l H \bullet dl = H_\theta \int_l dl = 2\pi r_1 H_\theta, \qquad \int_s J \bullet ds = I(t) \quad \text{and} \quad \partial D/\partial t = 0$$

we have

$$2\pi r_1 H_\theta = I(t)$$

or

$$2\pi r_1 dH_\theta/dt = dI/dt \tag{3.92}$$

(note that $\partial H/\partial t = \mathrm{d}H_\theta/dt$). Combining equations (3.91) and (3.92)

$$-\partial J_z/\partial r|_{r=r_1} = -(\mu\sigma)/(2\pi r_1)dI/dt \tag{3.93}$$

which is the required second boundary condition. In the subsequent programming, we take the current, $I(t)$, to be

$$I(t) = I_{\max}\sin(\omega_I t) \tag{3.94}$$

Also, the current density, $J(r, t)$ computed as the solution to equations (3.88), (3.89), (3.90) and (3.93) must satisfy the condition

$$I(t) = \int_0^{r_1}\int_0^{2\pi} rJ(r, t)\, dr\, d\theta = 2\pi\int_0^{r_1} rJ(r, t)\, dr \tag{3.95}$$

$I(t)$ from equation (3.95) should agree with $I(t)$ from equation (3.94). In the following program, 3.3a, the integral on the right-hand side of equation (3.95) is computed by *Simpson's rule* from $J(r,t)$ computed by the numerical method of lines solution of equations (3.88), (3.89), (3.90) and (3.93).

Subroutines INITAL, DERV, PRINT, and PLOT are listed in Program 3.3a for a 51-point grid in the radial coordinate, r.

The following points can be noted about Program 3.3a:

1. Homogeneous initial condition (3.89) is implemented in DO loop 2 in subroutine INITAL over a 51-point grid in r.

2. The derivative $\partial J/\partial r$ in equation (3.88) is computed by a call to subroutine DSS004 which has five-point centered approximations to be discussed subsequently:

```
C...
C...  DERIVATIVE J
C...             R
      CALL DSS004(0.0D0,R1,NR,C,CR)
```

The current, $J(r,t)$, in array C is differentiated to $\partial J/\partial r$ in array CR over the interval $0 \le r \le r_1$ defined on a spatial grid of NR (= 51) points. The basic difference between the finite difference approximations in DSS004 and DSS020 is that the former are centered while the latter are noncentered (biased upwind). Centered approximations are typically used in parabolic PDEs such as equation (3.88), for which there is no preferred direction, as distinguished from hyperbolic PDEs which have a preferred direction, typically due to convection (e.g., equation (3.11)). In other words, DSS004 does not require the sixth argument of DSS020 to specify a direction in r.

3. Boundary condition (3.90) is used to define the derivative $\partial J(0,t)/\partial r$ (=CR(1)=0):

```
C...
C...  BOUNDARY CONDITION AT R = 0
      NL=2
      CR(1)=0.0D0
```

NL=2 designates a Neumann boundary condition, that is, a boundary condition which defines the derivative of the dependent variable like $\partial J(0,t)/\partial r$.

4. Boundary condition (3.93) is used to define the derivative $\partial J(r_1,t)/\partial r$ (=CR(NR)), which is again Neumann (so NU=2):

```
C...
C...  BOUNDARY CONDITION AT R = R1
      NU=2
      CR(NR)=(MU*SIGMA)/(2.0D0*PI*R1)*CMAX*WI*DCOS(WI*T)
```

Note that we have used the derivative of the current, $dI(t)/dt = I_{\max}\omega_I \cos(\omega_I t)$, from equation (3.94).

5. The derivative $\partial^2 J/\partial r^2$ in equation (3.88) is computed by a call to subroutine

```
      SUBROUTINE INITAL
      IMPLICIT DOUBLE PRECISION (A-H,O-Z)
      PARAMETER(NR=51)
      COMMON/T/          T,      NSTOP,      NORUN
     1      /Y/     C(NR)
     2      /F/    CT(NR)
     3      /S/    CR(NR),   CRR(NR)
     4      /R/         R1,      R(NR),        DR,          PI,       CMAX,
     5                  MU,      SIGMA,        WI
     ,6      /I/         IP
      DOUBLE PRECISION MU
C...
C...  RADIAL GRID
C...
C...     CONDUCTOR RADIUS
         R1=1.0D0
C...
C...     GRID INCREMENT
         DR=R1/DFLOAT(NR-1)
C...
C...     GRID POINTS
         DO 1 I=1,NR
            R(I)=DFLOAT(I-1)*DR
1        CONTINUE
C...
C...  PI
         PI=4.0D0*DATAN(1.0D0)
C...
C...  MAXIMUM CONDUCTOR CURRENT
         CMAX=1.0D+03
C...
C...  PERMEABILITY
         MU=1.0D0
C...
C...  CONDUCTIVITY
         SIGMA=1.0D+01
C...
C...  CURRENT FREQUENCY
         WI=1.0D0
C...
C...  INITIAL CONDITION
         DO 2 I=1,NR
            C(I)=0.0D0
2        CONTINUE
C...
C...  INITIALIZE COUNTER FOR PRINTED AND PLOTTED SOLUTION
      IP=0
      RETURN
      END

      SUBROUTINE DERV
      IMPLICIT DOUBLE PRECISION (A-H,O-Z)
```

Program 3.3a Subroutines INITAL, DERV, PRINT, and PLOT for the Method of Lines Integration of Equations (3.88), (3.89), (3.90), and (3.93). *Continued next pages.*

```
      PARAMETER(NR=51)
      COMMON/T/          T,      NSTOP,      NORUN
     1      /Y/     C(NR)
     2      /F/    CT(NR)
     3      /S/    CR(NR),   CRR(NR)
     4      /R/         R1,      R(NR),        DR,          PI,       CMAX,
     5                  MU,      SIGMA,        WI
     6      /I/         IP
      DOUBLE PRECISION MU
C...
C...  DERIVATIVE J
C...               R
      CALL DSS004(0.0D0,R1,NR,C,CR)
C...
C...  BOUNDARY CONDITION AT R = 0
      NL=2
      CR(1)=0.0D0
C...
C...  BOUNDARY CONDITION AT R = R1
      NU=2
      CR(NR)=(MU*SIGMA)/(2.0D0*PI*R1)*CMAX*WI*DCOS(WI*T)
C...
C...  DERIVATIVE J
C...               RR
      CALL DSS044(0.0D0,R1,NR,C,CR,CRR,NL,NU)
C...
C...  PDE
      DO 1 I=1,NR
         IF(I.EQ.1)THEN
            CT(I)=(1.0D0/(MU*SIGMA))*2.0D0*CRR(I)
         ELSE
            CT(I)=(1.0D0/(MU*SIGMA))*(CRR(I)+1.0D0/R(I)*CR(I))
         END IF
1     CONTINUE
      RETURN
      END

      SUBROUTINE PRINT(NI,NO)
      IMPLICIT DOUBLE PRECISION (A-H,O-Z)
      PARAMETER(NR=51)
      COMMON/T/          T,      NSTOP,      NORUN
     1      /Y/     C(NR)
     2      /F/    CT(NR)
     3      /S/    CR(NR),   CRR(NR)
     4      /R/         R1,      R(NR),        DR,          PI,       CMAX,
     5                  MU,      SIGMA,        WI
     6      /I/         IP
      DOUBLE PRECISION MU
C...
C...  PRINT A HEADING FOR THE SOLUTION
```

Program 3.3a *Continued.*

```
      IP=IP+1
      IF(IP.EQ.1)THEN
         WRITE(NO,1)
1        FORMAT(7X,'T'
     1   ,1X,'J(0.0*r1,t)',1X,'J(0.2*r1,t)',1X,'J(0.4*r1,t)',
     2   /,9X,'J(0.6*r1,t)',1X,'J(0.8*r1,t)',1X,'J(1.0*r1,t)',
     3   /,9X,'       itot',1X,'       i(t)',1X,'       diff',
     4   /)
         END IF
C...
C...  INTEGRAL OF THE CURRENT DENSITY BY SIMPSON'S RULE
      CTOT=0.
      DO 2 I=2,NR-1,2
         CTOT=CTOT+(R(I-1)*C(I-1)+4.0D0*R(I)*C(I)+R(I+1)*C(I+1))
2     CONTINUE
      CTOT=2.0D0*PI*DR/3.0D0*CTOT
C...
C...  CONDUCTOR CURRENT AND DIFFERENCE OF CURRENTS
      CC=CMAX*DSIN(WI*T)
      DIFF=CTOT-CC
C...
C...  PRINT THE NUMERICAL SOLUTION
      NRI=10
      WRITE(NO,3)T,(C(I),I=1,NR,NRI),CTOT,CC,DIFF
3     FORMAT(F8.3,3F12.1,/,8X,3F12.1,/,8X,3F12.1,/)
C...
C...  STORE AND PLOT THE SOLUTION
      CALL PLOT
      RETURN
      END

      SUBROUTINE PLOT
      IMPLICIT DOUBLE PRECISION (A-H,O-Z)
      PARAMETER(NR=51)
      COMMON/T/          T,     NSTOP,     NORUN
     1      /Y/     C(NR)
     2      /F/    CT(NR)
     3      /S/    CR(NR),   CRR(NR)
     4      /R/         R1,     R(NR),        DR,          PI,       CMAX,
     5                  MU,     SIGMA,        WI
     6      /I/         IP
      DOUBLE PRECISION MU
C...
C...  INITIALIZE PLOT NUMBER
      DATA NCASE/0/
C...
C...  OPEN FILE FOR TOP DRAWER PLOTTING
      IF(IP.EQ.1)THEN
         OPEN(4,FILE='T.TOP',STATUS='NEW')
C...
```

Program 3.3a *Continued.*

```
C...      WRITE TOP DRAW FILE FOR PLOTTING CURRENT DENSITY
          WRITE(4,17)
17        FORMAT(' SET LIMITS X FROM 0 TO 1 Y FROM -600 TO 600',/,
     1          ' SET FONT DUPLEX')
C...
C...      SET THE CASE NUMBER FOR THE PLOT LABEL
          NCASE=NCASE+1
       END IF
C...
C...   WRITE THE SOLUTION IN FILE T.TOP FOR PLOTTING
       WRITE(4,14)(R(I),C(I),I=1,NR)
14     FORMAT(2F10.3)
       WRITE(4,16)
16     FORMAT(' JOIN 1')
C...
C...   AT THE END OF THE SOLUTION, LABEL THE PLOT
       IF(IP.EQ.5)THEN
          WRITE(4,18)NCASE
18        FORMAT(
     +    ' Title 4.5 9.5 "Figure ',I2,
     +    ': Current Density - J(r,t)"'
     +    ,/,' TITLE LEFT "J(r,t) (amps/cm**2)"'
     +    ,/,' TITLE BOTTOM "r (cm)"'
     +    ,/,'TITLE 3.5 0.40 '
     +       '"(mu)(sigma) = 10.0 sec/cm**2, wi = 1.0 1/sec"')
C...
C...      NEW PLOT
          WRITE(4,19)
19        FORMAT(' NEW FRAME')
       END IF
       RETURN

       END
```

Program 3.3a *Continued.*

DSS044 which has five-point centered approximations to be discussed subsequently:

```
C...
C...   DERIVATIVE J
C...               RR
       CALL DSS044(0.0D0,R1,NR,C,CR,CRR,NL,NU)
```

Subroutine DSS044 computes $\partial^2 J/\partial r^2$ in array CRR from J in array C, using the two end derivatives $\partial J(0,t)/\partial r$ and $\partial J(r_1,t)/\partial r$ in CR(1) and CR(NR) set previously.

6. The first derivative, $\partial J/\partial r$, and second derivative, $\partial^2 J/\partial r^2$, in equation (3.88) are now defined numerically in arrays CR and CRR, respectively, and we can therefore program equation (3.88). However, there is one detail we should consider first. At $r = 0$, the term $(1/r)(\partial J/\partial r)$ in equation (3.88) has the indeterminate form $0/0$ (due to boundary condition (3.90)). Thus, we must apply l'Hospital's rule

(differentiating the numerator, $\partial J/\partial r$, and denominator, r, with respect to r):

$$\lim_{r\to 0}\frac{1}{r}\frac{\partial J}{\partial r} = \lim_{r\to 0}\frac{1}{1}\frac{\partial^2 J}{\partial r^2} = \frac{\partial^2 J}{\partial r^2} \tag{3.96}$$

Equation (3.88) at $r = 0$ then becomes

$$\partial J/\partial t = 2/(\sigma\mu)\left\{\frac{\partial^2 J}{\partial r^2}\right\} \tag{3.97}$$

Thus, when we step along the grid in r, we use equation (3.88) except at $r = 0$, when we switch to equation (3.97). This special case for $r = 0$ (I=1) is programmed in DO loop 1 of subroutine DERV:

```
C...
C...  PDE
      DO 1 I=1,NR
         IF(I.EQ.1)THEN
            CT(I)=(1.0D0/(MU*SIGMA))*2.0D0*CRR(I)
         ELSE
            CT(I)=(1.0D0/(MU*SIGMA))*(CRR(I)+1.0D0/R(I)*CR(I))
         END IF
1     CONTINUE
```

Note again the close resemblance of the coding to the PDE, equation (3.88). Also, this coding illustrates the ease with which variable coefficients can be included in PDEs, in this case $1/r$ in the term $(1/r)(\partial J/\partial r)$. This is another example of how relatively straightforward a numerical solution can be computed compared with an analytical solution. The analytical solution of equation (3.88) with the given initial and boundary conditions, equations (3.89), (3.90), and (3.93), would be a relatively complicated infinite series of Bessel functions; additionally, the analytical solution would be relatively difficult to evaluate numerically, and depending on the rate of convergence of the Bessel series, may require many terms to achieve the same accuracy as with the numerical solution.

7. $I(t)$ given by equation (3.94) is computed in subroutine PRINT by Simpson's rule integration:

```
C...
C...  INTEGRAL OF THE CURRENT DENSITY BY SIMPSON'S RULE
      CTOT=0.
      DO 2 I=2,NR-1,2
         CTOT=CTOT+(R(I-1)*C(I-1)+4.0D0*R(I)*C(I)+R(I+1)*C(I+1))
2     CONTINUE
      CTOT=2.0D0*PI*DR/3.0D0*CTOT
```

The coding in DO loop 2 implements the *numerical quadrature* (numerical integration)

$$\begin{aligned} I(t) &= 2\pi\int_{10}^{r} rJ(r,t)\,dr \\ &\approx 2\pi(\Delta r/3)\left\{0J(0) + 4(\Delta r)J(\Delta r) + 2(2\Delta r)J(2\Delta r)\right. \\ &\quad \left. + 4(3\Delta r)J(3\Delta r) + \cdots + r_1 J(r_1)\right\} \end{aligned} \tag{3.98}$$

8. The current given by equation (3.94) is then computed and printed with the current calculated from equation (3.98), along with the difference between the two currents:

```
C...
C...  CONDUCTOR CURRENT AND DIFFERENCE OF CURRENTS
      CC=CMAX*DSIN(WI*T)
      DIFF=CTOT-CC
C...
C...  PRINT THE NUMERICAL SOLUTION
      NRI=10
      WRITE(NO,3)T,(C(I),I=1,NR,NRI),CTOT,CC,DIFF
3     FORMAT(F8.3,3F12.1,/,8X,3F12.1,/,8X,3F12.1,/)
```

Finally, the current profiles, $J(r,t)$ vs. r, for a series of times are plotted by a call to subroutine PLOT. Basically, PLOT creates a file with the numbers to be plotted that is then read by a plotting system. This arrangement for plotting is machine specific, but subroutine PLOT is listed in Program 3.3a to indicate how the plots to follow were created. Then, with this information, a similar arrangement can be used on other computers. Alternatively, the CALL PLOT statement in PRINT can be deleted or changed to a comment.

The integration of the 51 ODEs was performed with LSODE. The main program is similar to earlier main programs that call LSODE, e.g., Program 2.8b, so it is not listed here. The data file read by the main program is listed in Program 3.3b.

Two runs are programmed for 51 ODEs. In the first, the independent variable t in equation (3.88) is defined over the interval $0 \le t \le 1$ with an output interval of 0.25, so that there are five calls to PRINT and PLOT, and thus five spatial profiles, $J(r,t)$ vs. r, are printed and plotted. In the second case, the independent variable t in equation (3.88) is defined over the interval $0 \le t \le 2\pi$ with an output interval of $\pi/2$ (again there are five calls to PRINT and PLOT).

The output from Programs 3.2a and 3.2b is listed in Table 3.2. The following points can be noted about the output in Table 3.2:

1. The currents from equations (3.94) and (3.98) agree to four figures. Thus, the 51-point grid and the method of lines implemented on the grid apparently give a numerical solution to equations (3.88), (3.89), (3.90) and (3.93) of acceptable accuracy; the comparison of the two currents from equations (3.94) and (3.98), of course, only confirms the accuracy of the solution in the integral sense of equation (3.98) and not in the sense of point accuracy with respect to radial position r.

```
TRANSIENT CURRENT DENSITY
0.        1.0       0.25
   51                     0.00001
TRANSIENT CURRENT DENSITY
0.        6.2831854 1.5707964
   51                     0.00001
END OF RUNS
```

Program 3.3b Data File for the Solution of Equations (3.88), (3.89), (3.90) and (3.93).

Table 3.2 Numerical Output from Programs 3.2a and b. *Continued next page.*

```
RUN NO. -    1  TRANSIENT CURRENT DENSITY

INITIAL T -   0.000D+00

  FINAL T -   0.100D+01

  PRINT T -   0.250D+00

NUMBER OF DIFFERENTIAL EQUATIONS -  51

MAXIMUM INTEGRATION ERROR -  0.100D-04


      T J(0.0*r1,t) J(0.2*r1,t) J(0.4*r1,t)
        J(0.6*r1,t) J(0.8*r1,t) J(1.0*r1,t)
               itot        i(t)        diff

  0.000         0.0         0.0         0.0
                0.0         0.0         0.0
                0.0         0.0         0.0

  0.250         0.0         0.1         1.3
               14.1        85.1       300.9
              247.4       247.4         0.0

  0.500         1.9         4.6        19.1
               68.1       191.0       419.1
              479.4       479.4         0.0

  0.750        14.3        22.9        55.8
              132.7       275.5       485.6
              681.6       681.6         0.0

  1.000        41.5        55.5       102.6
              194.1       335.9       509.2
              841.5       841.5         0.0


LSODE COMPUTATIONAL STATISTICS

LAST STEP SIZE                          0.461D-01

LAST ORDER OF THE METHOD                        5

TOTAL NUMBER OF STEPS TAKEN                   129

NUMBER OF FUNCTION EVALUATIONS               1224

NUMBER OF JACOBIAN EVALUATIONS                 21


RUN NO. -    2  TRANSIENT CURRENT DENSITY
```

Table 3.2 *Continued.*

```
INITIAL T -  0.000D+00

  FINAL T -  0.628D+01

  PRINT T -  0.157D+01

NUMBER OF DIFFERENTIAL EQUATIONS -  51

MAXIMUM INTEGRATION ERROR -  0.100D-04


      T J(0.0*r1,t) J(0.2*r1,t) J(0.4*r1,t)
        J(0.6*r1,t) J(0.8*r1,t) J(1.0*r1,t)
               itot         i(t)        diff

  0.000         0.0          0.0         0.0
                0.0          0.0         0.0
                0.0          0.0         0.0

  1.571       137.8        155.9       208.9
              289.6        376.6       422.0
             1000.0       1000.0         0.0

  3.142       238.6        227.3       187.0
               98.7        -64.4      -327.7
                0.0          0.0         0.0

  4.712      -100.9       -124.2      -190.6
             -287.5       -387.0      -436.9
            -1000.0      -1000.0         0.0

  6.283      -234.9       -224.2      -185.2
              -98.5         63.4       326.2
                0.0          0.0         0.0


LSODE COMPUTATIONAL STATISTICS

LAST STEP SIZE                          0.700D-01

LAST ORDER OF THE METHOD                        4

TOTAL NUMBER OF STEPS TAKEN                   196

NUMBER OF FUNCTION EVALUATIONS               1977

NUMBER OF JACOBIAN EVALUATIONS                 34
```

2. The solution for the second case indicates that the current goes through the successive values $I(0) = 0$, $I(\pi/2) = 1000$, $I(\pi) = 0$, $I(3\pi/2) = -1000$, $I(2\pi) = 0$ which follow from equation (3.94) with $I_{\text{max}} = 1000$, $\omega_I = 1$.

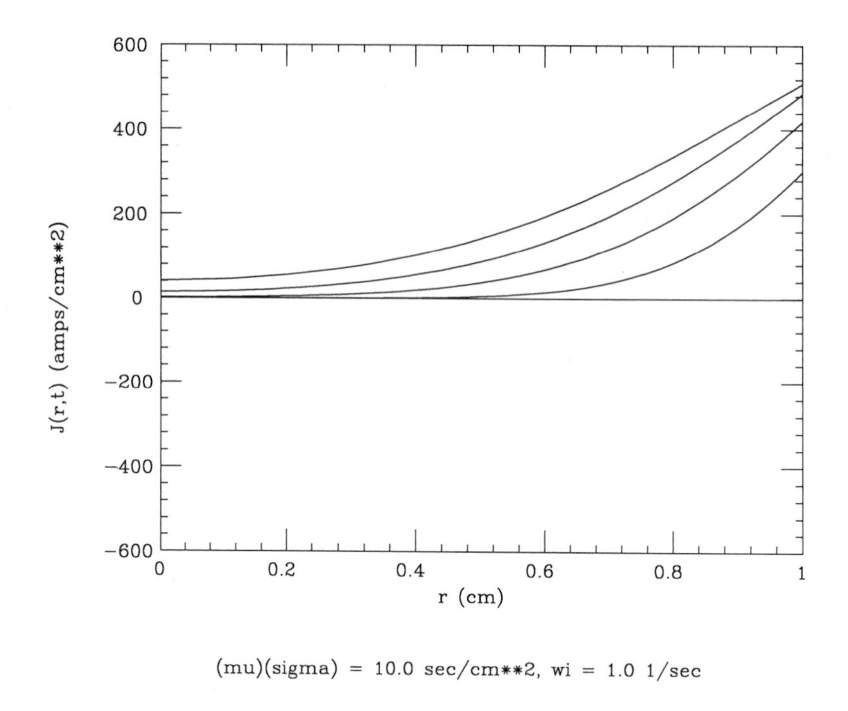

Figure 3.3 (a) Graphical Output of Programs 3.3a and b for `NORUN=1`.

3. The computational statistics indicate that `LSODE` computed the solution with modest effort using the stiff option, `MF=22`. Solutions with the nonstiff option, `MF=10`, were not attempted.

The graphical output for the two runs from subroutine `PLOT` in Program 3.3a is given in Figures 3.3a and b. The right-hand slope, $\partial J(r_1, t)/\partial r$, is a direct indication of boundary condition (3.93) with $I(t)$ given by equation (3.94).

In constructing Programs 3.3a and b, we used two spatial differentiation routines, `DSS004` and `DSS044`, to calculate the first and second derivatives, $\partial J/\partial r$ and $\partial^2 J/\partial r^2$, in equation (3.88). We now consider the differentiation formulas in these subroutines.

The first thing we should note is that these subroutines were applied to equation (3.88) which is a parabolic PDE describing the diffusion of the current density, $J(r, t)$, radially in the cylindrical conductor of Figure 3.2. This diffusion process is in distinction to the convection of Figure 3.1 modeled by the first-order hyperbolic PDE equation (3.11) in the sense that there is not a preferred direction and therefore we used centered rather than noncentered (biased upwind) approximations. In fact, a centered approximation was already put in subroutine `DSS020` which is used in `DSS004`. For a positive velocity, the five-point centered formula at grid point $i = 3$ in `DSS020` is

```
      UX( 3)=R4FDX*
     1(  +1.*U( 1)  -8.*U( 2)  +0.*U( 3)  +8.*U( 4)  -1.*U( 5))
```

and for a negative velocity, the five-point centered formula at grid point $i = N - 2$ is

```
      UX(N-2)=R4FDX*
     1(  +1.*U(N-4)  -8.*U(N-3)  +0.*U(N-2)  +8.*U(N-1)  -1.*U( N))
```

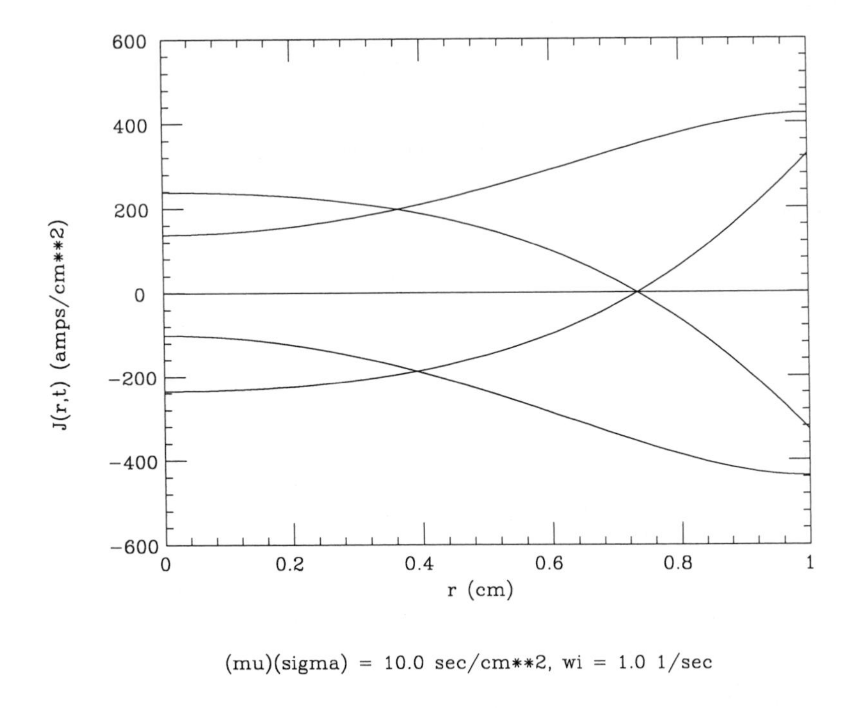

Figure 3.3 (b) Graphical Output of Programs 3.3a and b for `NORUN=2`.

Note that the weighting coefficients in these five-point formulas are 1, −8, 0, 8, −1, where the coefficient 0 is for the same point as the one for the derivative, e.g., in the first formula `0.*U(3)` and `UX(3)` both apply to the point $i = 3$. Thus, these formulas are centered in the sense that they use the same number of points on either side of the point where the first derivative, (e.g., `UX( 3)`) is calculated, and these points are located symmetrically with respect to the point for the derivative, for example, `-8.*U(2)` is paired symmetrically with `+8.*U(4)`, and `+1.*U(1)` is paired symmetrically with `-1.*U( 5))`.

We can note two other properties of these formulas: (a) the weighting coefficients are opposite in sign for the paired points, e.g, −8 and 8 for the points on either side of the central point, which in general is true for formulas to calculate odd order derivatives, like the first derivative `UX(3)` and (b) the coefficients sum to zero, as required to differentiate a constant to zero.

In summary, we use the centered five-point centered approximation with weighting coefficients $1, -8, 0, 8, -1$ at the interior points $i = 3, 4, \ldots, N-3, N-2$. Then, special formulas are required at the end points $i = 1, 2, N-1$, and N. All of these formulas are derived in the same way as equation (3.69), and we will go through the derivation for the centered formula after listing subroutine `DSS004` in Program 3.4a.

The comments in `DSS004` indicate that all of the weighting coefficients in the five differentiation formulas can be summarized in a 5×5 matrix. Note that the centered formula is represented by the third row in this matrix, and we now derive this formula to indicate the origin of the five weighting coefficients, $1, -8, 0, 8, -1$. Because we are interested in a centered approximation, we use the dependent variable $u(x)$ at x_{i-2}, x_{i-1}, x_i, x_{i+1}, and x_{i+2} in the approximation. This suggests writing a Taylor series for

```
      SUBROUTINE DSS004(XL,XU,N,U,UX)
C...
C...  SUBROUTINE DSS004 COMPUTES THE FIRST DERIVATIVE, U , OF A
C...                                                   X
C...  VARIABLE U OVER THE SPATIAL DOMAIN XL LE X LE XU FROM CLASSICAL
C...  FIVE-POINT, FOURTH-ORDER FINITE DIFFERENCE APPROXIMATIONS
C...
C...  ARGUMENT LIST
C...
C...     XL      LOWER BOUNDARY VALUE OF X (INPUT)
C...
C...     XU      UPPER BOUNDARY VALUE OF X (INPUT)
C...
C...     N       NUMBER OF GRID POINTS IN THE X DOMAIN INCLUDING THE
C...             BOUNDARY POINTS (INPUT)
C...
C...     U       ONE-DIMENSIONAL ARRAY CONTAINING THE VALUES OF U AT
C...             THE N GRID POINT POINTS FOR WHICH THE DERIVATIVE IS
C...             TO BE COMPUTED (INPUT)
C...
C...     UX      ONE-DIMENSIONAL ARRAY CONTAINING THE NUMERICAL
C...             VALUES OF THE DERIVATIVES OF U AT THE N GRID POINTS
C...             (OUTPUT)
C...
C...  THE WEIGHTING COEFFICIENTS CAN BE SUMMARIZED AS
C...
C...             -25   48  -36   16   -3
C...
C...              -3  -10   18   -6    1
C...
C...       1/12    1   -8    0    8   -1
C...
C...              -1    6  -18   10    3
C...
C...               3  -16   36  -48   25
C...
C...  WHICH ARE THE COEFFICIENTS REPORTED BY BICKLEY FOR N = 4, M =
C...  1, P = 0, 1, 2, 3, 4 (BICKLEY, W. G., FORMULAE FOR NUMERICAL
C...  DIFFERENTIATION, MATH. GAZ., VOL. 25, 1941.  NOTE - THE BICKLEY
C...  COEFFICIENTS HAVE BEEN DIVIDED BY A COMMON FACTOR OF TWO).
C...
      DIMENSION U(N),UX(N)
C...
C...  COMPUTE THE SPATIAL INCREMENT
      DX=(XU-XL)/FLOAT(N-1)
      R4FDX=1./(12.*DX)
      NM2=N-2
C...
C...  THE FIVE POINT FORMULAS HAVE BEEN FORMATTED SO THAT THE NUMERI-
C...  CAL WEIGHTING COEFFICIENTS CAN BE MORE EASILY ASSOCIATED WITH
```

Program 3.4a Subroutine DSS004 for the Fourth-Order Approximation of First-Order Derivatives. *Continued next page.*

```
C...   THE BICKLEY MATRIX ABOVE
       UX( 1)=R4FDX*
      1( -25.*U( 1) +48.*U( 2) -36.*U( 3) +16.*U( 4) -3.*U( 5))
       UX( 2)=R4FDX*
      1( -3.*U( 1) -10.*U( 2) +18.*U( 3) -6.*U( 4) +1.*U( 5))
       DO 1 I=3,NM2
       UX( I)=R4FDX*
      1( +1.*U(I-2) -8.*U(I-1) +0.*U( I) +8.*U(I+1) -1.*U(I+2))
1      CONTINUE
       UX(N-1)=R4FDX*
      1( -1.*U(N-4) +6.*U(N-3) -18.*U(N-2) +10.*U(N-1) +3.*U( N))
       UX( N)=R4FDX*
      1( 3.*U(N-4) -16.*U(N-3) +36.*U(N-2) -48.*U(N-1) +25.*U( N))
       RETURN
       END
```

Program 3.4a *Continued.*

grid points $i-2$, $i-1$, $i+1$, and $i+2$

$$u(x_{i-2}) = u(x_i) + (du(x_i)/dx)(-2\Delta x) + (1/2!)(d^2u(x_i)/dx^2)(-2\Delta x)^2 + \cdots \quad (3.99)$$

$$u(x_{i-1}) = u(x_i) + (du(x_i)/dx)(-\Delta x) + (1/2!)(d^2u(x_i)/dx^2)(-\Delta x)^2 + \cdots \quad (3.100)$$

$$u(x_{i+1}) = u(x_i) + (du(x_i)/dx)(\Delta x) + (1/2!)(d^2u(x_i)/dx^2)(\Delta x)^2 + \cdots \quad (3.101)$$

$$u(x_{i+2}) = u(x_i) + (du(x_i)/dx)(2\Delta x) + (1/2!)(d^2u(x_i)/dx^2)(2\Delta x)^2 + \cdots \quad (3.102)$$

Now, proceeding as we did with the four-point approximation of equation (3.69), we take a linear combination of equations (3.99) to (3.102) to keep the derivative of interest, $du(x_i)/dx$, and drop as many of the higher derivative terms as possible (to give maximum accuracy in the final differentiation formula). If we multiply equation (3.99) by a constant a, equation (3.100) by a constant b, etc.

$$\begin{aligned} au(x_{i-2}) = {} & au(x_i) + a(du(x_i)/dx)(-2\Delta x) \\ & + a(1/2!)(d^2u(x_i)/dx^2)(-2\Delta x)^2 + \cdots \end{aligned} \quad (3.103)$$

$$\begin{aligned} bu(x_{i-1}) = {} & bu(x_i) + b(du(x_i)/dx)(-\Delta x) \\ & + b(1/2!)(d^2u(x_i)/dx^2)(-\Delta x)^2 + \cdots \end{aligned} \quad (3.104)$$

$$\begin{aligned} cu(x_{i+1}) = {} & cu(x_i) + c(du(x_i)/dx)(\Delta x) \\ & + c(1/2!)(d^2u(x_i)/dx^2)(\Delta x)^2 + \cdots \end{aligned} \quad (3.105)$$

$$\begin{aligned} du(x_{i+2}) = {} & du(x_i) + d(du(x_i)/dx)(2\Delta x) \\ & + d(1/2!)(d^2u(x_i)/dx^2)(2\Delta x)^2 + \cdots \end{aligned} \quad (3.106)$$

then sum the resulting equations, we can retain the first derivative term, $du(x_i)/dx$ by imposing the condition

$$-2a - b + c + 2d = 1 \quad (3.107)$$

The 1 in the right-hand side of equation (3.107) ensures that when equations (3.103) to (3.106) are added with the numerical values of a, b, c, and d (still to be determined), the first derivative term will remain. Similarly, in order to eliminate the second derivative term, $d^2u(x_i)/dx^2$, we can impose the condition

$$4a + b + c + 4d = 0 \tag{3.108}$$

The 0 in the right-hand side of equation (3.108) ensures that when equations (3.103) to (3.106) are added with the numerical values of a, b, c, and d, the second derivative term will be eliminated.

In the same way, to eliminate the third- and fourth-order derivative terms, we can impose the conditions

$$-8a - b + c + 8d = 0 \tag{3.109}$$

$$16a + b + c + 16d = 0 \tag{3.110}$$

We now have four linear algebraic equations ((3.07) to (3.110)) for the four constants a to d. Simultaneous solution gives $a = 2/4!$, $b = -16/4!$, $c = 16/4!$, $d = -2/4!$. Then, if equations (3.103) to (3.106) are added with these numerical values for a to d, we obtain for the derivative of interest, $du(x_i)/dx$

$$du(x_i)/dx = 1/(4!\Delta x)(2u(x_{i-2}) - 16u(x_{i-1}) + 0u(x_i) + 16u(x_{i+1}) - 2u(x_{i+2})) + O(\Delta x^4) \tag{3.111}$$

Equation (3.111) is the result we seek, with the usual convention of 4! in the denominator rather than 12 as in subroutine DSS004 of Program 3.4a so that the weighting coeffcients in DSS004 are multiplied by 2 in equation (3.111); note that the five-point approximation of equation (3.111) is fourth-order correct. A similar procedure can be followed to derive the four noncentered approximations at grid points $i = 1, 2, N-1$, and N in subroutine DSS004.

Next, we consider the five-point centered approximations of subroutine DSS044 for calculating second-order spatial derivatives, like $\partial^2 J/\partial r^2$ in equation (3.88). The procedure is basically the same as for the derivation of equation (3.111). There is one additional complication, however. Whereas in the case of computing first derivatives, such as $du(x_i)/dx$ in equation (3.111), we did not have to be concerned with boundary conditions explicitly in the derivation or programming of the five-point approximations, the same is not true in the derivation and subsequent programming of the approximations for second derivatives. Now we must consider three possible types of boundary conditions that are defined along with the second derivative: (a) Dirichlet boundary conditions in which the dependent variable is specified at a boundary value of x; (b) Neumann boundary conditions in which the first-order spatial derivative of the dependent variable is specified at a boundary value of x, such as equations (3.90) and (3.93); and (c) a combination of Dirichlet and Neumann boundary conditions, called boundary conditions of the third type, in which the dependent variable and its first-order spatial derivative are both specified at a boundary value of x. Note that for the cases of (b) and (c), the spatial derivative that is specified is first order, since it is for a second-order spatial derivative; in general, for a derivative of order n, the highest order derivative in a boundary condition is of order $n-1$. Also, we will see that (c) can be handled in the same way as (b) when we are computing a method of lines solution to a PDE; thus we actually have to consider only (a) and (b).

We already have a procedure for implementing Dirichlet boundary conditions since they were used in the solution of equation (3.11). Specifically, in the derivative subroutine, such as DERV1, we merely set the dependent variable at the boundary. For example, in the case of equation (3.11), we used the statements in subroutine DERV1 of Program 3.1c:

```
C...
C...   BOUNDARY CONDITION
         U1(1)=1.0D+00-DEXP(-C2*T)
        U1T(1)=0.0D+00
```

to implement boundary condition (3.32)

$$u_1(0,t) = 1 - e^{-c_2 t}$$

Therefore, we basically require a new method only for Neumann boundary conditions.

The procedure for deriving an approximation to the second derivative $du^2(x_i)/dx^2$ is essentially the same as for the derivative $du(x_i)/dx$ of equation (3.111). Since we are interested in an approximation of $du^2(x_i)/dx^2$ at a boundary, for example, at $x = 0$, for which we can include a Dirichlet or Neumann boundary condition, we use $u(x)$ at the boundary and at the five interior points adjacent to the boundary. If we are considering the left boundary, then we use the Taylor series for $u(x_2)$ to $u(x_6)$

$$au(x_2) = au(x_1) + a(du(x_1)/dx)(\Delta x) + a(1/2!)(d^2u(x_1)/dx^2)(\Delta x)^2 + \cdots \tag{3.112}$$

$$bu(x_3) = bu(x_1) + b(du(x_1)/dx)(2\Delta x) + b(1/2!)(d^2u(x_1)/dx^2)(2\Delta x)^2 + \cdots \tag{3.113}$$

$$cu(x_4) = cu(x_1) + c(du(x_1)/dx)(3\Delta x) + c(1/2!)(d^2u(x_1)/dx^2)(3\Delta x)^2 + \cdots \tag{3.114}$$

$$du(x_5) = du(x_1) + d(du(x_1)/dx)(4\Delta x) + d(1/2!)(d^2u(x_1)/dx^2)(4\Delta x)^2 + \cdots \tag{3.115}$$

$$eu(x_6) = eu(x_1) + e(du(x_1)/dx)(5\Delta x) + e(1/2!)(d^2u(x_1)/dx^2)(5\Delta x)^2 + \cdots \tag{3.116}$$

To retain the second derivative, $d^2u(x_1)/dx^2$, we impose the condition

$$a + 4b + 9c + 16d + 25e = 2! \tag{3.117}$$

(the 2! is included to facilitate the algebra when equations (3.112) to (3.116) are subsequently added). To delete the first, third, fourth and fifth derivatives, we impose the conditions

$$a + 2b + 3c + 4d + 5e = 0 \tag{3.118}$$

$$a + 8b + 27c + 64d + 125e = 0 \tag{3.119}$$

$$a + 16b + 81c + 256d + 625e = 0 \tag{3.120}$$

Simultaneous solution of equations (3.116) to (3.120) gives $a = -308/4!$, $b = 428/4!$, $c = -312/4!$, $d = 122/4!$, $e = -20/4!$. Substitution of these values into equations (3.112) to (3.116), followed by the addition of these equations and the algebraic solution for $d^2u(x_1)/dx^2$, gives

$$\begin{aligned} du^2(x_1)/dx^2 &= 1/(4!\Delta x^2)(90u(x_1) - 308u(x_2) + 428u(x_3) \\ &\quad - 312u(x_4) + 122u(x_5) - 20u(x_6)) + O(\Delta x^4) \end{aligned} \tag{3.121}$$

Note that enough points are used in the approximation, that is, $u(x_1)$, $u(x_2)$, $u(x_3)$, $u(x_4)$, $u(x_5)$, and $u(x_6)$, to eliminate the derivatives in the Taylor series up to including the fifth derivative. Then the sixth derivative term remains, for example, $(1/6!)(d^6u(x_1)/dx^6)(\Delta x)^6$ in the first Taylor series (3.112). When this sixth-order term is divided by Δx^2 to obtain the formula for $du^2(x_1)/dx^2$, a Δx^4 term remains; thus the formula of equation (3.121) is fourth-order correct. Also, as before, the weighting coefficients sum to zero.

A similar analysis at the right boundary gives

$$du^2(x_N)/dx^2 = 1/(4!\Delta x^2)(90u(x_N) - 308u(x_{N-1}) + 428u(x_{N-2}) - 312u(x_{N-3}) + 122u(x_{N-4}) - 20u(x_{N-5})) + O(\Delta x^4) \quad (3.122)$$

We see that for the approximation of second-order (and generally, even order) derivatives, the weighting coefficients at the left and right boundaries are the same (they are not of opposite sign as in the case of odd order derivatives, as for example, with the $du(x_1)/dx$ (= `UX(1)`) and $du(x_N)/dx$ (= `UX(N)`) formulas in subroutine `DSS004` of Program 3.4a).

Equations (3.121) and (3.122) can be used when Dirichlet boundary conditions are specified for a second-order derivative in a PDE. We also need approximations for a second derivative at the boundaries with Neumann boundary conditions. For $du^2(x_1)/dx^2$ with $du(x_1)/dx$ included as a boundary condition, we take a linear combination of $u(x)$ at $x = x_2$, x_3, x_4, and x_5, plus $du(x)/dx$ at $x = x_1$

$$au(x_2) + bu(x_3) + cu(x_4) + du(x_5) + edu(x_1)/dx$$

To drop $du(x_1)/dx$,

$$a + 2b + 3c + 4d + e = 0 \quad (3.123)$$

Similarly, to drop $d^3u(x_1)/dx^3$, $d^4u(x_1)/dx^4$, and $d^5u(x_1)/dx^5$

$$a + 8b + 27c + 64d = 0 \quad (3.124)$$

$$a + 16b + 81c + 256d = 0 \quad (3.125)$$

$$a + 32b + 243c + 1024d = 0 \quad (3.126)$$

Finally, to retain $d^2u(x_1)/dx^2$

$$a + 4b + 9c + 16d = 2! \quad (3.127)$$

Solution of equations (3.123) to (3.127) gives $a = 192/4!$, $b = -72/4!$, $c = 64/(3(4!))$, $d = -3/4!$, $e = -100/4!$. Substitution of these values in the five Taylor series, followed by solution for $d^2u(x_2)/dx^2$, gives

$$du^2(x_1)/dx^2 = 1/(4!\Delta x^2)((-415/3)u(x_1) + 192u(x_2) - 72u(x_3) + (64/3)u(x_4) - 3u(x_5) - 100(du(x_1)/dx)\Delta x) + O(\Delta x^4) \quad (3.128)$$

Similarly, for $du^2(x_N)/dx^2$ with $du(x_N)/dx$ as a boundary condition, we take a linear combination of $u(x)$ at $x = x_{N-1}$, x_{N-2}, x_{N-3}, and x_{N-4}, plus $du(x)/dx$ at $x = x_N$

$$au(x_{N-1}) + bu(x_{N-2}) + cu(x_{N-3}) + du(x_{N-4}) + edu(x_N)/dx$$

Applying the conditions to drop $du(x_N)/dx$, $d^3u(x_N)/dx^3$, $d^4u(x_N)/dx^4$, and $d^5u(x_N)/dx^5$, while retaining $du^2(x_N)/dx^2$, we arrive finally at

$$du^2(x_N)/dx^2 = 1/(4!\Delta x^2)((-415/3)u(x_N) + 192u(x_{N-1}) - 72u(x_{N-2}) + (64/3)u(x_{N-3}) - 3u(x_{N-4}) + 100(du(x_N)/dx)\Delta x) + O(\Delta x^4) \quad (3.129)$$

Equations (3.128) and (3.129) can be used to calculate $du^2(x_i)/dx^2$ at grid points $i = 1$ and N with Neumann boundary conditions (since $du(x_i)/dx$ at $i = 1$ and N are also included).

Formulas for $du^2(x_i)/dx^2$, $i = 2, 3, \ldots N-2, N-1$, are derived in the same way as equations (3.121) and (3.122). The final results are

$$du^2(x_2)/dx^2 = 1/(4!\Delta x^2)(20u(x_1) - 30u(x_2) - 8u(x_3) + 28u(x_4) - 12u(x_5) + 2u(x_6)) + O(\Delta x^4) \quad (3.130)$$

$$du^2(x_{N-1})/dx^2 = 1/(4!\Delta x^2)(20u(x_N) - 30u(x_{N-1}) - 8u(x_{N-2}) + 28u(x_{N-3}) - 12u(x_{N-4}) + 2u(x_{N-5})) + O(\Delta x^4) \quad (3.131)$$

$$du^2(x_i)/dx^2 = 1/(4!\Delta x^2)(-2u(x_{i-2}) + 32u(x_{i-1}) - 60u(x_i) + 32u(x_{i+1}) - 2u(x_{i+2})) + O(\Delta x^4), \quad i = 3, \ldots, N-2 \quad (3.132)$$

The weighting coefficients in subroutine DSS044 have been reduced by a factor of two since the coefficient multiplying the right-hand side of each equation is 1/12 rather than $1/4! = 1/24$. Otherwise, the coding is a straightforward implementation of equations (3.121) and (3.122) and (3.128) to (3.132). Note in particular that arguments `NL` and `NU` are inputs to `DSS044` which specify either Dirichlet or Neumann boundary conditions at grid points $i = 1$ and N. `NL` and `NU` were used in the programming of boundary conditions (3.90) and (3.93) in subroutine `DERV` of Program 3.3a; since these are Neumann boundary conditions ($\partial J/\partial r$ is specified at $r = 0$ and $r = r_1$), `NL=NU=2`.

We conclude this chapter with another example of a PDE which is first order in time, t, and second order in the spatial variable, x, the *cubic Schrödinger equation* (CSE), mentioned briefly at the end of Chapter 1

$$iu_t + u_{xx} + q|u|^2u = 0 \quad (3.133)$$

$i = \sqrt{-1}$ and therefore $u(x,t)$ is complex. Equation (3.133) has the solution (for $q = 1$)

$$u(x,t) = \sqrt{2}e^{i(0.5x+0.75t)}\text{sech}(x-t) \quad (3.134)$$

It follows immediately from equation (3.134) that

$$|u(x,t)| = \sqrt{2}\ \text{sech}(x-t) \quad (3.135)$$

which is the equation for a *soliton*, initially centered at $x = 0$, traveling left to right with unit velocity and a maximum amplitude of $\sqrt{2}$ (without changing shape).

We now develop a numerical method of lines solution to equation (3.133) for comparison with the analytical solution, equation (3.134). We take as the initial condition for equation (3.133)

$$u(x,0) = \sqrt{2}\exp\{i(0.5x)\}\,\text{sech}(x) \quad (3.136)$$

which follows directly from equation (3.134). The boundary conditions are taken as

$$u(-\infty,t) = u(\infty,t) = 0 \quad (3.137)(3.138)$$

For a numerical solution on a spatial grid of finite length, we actually use

$$u(-30,t) = u(70,t) = 0 \quad (3.139)(3.140)$$

Equations (3.139) and (3.140) allow for movement of the soliton from left to right, yet for the interval of t used in the calculations, the soliton does not come close enough to the finite boundaries at $x = -30$ and $x = 70$ to depart significantly from a zero value, i.e., equations (3.139) and (3.140) and (3.137) and (3.138) are essentially equivalent.

```
      SUBROUTINE DSS044(XL,XU,N,U,UX,UXX,NL,NU)
C...
C...  SUBROUTINE DSS044 COMPUTES A FOURTH ORDER APPROXIMATION OF A
C...  SECOND ORDER DERIVATIVE, WITH OR WITHOUT THE NORMAL DERIVATIVE
C...  AT THE BOUNDARY.
C...
C...  ARGUMENT LIST
C...
C...     XL      LEFT VALUE OF THE SPATIAL INDEPENDENT VARIABLE (INPUT)
C...
C...     XU      RIGHT VALUE OF THE SPATIAL INDEPENDENT VARIABLE (INPUT)
C...
C...     N       NUMBER OF SPATIAL GRID POINTS, INCLUDING THE END
C...             POINTS (INPUT)
C...
C...     U       ONE-DIMENSIONAL ARRAY OF THE DEPENDENT VARIABLE TO BE
C...             DIFFERENTIATED (INPUT)
C...
C...     UX      ONE-DIMENSIONAL ARRAY OF THE FIRST DERIVATIVE OF U.
C...             THE END VALUES OF UX, UX(1) AND UX(N), ARE USED IN
C...             NEUMANN BOUNDARY CONDITIONS AT X = XL AND X = XU,
C...             DEPENDING ON THE ARGUMENTS NL AND NU (SEE THE DE-
C...             SCRIPTION OF NL AND NU BELOW)
C...
C...     UXX     ONE-DIMENSIONAL ARRAY OF THE SECOND DERIVATIVE OF U
C...             (OUTPUT)
C...
C...     NL      INTEGER INDEX FOR THE TYPE OF BOUNDARY CONDITION AT
C...             X = XL (INPUT).  THE ALLOWABLE VALUES ARE
C...
C...                1 - DIRICHLET BOUNDARY CONDITION AT X = XL
C...                    (UX(1) IS NOT USED)
C...
C...                2 - NEUMANN BOUNDARY CONDITION AT X = XL
C...                    (UX(1) IS USED)
C...
C...     NU      INTEGER INDEX FOR THE TYPE OF BOUNDARY CONDITION AT
C...             X = XU (INPUT).  THE ALLOWABLE VALUES ARE
C...
C...                1 - DIRICHLET BOUNDARY CONDITION AT X = XU
C...                    (UX(N) IS NOT USED)
C...
C...                2 - NEUMANN BOUNDARY CONDITION AT X = XU
C...                    (UX(N) IS USED)
C...
      REAL U(N), UX(N), UXX(N)
C...
C...  GRID SPACING
      DX=(XU-XL)/FLOAT(N-1)
C...
C...  1/(12*DX**2) FOR SUBSEQUENT USE
      R12DXS=1./(12.0E0*DX**2)
```

Program 3.4b Subroutine DSS044 for the Fourth-Order Approximation of Second-Order Derivatives. *Continued next pages.*

```
C...
C...  UXX AT THE LEFT BOUNDARY
C...
C...     WITHOUT UX (EQUATION (3.121))
         IF(NL.EQ.1)THEN
         UXX(1)=R12DXS*
     1                 (   45.0E0*U(1)
     2                   -154.0E0*U(2)
     3                   +214.0E0*U(3)
     4                   -156.0E0*U(4)
     5                    +61.0E0*U(5)
     6                    -10.0E0*U(6))
C...
C...     WITH UX (EQUATION (3.128))
         ELSE IF(NL.EQ.2)THEN
         UXX(1)=R12DXS*
     1                 (-415.0E0/6.0E0*U(1)
     2                          +96.0E0*U(2)
     3                          -36.0E0*U(3)
     4                   +32.0E0/3.0E0*U(4)
     5                    -3.0E0/2.0E0*U(5)
     6                          -50.0E0*UX(1)*DX)
         END IF
C...
C...  UXX AT THE RIGHT BOUNDARY
C...
C...     WITHOUT UX (EQUATION (3.122))
         IF(NU.EQ.1)THEN
         UXX(N)=R12DXS*
     1                 (   45.0E0*U(N  )
     2                   -154.0E0*U(N-1)
     3                   +214.0E0*U(N-2)
     4                   -156.0E0*U(N-3)
     5                    +61.0E0*U(N-4)
     6                    -10.0E0*U(N-5))
C...
C...     WITH UX (EQUATION (3.129))
         ELSE IF(NU.EQ.2)THEN
         UXX(N)=R12DXS*
     1                 (-415.0E0/6.0E0*U(N  )
     2                          +96.0E0*U(N-1)
     3                          -36.0E0*U(N-2)
     4                   +32.0E0/3.0E0*U(N-3)
     5                    -3.0E0/2.0E0*U(N-4)
     6                          +50.0E0*UX(N  )*DX)
         END IF
C...
C...  UXX AT THE INTERIOR GRID POINTS
C...
C...     I = 2 (EQUATION (3.130))
         UXX(2)=R12DXS*
```

Program 3.4b *Continued.*

```
     1                     (   10.0E0*U(1)
     2                        -15.0E0*U(2)
     3                         -4.0E0*U(3)
     4                        +14.0E0*U(4)
     5                         -6.0E0*U(5)
     6                         +1.0E0*U(6))
C...
C...      I = N-1 (EQUATION (3.131))
          UXX(N-1)=R12DXS*
     1                       ( 10.0E0*U(N  )
     2                        -15.0E0*U(N-1)
     3                         -4.0E0*U(N-2)
     4                        +14.0E0*U(N-3)
     5                         -6.0E0*U(N-4)
     6                         +1.0E0*U(N-5))
C...
C...      I = 3,..., N-2 (EQUATION (3.132))
          DO 1 I=3,N-2
          UXX(I)=R12DXS*
     1                     (   -1.0E0*U(I-2)
     2                        +16.0E0*U(I-1)
     3                        -30.0E0*U(I  )
     4                        +16.0E0*U(I+1)
     5                         -1.0E0*U(I+2))
1         CONTINUE
      RETURN
      END
```

Program 3.4b *Continued.*

To program equation (3.133), we write u as a *complex variable,*

$$u = v + iw \tag{3.141}$$

Substitution of u from equation (3.141) in equation (3.133), and separation into real and imaginary parts gives the two real-coupled PDEs

$$v_t + w_{xx} + Q(v^2 + w^2)w = 0 \tag{3.142}$$

$$w_t - v_{xx} - Q(v^2 + w^2)v = 0 \tag{3.143}$$

The initial conditions for equations (3.142) and (3.143) follow directly from equation (3.136)

$$v(x, 0) = \sqrt{2}\cos(0.5x)\text{sech}(x) \tag{3.144}$$

$$w(x, 0) = \sqrt{2}\sin(0.5x)\text{sech}(x) \tag{3.145}$$

Homogeneous Dirichlet boundary conditions follow from equations (3.139) and (3.140)

$$v(-30, t) = w(-30, t) = 0 \qquad (3.146)(3.147)$$

$$v(70, t) = w(70, t) = 0 \qquad (3.148)(3.149)$$

Equations (3.142) to (3.149) are the problem system. We now consider a method of lines code for the implementation of these equations.

Subroutine INITAL for initial conditions (3.144) and (3.145) is listed in Program 3.5a. We can note the following points about subroutine INITAL:

1. A grid of 400 intervals (401 points) is defined in x. This relatively large number of points is used in order to define the sharp spatial variations of the solitons given by equation (3.135), which we will observe in a subsequent plot of the solution.

2. The COMMON block has a series of arrays relating to equations (3.142) to (3.149):

```
      COMMON/T/          T,      NSTOP,      NORUN
     1      /Y/   V(0:NG),    W(0:NG)
     2      /F/  VT(0:NG),   WT(0:NG)
     3      /S/  VX(0:NG),   WX(0:NG), VXX(0:NG), WXX(0:NG),V2W2(0:NG),
     4            X(0:NG)
     5      /C/         XL,          XR,          Q,      ROOT2,        SSET
     6      /I/         IP
```

(2.1) $v(x,t)$ and $w(x,t)$ in equations (3.142) to (3.149) are in arrays V(0:NG) and W(0:NG), respectively.

(2.2) The temporal derivatives v_t and w_t given by equations (3.142) and (3.143) are in arrays VT(0:NG) and WT(0:NG), thus the temporal integration of derivatives in COMMON/F/ to dependent variables in COMMON/Y/ follows the format used in the preceding programs.

(2.3) The first-order spatial derivatives v_x and w_x are in arrays VX(0:NG) and WX(0:NG), and the second-order spatial derivatives v_{xx} and w_{xx} are in arrays VXX(0:NG) and WXX(0:NG).

(2.4) The absolute value of u, given by $u = (v^2 + w^2)^{1/2}$, is in array V2W2(0:NG).

(2.5) The values of x along the 401-point grid are in array X(0:NG).

3. The boundary values of x in equations (3.146) to (3.149) are defined by the programming

```
C...
C...  LOWER AND UPPER LIMITS OF X
      XL=-30.0D0
      XR= 70.0D0
```

4. Initial conditions (3.144) and (3.145) are defined in DO loop 10, along with $|u|$

```
      DO 10 I=0,NG
C...
C...     INITIAL UNIFORM GRID
         X(I)=XL+DFLOAT(I)*STEP
C...
C...     SECH(X)
         SCH=2.0D0/(DEXP(X(I))+DEXP(-X(I)))
C...
C...     REAL PART
         V(I)=ROOT2*DCOS(0.5D0*X(I))*SCH
C...
C...     IMAGINARY PART
         W(I)=ROOT2*DSIN(0.5D0*X(I))*SCH
C...
```

```
      SUBROUTINE INITAL
      IMPLICIT DOUBLE PRECISION (A-H,O-Z)
      PARAMETER (NG=400)
      COMMON/T/           T,      NSTOP,      NORUN
     1      /Y/   V(0:NG),    W(0:NG)
     2      /F/  VT(0:NG),  WT(0:NG)
     3      /S/  VX(0:NG),  WX(0:NG), VXX(0:NG), WXX(0:NG),V2W2(0:NG),
     4            X(0:NG)
     5      /C/        XL,        XR,         Q,      ROOT2,      SSET
     6      /I/        IP
C...
C...  LOWER AND UPPER LIMITS OF X
      XL=-30.0D0
      XR= 70.0D0
C...
C...  Q IN CUBIC SCHRODINGER EQUATION
      Q=1.0D0
C...
C...  PRECOMPUTE 2**0.5
      ROOT2=DSQRT(2.0D0)
C...
C...  GRID SPACING
      STEP=(XR-XL)/DFLOAT(NG)
C...
C...  INITIAL REAL, IMAGINARY PARTS (V(X,T), W(X,T)), ABSOLUTE VALUE OF
C...
C...     U(X,0) = U0(X) = (2**0.5)EXP(0.5*I*X)SECH(X), I = (-1)**0.5
C...
C...  OVER XL GE X LE XU
C...
      DO 10 I=0,NG
C...
C...     INITIAL UNIFORM GRID
         X(I)=XL+DFLOAT(I)*STEP
C...
C...     SECH(X)
         SCH=2.0D0/(DEXP(X(I))+DEXP(-X(I)))
C...
C...     REAL PART
         V(I)=ROOT2*DCOS(0.5D0*X(I))*SCH
C...
C...     IMAGINARY PART
         W(I)=ROOT2*DSIN(0.5D0*X(I))*SCH
C...
C...     ABSOLUTE VALUE
         V2W2(I)=DSQRT(V(I)**2+W(I)**2)
10    CONTINUE
C...
C...  INITIAL DERIVATIVES
      CALL DERV
C...
```

Program 3.5a Subroutine INITAL for Initial Conditions (3.144) and (3.145). *Continued next page.*

```
C...  INITIALIZE COUNTER FOR PLOTTED SOLUTION, CUMULATIVE SUM OF SQUARES
C...  OF ERRORS
      IP=0
      SSET=0.0D0
      RETURN
      END
```

Program 3.5a *Continued.*

```
C...       ABSOLUTE VALUE
           V2W2(I)=DSQRT(V(I)**2+W(I)**2)
10       CONTINUE
```

STEP used in DO loop 10 is the grid spacing computed as STEP=(XR-XL)/DFLOAT(NG).

5. Finally, the initial derivatives VT(0:NG) and WT(0:NG) are computed by a call to DERV, and a counter for the plotted solution, IP, and the sum of squares of the errors (differences between the numerical and analytical solutions), SSET, are initialized.

Subroutine DERV, which computes the temporal derivatives v_t and w_t given by equations (3.142) and (3.143), is listed in Program 3.5b. We can note the following points about subroutine DERV:

1. The derivative v_{xx} in equation (3.143) is computed by a call to subroutine DSS044

```
C...  V
C...   XX
           NL=1
           NU=1
           CALL DSS044(XL,XR,NG+1,V,VX,VXX,NL,NU)
```

Note that we do not program boundary conditions (3.146) and (3.148) other than to specify that they are Dirichlet (NL=1, NU=1). This is possible since the boundary values are defined in DO loop 10 in subroutine INITAL, that is V(0) ~ 0, V(NG) ~ 0 through the statement

```
C...
C...       REAL PART
           V(I)=ROOT2*DCOS(0.5D0*X(I))*SCH
```

(note that SCH ~ 0 at $x = -30$ and 70) and they are not permitted to deviate from these initial values by setting the boundary temporal derivatives to zero in DERV

```
           VT(0)=0.0D0
           VT(NG)=0.0D0
```

Thus, this is an alternative way to define constant boundary conditions, that is, the constant values are set in INITAL, then their temporal derivatives are set to zero in DERV so that the initial values are not changed. Another possibility, of

```
      SUBROUTINE DERV
      IMPLICIT DOUBLE PRECISION (A-H,O-Z)
      PARAMETER (NG=400)
      COMMON/T/          T,      NSTOP,      NORUN
     1      /Y/   V(0:NG),   W(0:NG)
     2      /F/  VT(0:NG),  WT(0:NG)
     3      /S/  VX(0:NG),  WX(0:NG), VXX(0:NG), WXX(0:NG),V2W2(0:NG),
     4            X(0:NG)
     5      /C/        XL,         XR,          Q,      ROOT2,        SSET
     6      /I/        IP
C...
C...  PDES
C...
C...  V
C...   XX
        NL=1
        NU=1
        CALL DSS044(XL,XR,NG+1,V,VX,VXX,NL,NU)
C...
C...  W
C...   XX
        NL=1
        NU=1
        CALL DSS044(XL,XR,NG+1,W,WX,WXX,NL,NU)
C...
C...  ODES AT THE BOUNDARIES
        VT(0)=0.0D0
        WT(0)=0.0D0
        VT(NG)=0.0D0
        WT(NG)=0.0D0
C...
C...  ODES AT THE INTERIOR POINTS
      DO 10 I=1,NG-1
C...
C...     V**2 + W**2
         V2W2(I)=V(I)**2+W(I)**2
C...
C...     VT
         VT(I)=-WXX(I)-Q*V2W2(I)*W(I)
C...
C...     WT
         WT(I)= VXX(I)+Q*V2W2(I)*V(I)
10    CONTINUE
      RETURN
      END
```

Program 3.5b Subroutine DERV for Equations (3.142) and (3.143).

course, is to set the boundary values of the dependent variable in DERV as we did before (e.g., in DERV1 of Program 3.1c), but this may cause problems with some implicit ODE integrators (as discussed in paragraph (1) after Program 3.1c); in other words, when using these implicit integrators, the dependent variables should not be set in DERV, but rather, only their temporal derivatives should be set which are then inputs to the ODE integrator (through COMMON/F/), and from which the boundary dependent variables are computed by the ODE integrator (and are returned through COMMON/Y/). However, explicit integrator RKF45, which was used for this problem of 802 ODEs does not have this limitation, and the boundary dependent variables can be set in DERV.

2. The same procedure is repeated to compute derivative w_{xx} in equation (3.142) (by a call to subroutine DSS044).

3. The ODEs which approximate equations (3.142) and (3.143) are programmed over the interior grid points $i = 1, 2, \ldots, 399$ in DO loop 10

```
C...
C...  ODES AT THE INTERIOR POINTS
      DO 10 I=1,NG-1
C...
C...     V**2 + W**2
         V2W2(I)=V(I)**2+W(I)**2
C...
C...     VT
         VT(I)=-WXX(I)-Q*V2W2(I)*W(I)
C...
C...     WT
         WT(I)= VXX(I)+Q*V2W2(I)*V(I)
10    CONTINUE
      RETURN
      END
```

Note again the close resemblance of the PDEs and their method of lines programming. Also, note that all 802 ODEs in COMMON/F/ have now been programmed, which is essential before the return from DERV.

Subroutine PRINT in Program 3.5c prints the numerical and analytical solutions and calls subroutine PLOT1 to plot the soliton obtained from the numerical solution.

We can note the following points about subroutine PRINT:

1. After a heading for the numerical solution is printed via FORMAT 2, the absolute value of the numerical solution, $|u| = (v^2 + w^2)^{1/2}$, and the absolute value of the analytical solution given by equation (3.135) (computed in function EXACT), and the difference, DIFF, between these two absolute values are computed in the neighborhood of the peak of the soliton; this neighborhood is defined by $|x - t| < 2$ in equation (3.135)

```
      SSE=0.0D0
      DO 3 I=1,NG
         IF(ABS(X(I)-T).LT.2.0D0)THEN
            SRN=DSQRT(V2W2(I))
            SRE=DSQRT(EXACT(X(I),T))
            DIFF=SRN-SRE
            WRITE(NO,4)I,T,X(I),X(I)-T,SRN,SRE,DIFF
4           FORMAT(I6,F7.2,F7.1,F9.3,2F10.5,D13.3)
            SSE=SSE+DIFF**2
         END IF
3     CONTINUE
```

```
      SUBROUTINE PRINT(NI,NO)
      IMPLICIT DOUBLE PRECISION (A-H,O-Z)
      PARAMETER (NG=400)
      COMMON/T/          T,      NSTOP,      NORUN
     1      /Y/  V(0:NG),   W(0:NG)
     2      /F/  VT(0:NG),  WT(0:NG)
     3      /S/  VX(0:NG),  WX(0:NG), VXX(0:NG), WXX(0:NG),V2W2(0:NG),
     4             X(0:NG)
     5      /C/        XL,         XR,           Q,      ROOT2,       SSET
     6      /I/        IP
C...
C...  INCREMENT THE COUNTER FOR THE PLOTTED OUTPUT
      IP=IP+1
C...
C...  MONITOR PROGRESS OF THE SOLUTION
      WRITE(*,1)NORUN,IP,T
1     FORMAT(' NORUN = ',I2,'  IP = ',I3,'  TIME = ',F8.1)
C...
C...  WRITE THE NUMERICAL AND ANALYTICAL SOLUTIONS AND THE DIFFERENCE
C...  BETWEEN THE TWO IN THE NEIGHBORHOOD OF THE SOLITON PEAK.  COMPUTE
C...  THE SUM OF SQUARES OF ERRORS
      WRITE(NO,2)
2     FORMAT(5X,'I',6X,'T',3X,'X(I)',3X,'X(I)-T',3X,'ABS(UN)'
     +       3X,'ABS(UE)',9X,'DIFF')
      SSE=0.0D0
      DO 3 I=1,NG
         IF(ABS(X(I)-T).LT.2.0D0)THEN
            SRN=DSQRT(V2W2(I))
            SRE=DSQRT(EXACT(X(I),T))
            DIFF=SRN-SRE
            WRITE(NO,4)I,T,X(I),X(I)-T,SRN,SRE,DIFF
4           FORMAT(I6,F7.2,F7.1,F9.3,2F10.5,D13.3)
            SSE=SSE+DIFF**2
         END IF
3     CONTINUE
C...
C...  WRITE THE SUM OF SQUARES OF ERRORS
      WRITE(NO,5)SSE
5     FORMAT(/,' SSE = ',D11.3,//)
C...
C...  PLOT THE SOLUTION
      CALL PLOT1
C...
C...  ACCUMULATE THE SUM OF SQUARES OF ERRORS FOR OUTPUT AT THE END
C...  OF THE RUN
      SSET=SSET+SSE
      IF(IP.EQ.7)THEN
         WRITE(NO,6)SSET
6        FORMAT(//,' CUMULATIVE SSE = ',D11.3,//)
      END IF
      RETURN
      END
```

Program 3.5c Subroutine PRINT to Print and Plot the Solution of Equations (3.142)–(3.149).

Also, the running sum of squares of the difference between the numerical and analytical values of $|u|$, SSE, is computed for subsequent printing via FORMAT 5.

2. During each call to PRINT, the soliton is plotted by a call to subroutine PLOT1. Again, subroutine PLOT1 writes a file which is then read by a plotting system to produce the plot. This system is computer specific, but we include subroutine PLOT1 in Program 3.5d to indicate the source of the plotting in Figure 3.4.

3. Finally, at the end of the solution (the seventh call to PRINT, for which $IP = 7$), the sum of the squares of the differences between the numerical and analytical values of $|u|$ is printed via FORMAT 6.

Subroutine PLOT1 is listed in Program 3.5d. We can note the following points about subroutine PLOT1:

1. At the beginning of the solution, a plotting file, T.TOP, is opened and the scaling of the axes is defined

```
C...
C...    WRITE A FILE FOR TOP DRAWER PLOTTING
        IF(IP.EQ.1)THEN
           OPEN(2,FILE='T.TOP',STATUS='NEW')
           WRITE(2,12)
12         FORMAT(' SET LIMITS X FROM -30 TO 70 Y FROM 0. TO 1.5',/,
     1            ' SET FONT DUPLEX')
        END IF
```

2. During each call to PRINT, and therefore PLOT1, the numerical solution is written onto the plotting file as $|u|$ vs. x

```
C...
C...    WRITE THE CURRENT SOLUTION
        DO 16 I=1,NG
           SR=SQRT(V2W2(I))
C...
C...       USE THE FOLLOWING STATEMENT TO PLOT THE EXACT SOLUTION
C...       SR=DSQRT(EXACT(X(I),T))
           WRITE(2,11)X(I),SR
11         FORMAT(F10.3,F10.5)
16      CONTINUE
        WRITE(2,13)
13      FORMAT(' JOIN 1')
```

The 401 points of each solution are then joined to form a continuous curve by FORMAT 13.

```
      SUBROUTINE PLOT1
      IMPLICIT DOUBLE PRECISION (A-H,O-Z)
      PARAMETER (NG=400)
      COMMON/T/          T,      NSTOP,      NORUN
     1      /Y/   V(0:NG),   W(0:NG)
     2      /F/  VT(0:NG),  WT(0:NG)
     3      /S/  VX(0:NG),  WX(0:NG), VXX(0:NG), WXX(0:NG),V2W2(0:NG),
     4             X(0:NG)
     5      /C/         XL,         XR,         Q,     ROOT2,       SSET
     6      /I/         IP
C...
C...  WRITE A FILE FOR TOP DRAWER PLOTTING
      IF(IP.EQ.1)THEN
         OPEN(2,FILE='T.TOP',STATUS='NEW')
         WRITE(2,12)
12       FORMAT(' SET LIMITS X FROM -30 TO 70 Y FROM 0. TO 1.5',/,
     1          ' SET FONT DUPLEX')
      END IF
C...
C...  WRITE THE CURRENT SOLUTION
      DO 16 I=1,NG
         SR=SQRT(V2W2(I))
C...
C...     USE THE FOLLOWING STATEMENT TO PLOT THE EXACT SOLUTION
C...     SR=DSQRT(EXACT(X(I),T))
         WRITE(2,11)X(I),SR
11       FORMAT(F10.3,F10.5)
16    CONTINUE
      WRITE(2,13)
13    FORMAT(' JOIN 1')
C...
C...  WRITE THE PLOT LABELS
      IF(IP.EQ.7)THEN
      WRITE(2,14)NORUN
14    FORMAT(
     +' SET FONT DUPLEX'
     +,/,' TITLE 3.0 9.5
     + "Figure ',I2,'a: Solitons of the Cubic Schrodinger Equation"'
     +,/,' TITLE LEFT "          |U| = (V(x,t)**2 + W(x,t)**2)**0.5"'
     +,/,' TITLE BOTTOM "x"'
     +,/,' TITLE 4.5 0.75 "|U| of iU  + U   + Q(|U|**2)U = 0 vs x"'
     +,/,' TITLE 4.5 0.65 "          t     xx"'
     +,/,' TITLE 4.0 0.30 "t = 0, 5,..., 30, Q = 1, NG = 400,
     + Uniform Grid"')
C...
C...  NEXT PLOT
      WRITE(2,15)
15    FORMAT('NEW FRAME')
      END IF
      RETURN
      END
```

Program 3.5d Subroutine PLOT1 to Write the File that is Plotted in Figure 3.4.

3. At the end of the solution, the plot is labeled

```
C...
C...  WRITE THE PLOT LABELS
      IF(IP.EQ.7)THEN
      WRITE(2,14)NORUN
14    FORMAT(
     +' SET FONT DUPLEX'
     +,/,' TITLE 3.0 9.5
     + "Figure ',I2,'a: Solitons of the Cubic Schrodinger Equation"'
     +,/,' TITLE LEFT "          |U| = (V(x,t)**2 + W(x,t)**2)**0.5"'
     +,/,' TITLE BOTTOM "x"'
     +,/,' TITLE 4.5 0.75 "|U| of iU  + U   + Q(|U|**2)U = 0 vs x"'
     +,/,' TITLE 4.5 0.65 "           t     xx"'
     +,/,' TITLE 4.0 0.30 "t = 0, 5,..., 30, Q = 1, NG = 400,
     + Uniform Grid"')
```

Function EXACT in Program 3.5e performs a straightforward calculation of $|u(x,t)|^2$ from equation (3.135).

The main program that calls subroutines INITAL, PRINT, and RKF45, and indirectly DERV via RKF45, is similar to the several main programs discussed earlier, e.g., Program 2.2a, and therefore it is not listed here. The only additional requirement for this main program is to increase the work array used by RKF45 to accommodate the 802 ODEs.

The data file read by the main program is listed in Program 3.5f. Note that time, t, is defined over the interval $0 \le t \le 30$ with an output interval of 5, thus subroutine PRINT in Program 3.5c is called seven times (including $t = 0$).

The output from Programs 3.5a to f is listed in Table 3.3. The graphical output of the solution produced by subroutine PLOT1 appears in Figure 3.4. We can note the following about the numerical and graphical output of Table 3.3 and Figure 3.4:

1. Equations (3.141) to (3.149) are a stringent test problem since the traveling soliton has a very sharp spatial variation as indicated in Figure 3.4 (the solitons moving left to right start at $t = 0$ (centered at $x = 0, 5, 10, \ldots, 30$)). Also, the peak of the soliton, where the curvature is most extreme should have a value of $\sqrt{2} = 1.41412$; this peak value taken from Table 3.3 is listed below for several values of time, t

$$|u(x = 10, t = 10)| = 1.41522$$
$$|u(x = 20, t = 20)| = 1.41523$$
$$|u(x = 30, t = 30)| = 1.41499$$

```
      DOUBLE PRECISION FUNCTION EXACT(X,T)
      IMPLICIT DOUBLE PRECISION (A-H,O-Z)
      EXACT=2.0D0*(2.0D0/(DEXP(X-T)+DEXP(-(X-T))))**2
      RETURN
      END
```

Program 3.5e Function EXACT for the Exact Solution of Equation (3.135).

```
CUBIC SCHRODINGER EQUATION ON A UNIFORM GRID, DSS044, DOUBLE PRECISION
0.          30.        5.
  802                        0.00001
END OF RUNS
```

Program 3.5f Data File for the Solution of Equations (3.142) to (3.149).

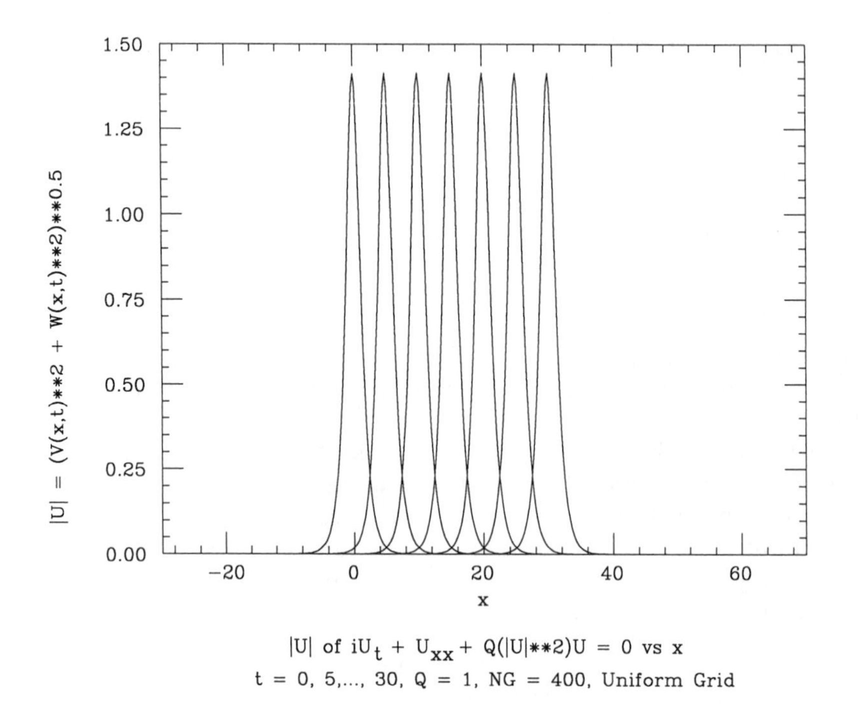

Figure 3.4 Graphical Output of Programs 3.5a to f.

The close agreement of these peak values with the exact value, 1.41412, as well as the generally good agreement between the numerical and analytical solutions in Table 3.3 indicates that the method of lines solution has acceptable accuracy; in other words, the fourth-order formulas in DSS044 performed satisfactorily on the 401-point grid. In fact, when the exact solution of equation (3.135) is plotted in the same way, the differences between the plotted numerical and analytical solutions are imperceptible.

In summary, an accurate method of lines solution for the CSE could be computed using library routines plus a series of straightforward routines to define the problem system equations; also, the computer run times were modest considering that a solution to 802 nonlinear ODEs was computed. Thus we advocate the use of the method of lines as an efficient and reliable procedure for the solution of PDEs. To explore this idea further, we go to the next chapter where we again consider PDEs first order in time, but we consider some extensions and variants of the methods discussed in this chapter.

Table 3.3 Numerical Output of Programs 3.5a to f. *Continued next pages.*

```
RUN NO. -   1  CUBIC SCHRODINGER EQUATION ON A UNIFORM GRID

INITIAL T -  0.000D+00

  FINAL T -  0.300D+02

  PRINT T -  0.500D+01

NUMBER OF DIFFERENTIAL EQUATIONS - 802

INTEGRATION ALGORITHM - RKF45

MAXIMUM INTEGRATION ERROR -  0.100D-04

     I       T    X(I)    X(I)-T   ABS(UN)    ABS(UE)          DIFF
   113    0.00    -1.8    -1.750   0.47710    0.47710     0.694D-17
   114    0.00    -1.5    -1.500   0.60118    0.60118     0.000D+00
   115    0.00    -1.3    -1.250   0.74889    0.74889     0.278D-16
   116    0.00    -1.0    -1.000   0.91649    0.91649     0.139D-16
   117    0.00    -0.8    -0.750   1.09232    1.09232     0.278D-16
   118    0.00    -0.5    -0.500   1.25415    1.25415     0.000D+00
   119    0.00    -0.3    -0.250   1.37114    1.37114     0.000D+00
   120    0.00     0.0     0.000   1.41421    1.41421     0.000D+00
   121    0.00     0.3     0.250   1.37114    1.37114     0.000D+00
   122    0.00     0.5     0.500   1.25415    1.25415     0.000D+00
   123    0.00     0.8     0.750   1.09232    1.09232     0.278D-16
   124    0.00     1.0     1.000   0.91649    0.91649     0.139D-16
   125    0.00     1.3     1.250   0.74889    0.74889     0.278D-16
   126    0.00     1.5     1.500   0.60118    0.60118     0.000D+00
   127    0.00     1.8     1.750   0.47710    0.47710     0.694D-17

SSE =    0.356D-32


     I       T    X(I)    X(I)-T   ABS(UN)    ABS(UE)          DIFF
   133    5.00     3.3    -1.750   0.47769    0.47710     0.587D-03
   134    5.00     3.5    -1.500   0.60189    0.60118     0.715D-03
   135    5.00     3.8    -1.250   0.74974    0.74889     0.857D-03
   136    5.00     4.0    -1.000   0.91750    0.91649     0.102D-02
   137    5.00     4.3    -0.750   1.09351    1.09232     0.119D-02
   138    5.00     4.5    -0.500   1.25549    1.25415     0.134D-02
   139    5.00     4.8    -0.250   1.37244    1.37114     0.129D-02
   140    5.00     5.0     0.000   1.41503    1.41421     0.817D-03
   141    5.00     5.3     0.250   1.37105    1.37114    -0.954D-04
   142    5.00     5.5     0.500   1.25310    1.25415    -0.106D-02
   143    5.00     5.8     0.750   1.09065    1.09232    -0.167D-02
   144    5.00     6.0     1.000   0.91463    0.91649    -0.186D-02
   145    5.00     6.3     1.250   0.74713    0.74889    -0.176D-02
   146    5.00     6.5     1.500   0.59966    0.60118    -0.152D-02
   147    5.00     6.8     1.750   0.47585    0.47710    -0.125D-02

SSE =    0.225D-04
```

Table 3.3 *Continued.*

I	T	X(I)	X(I)-T	ABS(UN)	ABS(UE)	DIFF
153	10.00	8.3	-1.750	0.47845	0.47710	0.135D-02
154	10.00	8.5	-1.500	0.60288	0.60118	0.170D-02
155	10.00	8.8	-1.250	0.75096	0.74889	0.207D-02
156	10.00	9.0	-1.000	0.91890	0.91649	0.241D-02
157	10.00	9.3	-0.750	1.09496	1.09232	0.264D-02
158	10.00	9.5	-0.500	1.25677	1.25415	0.262D-02
159	10.00	9.8	-0.250	1.37328	1.37114	0.214D-02
160	10.00	10.0	0.000	1.41522	1.41421	0.100D-02
161	10.00	10.3	0.250	1.37053	1.37114	-0.610D-03
162	10.00	10.5	0.500	1.25203	1.25415	-0.212D-02
163	10.00	10.8	0.750	1.08928	1.09232	-0.304D-02
164	10.00	11.0	1.000	0.91318	0.91649	-0.331D-02
165	10.00	11.3	1.250	0.74574	0.74889	-0.314D-02
166	10.00	11.5	1.500	0.59841	0.60118	-0.276D-02
167	10.00	11.8	1.750	0.47477	0.47710	-0.233D-02

SSE = 0.823D-04

I	T	X(I)	X(I)-T	ABS(UN)	ABS(UE)	DIFF
173	15.00	13.3	-1.750	0.47928	0.47710	0.218D-02
174	15.00	13.5	-1.500	0.60393	0.60118	0.275D-02
175	15.00	13.8	-1.250	0.75223	0.74889	0.334D-02
176	15.00	14.0	-1.000	0.92032	0.91649	0.383D-02
177	15.00	14.3	-0.750	1.09638	1.09232	0.406D-02
178	15.00	14.5	-0.500	1.25799	1.25415	0.384D-02
179	15.00	14.8	-0.250	1.37406	1.37114	0.291D-02
180	15.00	15.0	0.000	1.41533	1.41421	0.111D-02
181	15.00	15.3	0.250	1.36991	1.37114	-0.123D-02
182	15.00	15.5	0.500	1.25084	1.25415	-0.331D-02
183	15.00	15.8	0.750	1.08783	1.09232	-0.449D-02
184	15.00	16.0	1.000	0.91173	0.91649	-0.476D-02
185	15.00	16.3	1.250	0.74443	0.74889	-0.445D-02
186	15.00	16.5	1.500	0.59726	0.60118	-0.392D-02
187	15.00	16.8	1.750	0.47378	0.47710	-0.332D-02

SSE = 0.181D-03

I	T	X(I)	X(I)-T	ABS(UN)	ABS(UE)	DIFF
193	20.00	18.3	-1.750	0.48033	0.47710	0.323D-02
194	20.00	18.5	-1.500	0.60510	0.60118	0.392D-02
195	20.00	18.8	-1.250	0.75351	0.74889	0.462D-02
196	20.00	19.0	-1.000	0.92170	0.91649	0.521D-02
197	20.00	19.3	-0.750	1.09778	1.09232	0.545D-02
198	20.00	19.5	-0.500	1.25917	1.25415	0.502D-02
199	20.00	19.8	-0.250	1.37470	1.37114	0.356D-02
200	20.00	20.0	0.000	1.41523	1.41421	0.101D-02
201	20.00	20.3	0.250	1.36911	1.37114	-0.203D-02

Table 3.3 *Continued.*

```
  202  20.00   20.5    0.500   1.24958   1.25415   -0.457D-02
  203  20.00   20.8    0.750   1.08638   1.09232   -0.594D-02
  204  20.00   21.0    1.000   0.91032   0.91649   -0.617D-02
  205  20.00   21.3    1.250   0.74319   0.74889   -0.570D-02
  206  20.00   21.5    1.500   0.59624   0.60118   -0.494D-02
  207  20.00   21.8    1.750   0.47294   0.47710   -0.416D-02

SSE =    0.315D-03


    I       T   X(I)   X(I)-T  ABS(UN)   ABS(UE)         DIFF
  213  25.00   23.3   -1.750   0.48134   0.47710    0.424D-02
  214  25.00   23.5   -1.500   0.60634   0.60118    0.516D-02
  215  25.00   23.8   -1.250   0.75495   0.74889    0.606D-02
  216  25.00   24.0   -1.000   0.92322   0.91649    0.673D-02
  217  25.00   24.3   -0.750   1.09917   1.09232    0.685D-02
  218  25.00   24.5   -0.500   1.26018   1.25415    0.603D-02
  219  25.00   24.8   -0.250   1.37515   1.37114    0.401D-02
  220  25.00   25.0    0.000   1.41506   1.41421    0.847D-03
  221  25.00   25.3    0.250   1.36838   1.37114   -0.276D-02
  222  25.00   25.5    0.500   1.24843   1.25415   -0.572D-02
  223  25.00   25.8    0.750   1.08501   1.09232   -0.731D-02
  224  25.00   26.0    1.000   0.90889   0.91649   -0.759D-02
  225  25.00   26.3    1.250   0.74185   0.74889   -0.703D-02
  226  25.00   26.5    1.500   0.59509   0.60118   -0.608D-02
  227  25.00   26.8    1.750   0.47207   0.47710   -0.503D-02

SSE =    0.490D-03


    I       T   X(I)   X(I)-T  ABS(UN)   ABS(UE)         DIFF
  233  30.00   28.3   -1.750   0.48226   0.47710    0.516D-02
  234  30.00   28.5   -1.500   0.60745   0.60118    0.628D-02
  235  30.00   28.8   -1.250   0.75625   0.74889    0.737D-02
  236  30.00   29.0   -1.000   0.92467   0.91649    0.818D-02
  237  30.00   29.3   -0.750   1.10062   1.09232    0.830D-02
  238  30.00   29.5   -0.500   1.26138   1.25415    0.723D-02
  239  30.00   29.8   -0.250   1.37580   1.37114    0.466D-02
  240  30.00   30.0    0.000   1.41499   1.41421    0.780D-03
  241  30.00   30.3    0.250   1.36765   1.37114   -0.350D-02
  242  30.00   30.5    0.500   1.24724   1.25415   -0.691D-02
  243  30.00   30.8    0.750   1.08362   1.09232   -0.870D-02
  244  30.00   31.0    1.000   0.90751   0.91649   -0.897D-02
  245  30.00   31.3    1.250   0.74059   0.74889   -0.830D-02
  246  30.00   31.5    1.500   0.59397   0.60118   -0.721D-02
  247  30.00   31.8    1.750   0.47109   0.47710   -0.601D-02

SSE =    0.704D-03

CUMULATIVE SSE =    0.179D-02
```

References

Bateman, H. (1932). *Partial Differential Equations of Mathematical Physics.* Cambridge University Press, Cambridge.

Beyer, W. H. (1978). *CRC Handbook of Mathematical Sciences*, 5th ed., CRC Press, Inc., West Palm Beach, FL.

Carslaw, H. S. and J. C. Jaeger (1959). *Conduction of Heat in Solids*, Oxford Science Publications and Oxford University Press, Clarendon Press, Oxford, Second Edition.

Press, W. H., B. P. Flannery, S. A. Teukolsy and W. T. Vetterling (1986). *Numerical Recipes*, Cambridge University Press, Cambridge.

Schiesser, W. E. (1991). *The Numerical Method of Lines Integration of Partial Differential Equations*. Academic Press, San Diego.

Strang, G. (1986). *Introduction to Applied Mathematics*, Wellesley-Cambridge Press, Wellesley, MA.

chapter four

Partial Differential Equations First Order in Time (continued)

In this chapter, we continue the discussion of the method of lines solution of PDEs which are first order in time. In addition to considering other forms of PDEs first order in time, we also consider alternate methods for algebraically approximating the spatial derivatives; that is, previously we used *finite differences*, but other well-established methods can be used such as *finite elements*, *finite volumes*, and *weighted residuals*.

4.1 PDEs with First- and Second-Order Spatial Derivatives

We start by considering the advection equation, previously numbered (3.55)

$$\frac{\partial u}{\partial t} + v\frac{\partial u}{\partial x} = 0 \tag{4.1}$$

which is first order in space. Equation (4.1) models only convection with a constant velocity, v. However, in most physical situations, convection is not the only characteristic which must be considered; that is, all elements of the fluid will not flow at the same uniform velocity, v. For example, some *diffusion* or *dispersion* will also take place in a physical setting, which is typically characterized by a *diffusivity* or *dispersion coefficient*, D; then equation (4.1) becomes

$$\frac{\partial u}{\partial t} + v\frac{\partial u}{\partial x} = D\frac{\partial^2 u}{\partial x^2} \tag{4.2}$$

which is a *convective diffusion equation*, or a *hyperbolic parabolic equation*; it is hyperbolic due to the convective term, $v\partial u/\partial x$, and it is parabolic due to the diffusive term, $D\partial^2 u/\partial x^2$. This diffusive term occurs in the following way. If in addition to convection of heat into and out of the incremental section of length Δx in Figure 3.1, we consider diffusion (or conduction) into and out of the incremental section, then we have a difference of two additional terms in equation (3.1)

$$\begin{aligned} A_c\Delta x\rho C_p\frac{\partial u_1}{\partial t} &= vA_c\rho C_p u_1|_x - vA_c\rho C_p u_1|_{x+\Delta x} + A_h\Delta x U(u_2 - u_1) \\ &\quad - A_c D\rho C_p \left.\frac{\partial u_1}{\partial x}\right|_x - \left\{-A_c D\rho C_p \left.\frac{\partial u_1}{\partial x}\right|_{x+\Delta x}\right\} \end{aligned} \tag{4.3}$$

where we have used *Fourier's first law* (*Fick's first law*) for the heat flux, q, due to the gradient $\partial u_1/\partial x$

$$q = -A_c D \rho C_p \frac{\partial u_1}{\partial x} \tag{4.4}$$

The minus in equation (4.4) indicates the flux, q, is in the direction of decreasing u_1. If we now divide equation (4.3) by $A_c \Delta x \rho C_p$ and take the limit $\Delta x \to 0$, we obtain equation (4.2) for the case of no heat transfer ($U = 0$). Note that the diffusivity, D, has the units m^2/s, which is clear from equation (4.2). Also, for just a diffusive system (no convection, or $v = 0$), equation (4.2) reduces to

$$\frac{\partial u}{\partial t} = D \frac{\partial^2 u}{\partial x^2} \tag{4.5}$$

which is *Fourier's second law* (*Fick's second law*).

Since equation (4.2) is first order in t and second order in x, we require one initial condition and two boundary conditions. The initial condition is straightforward

$$u(x,0) = f(x) \tag{4.6}$$

Boundary conditions for equation (4.2) have been discussed at considerable length. We merely indicate some possibilities:

1. If the entering value of u at $x = 0$ is a prescribed function of t, $g(t)$, we have as the left-end boundary condition

$$u(0,t) = g(t) \tag{4.7}$$

2. If only accumulation and convection take place at the right boundary, $x = x_r$, we have as the right-end boundary condition

$$\frac{\partial u(x_r,t)}{\partial t} + v \frac{\partial u(x_r,t)}{\partial x} = 0 \tag{4.8}$$

Equation (4.8) is an acceptable boundary condition for equation (4.2) since it is only first order in x (it must be at least one order lower than the PDE for which it is a boundary condition).

Equations (4.2), (4.6), (4.7) and (4.8) constitute a complete convective diffusion problem, and a solution can be computed by the method of lines; for example, the convective derivative, $\partial u/\partial x$, can be computed by the five-point biased upwind approximations in `DSS020`, and the diffusive derivative, $\partial^2 u/\partial x^2$, can be computed by the five-point centered approximations in `DSS044`. In fact this has been done [Silebi et al. (1992)] for a more general form of equation (4.2)

$$\frac{\partial u}{\partial t} + v \frac{\partial u}{\partial x} = D \frac{\partial^2 u}{\partial x^2} - r(u,x,t) \tag{4.9}$$

where $r(u,x,t)$ is typically a volumetric rate of consumption due to chemical reaction, for example.

Equations (4.2), (4.6), (4.7) and (4.8) are all linear, and therefore in principle, can be solved analytically. In practice this could prove to be difficult, but a numerical solution is straightforward. If $r(u,x,t)$ is nonlinear in u, which is often the case for chemical reactions, equations (4.6), (4.7), (4.8) and (4.9) are essentially impossible to

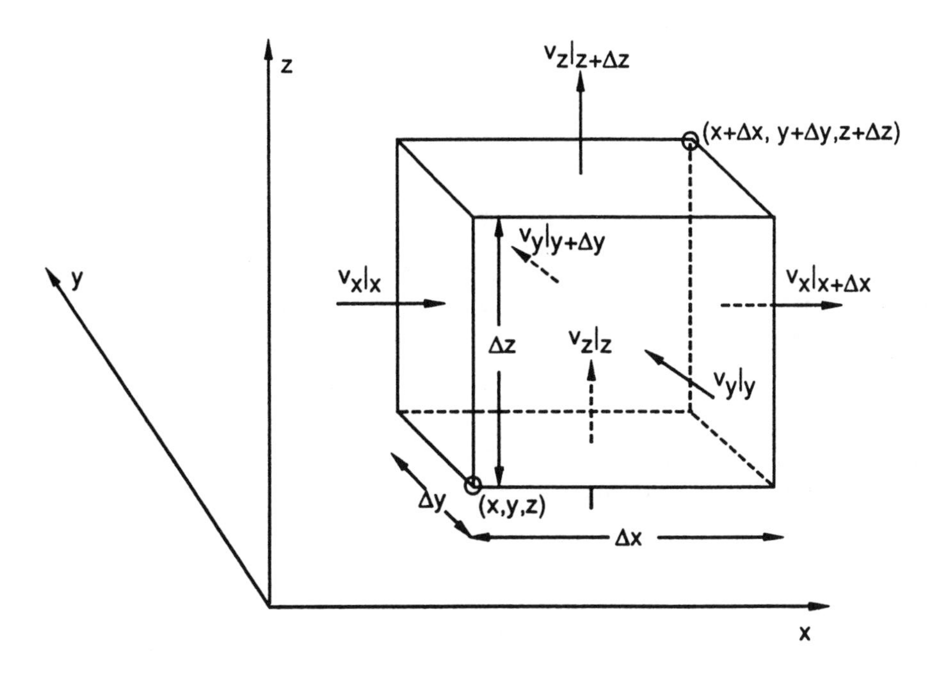

Figure 4.1 Incremental Volume of Fluid in Cartesian Coordinates.

solve analytically, and a numerical procedure must be followed. We have, however, a numerical method of lines procedure available which should accommodate essentially any form of $r(u, x, t)$. Thus, convective diffusion equations, at least in one spatial dimension, should be amenable to numerical solution by the methods discussed in Chapter 3.

Other sources of nonlinearity (in addition to $r(u, x, t)$ in equation (4.9)) in convective diffusion problems are the convective and diffusion terms themselves. We now consider a hyperbolic parabolic PDE with a nonlinear convective term followed by a PDE with both nonlinear convective and diffusion terms. The start of the problem formulation is the analysis of the flowing system indicated in Figure 4.1.

If a *mass (continuity) balance* is written for the incremental volume of Figure 4.1, we have

$$\begin{aligned} \Delta x \Delta y \Delta z \frac{\partial \rho}{\partial t} = & \; \Delta y \Delta z v_x \rho|_x - \Delta y \Delta z v_x \rho|_{x+\Delta x} \\ & + \Delta x \Delta z v_y \rho|_y - \Delta x \Delta z v_y \rho|_{y+\Delta y} \\ & + \Delta x \Delta y v_z \rho|_z - \Delta x \Delta y v_z \rho|_{z+\Delta z} \end{aligned} \tag{4.10}$$

where ρ = fluid density (kg/m^3); x, y, x = Cartesian spatial coordinates (m); Δx, Δy, Δz = incremental lengths in the x, y, and z directions (m); t = time (s); and v_x, v_y, v_z = components of fluid velocity in the x, y, and z directions (m/s).

The term $\Delta x \Delta y \Delta z \partial \rho / \partial t$ has the net units of kg/s for the mass accumulating in the incremental volume of Figure 4.1 per unit time. Similarly, the right-hand convective terms each have the units of kg/s for the mass per unit time flowing into or out of the incremental volume in each of the three directions.

Division of equation (4.10) by $\Delta x \Delta y \Delta z$ gives

$$\begin{aligned}\frac{\partial \rho}{\partial t} = &-\frac{v_x\rho|_{x+\Delta x} - v_x\rho|_x}{\Delta x}\\ &-\frac{v_y\rho|_{y+\Delta y} - v_y\rho|_y}{\Delta y}\\ &-\frac{v_z\rho|_{z+\Delta z} - v_z\rho|_z}{\Delta z}\end{aligned} \tag{4.11}$$

In the limit as $\Delta x \to 0$, $\Delta y \to 0$, $\Delta z \to 0$, the right-hand side divided differences in equation (4.11) beome partial derivatives, and we arrive at

$$\frac{\partial \rho}{\partial t} = -\frac{\partial(\rho v_x)}{\partial x} - \frac{\partial(\rho v_y)}{\partial y} - \frac{\partial(\rho v_z)}{\partial z} \tag{4.12}$$

Equation (4.12) is *the equation of continuity* in Cartesian coordinates. It can be written more generally for any coordinate system as

$$\frac{\partial \rho}{\partial t} = -\nabla \cdot (\rho \bar{v}) = -\rho \nabla \cdot \bar{v} - \bar{v} \cdot \nabla \rho \tag{4.13}$$

∇ is a *vector differential operator*, defined in Cartesian coordinates as

$$\nabla = \bar{\imath}\frac{\partial}{\partial x} + \bar{\jmath}\frac{\partial}{\partial y} + \bar{k}\frac{\partial}{\partial z} \tag{4.14}$$

∇ can be applied to a vector field or a scalar field, for which it is called the *divergence operator* or the *gradient operator*, respectively. We consider briefly these two cases.

1. *Divergence Operator*. For a vector field, with scalar Cartesian components v_x, v_y, and v_z,

$$\bar{v} = \bar{\imath}v_x + \bar{\jmath}v_y + \bar{k}v_z \tag{4.15}$$

the term $\nabla \cdot (\rho v)$ in equation (4.13) becomes

$$\nabla \cdot (\rho \bar{v}) = \frac{\partial(\rho v_x)}{\partial x} + \frac{\partial(\rho v_y)}{\partial y} + \frac{\partial(\rho v_z)}{\partial z} \tag{4.16}$$

which follows from equations (4.14) and (4.15) and the definition of the dot product between two vectors $\bar{a}$ and $\bar{b}$ and common angle θ defined as

$$\bar{a} \cdot \bar{b} = |\bar{a}||\bar{b}|\cos(\theta) \tag{4.17}$$

It follows immediately from (4.17) that

$$\bar{a} \cdot \bar{a} = |\bar{a}||\bar{a}| = |\bar{a}|^2$$

Also, if $\bar{a}$ and $\bar{b}$ are orthogonal unit vectors,

$$\bar{a} \cdot \bar{b} = |1||1|\cos(\theta) = \delta_{ij}$$

where δ_{ij} is the Kronecker delta

$$\delta_{ij} = \begin{matrix} 1, & i = j \\ 0, & i \neq j \end{matrix}$$

2. *Gradient Operator*. For a scalar field, $\rho = \rho(x, y, z)$, the term $\nabla\rho$ (in equation (4.13)) becomes from equation (4.14)

$$\nabla\rho = \bar{\imath}\frac{\partial\rho}{\partial x} + \bar{\jmath}\frac{\partial\rho}{\partial y} + \bar{k}\frac{\partial\rho}{\partial z} \tag{4.18}$$

For an incompressible fluid, $\rho = \text{const}$, $\nabla\rho = 0$ from equation (4.18) and therefore from equation (4.13),

$$\nabla \cdot \bar{v} = \bar{0} \tag{4.19}$$

Equation (4.19) is the *equation of continuity for an incompressible fluid.*

Finally, we note that the successive application of the divergence and gradient operators to a scalar field defines the *Laplacian of the scalar field*, ∇^2. In Cartesian coordinates, the Laplacian of the scalar field, $\phi = \phi(x, y, z)$ (from equations (4.14) and (4.18)), is

$$\begin{aligned}\nabla^2\phi = \nabla\cdot\nabla\phi &= \left\{\bar{\imath}\frac{\partial}{\partial x} + \bar{\jmath}\frac{\partial}{\partial y} + \bar{k}\frac{\partial}{\partial z}\right\}\cdot\left\{\bar{\imath}\frac{\partial\phi}{\partial x} + \bar{\jmath}\frac{\partial\phi}{\partial y} + \bar{k}\frac{\partial\phi}{\partial z}\right\} \\ &= \frac{\partial^2\phi}{\partial x^2} + \frac{\partial^2\phi}{\partial y^2} + \frac{\partial^2\phi}{\partial z^2}\end{aligned} \tag{4.20}$$

As an example application of the Laplacian operator, if ϕ in equation (4.20) is considered as the temperature, u, in equation (4.5), then the derivative $\partial^2u/\partial x^2$ in equation (4.5) is just the first term of the Laplacian of u. In other words, equation (4.5) could be generalized to three dimensions as $\partial u/\partial t = D\nabla^2u$.

Equation (4.13) has two dependent variables, ρ and v, so that a second equation is required, which is the *momentum balance* for the fluid. We consider now a momentum balance for the special case of *negligible viscous effects*. For the fluid in the incremental volume of Figure 4.1, the momentum balance is a vector equation with components in the three directions, x, y, and z. For the x component, we have:

$$\begin{aligned}\Delta x\Delta y\Delta z\frac{\partial(\rho v_x)}{\partial t} = \; & \Delta y\Delta z v_x(\rho v_x)|_x - \Delta y\Delta z v_x(\rho v_x)|_{x+\Delta x} \\ & +\Delta x\Delta z v_y(\rho v_x)|_y - \Delta x\Delta z v_y(\rho v_x)|_{y+\Delta y} \\ & +\Delta x\Delta y v_z(\rho v_x)|_z - \Delta x\Delta y v_z(\rho v_x)|_{z+\Delta z} \\ & +\Delta y\Delta z P|_x - \Delta y\Delta z P|_{x+\Delta x} \\ & +\Delta x\Delta y\Delta z(\rho g_x)\end{aligned} \tag{4.21}$$

Each of the terms in equation (4.21) requires some explanation.

1. $\Delta x\Delta y\Delta z\dfrac{\partial(\rho v_x)}{\partial t}$

   ρv_x is the momentum of the fluid due to the x-component of velocity, v_x, per unit volume of the fluid. Thus, this term is the accumulation of the x-component of momentum of the fluid in the incremental volume of Figure 4.1. In other words, it is the time rate of change of momentum in Newton's second law,

$$dm/dt = \sum_i F_i \tag{4.22}$$

   where m denotes momentum. This term has the units of $kg\text{-}m/s^2 = N$ (Newtons).

2. $\Delta y \Delta z v_x(\rho v_x)|_x - \Delta y \Delta z v_x(\rho v_x)|_{x+\Delta x}$

   The volumetric flow rate of the fluid through the face of the incremental volume in Figure 4.1 with area $\Delta y \Delta z$ is $\Delta y \Delta z v_x$. Thus, this term is the net rate of flow of the x-component of momentum into (or out of) the incremental volume in the x direction. Again, it has the units of $kg\text{-}m/s^2 = N$, and constitutes a force in the right-hand side of equation (4.22).

3. $\Delta x \Delta z v_y(\rho v_x)|_y - \Delta x \Delta z v_y(\rho v_x)|_{y+\Delta y}$

   The fluid can have an x-component of momentum, (ρv_x), while flowing through the face with area $\Delta x \Delta z$. Thus, this term is the net rate of flow of the x-component of momentum into (or out of) the incremental volume in the y-direction, and, of course, it has the units of $kg\text{-}m/s^2 = N$; that is, it is a force in the x-direction.

4. $\Delta x \Delta y v_z(\rho v_x)|_z - \Delta x \Delta y v_z(\rho v_x)|_{z+\Delta z}$

   As with the preceding term, the fluid can have an x-component of momentum, (ρv_x), while flowing through the face with area $\Delta x \Delta y$. Thus, this term is the net rate of flow of the x-component of momentum into (or out of) the incremental volume in the z-direction; it has the units of $kg\text{-}m/s^2 = N$, and it is a force in the x-direction.

5. $\Delta y \Delta z P|_x - \Delta y \Delta z P|_{x+\Delta x}$

   The fluid pressure, P, exerts a force in the x-direction on the face with area $\Delta y \Delta z$. Thus, this term is the net pressure force in the x-direction, with the units $(m^2)(N/m^2) = N$.

6. $\Delta x \Delta y \Delta z(\rho g_x)$

   A gravitational force in the x-direction acting on the mass in the incremental volume of Figure 4.1 is given in terms of the x-component of gravity, g_x. The units of this term are $(m^3)(kg/m^3)(m/s^2) = kg\text{-}m/s^2 = N$.

If equation (4.21) is divided by the incremental volume, $\Delta x \Delta y \Delta z$,

$$\begin{aligned}\frac{\partial(\rho v_x)}{\partial t} = &-\frac{v_x(\rho v_x)|_{x+\Delta x} - v_x(\rho v_x)|_x}{\Delta x}\\ &-\frac{v_y(\rho v_x)|_{y+\Delta y} - v_y(\rho v_x)|_y}{\Delta y}\\ &-\frac{v_z(\rho v_x)|_{z+\Delta z} - v_z(\rho v_x)|_z}{\Delta z}\\ &-\frac{P|_{x+\Delta x} - P|_x}{\Delta x} + \rho g_x \end{aligned} \tag{4.23}$$

We recognize a series of partial derivatives when $\Delta x \to 0$, $\Delta y \to 0$, $\Delta z \to 0$. Equation (4.23) then becomes in the limit

$$\frac{\partial(\rho v_x)}{\partial t} = -\frac{\partial\{v_x(\rho v_x)\}}{\partial x} - \frac{\partial\{v_y(\rho v_x)\}}{\partial y} - \frac{\partial\{v_z(\rho v_x)\}}{\partial z} - \frac{\partial P}{\partial x} + \rho g_x \tag{4.24}$$

which is the x-component of the momentum balance for the fluid in the incremental volume of Figure 4.1.

All three component momentum balances can be summarized by the vector equation

$$\frac{\partial(\rho \bar{v})}{\partial t} = -[\nabla \cdot \rho \bar{v}\bar{v}] - \nabla P + \rho \bar{g} \tag{4.25}$$

where in Cartesian coordinates

$$\bar{v} = \bar{\imath}v_x + \bar{\jmath}v_y + \bar{k}v_z \tag{4.26}$$

$$\bar{g} = \bar{\imath}g_x + \bar{\jmath}g_y + \bar{k}g_z \tag{4.27}$$

$$\nabla P = \bar{\imath}\frac{\partial P}{\partial x} + \bar{\jmath}\frac{\partial P}{\partial y} + \bar{k}\frac{\partial P}{\partial z} \tag{4.28}$$

$$\bar{v}\bar{v} = \begin{bmatrix} v_x v_x & v_x v_y & v_x v_z \\ v_y v_x & v_y v_y & v_y v_z \\ v_z v_x & v_z v_y & v_z v_z \end{bmatrix} \tag{4.29}$$

$$\begin{aligned}
[\nabla \cdot \rho\bar{v}\bar{v}] &= \left\{\bar{\imath}\frac{\partial}{\partial x} + \bar{\jmath}\frac{\partial}{\partial y} + \bar{k}\frac{\partial}{\partial z}\right\} \cdot \rho \begin{bmatrix} v_x v_x & v_x v_y & v_x v_z \\ v_y v_x & v_y v_y & v_y v_z \\ v_z v_x & v_z v_y & v_z v_z \end{bmatrix} \\
&= \bar{\imath}\left\{\frac{\partial(\rho v_x v_x)}{\partial x} + \frac{\partial(\rho v_y v_x)}{\partial y} + \frac{\partial(\rho v_z v_x)}{\partial z}\right\} \\
&\quad + \bar{\jmath}\left\{\frac{\partial(\rho v_x v_y)}{\partial x} + \frac{\partial(\rho v_y v_y)}{\partial y} + \frac{\partial(\rho v_z v_y)}{\partial z}\right\} \\
&\quad + \bar{k}\left\{\frac{\partial(\rho v_x v_z)}{\partial x} + \frac{\partial(\rho v_y v_z)}{\partial y} + \frac{\partial(\rho v_z v_z)}{\partial z}\right\}
\end{aligned} \tag{4.30}$$

$\bar{v}\bar{v}$ given by equation (4.29) is a *second-order tensor* with nine components (scalars and vectors can be considered as zeroth- and first-order tensors, with one and three components, respectively). The $\bar{\imath}$ component of equation (4.25) is equation (4.24). Equations (4.13) and (4.25), which are the *general continuity* and *momentum balances* for an inviscid (zero viscosity) fluid, can be written in other coordinate systems for which the various differential operators, that is, $\nabla\cdot$, ∇, (and subsequently) ∇^2, are defined.

Also, the fluid pressure must be available to use in the term ∇P in equation (4.25). Generally, the fluid pressure is computed from an equation of state, e.g.,

$$P = P(\rho, T) \tag{4.31}$$

T is given for an isothermal system, or, more generally, is computed from an energy balance for a nonisothermal system. Alternatively, the pressure might be computed from an equation of state of the form

$$P = P(\rho, u) \tag{4.32}$$

where u is the fluid internal energy computed from an energy balance.

To illustrate how the continuity and momentum equations can be combined, we consider first equation (4.24)

$$\frac{\partial(\rho v_x)}{\partial t} = -\frac{\partial\{v_x(\rho v_x)\}}{\partial x} - \frac{\partial\{v_y(\rho v_x)\}}{\partial y} - \frac{\partial\{v_z(\rho v_x)\}}{\partial z} - \frac{\partial P}{\partial x} + \rho g_x$$

or

$$\begin{aligned}
\rho\frac{\partial v_x}{\partial t} + v_x\left\{\frac{\partial \rho}{\partial t} + \frac{\partial(\rho v_x)}{\partial x} + \frac{\partial(\rho v_y)}{\partial y} + \frac{\partial(\rho v_z)}{\partial z}\right\} \\
= -\rho v_x\frac{\partial v_x}{\partial x} - \rho v_y\frac{\partial v_x}{\partial y} - \rho v_z\frac{\partial v_x}{\partial z} - \frac{\partial P}{\partial x} + \rho g_x
\end{aligned} \tag{4.33}$$

where the term in brackets is zero from equation (4.12). Equation (4.33) can be rearranged to

$$\rho\left\{\frac{\partial v_x}{\partial t}+v_x\frac{\partial v_x}{\partial x}+v_y\frac{\partial v_x}{\partial y}+v_z\frac{\partial v_x}{\partial z}\right\}=-\frac{\partial P}{\partial x}+\rho g_x \tag{4.34}$$

or

$$\rho\frac{Dv_x}{Dt}=-\frac{\partial P}{\partial x}+\rho g_x \tag{4.35}$$

where D/Dt is the *substantial derivative*, which, in Cartesian coordinates, is

$$\frac{D}{Dt}=\frac{\partial}{\partial t}+v_x\frac{\partial}{\partial x}+v_y\frac{\partial}{\partial y}+v_z\frac{\partial}{\partial z} \tag{4.36}$$

Equation (4.35) generalized to a vector equation is

$$\rho\frac{D\bar{v}}{Dt}=-\nabla P+\rho\bar{g} \tag{4.37}$$

which is the famous *Euler equation* of fluid mechanics. If viscous forces are included in the preceding analysis, then equation (4.37) becomes

$$\rho\frac{D\bar{v}}{Dt}=-\nabla P-\mu\nabla^2\bar{v}+\rho\bar{g} \tag{4.38}$$

where μ is the viscosity for an incompressible Newtonian fluid. Equation (4.38) is the *Navier Stokes equation* of fluid mechanics. $\nabla^2\bar{v}$ is the Laplacian of the vector field $\bar{v}$, which in Cartesian coordinates is

$$\nabla^2\bar{v}=\bar{\imath}\nabla^2 v_x+\bar{\jmath}\nabla^2 v_y+\bar{k}\nabla^2 v_z \tag{4.39}$$

where the three right-hand terms are the Laplacian of a scalar, e.g., the $\bar{\imath}$ component is

$$\nabla^2 v_x=\frac{\partial^2 v_x}{\partial x^2}+\frac{\partial^2 v_x}{\partial y^2}+\frac{\partial^2 v_x}{\partial z^2} \tag{4.40}$$

If the $\bar{\imath}$ component of equation (4.38) is written in one dimension, and pressure and gravitational effects are neglected, we obtain

$$\frac{\partial v_x}{\partial t}=-v_x\frac{\partial v_x}{\partial x}+(\mu/\rho)\frac{\partial^2 v_x}{\partial x^2} \tag{4.41}$$

Note that the *kinematic viscosity*, (μ/ρ), has the units of $(kg/m\text{-}s)(m^3/kg)=m^2/s$, i.e., the units of a diffusivity.

Equation (4.41) is the *one-dimensnional Burgers' equation*, and is the starting point for the next PDE example. Burgers' equation is a standard test problem for PDE numerical methods for the following reasons:

1. It is a nonlinear equation with known analytical solutions (the solutions depend on the particular initial and boundary conditions).

2. It can be made increasingly difficult to solve numerically by the particular choice of initial conditions and the value of μ/ρ. Concerning this latter point, note that:

   (2.1) For small μ/ρ, Burgers' equation is strongly (first-order) hyperbolic, while for large μ/ρ it is strongly parabolic; for the hyperbolic case, the solution can exhibit steep moving fronts which are difficult to resolve numerically.

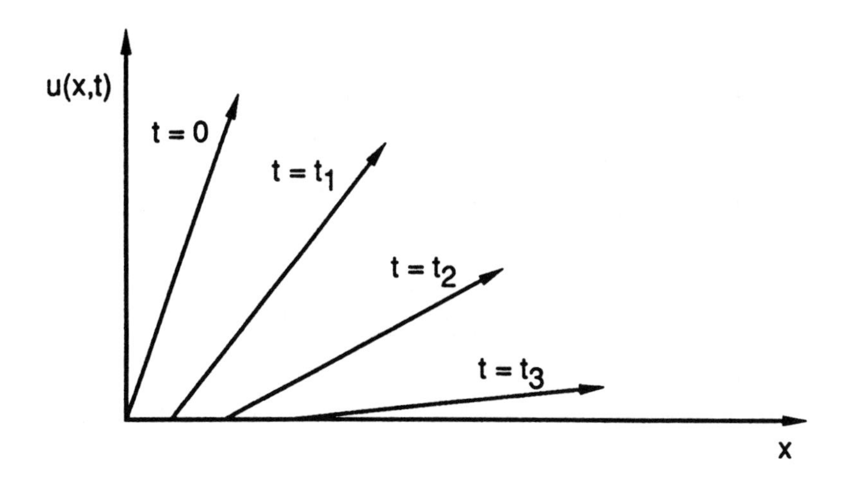

Figure 4.2 (a) Evolution of the Velocity Profiles from equation (4.42) for $v_x(x,t)$ Increasing with Increasing x.

(2.2) For $\mu/\rho = 0$, we have (from equation (4.41))

$$\frac{\partial v_x}{\partial t} = -v_x \frac{\partial v_x}{\partial x} \tag{4.42}$$

which is the *inviscid Burgers' equation*. Note that it resembles the advection equation, equation (4.1) with regard to the number and types of derivatives. The essential difference between equations (4.1) and (4.42) is that for the latter, the dependent variable is the velocity, $v_x(x,t)$, (rather than u as in equation (4.1)) and the convective term, $-v_x \partial v_x/\partial x = -\partial((1/2)v_x{}^2)/\partial x$ is nonlinear in v_x.

Also, as we observed with the advection equation, equation (4.42) defines the way the velocity, v_x, "flows" or moves, say left to right, with velocity v_x. This interpretation of equation (4.42) leads to two important cases. If the velocity, v_x, is an increasing function of x, then the velocity will flow faster at larger values of x, as depicted in Figure 4.2a, and the *velocity profile will flatten with time*.

A simple example of this case corresponds to the initial and boundary conditions for equation (4.42)

$$v_x(x,0) = x, v_x(0,t) = 0 \tag{4.43)(4.44}$$

for which the analytical solution (to equations (4.42), (4.43) and (4.44)) is

$$v_x(x,t) = \frac{x}{1+t} \tag{4.45}$$

Note that initially, $v_x(x,0) = x$, that is, an increasing function of x as depicted in Figure 4.2a; thus the velocity profiles flatten with time and, in fact, $v_x(x,\infty) = 0$ for finite x.

However, if at some point in time the velocity, v_x, is a decreasing function of x, then the velocity will flow faster at smaller values of x, as depicted in Figure 4.2b, and *the velocity profile will sharpen with time*. If this front sharpening or steepening process continues long enough, a shock (or discontinuity) can develop.

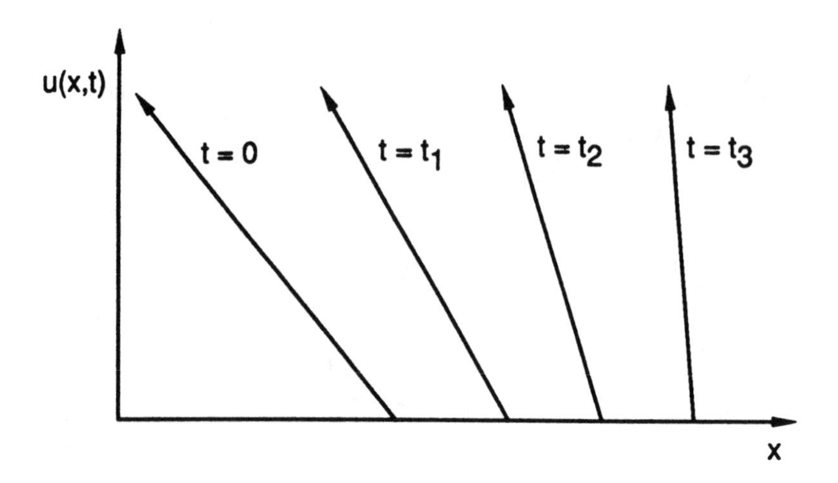

Figure 4.2 (b) Evolution of the Velocity Profiles from Equation (4.42) for $v_x(x,t)$ Decreasing with Increasing x.

We now consider a case of Burgers' equation, (4.41), for which front sharpening takes place. The essential difference between equations (4.41) and (4.42) is that the former has the viscous term, $(\mu/\rho)(\partial^2 v_x/\partial x^2)$, which tends to reduce front sharpening through diffusion. In other words, the progression of front sharpening, possibly toward a shock, will not occur as rapidly for equation (4.41) as for equation (4.42). The initial and boundary conditions we will use for Burgers' equation, equation (4.41), are given by [Madsen et al. (1976)]

$$v_x(x,t) = \phi(x,t) = \frac{0.1e^{-a} + 0.5e^{-b} + e^{-c}}{e^{-a} + e^{-b} + e^{-c}} \tag{4.46}$$

where

$$a = (0.05/\mu)(x - 0.5 + 4.95t)$$

$$b = (0.25/\mu)(x - 0.5 + 0.75t)$$

$$c = (0.5/\mu)(x - 0.375)$$

where we now use μ in place of μ/ρ. $\phi(x,t)$ of equation (4.46) is an analytical solution to equation (4.41), and the initial and boundary conditions for equation (4.41) are taken to be consistent with equation (4.46)

$$v_x(x,0) = \phi(x,0) \tag{4.47}$$

$$v_x(0,t) = \phi(0,t), \qquad v_x(1,y) = \phi(1,t) \qquad (4.48)(4.49)$$

We now consider the method of lines programming of equation (4.41) and equations (4.47) to (4.49). Also, in the programming and associated discussion, we follow the usual convention of calling the dependent variable of this PDE problem $u(x,t)$ rather than $v_x(x,t)$. Subroutine `INITAL` for the method of lines solution of equations (4.41) and (4.47) to (4.49) on a 201-point grid is given in Program 4.1a.

```
      SUBROUTINE INITAL
      IMPLICIT DOUBLE PRECISION (A-H,O-Z)
      PARAMETER (NX=201)
      COMMON/T/          T,      NFIN,      NORUN
     1      /Y/     U(NX)
     2      /F/    UT(NX)
     3      /S/    UX(NX),   UXX(NX),      X(NX)
     4      /C/       VIS
C...
C...  PROBLEM PARAMETERS
      VIS=0.003D+00
C...
C...  SPATIAL GRID AND THE INITIAL CONDITION
      DO 1 I=1,NX
         X(I)=DFLOAT(I-1)/DFLOAT(NX-1)
         U(I)=PHI(X(I),0.D+00)
1     CONTINUE
C...
C...  INITIAL DERIVATIVES
      CALL DERV
      RETURN
      END
```

Program 4.1a Subroutine INITAL for Initial Condition (4.47).

We can note the following points about subroutine INITAL:

1. $u(x,t)$ (rather than $v_x(x,t)$), $\partial u/\partial t$, $\partial u/\partial x$, and $\partial u^2/\partial x^2$ are in arrays U, UT, UX, and UXX, respectively.

2. μ is given the value 0.003 (as variable VIS) which is small enough to produce a strongly hyperbolic problem with a steep moving front.

3. The uniform spatial grid on 201 points for $0 \le x \le 1$ is in array X.

4. The initial condition of equation (4.47) is set in DO loop 1 by calling function PHI which implements equation (4.46). This initial condition decreases with increasing x so the solution to equation (4.41) has a front-sharpening characteristic, i.e., the calculation of the solution becomes increasingly difficult with increasing t.

Function PHI is listed in Program 4.1b.
We can note the following points about function PHI:

1. μ is passed from subroutine INITAL (where is it set to a value of 0.003) to funtion PHI through COMMON/C/ so it can be used in the calculation of $u(x,t)$ according to equation (4.46).

2. PHI was programmed to run on a short word length (32-bit) computer, and as t (the second argument, T) in $\phi(x,t)$ increased beyond $t = 0.9$, the exponential functions in the numerator and denominator of equation (4.46) began to overflow. Therefore, for $t \ge 0.9$, the right-hand side of equation (4.46) was multiplied by e^c/e^c to improve the scaling of the exponentials. However, computational problems developed for small t (i.e., $t < 0.5$) with this scaling; thus the original form of equation (4.46) was retained for $t < 0.9$.

```
      DOUBLE-PRECISION FUNCTION PHI(X,T)
C...
C...  FUNCTION PHI(T,X) COMPUTES THE EXACT SOLUTION FOR COMPARISON
C...  WITH THE NUMERICAL SOLUTION.  IT IS ALSO USED TO DEFINE THE
C...  INITIAL AND BOUNDARY CONDITIONS FOR THE NUMERICAL SOLUTION
C...
      IMPLICIT DOUBLE PRECISION (A-H,O-Z)
      COMMON/C/       VIS
C...
C...  ANALYTICAL SOLUTION TO BURGERS' EQUATION
      A=(0.05D+00/VIS)*(X-0.5D+00+4.95D+00*T)
      B=(0.25D+00/VIS)*(X-0.5D+00+0.75D+00*T)
      C=( 0.5D+00/VIS)*(X-0.375D+00)
      IF(T.LT.0.9D+00)THEN
         EA=DEXP(-A)
         EB=DEXP(-B)
         EC=DEXP(-C)
         PHI=(0.1D+00*EA+0.5D+00*EB+EC)/(EA+EB+EC)
      ELSE
         ECMA=DEXP(C-A)
         ECMB=DEXP(C-B)
         PHI=(0.1D+00*ECMA+0.5D+00*ECMB+1.0D+00)/(ECMA+ECMB+1.0D+00)
      END IF
      RETURN
      END
```

Program 4.1b Function PHI for the Analytical Solution of Equation (4.46).

With initial condition (4.47) programmed, we can proceed with the method of lines calculation of $\partial u/\partial t$ from equation (4.41), which is implemented in subroutine DERV in Program 4.1c.

We can note the following points about subroutine DERV:

1. Boundary conditions (4.48) and (4.49) are first programmed by a call to function PHI for $x = 0$ and $x = 1$

```
C...
C...  BOUNDARY CONDITIONS
      U(1) =PHI(X(1) ,T)
      U(NX)=PHI(X(NX),T)
      UT(1) =0.D+00
      UT(NX)=0.D+00
```

   Also, the temporal derivatives $\partial u(0,t)/\partial t$ and $\partial u(1,t)/\partial t$ are set to zero to prevent the ODE integrator from moving the boundary values, $u(0,t)$ and $u(1,t)$, from their values given by equation (4.46).

2. The derivative $\partial u/\partial x$ in equation (4.41) is then computed by a call to subroutine DSS020

```
C...
C...  UX
      CALL DSS020(0.0D0,1.0D0,NX,U,UX,1.0D0)
```

```
      SUBROUTINE DERV
      IMPLICIT DOUBLE PRECISION (A-H,O-Z)
      PARAMETER (NX=201)
      COMMON/T/          T,      NFIN,     NORUN
     1      /Y/     U(NX)
     2      /F/    UT(NX)
     3      /S/    UX(NX),   UXX(NX),      X(NX)
     4      /C/       VIS
C...
C...  BOUNDARY CONDITIONS
      U(1) =PHI(X(1) ,T)
      U(NX)=PHI(X(NX),T)
      UT(1) =0.D+00
      UT(NX)=0.D+00
C...
C...  UX
      CALL DSS020(0.0D0,1.0D0,NX,U,UX,1.0D0)
C...
C...  UXX
      NL=1
      NU=1
      CALL DSS044(0.0D0,1.0D0,NX,U,UX,UXX,NL,NU)
C...
C...  BURGERS' EQUATION
      DO 2 I=2,NX-1
         UT(I)=VIS*UXX(I)-U(I)*UX(I)
2     CONTINUE
      RETURN
      END
```

Program 4.1c Subroutine DERV for Equation (4.41).

Note that the last argument of DSS020 is 1.0D0 so that the five-point biased upwind approximations are used corresponding to a positive velocity. This was done since $u(x,t)$ from equation (4.41) is positive for all x and t. In other words, the convective term $-u\partial u/\partial x$ has a positive velocity.

3. The derivative $\partial^2 u/\partial x^2$ in equation (4.41) is next computed by a call to subroutine DSS044, with the specification of Dirichlet boundary conditions according to equations (4.48) and (4.49)

```
C...
C...  UXX
      NL=1
      NU=1
      CALL DSS044(0.0D0,1.0D0,NX,U,UX,UXX,NL,NU)
```

4. Burgers' equation, (4.41), is finally programmed in DO loop 2 for grid points 2 to 200

```
C...
C...  BURGERS' EQUATION
      DO 2 I=2,NX-1
```

```
      UT(I)=VIS*UXX(I)-U(I)*UX(I)
2     CONTINUE
      RETURN
      END
```

Again, we note two of the principal advantages of the numerical method of lines: (a) the straightforward solution of nonlinear problems and (b) the close resemblance of the programming to the PDE(s).

This essentially completes the programming of equations (4.41) and (4.47) to (4.49). Basically, all that remains is to display the output in a form that can be easily interpreted. This is not necessarily a trivial task as the number and complexity of the PDEs increases. Here we have a solution over 201 points; thus some graphical output is required to visualize the numerical solution and how it compares with the analytical solution. Numerical and graphical output is produced by subroutine PRINT, listed in Program 4.1d.

We can note the following points about subroutine PRINT:

1. The numerical solution in array U is stored in array UN for subsequent plotting by a call to subroutione PLOT1

```
C...
C...  STORE THE NUMERICAL SOLUTION FOR SUBSEQUENT PLOTTING VIA
C...  SUBROUTINE PLOT1
      DO 10 I=1,NX
         UN(IP,I)=U(I)
10    CONTINUE
```

   The index IP has the values of 1 to 6 corresponding to six calls to PRINT for $t = 0, 0.2, 0.4, \ldots, 1.0$.

2. During each of the six calls to PRINT, the analytical solution given by equation (4.46), UANAL, and the difference between the numerical and analytical solutions, ERROR, are computed and printed

```
      UANAL=PHI(X(I),T)
      ERROR=U(I)-UANAL
```

3. Also, during the six calls to PRINT, the diffusive and convective terms in equation (4.41), $\mu\partial^2 u/\partial x^2$ and $-u\partial u/\partial x$, are computed and printed (as DIFF and CONV).

```
      DIFF=VIS*UXX(I)
      CONV=-U(I)*UX(I)
```

   This programming illustrates an important technique in the numerical study of PDEs, the printing of all of the terms in the PDE(s) to observe the magnitudes and relative contributions of the terms. This procedure is useful for detecting programming errors (e.g., a term may have a magnitude which is not consistent with its physical meaning) and for understanding the relative contributions of the terms (e.g., in the case of equation (4.41), the relative contributions of the diffusive term, $\mu\partial^2 u/\partial x^2$, and the convective term, $-u\partial u/\partial x$).

```
      SUBROUTINE PRINT(NI,NO)
      IMPLICIT DOUBLE PRECISION (A-H,O-Z)
      PARAMETER (NX=201)
      COMMON/T/          T,      NFIN,     NORUN
     1      /Y/     U(NX)
     2      /F/    UT(NX)
     3      /S/    UX(NX),   UXX(NX),     X(NX)
     4      /C/       VIS
     5      /P/  UN(6,NX),        IP
C...
C...  INITIALIZE A COUNTER FOR THE PLOTTING
      DATA IP/0/
C...
C...  MONITOR THE SOLUTION
      IP=IP+1
      WRITE(*,*)IP,T
C...
C...  STORE THE NUMERICAL SOLUTION FOR SUBSEQUENT PLOTTING VIA SUBROU-
C...  TINE PLOT1
      DO 10 I=1,NX
         UN(IP,I)=U(I)
10    CONTINUE
C...
C...  PRINT THE NUMERICAL AND ANALYTICAL SOLUTIONS, THE DIFFERENCE
C..   BETWEEN THE SOLUTIONS AND THE TERMS IN BURGERS' EQUATION
      WRITE(NO,2)T
2     FORMAT(1H ,//,5H T = ,F5.2,/,
     1 9X,1HX,5X,5HU NUM,4X,6HU ANAL,5X,5HERROR,3X,7HVIS*UXX,5X,5H-U*UX,
     2 8X,2HUT)
      DO 3 I=1,NX
         UANAL=PHI(X(I),T)
         ERROR=U(I)-UANAL
         DIFF=VIS*UXX(I)
         CONV=-U(I)*UX(I)
         WRITE(NO,4)X(I),U(I),UANAL,ERROR,DIFF,CONV,UT(I)
4        FORMAT(F10.3,6F10.5)
3     CONTINUE
C...
C...  PLOT THE SOLUTION
      CALL PLOT1
C...
C...  REINITIALIZE THE COUNTER FOR PLOTTING
      IF(IP.EQ.6)IP=0
      RETURN
      END
```

Program 4.1d Subroutine PRINT to Display the Numerical and Analytical Solutions of Equations (4.41) and (4.47) to (4.49).

4. During each call to PRINT, a call to PLOT1 to write the numerical and analytical solutions to a file is then sent to a plotting system. This approach to plotting was illustrated in Program 3.5d, and therefore PLOT1 is not listed here.

```
MADSEN, ET AL, LAPIDUS, ET AL, NUM METH DIFF SYS, PP. 236-237
0.         1.0        0.2
  201                     0.00001
END OF RUNS
```

Program 4.1e Data File for Equations (4.41) and (4.47) to (4.49).

The data file for Programs 4.1a to d is listed in Program 4.1e.

201 ODEs are specified to be integrated with an error tolerance of 0.00001. The integration was performed by subroutine RKF45 called by a main program similar to Program 2.2a; thus it is not listed here.

Abbreviated numerical and graphical output from Programs 4.1a to e are given in Table 4.1 and Figure 4.3.

We can note the following points about the numerical and analytical solutions to Equations (4.41) and (4.47) to (4.49):

1. The graphical output of Figure 4.3 demonstrates the front-steepening characteristic of the solutions with increasing t as discussed previously (see Figure 4.2b); (the curves from left to right are for $t = 0, 0.2, 0.4, \ldots, 1$). Note that the initial condition $u(x, 0)$ decreases with increasing x, which initiates the front steepening since the velocity at small x is greater than the velocity at larger x. Figure 4.3 suggests that if the solution is allowed to continue beyond $t = 1$, the front would steepen to the point where it could not be resolved spatially with the 201-point grid.

2. The output in Table 4.1 indicates that the numerical and analytical solutions agree to about three figures where the solution has the most rapid spatial variation (at $t = 1$) and only about 10 numerical solution points occur along the front out of a total of 201 points. Thus, this region of sharp spatial variation has the poorest numerical resolution of the solution, as illustrated by the following portion of the numerical and analytical solutions at $t = 1$ (from Table 4.1)

```
0.890   0.96310   0.96356  -0.00046  -2.23238   5.18801   2.95563
0.895   0.92486   0.92622  -0.00136  -3.86818   9.57409   5.70591
0.900   0.85449   0.85695  -0.00246  -5.44862  15.47039  10.02176
0.905   0.74003   0.74286  -0.00283  -5.09351  20.20669  15.11318
0.910   0.58536   0.58739  -0.00202  -1.21970  19.52242  18.30271
0.915   0.42125   0.42237  -0.00112   3.81360  13.18044  16.99404
0.920   0.28701   0.28780  -0.00079   5.80658   6.42348  12.23006
0.925   0.19913   0.19972  -0.00059   4.68122   2.60082   7.28204
0.930   0.14971   0.15005  -0.00034   2.83677   1.03251   3.86929
0.935   0.12422   0.12437  -0.00016   1.50290   0.43456   1.93746
```

As Figure 4.3 indicates, these differences between the numerical and analytical solutions are essentially imperceptible when compared graphically. Outside of this region of sharp spatial variation, the numerical and analytical solutions agree to better than three figures. However, if better resolution is required, the number of grid points can easily be increased (recall the 401-point grid for the cubic Schrödinger equation or Programs 3.5a to f). Note also that at $t = 0.905$, the diffusive and convective terms of equation (4.41) have the relatively large values of $\mu \partial^2 u/\partial x^2 = -5.09351$ and $-u \partial u/\partial x = 20.20669$; the relative magnitudes of these two terms indicates that equation (4.41), for the parameters chosen, particularly $\mu = 0.003$, is more hyperbolic than parabolic.

Table 4.1 Abbreviated Numerical Output from Programs 4.1a to e. *Continued next pages.*

```
RUN NO. -    1  MADSEN, ET AL, LAPIDUS, ET AL, NUM METH DIFF SYS,
                PP. 236-237

INITIAL T -  0.000D+00

  FINAL T -  0.100D+01

  PRINT T -  0.200D+00

NUMBER OF DIFFERENTIAL EQUATIONS - 201

INTEGRATION ALGORITHM - RKF45

MAXIMUM INTEGRATION ERROR -  0.100D-04


T =  0.00
```

X	U NUM	U ANAL	ERROR	VIS*UXX	-U*UX	UT
0.000	1.00000	1.00000	0.00000	0.00000	0.00000	0.00000
0.005	1.00000	1.00000	0.00000	0.00000	0.00000	0.00000
0.010	1.00000	1.00000	0.00000	0.00000	0.00000	0.00000
0.015	1.00000	1.00000	0.00000	0.00000	0.00000	0.00000
0.020	1.00000	1.00000	0.00000	0.00000	0.00000	0.00000
0.025	1.00000	1.00000	0.00000	0.00000	0.00000	0.00000
		.			.	
		.			.	
		.			.	
0.200	0.99237	0.99237	0.00000	-0.15182	0.62210	0.47028
0.205	0.98851	0.98851	0.00000	-0.22312	0.92518	0.70206
0.210	0.98278	0.98278	0.00000	-0.32265	1.36239	1.03975
0.215	0.97433	0.97433	0.00000	-0.45538	1.97721	1.52183
0.220	0.96207	0.96207	0.00000	-0.61975	2.80935	2.18960
0.225	0.94464	0.94464	0.00000	-0.79891	3.87303	3.07413
0.230	0.92057	0.92057	0.00000	-0.94984	5.12075	4.17091
0.235	0.88865	0.88865	0.00000	-0.99956	6.40336	5.40380
0.240	0.84853	0.84853	0.00000	-0.86573	7.46166	6.59593
0.245	0.80134	0.80134	0.00000	-0.51113	7.99704	7.48591
0.250	0.75000	0.75000	0.00000	0.00000	7.82085	7.82085
0.255	0.69866	0.69866	0.00000	0.51113	6.98063	7.49176
0.260	0.65147	0.65147	0.00000	0.86573	5.73751	6.60324
0.265	0.61135	0.61135	0.00000	0.99955	4.40904	5.40859
0.270	0.57943	0.57943	0.00000	0.94984	3.22262	4.17246
0.275	0.55536	0.55536	0.00000	0.79890	2.27485	3.07375
0.280	0.53793	0.53793	0.00000	0.61975	1.56891	2.18866
0.285	0.52567	0.52567	0.00000	0.45537	1.06562	1.52099
0.290	0.51722	0.51722	0.00000	0.32264	0.71653	1.03917
0.295	0.51149	0.51149	0.00000	0.22311	0.47861	0.70172
0.300	0.50763	0.50763	0.00000	0.15181	0.31828	0.47009
		.			.	
		.			.	
		.			.	

Table 4.1 *Continued.*

0.500	0.30000	0.30000	0.00000	0.00000	2.00099	2.00099
0.505	0.26697	0.26697	0.00000	0.21397	1.73230	1.94627
0.510	0.23570	0.23570	0.00000	0.38409	1.40954	1.79363
0.515	0.20758	0.20758	0.00000	0.48437	1.08849	1.57286
0.520	0.18344	0.18344	0.00000	0.51306	0.80742	1.32049
0.525	0.16355	0.16355	0.00000	0.48628	0.58250	1.06878
0.530	0.14768	0.14768	0.00000	0.42654	0.41322	0.83976
0.535	0.13536	0.13536	0.00000	0.35387	0.29070	0.64457
0.540	0.12599	0.12599	0.00000	0.28194	0.20400	0.48594
0.545	0.11897	0.11897	0.00000	0.21812	0.14328	0.36140
0.550	0.11378	0.11378	0.00000	0.16517	0.10090	0.26608
0.555	0.10997	0.10997	0.00000	0.12316	0.07128	0.19444
0.560	0.10719	0.10719	0.00000	0.09081	0.05050	0.14132
0.565	0.10518	0.10518	0.00000	0.06642	0.03588	0.10230
0.570	0.10373	0.10373	0.00000	0.04830	0.02554	0.07384
0.575	0.10268	0.10268	0.00000	0.03498	0.01821	0.05319
0.580	0.10192	0.10192	0.00000	0.02525	0.01300	0.03826
0.585	0.10138	0.10138	0.00000	0.01819	0.00929	0.02749
0.590	0.10099	0.10099	0.00000	0.01309	0.00665	0.01974
0.595	0.10071	0.10071	0.00000	0.00940	0.00476	0.01416
0.600	0.10051	0.10051	0.00000	0.00675	0.00340	0.01016
		.			.	
		.			.	
		.			.	
0.975	0.10000	0.10000	0.00000	0.00000	0.00000	0.00000
0.980	0.10000	0.10000	0.00000	0.00000	0.00000	0.00000
0.985	0.10000	0.10000	0.00000	0.00000	0.00000	0.00000
0.990	0.10000	0.10000	0.00000	0.00000	0.00000	0.00000
0.995	0.10000	0.10000	0.00000	0.00000	0.00000	0.00000
1.000	0.10000	0.10000	0.00000	0.00000	0.00000	0.00000

T = 0.20

X	U NUM	U ANAL	ERROR	VIS*UXX	-U*UX	UT
0.000	1.00000	1.00000	0.00000	0.00000	0.00000	0.00000
0.005	1.00000	1.00000	0.00000	0.00000	0.00000	0.00000
0.010	1.00000	1.00000	0.00000	0.00000	0.00000	0.00000
0.015	1.00000	1.00000	0.00000	0.00000	0.00000	0.00000
0.020	1.00000	1.00000	0.00000	0.00000	0.00000	0.00000
0.025	1.00000	1.00000	0.00000	0.00000	0.00000	0.00000
		.			.	
		.			.	
		.			.	
0.400	0.75015	0.74999	0.00015	-0.00980	7.81225	7.80245
0.405	0.69882	0.69865	0.00018	0.50076	6.98586	7.48662
0.410	0.65157	0.65145	0.00012	0.86148	5.75049	6.61197
0.415	0.61136	0.61132	0.00003	1.00204	4.42109	5.42313
0.420	0.57936	0.57940	-0.00004	0.95467	3.22935	4.18402
0.425	0.55525	0.55531	-0.00007	0.80233	2.27705	3.07938

Table 4.1 *Continued.*

```
      0.430   0.53780   0.53786  -0.00006   0.62055   1.56925   2.18980
      0.435   0.52553   0.52557  -0.00004   0.45415   1.06637   1.52053
      0.440   0.51707   0.51709  -0.00002   0.32020   0.71900   1.03920
      0.445   0.51130   0.51130  -0.00001   0.21988   0.48352   0.70340
      0.450   0.50738   0.50737   0.00000   0.14776   0.32624   0.47399
      0.455   0.50470   0.50469   0.00001   0.09698   0.22283   0.31981
      0.460   0.50285   0.50284   0.00001   0.06136   0.15647   0.21782
      0.465   0.50151   0.50150   0.00001   0.03595   0.11589   0.15185
      0.470   0.50048   0.50047   0.00001   0.01703   0.09385   0.11088
      0.475   0.49959   0.49958   0.00001   0.00176   0.08600   0.08776
      0.480   0.49872   0.49871   0.00001  -0.01205   0.09014   0.07808
      0.485   0.49775   0.49774   0.00001  -0.02627   0.10576   0.07949
      0.490   0.49656   0.49655   0.00001  -0.04257   0.13379   0.09121
      0.495   0.49501   0.49500   0.00001  -0.06268   0.17647   0.11379
      0.500   0.49293   0.49293   0.00000  -0.08841   0.23735   0.14894
                          .                             .
                          .                             .
                          .                             .

      0.975   0.10000   0.10000   0.00000   0.00000   0.00000   0.00000
      0.980   0.10000   0.10000   0.00000   0.00000   0.00000   0.00000
      0.985   0.10000   0.10000   0.00000   0.00000   0.00000   0.00000
      0.990   0.10000   0.10000   0.00000   0.00000   0.00000   0.00000
      0.995   0.10000   0.10000   0.00000   0.00000   0.00000   0.00000
      1.000   0.10000   0.10000   0.00000   0.00000   0.00000   0.00000
        .
        .
        .

  T =  0.40
        .
        .
        .

  T =  0.60
        .
        .
        .

  T =  0.80
        .
        .
        .

  T =  1.00

          X     U NUM    U ANAL     ERROR   VIS*UXX     -U*UX        UT
      0.000   1.00000   1.00000   0.00000   0.00000   0.00000   0.00000
      0.005   1.00000   1.00000   0.00000   0.00000   0.00000   0.00000
      0.010   1.00000   1.00000   0.00000   0.00000   0.00000   0.00000
      0.015   1.00000   1.00000   0.00000   0.00000   0.00000   0.00000
```

Table 4.1 *Continued.*

0.020	1.00000	1.00000	0.00000	0.00000	0.00000	0.00000
0.025	1.00000	1.00000	0.00000	0.00000	0.00000	0.00000
		.			.	
		.			.	
		.			.	
0.890	0.96310	0.96356	-0.00046	-2.23238	5.18801	2.95563
0.895	0.92486	0.92622	-0.00136	-3.86818	9.57409	5.70591
0.900	0.85449	0.85695	-0.00246	-5.44862	15.47039	10.02176
0.905	0.74003	0.74286	-0.00283	-5.09351	20.20669	15.11318
0.910	0.58536	0.58739	-0.00202	-1.21970	19.52242	18.30271
0.915	0.42125	0.42237	-0.00112	3.81360	13.18044	16.99404
0.920	0.28701	0.28780	-0.00079	5.80658	6.42348	12.23006
0.925	0.19913	0.19972	-0.00059	4.68122	2.60082	7.28204
0.930	0.14971	0.15005	-0.00034	2.83677	1.03251	3.86929
0.935	0.12422	0.12437	-0.00016	1.50290	0.43456	1.93746
0.940	0.11164	0.11169	-0.00006	0.74944	0.19309	0.94254
0.945	0.10555	0.10557	-0.00001	0.36370	0.08859	0.45229
0.950	0.10265	0.10265	0.00000	0.17432	0.04134	0.21565
0.955	0.10126	0.10126	0.00000	0.08307	0.01946	0.10253
0.960	0.10060	0.10060	0.00000	0.03949	0.00920	0.04870
0.965	0.10029	0.10028	0.00000	0.01876	0.00437	0.02313
0.970	0.10014	0.10014	0.00000	0.00891	0.00208	0.01099
0.975	0.10007	0.10007	0.00000	0.00424	0.00099	0.00523
0.980	0.10003	0.10003	0.00000	0.00202	0.00048	0.00249
0.985	0.10002	0.10002	0.00000	0.00097	0.00023	0.00119
0.990	0.10001	0.10001	0.00000	0.00046	0.00011	0.00058
0.995	0.10000	0.10000	0.00000	0.00023	0.00005	0.00028
1.000	0.10000	0.10000	0.00000	-0.00007	0.00003	0.00000

Another very effective way to improve spatial resolution (in addition to adding more grid points), is to concentrate the grid points in the regions of sharpest spatial variation. This presupposes that: (a) we know in advance where the solution has sharp spatial variations in order to concentrate the grid points or we can program the numerical solution so that it detects these regions of sharp spatial variation and places the additional grid points automatically (these are called adaptive grids [Schiesser (1991)] and (b) we have approximations for spatial derivatives which can be applied to nonuniform grids, since now the grid points will not be equally spaced. These two issues, defining a nonuinform grid and approximating spatial derivatives on the nonuniform grid, have been studied extensively, and we cannot describe here all that has been developed. Rather, we present a few basic ideas to illustrate what might be done with the understanding that extensive literature is available containing a spectrum of approaches.

We will not actually consider the first issue of defining a nonuniform grid in which the grid points are concentrated where they are needed to achieve acceptable spatial resolution since the details of implementation are closely tied to the characteristics of a particular problem system; or, in the case of an adaptive grid which automatically places the grid points where they are required, the details of implementation are sufficiently complex to be beyond the scope of the present discussion. Rather, we concentrate on the second issue of developing approximations for spatial derivatives that can be applied to nonuniform grids.

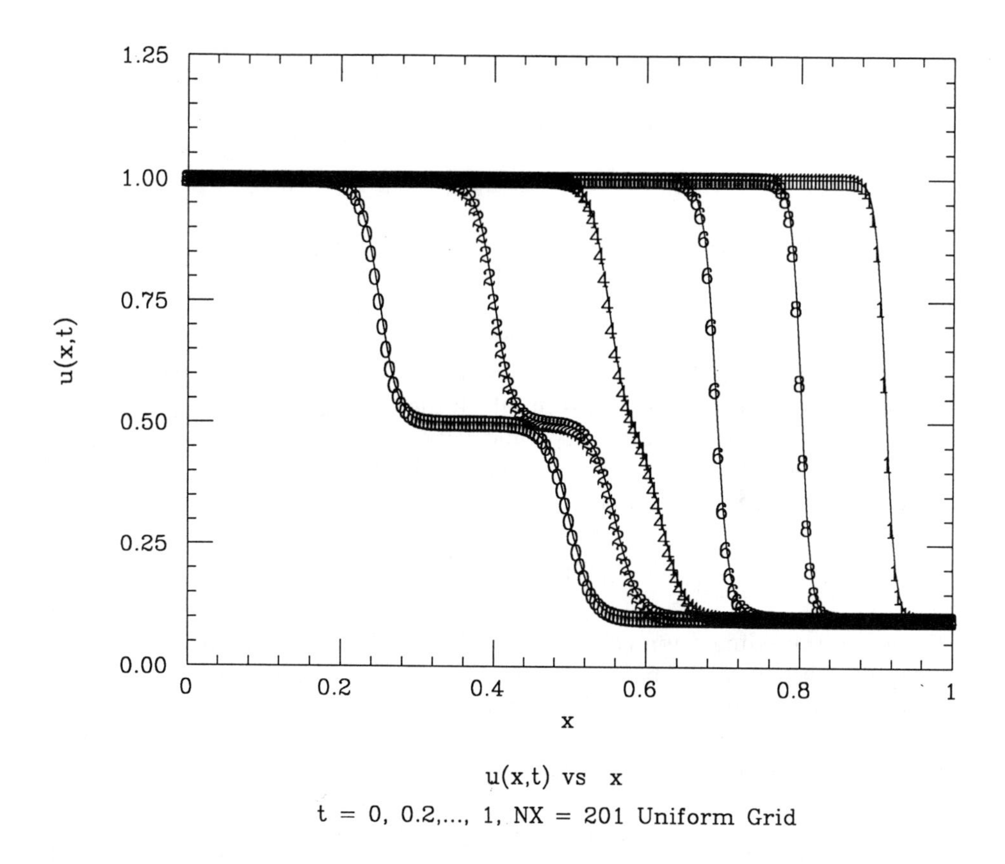

Figure 4.3 Numerical (Points) and Analytical (Solid Curves) Solutions of Equations (4.41) and (4.47) to (4.49) from Programs 4.1a to e.

4.2 Nonuniform Spatial Grids

We will consider three basic approaches to the development of approximations for spatial derivatives on nonuniform grids:

1. The first approach is based on the use of the Taylor series, along the lines discussed in Chapter 3, starting with equation (3.62). For example, if we require an approximation for a first-order derivative, we could write a Taylor series for u_{i+1} as

$$u(x_{i+1}) = u(x_i) + (du(x_i)/dx)(\Delta x_1) + (1/2!)(d^2u(x_i)/dx^2)(\Delta x_1)^2 + \cdots \quad (4.50)$$

where $\Delta x_1 = x_{i+1} - x_i$. Similarly, we could write an approximation for $u(x_{i-1})$ as

$$u(x_{i-1}) = u(x_i) + (du(x_i)/dx)(-\Delta x_2) + (1/2!)(d^2u(x_i)/dx^2)(-\Delta x_2)^2 + \cdots \quad (4.51)$$

where $\Delta x_2 = x_i - x_{i-1}$.

If equation (4.50) is now multiplied by $1/\Delta x_1{}^2$ and equation (4.51) is multiplied by $-1/\Delta x_2{}^2$ and the two resulting equations are added (in order to drop out the second derivative terms)

$$1/\Delta x_1^2\,(u(x_{i+1}) - u(x_i)) = (du(x_i)/dx)(1/\Delta x_1) + \cdots$$

$$-1/\Delta x_2^2\,(u(x_{i-1}) - u(x_i)) = (du(x_i)/dx)(1/\Delta x_2) + \cdots$$

or

$$du(x_i)/dx = \frac{1}{1/\Delta x_1 + 1/\Delta x_2}\left\{1/\Delta x_1{}^2\,(u(x_{i+1}) - u(x_i)) - 1/\Delta x_2{}^2\,(u(x_{i-1}) - u(x_i))\right\} \tag{4.52}$$

Equation (4.52) could then be used to approximate a first-order spatial derivative in a PDE over a nonuniform grid in which the basic grid spacings are Δx_1 and Δx_2; and, of course, Δx_1 and Δx_2 could vary in magnitude along the grid. As expected, when $\Delta x_1 = \Delta x_2 = \Delta x$, equation (4.52) reduces to the usual second-order, centered approximation

$$du(x_i)/dx = \frac{\{u(x_{i+1}) - u(x_{i-1})\}}{2\Delta x} \tag{4.53}$$

2. While the preceding approach based on the Taylor series has the advantage of using established mathematics, it is in some cases difficult to use in the derivation of approximations for particular spatial derivatives. For these, a more intuitive approach is frequently used. For example, consider an approximation for $(1/x^a)\partial\left(x^a\mu(u)\partial u/\partial x\right)/\partial x$. We could write

$$\frac{1}{x^a}\frac{\partial\left(x^a\mu(u)\partial u/\partial x\right)}{\partial x} \approx \left(\frac{1}{x_i^a}\right)\left\{\frac{\dfrac{\left(x_{i+1/2}^a\right)\left(\mu\left(u_{i+1/2}\right)\left(u_{i+1}-u_i\right)\right)}{\Delta x_{i+1/2}} - \dfrac{\left(x_{i-1/2}^a\right)\left(\mu\left(u_{i-1/2}\right)\left(u_i-u_{i-1}\right)\right)}{\Delta x_{i-1/2}}}{\Delta x_i}\right\} \tag{4.54}$$

where

$$x_{i+1/2}^a = \left\{\frac{x_{i+1}+x_i}{2}\right\}^a, \qquad x_{i-1/2}^a = \left\{\frac{x_{i-1}+x_i}{2}\right\}^a$$

$$\mu(u_{i+1/2}) = \frac{\mu(u_{i+1}) + \mu(u_i)}{2}, \qquad \mu(u_{i-1/2}) = \frac{\mu(u_i) + \mu(u_{i-1})}{2}$$

$$\Delta x_{i+1/2} = x_{i+1} - x_i, \qquad \Delta x_{i-1/2} = x_i - x_{i-1}$$

$$\Delta x_i = (\Delta x_{i+1/2} + \Delta x_{i-1/2})/2 = (x_{i+1} - x_{i-1})/2$$

The approximation of equation (4.54) is based on three points, x_{i-1}, x_i, and x_{i+1} (which can be unequally spaced), for which the corresponding values of the dependent variable are u_{i-1}, u_i, and $u_{i+1/2}$. Several important special cases are included:

(a) $\mu =$ const (the diffusivity does not depend on the dependent variable), for which equation (4.54) reduces to

$$\frac{\mu}{x^a}\frac{\partial\left(x^a\partial u/\partial x\right)}{\partial x} \approx \left(\frac{\mu}{x_i^a}\right)\left\{\frac{\dfrac{(x_{i+1/2}^a)(u_{i+1} - u_i)}{\Delta x_{i+1/2}} - \dfrac{(x_{i-1/2}^a)(u_i - u_{i-1})}{\Delta x_{i-1/2}}}{\Delta x_i}\right\} \tag{4.55}$$

(b) Case (a) plus equal spacing so that $\Delta x_{i+1/2} = \Delta x_i = \Delta x_{i-1/2} = \Delta x$, and equation (4.55) reduces to

$$\frac{\mu}{x^a}\frac{\partial\left(x^a \partial u/\partial x\right)}{\partial x} \approx \left(\frac{\mu}{x^a{}_i}\right)\left\{\frac{\frac{(x^a_{i+1/2})(u_{i+1}-u_i)}{\Delta x} - \frac{(x^a_{i-1/2})(u_i-u_{i-1})}{\Delta x}}{\Delta x}\right\} \tag{4.56}$$

(c) Case (b) with $a = 2$ gives an approximation of the linear, radial diffusion term in spherical coordinates (with $r = x$)

$$\frac{\mu}{r^2}\frac{\partial\left(r^2 \partial u/\partial r\right)}{\partial r} = \mu\left\{\frac{\partial^2 u}{\partial r^2} + \frac{2}{r}\frac{\partial u}{\partial r}\right\}$$

(d) Case (b) with $a = 1$ gives an approximation of the linear, radial diffusion term in cylindrical coordinates (with $r = x$)

$$\frac{\mu}{r}\frac{\partial\left(r \partial u/\partial r\right)}{\partial r} = \mu\left\{\frac{\partial^2 u}{\partial r^2} + \frac{1}{r}\frac{\partial u}{\partial r}\right\}$$

(e) Case (b) with $a = 0$ gives an approximation of the linear diffusion term in Cartesian coordinates

$$\mu\frac{\partial\left(\partial u/\partial x\right)}{\partial x} \approx \mu\left\{\frac{\frac{(u_{i+1}-u_i)}{\Delta x} - \frac{(u_i-u_{i-1})}{\Delta x}}{\Delta x}\right\} = \mu\left\{\frac{u_{i+1}-2u_i+u_{i-1}}{\Delta x^2}\right\} \tag{4.57}$$

Equation (4.57) is the well-known three-point, second-order, centered approximation for a second derivative.

3. A third approach to the development of approximations for spatial derivatives on nonuniform grids is to use *established functional approximation methods* (or develop a new functional approximation method). In other words, if the solution to a PDE can be approximated on a nonuniform grid by a known function, this function can then be differentiated with respect to the spatial independent variable(s) to obtain the required approximations to the spatial derivatives.

To illustrate this approach, we could consider the approximation of a one-dimensional PDE solution by a *cubic spline*

$$u(x,t) \approx u(x_i,t) + a_1(t)(x-x_i) + a_2(t)(x-x_i)^2 + a_3(t)(x-x_i)^3 \tag{4.58}$$

where a_1 to a_3 are the coefficients of the cubic spline and $u(x_i,t)$ is the PDE dependent variable evaluated at grid point $x = x_i$. Then the required partial derivatives for a second-order PDE, such as equation (4.41) are

$$\frac{\partial u(x,t)}{\partial x} \approx a_1(t) + 2a_2(t)(x-x_i) + 3a_3(t)(x-x_i)^2 \tag{4.59}$$

$$\frac{\partial^2 u(x,t)}{\partial x^2} \approx 2a_2(t) + 6a_3(t)(x-x_i) \tag{4.60}$$

If $u(x,t)$ is known at some t (from a method of lines solution), a_1 to a_3 can be evaluated according to equation (4.58). Then the derivatives $\partial u(x,t)/\partial x$ and $\partial^2 u(x,t)/\partial x^2$ can be computed from equations (4.59) and (4.60) to continue the method of lines solution in t.

```
      SUBROUTINE INITAL
      IMPLICIT DOUBLE PRECISION (A-H,O-Z)
      PARAMETER (NX=201)
      COMMON/T/           T,         NFIN,       NORUN
     1      /Y/      U(NX)
     2      /F/     UT(NX)
     3      /S/     UX(NX),    UXX(NX),       X(NX)
     4      /C/        VIS
     5      /A/     A1(NX),     A2(NX),     A3(NX)
C...
C...  PROBLEM PARAMETERS
      VIS=0.003D+00
C...
C...  SPATIAL GRID AND THE INITIAL CONDITION
      DO 1 I=1,NX
         X(I)=DFLOAT(I-1)/DFLOAT(NX-1)
         U(I)=PHI(X(I),0.D+00)
1     CONTINUE
C...
C...  INITIAL DERIVATIVES
      CALL DERV
      RETURN
      END
```

Program 4.2a Subroutine INITAL for the Cubic Spline Solution of Equations (4.41) and (4.47) to (4.49).

A cubic spline offers several advantages: (a) because of the way in which the spline is constructed, it can be used on a nonuniform grid in x; (b) also, the cubic spline coefficients, a_1 to a_3, are selected to ensure continuity in u, du/dx $(= \partial u(x,t)/\partial x$ in a PDE problem) and d^2u/dx^2 $(= \partial^2 u(x,t)/\partial x^2)$ at the grid points $x = x_i$; and (c) since the cubic spline is a well-established approach to functional approximation, library subroutines are readily available to compute the spline coefficients.

To illustrate these ideas, we return to Burgers' equation, (4.41), with initial and boundary conditions (4.47) to (4.49). Subroutines INITAL, DERV, and PRINT closely parallel those of Programs 4.1a, 4.1b, and 4.1d. Subroutine INITAL is listed in Program 4.2a.

The only new detail of subroutine INITAL is the additional COMMON/A/ to store the spline coefficients, a_1, a_2, and a_3, in equation (4.58):

```
     5      /A/     A1(NX),     A2(NX),     A3(NX)
```

Subroutine DERV in Program 4.2b is also similar to Program 4.1b.

In place of using subroutines DSS020 and DSS044 to calculate the first- and second-order derivatives $\partial u/\partial x$ and $\partial^2 u/\partial x^2$ of Program 4.1b, we now call subroutine DSS038 twice to calculate these derivatives:

```
C...
C...  UX
      CALL DSS038(NX,X,U ,UX ,A1,A2,A3)
C...
C...  UXX
      CALL DSS038(NX,X,UX,UXX,A1,A2,A3)
```

```
      SUBROUTINE DERV
      IMPLICIT DOUBLE PRECISION (A-H,O-Z)
      PARAMETER (NX=201)
      COMMON/T/          T,       NFIN,      NORUN
     1      /Y/      U(NX)
     2      /F/     UT(NX)
     3      /S/     UX(NX),   UXX(NX),     X(NX)
     4      /C/        VIS
     5      /A/     A1(NX),    A2(NX),    A3(NX)
C...
C...  BOUNDARY CONDITIONS
      U(1) =PHI(X(1) ,T)
      U(NX)=PHI(X(NX),T)
      UT(1) =0.D+00
      UT(NX)=0.D+00
C...
C...  UX
      CALL DSS038(NX,X,U ,UX ,A1,A2,A3)
C...
C...  UXX
      CALL DSS038(NX,X,UX,UXX,A1,A2,A3)
C...
C...  BURGERS' EQUATION
      DO 2 I=2,NX-1
         UT(I)=VIS*UXX(I)-U(I)*UX(I)
2     CONTINUE
      RETURN
      END
```

Program 4.2b Subroutine DERV for the Cubic Spline Solution of Equations (4.41) and (4.47) to (4.49).

In the first call, u (array U) is differentiated to $\partial u/\partial x$ (array UX), and in the second call, $\partial u/\partial x$ is differentiated to $\partial^2 u/\partial x^2$ (array UXX). DSS038, in turn, merely calls a library routine for cubic splines to compute the spline coefficients for U in the first call and UX in the second call. Once these spline coefficients are evaluated, they can be used in equation (4.59) to compute a first-order spatial derivative that is then returned from DSS038 as the fourth argument. There is one important difference between DSS038 and subroutines DSS020 and DSS044. The spatial grid in x (array X) is passed to DSS038 as its second argument, and this grid, which is used in the calculation of the spline coefficients, does not have to be uniformly spaced. Thus, the grid points can be selected to concentrate them in regions of rapid variation of $u(x,t)$ with x, e.g., the steep fronts of Figure 4.3. This was not done in the present example, since the grid in x is uniform, as defined in DO loop 1 of subroutine INITAL. However, a nonuniform grid can be used to good advantage to resolve sharp spatial variations in the solution, and this programming indicates that the construction and use of a nonuniform grid is straightforward (the only requirement is to store the grid points in array X).

Subroutine DSS038, and the cubic spline routine it calls, SPLINE, are listed in Program 4.2c. We can note the following points about DSS038:

1. Subroutine SPLINE is called first to compute the cubic spline coefficients for array U defined on grid X with N points. The computed coefficients are returned from

```
      SUBROUTINE DSS038(N,X,U,UX,B,C,D)
C...
C...  SUBROUTINE DSS038 IS A NUMERICAL DIFFERENTIATION ROUTINE BASED ON
C...  THE CUBIC SPLINE OF FORSYTHE, G. E., M. A. MALCOLM, AND C. B.
C...  MOLER, COMPUTER METHODS FOR MATHEMATICAL COMPUTATIONS, PRENTICE-
C...  HALL, ENGLEWOOD CLIFFS, NJ, 1977, PP. 76-79.
C...
C...  ARGUMENT LIST
C...
C...     N       NUMBER OF SPATIAL GRID POINTS (INPUT)
C...
C...     X       ONE-DIMENSIONAL ARRAY CONTAINING THE VALUES OF THE
C...             SPATIAL (BOUNDARY VALUE) INDEPENDENT VARIABLE (INPUT)
C...
C...     U       ONE-DIMENSIONAL ARRAY CONTAINING THE VALUES OF THE
C...             DEPENDENT VARIABLE (INPUT)
C...
C...     UX      ONE-DIMENSIONAL ARRAY CONTAINING THE DERIVATIVE OF U
C...             WITH RESPECT TO X (OUTPUT)
C...
C...     B,C,D   ONE-DIMENSIONAL ARRAYS CONTAINING THE CUBIC SPLINE CO-
C...             EFFICIENTS (OUTPUT)
C...
      IMPLICIT DOUBLE PRECISION (A-H,O-Z)
C...
C...  VARIABLE DIMENSION THE ARRAYS PASSED AS ARGUMENTS
      DIMENSION  X(N), U(N), UX(N), B(N), C(N), D(N)
C...
C...  CALL SUBROUTINE SPLINE TO COMPUTE THE SPLINE COEFFICIENTS
      CALL SPLINE(N,X,U,B,C,D)
C...
C...  THE NUMERICAL DERIVATIVE IS OBTAINED BY DIFFERENTIATION OF THE
C...  CUBIC SPLINE
C...
C...     S(X) = U(I) + B(I)*(X - X(I)) + C(I)*(X - X(I))**2
C...
C...                 + D(I)*(X - X(I))**3
C...  THUS
C...
C...     UX(I) = DS(X(I))/DX = B(I)
C...
      DO 1 I=1,N
      UX(I)=B(I)
1     CONTINUE
      RETURN
      END

      SUBROUTINE SPLINE (N, X, Y, B, C, D)
C...
C...  THIS SUBROUTINE APPEARS IN
C...
```

Program 4.2c Subroutines DSS038 and SPLINE for the Cubic Spline Calculation of First-Order Spatial Derivatives. *Continued next pages.*

```
C...      FORSYTHE, G. E., M. A. MALCOLM AND C. B. MOLER, COMPUTER
C...      METHODS FOR MATHEMATICAL COMPUTATIONS, PRENTICE-HALL, INC.,
C...      ENGLEWOOD CLIFFS, NJ, 1977, PP. 76-78
C...
      INTEGER N
      DOUBLE PRECISION X(N), Y(N), B(N), C(N), D(N)
C
C  THE COEFFICIENTS B(I), C(I), AND D(I), I=1,2,...,N ARE COMPUTED
C  FOR A CUBIC INTERPOLATING SPLINE
C
C    S(X) = Y(I) + B(I)*(X-X(I)) + C(I)*(X-X(I))**2 + D(I)*(X-X(I))**3
C
C    FOR  X(I) .LE. X .LE. X(I+1)
C
C  INPUT..
C
C    N = THE NUMBER OF DATA POINTS OR KNOTS (N.GE.2)
C    X = THE ABSCISSAS OF THE KNOTS IN STRICTLY INCREASING ORDER
C    Y = THE ORDINATES OF THE KNOTS
C
C  OUTPUT..
C
C    B, C, D  = ARRAYS OF SPLINE COEFFICIENTS AS DEFINED ABOVE.
C
C  USING  P  TO DENOTE DIFFERENTIATION,
C
C    Y(I) = S(X(I))
C    B(I) = SP(X(I))
C    C(I) = SPP(X(I))/2
C    D(I) = SPPP(X(I))/6  (DERIVATIVE FROM THE RIGHT)
C
C  THE ACCOMPANYING FUNCTION SUBPROGRAM  SEVAL  CAN BE USED
C  TO EVALUATE THE SPLINE.
C
C
      INTEGER NM1, IB, I
      DOUBLE PRECISION T
C
      NM1 = N-1
      IF ( N .LT. 2 ) RETURN
      IF ( N .LT. 3 ) GO TO 50
C
C  SET UP TRIDIAGONAL SYSTEM
C
C  B = DIAGONAL, D = OFFDIAGONAL, C = RIGHT-HAND SIDE.
C
      D(1) = X(2) - X(1)
      C(2) = (Y(2) - Y(1))/D(1)
      DO 10 I = 2, NM1
         D(I) = X(I+1) - X(I)
         B(I) = 2.D+00*(D(I-1) + D(I))
         C(I+1) = (Y(I+1) - Y(I))/D(I)
         C(I) = C(I+1) - C(I)
```

Program 4.2c *Continued.*

```
   10 CONTINUE
C
C  END CONDITIONS.  THIRD DERIVATIVES AT  X(1)  AND  X(N)
C  OBTAINED FROM DIVIDED DIFFERENCES
C
      B(1) = -D(1)
      B(N) = -D(N-1)
      C(1) = 0.D+00
      C(N) = 0.D+00
      IF ( N .EQ. 3 ) GO TO 15
      C(1) = C(3)/(X(4)-X(2)) - C(2)/(X(3)-X(1))
      C(N) = C(N-1)/(X(N)-X(N-2)) - C(N-2)/(X(N-1)-X(N-3))
      C(1) = C(1)*D(1)**2/(X(4)-X(1))
      C(N) = -C(N)*D(N-1)**2/(X(N)-X(N-3))
C
C  FORWARD ELIMINATION
C
   15 DO 20 I = 2, N
         T = D(I-1)/B(I-1)
         B(I) = B(I) - T*D(I-1)
         C(I) = C(I) - T*C(I-1)
   20 CONTINUE
C
C  BACK SUBSTITUTION
C
      C(N) = C(N)/B(N)
      DO 30 IB = 1, NM1
         I = N-IB
         C(I) = (C(I) - D(I)*C(I+1))/B(I)
   30 CONTINUE
C
C  C(I) IS NOW THE SIGMA(I) OF THE TEXT
C
C  COMPUTE POLYNOMIAL COEFFICIENTS
C
      B(N) = (Y(N) - Y(NM1))/D(NM1) + D(NM1)*(C(NM1) + 2.D+00*C(N))
      DO 40 I = 1, NM1
         B(I) = (Y(I+1) - Y(I))/D(I) - D(I)*(C(I+1) + 2.D+00*C(I))
         D(I) = (C(I+1) - C(I))/D(I)
         C(I) = 3.D+00*C(I)
   40 CONTINUE
      C(N) = 3.D+00*C(N)
      D(N) = D(N-1)
      RETURN
C
   50 B(1) = (Y(2)-Y(1))/(X(2)-X(1))
      C(1) = 0.D+00
      D(1) = 0.D+00
      B(2) = B(1)
      C(2) = 0.D+00
      D(2) = 0.D+00
      RETURN
      END
```

Program 4.2c *Continued.*

SPLINE in arrays B, C, and D (A1, A2, and A3 in calling program DERV):

```
C...
C...  CALL SUBROUTINE SPLINE TO COMPUTE THE SPLINE COEFFICIENTS
      CALL SPLINE(N,X,U,B,C,D)
```

2. From equation (4.59) we have

$$\frac{\partial u(x_i,t)}{\partial x} \approx a_1(t) \tag{4.61}$$

Equation (4.61) is programmed in DO loop 1 of DSS038 (spline coefficient a_1 is stored in array B)

```
      DO 1 I=1,N
      UX(I)=B(I)
1     CONTINUE
```

We can note the following points about subroutine SPLINE, which is taken from Forsythe et al. (1977):

1. The cubic spline coefficients a_1 to a_3 are evaluated at all of the grid points, $x = x_i$, $i = 1, 2, \ldots 201$ (3×201 coefficients) at each point in time, t, when DERV is called. Thus these coefficients are evaluated many times during the course of computing the method of lines solution.

2. Each time SPLINE is called from DSS038, a tridiagonal system of linear algebraic equations is solved for the 3×201 spline coefficients, which is done efficiently by *Thomas' method*, a form of Gauss elimination or row reduction structured specifically for tridiagonal systems.

3. In general, cubic splines have two coefficients at the end points, $x = x_1$ and $x = x_n$ ($n = 201$ for this application), which can be selected arbitrarily. In the case of SPLINE, these coefficients are selected so that the third derivative, d^3u/dx^3, at the end points is approximated by divided differences. This is different than for the *natural cubic splines*, for which the coefficients are selected by setting the second derivatives at the end points to zero.

4. The solution of a set of tridiagonal equations is substantially more computer intensive than the use of the weighted sums in subroutines DSS020 and DSS044. This is the "price" that is paid for having the possible use of a nonuniform grid (at least as implemented with cubic splines; other approximations on nonuniform grids may be computationally more efficient).

Subroutine PRINT in Program 4.2d is essentially the same as in Program 4.1d. Subroutine PLOT1 and the data file for the cubic spline solution is the same as for Programs 4.1d and e. The main program, which calls RKF45, is not listed since it is similar to previous main programs, such as Program 2.2a.

The numerical and graphical output from Programs 4.2a to d are given in Table 4.2 and Figure 4.4. The numerical solution computed by cubic splines is generally more

```
      SUBROUTINE PRINT(NI,NO)
      IMPLICIT DOUBLE PRECISION (A-H,O-Z)
      PARAMETER (NX=201)
      COMMON/T/          T,       NFIN,     NORUN
     1      /Y/    U(NX)
     2      /F/   UT(NX)
     3      /S/   UX(NX),   UXX(NX),     X(NX)
     4      /C/        VIS
     5      /A/    A1(NX),   A2(NX),   A3(NX)
     6      /P/  UN(6,NX),        IP
C...
C...  INITIALIZE A COUNTER FOR THE PLOTTING
      DATA IP/0/
C...
C...  MONITOR THE SOLUTION
      IP=IP+1
      WRITE(*,*)IP,T
C...
C...  STORE THE NUMERICAL SOLUTION FOR SUBSEQUENT PLOTTING VIA SUBROU-
C...  TINE PLOT1
      DO 10 I=1,NX
         UN(IP,I)=U(I)
10    CONTINUE
C...
C...  PRINT THE NUMERICAL AND ANALYTICAL SOLUTIONS, THE DIFFERENCE
C..   BETWEEN THE SOLUTIONS AND THE TERMS IN BURGERS' EQUATION
      WRITE(NO,2)T
2     FORMAT(1H ,//,5H T = ,F5.2,//,
     1 9X,1HX,5X,5HU NUM,4X,6HU ANAL,5X,5HERROR,3X,7HVIS*UXX,5X,5H-U*UX,
     2 8X,2HUT)
      DO 3 I=1,NX
         UANAL=PHI(X(I),T)
         ERROR=U(I)-UANAL
         DIFF=VIS*UXX(I)
         CONV=-U(I)*UX(I)
         WRITE(NO,4)X(I),U(I),UANAL,ERROR,DIFF,CONV,UT(I)
4        FORMAT(F10.3,6F10.5)
3     CONTINUE
C...
C...  PLOT THE SOLUTION
      CALL PLOT1
C...
C...  REINITIALIZE THE COUNTER FOR PLOTTING
      IF(IP.EQ.6)IP=0
      RETURN
      END
```

Program 4.2d Subroutine `PRINT` to Display the Cubic Spline Numerical and Analytical Solutions of Equations (4.41) and (4.47) to (4.49).

Table 4.2 Abbreviated Numerical Output from Programs 4.2a to d. *Continued next pages.*

```
RUN NO. -    1  MADSEN, ET AL, LAPIDUS, ET AL, NUM METH DIFF SYS,
                PP. 236-237

INITIAL T -  0.000D+00

  FINAL T -  0.100D+01

  PRINT T -  0.200D+00

NUMBER OF DIFFERENTIAL EQUATIONS - 201

INTEGRATION ALGORITHM - RKF45

MAXIMUM INTEGRATION ERROR -  0.100D-04


T =  0.00

        X     U NUM    U ANAL     ERROR   VIS*UXX     -U*UX        UT
    0.000   1.00000   1.00000   0.00000   0.00000   0.00000   0.00000
    0.005   1.00000   1.00000   0.00000   0.00000   0.00000   0.00000
    0.010   1.00000   1.00000   0.00000   0.00000   0.00000   0.00000
    0.015   1.00000   1.00000   0.00000   0.00000   0.00000   0.00000
    0.020   1.00000   1.00000   0.00000   0.00000   0.00000   0.00000
    0.025   1.00000   1.00000   0.00000   0.00000   0.00000   0.00000
                .                                       .
                .                                       .
                .                                       .

    0.975   0.10000   0.10000   0.00000   0.00000   0.00000   0.00000
    0.980   0.10000   0.10000   0.00000   0.00000   0.00000   0.00000
    0.985   0.10000   0.10000   0.00000   0.00000   0.00000   0.00000
    0.990   0.10000   0.10000   0.00000   0.00000   0.00000   0.00000
    0.995   0.10000   0.10000   0.00000   0.00000   0.00000   0.00000
    1.000   0.10000   0.10000   0.00000   0.00000   0.00000   0.00000


T =  0.20

        X     U NUM    U ANAL     ERROR   VIS*UXX     -U*UX        UT
    0.000   1.00000   1.00000   0.00000   0.00004   0.00002   0.00000
    0.005   1.00000   1.00000   0.00000   0.00001  -0.00001  -0.00001
    0.010   1.00000   1.00000   0.00000  -0.00002   0.00000  -0.00002
    0.015   1.00000   1.00000   0.00000   0.00001   0.00002   0.00003
    0.020   1.00000   1.00000   0.00000   0.00002  -0.00002   0.00001
    0.025   1.00000   1.00000   0.00000  -0.00003  -0.00001  -0.00005
                .                                       .
                .                                       .
                .                                       .

    0.400   0.74997   0.74999  -0.00002   0.00122   7.81476   7.81599
    0.405   0.69861   0.69865  -0.00004   0.51331   6.97069   7.48400
    0.410   0.65142   0.65145  -0.00003   0.86683   5.73066   6.59749
```

Table 4.2 *Continued.*

0.415	0.61131	0.61132	-0.00001	0.99898	4.40848	5.40746
0.420	0.57940	0.57940	0.00000	0.94840	3.22654	4.17493
0.425	0.55532	0.55531	0.00001	0.79742	2.28014	3.07756
0.430	0.53787	0.53786	0.00001	0.61844	1.57393	2.19237
0.435	0.52558	0.52557	0.00001	0.45403	1.07029	1.52432
0.440	0.51709	0.51709	0.00001	0.32095	0.72154	1.04249
0.445	0.51130	0.51130	0.00000	0.22075	0.48486	0.70561
0.450	0.50737	0.50737	0.00000	0.14845	0.32677	0.47521
0.455	0.50469	0.50469	0.00000	0.09742	0.22289	0.32032
0.460	0.50284	0.50284	0.00000	0.06160	0.15631	0.21790
0.465	0.50150	0.50150	0.00000	0.03606	0.11564	0.15170
0.470	0.50047	0.50047	0.00000	0.01706	0.09360	0.11065
0.475	0.49958	0.49958	0.00000	0.00175	0.08577	0.08752
0.480	0.49871	0.49871	0.00000	-0.01207	0.08994	0.07787
0.485	0.49774	0.49774	0.00000	-0.02628	0.10559	0.07931
0.490	0.49655	0.49655	0.00000	-0.04256	0.13362	0.09106
0.495	0.49500	0.49500	0.00000	-0.06264	0.17628	0.11364
0.500	0.49292	0.49293	0.00000	-0.08831	0.23710	0.14878
	.				.	
	.				.	
	.				.	
0.975	0.10000	0.10000	0.00000	0.00000	0.00000	0.00000
0.980	0.10000	0.10000	0.00000	0.00000	0.00000	0.00000
0.985	0.10000	0.10000	0.00000	0.00000	0.00000	0.00000
0.990	0.10000	0.10000	0.00000	0.00000	0.00000	0.00000
0.995	0.10000	0.10000	0.00000	0.00000	0.00000	0.00000
1.000	0.10000	0.10000	0.00000	0.00000	0.00000	0.00000

.
.
.

T = 0.40

.
.
.

T = 0.60

.
.
.

T = 0.80

.
.
.

T = 1.00

Table 4.2 Continued.

X	U NUM	U ANAL	ERROR	VIS*UXX	-U*UX	UT
0.000	1.00000	1.00000	0.00000	0.00001	0.00001	0.00000
0.005	1.00000	1.00000	0.00000	0.00000	0.00000	0.00000
0.010	1.00000	1.00000	0.00000	0.00000	0.00000	0.00000
0.015	1.00000	1.00000	0.00000	0.00000	0.00000	0.00000
0.020	1.00000	1.00000	0.00000	0.00000	0.00000	0.00000
0.025	1.00000	1.00000	0.00000	0.00000	0.00000	0.00000
	.				.	
	.				.	
	.				.	
0.890	0.96349	0.96356	-0.00007	-2.16966	5.03116	2.86150
0.895	0.92629	0.92622	0.00007	-3.81704	9.40901	5.59198
0.900	0.85709	0.85695	0.00014	-5.54260	15.41790	9.87530
0.905	0.74349	0.74286	0.00062	-5.34819	20.49531	15.14712
0.910	0.58730	0.58739	-0.00009	-1.19921	19.74440	18.54520
0.915	0.42176	0.42237	-0.00061	3.97605	13.07189	17.04795
0.920	0.28761	0.28780	-0.00020	5.81309	6.39069	12.20378
0.925	0.19982	0.19972	0.00011	4.64229	2.65521	7.29750
0.930	0.15016	0.15005	0.00011	2.83678	1.06589	3.90268
0.935	0.12442	0.12437	0.00004	1.51623	0.44322	1.95945
0.940	0.11169	0.11169	0.00000	0.75934	0.19336	0.95270
0.945	0.10556	0.10557	-0.00001	0.36823	0.08741	0.45564
0.950	0.10263	0.10265	-0.00001	0.17594	0.04036	0.21630
0.955	0.10125	0.10126	-0.00001	0.08349	0.01884	0.10233
0.960	0.10059	0.10060	-0.00001	0.03947	0.00886	0.04833
0.965	0.10028	0.10028	0.00000	0.01867	0.00418	0.02284
0.970	0.10013	0.10014	0.00000	0.00881	0.00198	0.01079
0.975	0.10006	0.10007	0.00000	0.00418	0.00094	0.00511
0.980	0.10003	0.10003	0.00000	0.00197	0.00045	0.00242
0.985	0.10002	0.10002	0.00000	0.00094	0.00021	0.00115
0.990	0.10001	0.10001	0.00000	0.00045	0.00010	0.00055
0.995	0.10000	0.10000	0.00000	0.00021	0.00005	0.00026
1.000	0.10000	0.10000	0.00000	0.00011	0.00003	0.00000

accurate than that for finite differences. For example, for $t = 1$, we have at $x = 0.910$ (approximately the middle of the steepest fronts in Figures 4.3 and 4.4):

```
        X       U NUM      U ANAL      ERROR     VIS*UXX       -U*UX          UT

Finite differences (Table 4.1):

    0.910    0.58536    0.58739   -0.00202   -1.21970   19.52242   18.30271

Cubic splines (Table 4.2):

    0.910    0.58730    0.58739   -0.00009   -1.19921   19.74440   18.54520
```

The ratio of the errors is $(-0.00202)/(-0.00009) \approx 22.4$.

We also should note that the second derivative in equation (4.41), u_{xx}, could be calculated directly from the cubic spline implemented in subroutine SPLINE; in other words, it is not necessary to use stagewise differentiation, with two calls to DSS038

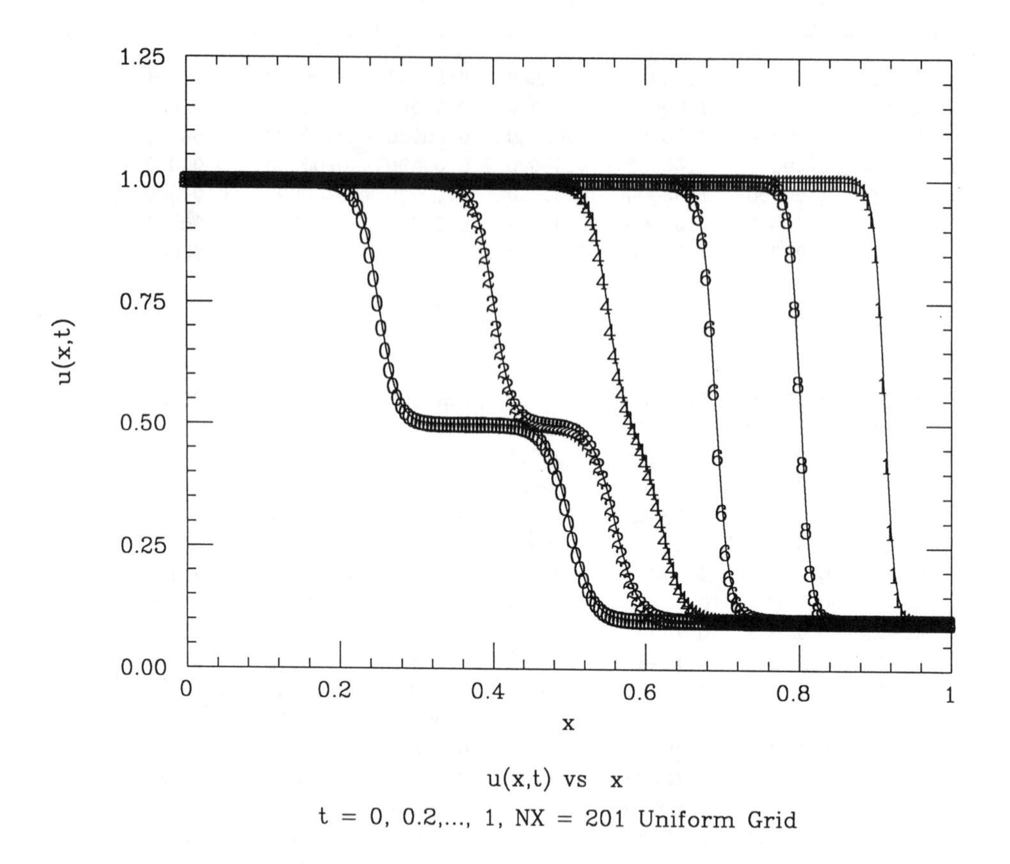

Figure 4.4 Numerical (Points) and Analytical (Solid Curves) Solutions of Equations (4.41) and (4.47) to (4.49) from Programs 4.2a to d.

in DERV of Program 4.2b. Rather, u_x and u_{xx} for equation (4.41) are available directly from the cubic spline. Equation (4.60) gives

$$\frac{\partial^2 u(x_i, t)}{\partial x^2} \approx 2a_2(t) \tag{4.62}$$

which can be used in combination with equation (4.61).

These ideas are easily put into a spatial differentiation routine, for example, subroutine DSS040 in Program 4.2e. Note that just one additional line of code is required in DO loop 1 to obtain the second derivative from equation (4.62) (in array UXX)

```
      DO 1 I=1,N
      UX(I)=B(I)
      UXX(I)=2.D+00*C(I)
1     CONTINUE
```

Subroutine DSS040 gave a solution with accuracy comparable to subroutine DSS038, and the computational effort is only about half as great (one call to DSS040 vs. two calls to DSS038).

Again, we point out that either DSS038 or DSS040 can be used on a nonuniform grid via array X, which could be particularly useful for a problem with a solution that

```
      SUBROUTINE DSS040(N,X,U,UX,UXX,B,C,D)
C...
C...  SUBROUTINE DSS040 IS A NUMERICAL DIFFERENTIATION ROUTINE BASED ON
C...  THE CUBIC SPLINE OF FORSYTHE, G. E., M. A. MALCOLM, AND C. B.
C...  MOLER, COMPUTER METHODS FOR MATHEMATICAL COMPUTATIONS, PRENTICE-
C...  HALL, ENGLEWOOD CLIFFS, N.J., 1977, PP. 76-79.
C...
C...  ARGUMENT LIST
C...
C...     N       NUMBER OF SPATIAL GRID POINTS (INPUT)
C...
C...     X       ONE-DIMENSIONAL ARRAY CONTAINING THE VALUES OF THE
C...             SPATIAL (BOUNDARY-VALUE) INDEPENDENT VARIABLE (INPUT)
C...
C...     U       ONE-DIMENSIONAL ARRAY CONTAINING THE VALUES OF THE
C...             DEPENDENT VARIABLE (INPUT)
C...
C...     UX      ONE-DIMENSIONAL ARRAY CONTAINING THE FIRST DERIVATIVE
C...             OF U WITH RESPECT TO X (OUTPUT)
C...
C...     UXX     ONE-DIMENSIONAL ARRAY CONTAINING THE SECOND DERIVATIVE
C...             OF U WITH RESPECT TO X (OUTPUT)
C...
C...     B,C,D   ONE-DIMENSIONAL ARRAYS CONTAINING THE CUBIC SPLINE CO-
C...             EFFICIENTS (OUTPUT)
C...
      IMPLICIT DOUBLE PRECISION (A-H,O-Z)
C...
C...  VARIABLE DIMENSION THE ARRAYS PASSED AS ARGUMENTS
      DIMENSION   X(N),  U(N), UX(N),UXX(N),  B(N),  C(N),  D(N)
C...
C...  CALL SUBROUTINE SPLINE TO COMPUTE THE SPLINE COEFFICIENTS
      CALL SPLINE(N,X,U,B,C,D)
C...
C...  THE NUMERICAL DERIVATIVES, UX AND UXX, ARE COMPUTED BY DIFFERENT-
C...  IATION OF THE CUBIC SPLINE
C...
C...     S(X) = U(I) + B(I)*(X - X(I)) + C(I)*(X - X(I))**2
C...
C...                 + D(I)*(X - X(I))**3
C...  THUS
C...
C...     UX(I) = DS(X(I))/DX = B(I)
C...
C...     UXX(I) = D(DS(X(I))/DX)/DX = 2*C(I)
C...
      DO 1 I=1,N
      UX(I)=B(I)
      UXX(I)=2.D+00*C(I)
1     CONTINUE
      RETURN
      END
```

Program 4.2e Subroutine DSS040 for the Cubic Spline Calculation of First- and Second-Order Spatial Derivatives.

has sharp spatial variation. Thus, the additional effort of solving systems of tridiagonal equations in SPLINE might be worth the additional computational effort (relative to finite differences) by producing a solution of better accuracy, as demonstrated by the preceding example, and additionally, might give substantially better spatial resolution through the use of a nonuniform grid.

Finally, we should note that this basic procedure of using a method for functional approximation, in this case cubic splines, can be applied generally in the method of lines solution of PDEs. The approximation of the PDE solution might be by finite elements, finite volumes, or weighted residual methods, for example, and we will consider each of these approaches in subsequent examples. Any functional approximation which represents the PDE solution with acceptable accuracy, and can be differentiated or can represent the PDE spatial derivatives in some fashion, can be used in the method of lines. This means that the method of lines is a flexible and open-ended approach to the numerical solution of PDEs that is directed essentially by the background, experience, and ingenuity of the analyst; that is, the method of lines is not a precisely defined algorithm or prescription for the numerical solution of PDEs.

One essential feature of the method of lines solution of PDEs that we have not considered thus far is the choice of the number of points in the spatial grid. This is an aspect of the numerical solution that must be considered for each new problem since a general answer to the question of how many points to use cannot be given; i.e., the required number of points is problem specific. The basic concept we follow is to use enough points to spatially resolve the solution with acceptable accuracy. In other words, the solution cannot have a large variation within one spatial interval. If we consider the solution in Figures 4.3 and 4.4, for example, we see the front at $t = 1$ is represented by approximately 10 solution points, which is about the minimum number that can accurately resolve the front. This requirement in turn sets the overall number of grid points at approximately 201; substantially fewer grid points would lead to solutions with poor resolution of the front. This might be manifested by oscillations or other unrealistic numerical effects. In the next example, we will see that the solution can be resolved accurately with just 11 grid points. Thus, the number of required grid points is quite problem specific.

This suggests, also, that some experimentation in formulating a method of lines solution is generally required. The preceding discussion implies that we know something about the spatial variation of the solution in order to select the number of grid points, but generally we are attempting the solution of a new problem and therefore we do not have the solution to guide us in the selection of the number of grid points. Thus, we might have to start with an assumed number of points, hopefully, guided by some physical insights into the nature of the solution. An initial execution of the resulting program usually indicates if we have made a good choice, particularly if the number of grid points is too small to resolve the solution spatially; that is, in this case we will usually see numerical distortions that are not physically realistic, and we can then execute the program with a larger number of grid points. Generally an overestimation of the required number of grid points will cause only two relatively minor adverse effects: (a) the computer run time may be larger than required for an optimum choice of the number of grid points and (b) generally, as the number of grid points is increased, the stiffness of the resulting system of ODEs that approximates the problem PDE(s) also increases, in which case an implicit ODE integrator may be required. This latter point (of increased stiffness) is rather dramatically illustrated by the next problem we consider, a hyperbolic parabolic problem with nonlinear convection and diffusion terms, the *modified Burgers' equation* discussed by Mickens (1991)

$$\frac{\partial u}{\partial t} = -u\frac{\partial u}{\partial x} + u\frac{\partial^2 u}{\partial x^2} \tag{4.63}$$

```
      SUBROUTINE INITAL
      IMPLICIT DOUBLE PRECISION (A-H,O-Z)
      PARAMETER (NX=11)
      COMMON/T/          T,      NFIN,      NORUN
     1      /Y/     U(NX)
     2      /F/    UT(NX)
     3      /S/    UX(NX),   UXX(NX),     X(NX)
     4      /C/          A,         B,          C,          D
C...
C...  CONSTANTS IN THE ANALYTICAL SOLUTION
      A= 1.0D+00
      B= 1.0D+00
      C= 1.0D+00
      D= 1.0D+00
C...
C...  SPATIAL GRID AND THE INITIAL CONDITION
      DO 1 I=1,NX
         X(I)=DFLOAT(I-1)/DFLOAT(NX-1)
         CALL UE(X(I),T,U(I))
1     CONTINUE
C...
C...  INITIAL DERIVATIVES
      CALL DERV
      RETURN
      END
```

Program 4.3a Subroutine `INITAL` for the Solution of Equation (4.63).

Note that now the diffusion term, $u\partial^2 u/\partial x^2$, has a diffusivity which is proportional to the dependent variable. Thus, as the dependent variable increases, therefore also increasing the velocity through the convective term, $-u\partial u/\partial x$, and possibly producing a front-sharpening condition (see Figure 4.2b), the diffusion term also increases which tends to offset the effect of the convective term. Therefore, we might expect that equation (4.63) has solutions that are more easily computed than those of the Burgers' equation (4.41). In fact, we will see that an accurate solution to equation (4.63) is possible with an 11-point grid, in contrast with the 201-point grid of Programs 4.1a–e and 4.2a–e for the Burgers' equation.

An analytical solution to equation (4.63) is given by Mickens (1991) (the reader should confirm this solution)

$$u(x,t) = \frac{a + be^x + cx}{ct + d} \tag{4.64}$$

We then use initial and boundary conditions for equation (4.63) that are consistent with equation (4.64). Note that the right-hand side of equation (4.64) is well behaved provided we avoid a choice of c and d that causes the denominator, $ct + d$, to go to zero for some t. This smoothness is the reason we can compute an accurate solution to equation (4.63) with only 11 grid points.

Subroutine `INITAL` for the method of lines solution of equation (4.63) is listed in Program 4.3a. We can note the following points about subroutine `INITAL`:

1. The COMMON area is similar to those of previous programs, e.g., Programs 4.1a and 4.2a.

2. The solution is computed for $a = b = c = d = 1$ in equation (4.64), for which the programming is

```
C...
C...  CONSTANTS IN THE ANALYTICAL SOLUTION
      A= 1.0D+00
      B= 1.0D+00
      C= 1.0D+00
      D= 1.0D+00
```

3. The initial condition for equation (4.63) is set in DO loop 1, using function UE which implements equation (4.64), in this case, for $t = 0$:

```
C...
C...  SPATIAL GRID AND THE INITIAL CONDITION
      DO 1 I=1,NX
         X(I)=DFLOAT(I-1)/DFLOAT(NX-1)
         CALL UE(X(I),T,U(I))
1     CONTINUE
```

Subroutine DERV for equation (4.63) is given in Program 4.3b. We can note the following points about subroutine DERV:

1. Two cases are programmed (NORUN=1 and 2). For the first case, Dirichlet boundary conditions are programmed as

```
C...
C...     CASE 1 - DIRICHLET BOUNDARY CONDITIONS
         IF(NORUN.EQ.1)THEN
            NL=1
            NU=1
            CALL UE(X( 1),T,U( 1))
            CALL UE(X(NX),T,U(NX))
         END IF
```

where the function UE implements equation (4.64), with $x = 0$ and $x = 1$ for the two calls to UE:

```
      SUBROUTINE UE(X,T,U)
C...
C...  SUBROUTINE EXACT COMPUTES THE EXACT SOLUTION FOR COMPARISON
C...  WITH THE NUMERICAL SOLUTION.  IT IS ALSO USED TO DEFINE THE
C...  INITIAL AND BOUNDARY CONDITIONS FOR THE NUMERICAL SOLUTION
C...
      IMPLICIT DOUBLE PRECISION (A-H,O-Z)
      COMMON/C/          A,          B,          C,          D
C...
C...  ANALYTICAL SOLUTION TO MODIFIED BURGERS' EQUATION
      U=(A+B*DEXP(X)+C*X)/(C*T+D)
      RETURN
      END
```

```
      SUBROUTINE DERV
      IMPLICIT DOUBLE PRECISION (A-H,O-Z)
      PARAMETER (NX=11)
      COMMON/T/          T,       NFIN,      NORUN
     1      /Y/     U(NX)
     2      /F/    UT(NX)
     3      /S/    UX(NX),   UXX(NX),      X(NX)
     4      /C/          A,          B,          C,          D
C...
C... BOUNDARY CONDITIONS
C...
C...      CASE 1 - DIRICHLET BOUNDARY CONDITIONS
         IF(NORUN.EQ.1)THEN
            NL=1
            NU=1
            CALL UE(X( 1),T,U( 1))
            CALL UE(X(NX),T,U(NX))
         END IF
C...
C...  UX
      CALL DSS020(0.0D0,1.0D0,NX,U,UX,1.0D+00)
C...
C...      CASE 2 - NEUMANN BOUNDARY CONDITIONS
         IF(NORUN.EQ.2)THEN
            NL=2
            NU=2
            CALL UXE(X( 1),T,UX( 1))
            CALL UXE(X(NX),T,UX(NX))
         END IF
C...
C...  UXX
      CALL DSS044(0.0D0,1.0D0,NX,U,UX,UXX,NL,NU)
C...
C...  MODIFIED BURGERS' EQUATION
      DO 2 I=1,NX
         UT(I)=U(I)*UXX(I)-U(I)*UX(I)
2     CONTINUE
      RETURN
      END
```

Program 4.3b Subroutine `DERV` for the Solution of Equation (4.63).

The convective derivative, $\partial u/\partial x$, in equation (4.63) is then computed by five-point biased upwind approximations, for flow in the positive x-direction (for both cases)

```
C...
C...  UX
      CALL DSS020(0.0D0,1.0D0,NX,U,UX,1.0D+00)
```

2. The Neumann boundary conditions at $x = 0$ and $x = 1$ from equation (4.64) are programmed as

```
C...
C...      CASE 2 - NEUMANN BOUNDARY CONDITIONS
```

```
      IF(NORUN.EQ.2)THEN
         NL=2
         NU=2
         CALL UXE(X( 1),T,UX( 1))
         CALL UXE(X(NX),T,UX(NX))
      END IF
```

where the function UXE provides the derivative, u_x, from equation (4.64)

```
      SUBROUTINE UXE(X,T,UX)
C...
C...  SUBROUTINE EXACT COMPUTES THE SPATIAL DERIVATIVE OF THE
C...  EXACT SOLUTION FOR COMPARISON WITH THE NUMERICAL SOLUTION.
C...  IT IS ALSO USED TO DEFINE THE BOUNDARY CONDITIONS FOR THE
C...  NUMERICAL SOLUTION
C...
      IMPLICIT DOUBLE PRECISION (A-H,O-Z)
      COMMON/C/          A,          B,          C,          D
C...
C...  ANALYTICAL DERIVATIVE FOR MODIFIED BURGERS' EQUATION
      UX=(B*DEXP(X)+C)/(C*T+D)
      RETURN
      END
```

3. The diffusive second derivative, $\partial^2 u/\partial x^2$, in equation (4.63) is then computed from the fourth-order centered approximations of DSS044:

```
C...
C...  UXX
      CALL DSS044(0.0D0,1.0D0,NX,U,UX,UXX,NL,NU)
```

4. The PDE, equation (4.63), is then implemented in DO loop 2 over the grid from $x = 0$ to $x = 1$:

```
C...
C...  MODIFIED BURGERS' EQUATION
      DO 2 I=1,NX
         UT(I)=U(I)*UXX(I)-U(I)*UX(I)
2     CONTINUE
```

Again, we note the close resemblance of the method of lines programming of a PDE with the PDE itself, and the ease with which nonlinearities can be accommodated.

Subroutine PRINT, listed in Program 4.3c, is very similar to the two preceding subroutines PRINT for Burgers' equation (Programs 4.1d and 4.2d). As with Burgers' equation, we compute and print the convective and diffusive terms in equation (4.63), $-u\partial u/\partial x$ and $u\partial^2 u/\partial x^2$. Again, we do not list subroutine PLOT1 since it just writes a file for plotting as discussed previously (e.g., see Program 3.5d).

The data file read by the main program is listed in Program 4.3d. Two sets of data are provided for the two runs programmed in subroutine DERV. The integration is by RKF45, called by a main program which is not listed, but is very similar to Program 2.2a.

Abbreviated numerical output and graphical output from Programs 4.3a to d are given in Table 4.3 and Figure 4.5.

```
      SUBROUTINE PRINT(NI,NO)
      IMPLICIT DOUBLE PRECISION (A-H,O-Z)
      PARAMETER (NX=11)
      COMMON/T/          T,       NFIN,      NORUN
     1      /Y/     U(NX)
     2      /F/    UT(NX)
     3      /S/    UX(NX),   UXX(NX),       X(NX)
     4      /C/          A,         B,          C,          D
     5      /P/  UN(6,NX),        IP
C...
C...  INITIALIZE A COUNTER FOR THE PLOTTING
      DATA IP/0/
C...
C...  MONITOR THE SOLUTION
      IP=IP+1
      WRITE(*,*)NORUN,IP,T
C...
C...  STORE THE NUMERICAL SOLUTION FOR SUBSEQUENT PLOTTING VIA SUBROU-
C...  TINE PLOT1
      DO 10 I=1,NX
         UN(IP,I)=U(I)
10    CONTINUE
C...
C...  PRINT THE NUMERICAL AND ANALYTICAL SOLUTIONS, THE DIFFERENCE
C..   BETWEEN THE SOLUTIONS AND THE TERMS IN BURGERS' EQUATION
      WRITE(NO,2)T
2     FORMAT(1H ,//,5H T = ,F5.2,/,
     1 9X,1HX,5X,5HU NUM,4X,6HU ANAL,5X,5HERROR,5X,5HU*UXX,5X,5H-U*UX,
     2 8X,2HUT)
      DO 3 I=1,NX
         CALL UE(X(I),T,UANAL)
         ERROR=U(I)-UANAL
         DIFF= U(I)*UXX(I)
         CONV=-U(I)*UX( I)
         WRITE(NO,4)X(I),U(I),UANAL,ERROR,DIFF,CONV,UT(I)
4        FORMAT(F10.3,6F10.5)
3     CONTINUE
C...
C...  PLOT THE SOLUTION
      CALL PLOT1
C...
C...  REINITIALIZE THE COUNTER FOR PLOTTING
      IF(IP.EQ.6)IP=0
      RETURN
      END
```

Program 4.3c Subroutine `PRINT` to Display the Numerical and Analytical Solutions for Equation (4.63).

The numerical and analytical solutions agree to four figures or better, even with only 11 grid points in x. This agreement is due in large part to the smooth solution of equation (4.64). In particular, the denominator $ct + d$ is well behaved with $c = d = 1$ and $t \geq 0$. Also, Figure 4.5 indicates there would be essentially no benefit in

```
Mickens, R. E., Trans SCS, pp 209-227, 1991
0.          5.0        1.0
   11                      0.000001
Mickens, R. E., Trans SCS, pp 209-227, 1991
0.          5.0        1.0
   11                      0.000001
END OF RUNS
```

Program 4.3d Data File for Equation (4.63).

using a nonuniform grid, such as could be implemented via subroutine DSS038 or DSS040. Thus, some knowledge of the solution is helpful in selecting a grid and the associated algebraic approximations of the PDE spatial derivatives when formulating a new method of lines code.

Finally, we should mention that one reason why only 11 grid points were used is that with more grid points, the stiffness of the ODEs increased very sharply. This became evident through the run times with increasing numbers of grid points. In fact, a run with 201 points, which admittedly, is "overkill" for this problem, resulted in a computer run that seemed interminable. Fortunately, such a large number of grid point points was not required for the particular choice of a, b, c, and d. If for other conditions, a sharp spatial front occurred that required a relatively large number of grid points to achieve an acceptable spatial resolution, the use of an implicit ODE integrator might have been required.

4.3 *PDEs with Mixed Partial Derivatives*

Scientific and engineering problems can lead to PDEs with *mixed (or cross) partial derivatives*. We now consider an example of how mixed partial dervivatives can be accommodated within the numerical method of lines. The PDE is

$$\frac{\partial u}{\partial t} = \frac{\partial^2 u}{\partial x^2} - 3\frac{\partial^2 u}{\partial x \partial y} + \frac{\partial^2 u}{\partial y^2} \tag{4.65}$$

Equation (4.65) has a solution

$$u(x,y,t) = e^x e^y e^{-t} \tag{4.66}$$

We then take the initial and boundary conditions to be consistent with equation (4.66). Thus, the initial condition for equation (4.65) is

$$u(x,y,0) = e^x e^y \tag{4.67}$$

Dirichlet boundary conditions are defined on the unit square, $0 \le x \le 1$, $0 \le y \le 1$, as

$$u(0,y,t) = e^y e^{-t}, \qquad u(1,y,t) = e^1 e^y e^{-t} \qquad (4.68)(4.69)$$

$$u(x,0,t) = e^x e^{-t}, \qquad u(x,1,t) = e^x e^1 e^{-t} \qquad (4.70)(4.71)$$

Various approaches within the method of lines can be considered for the approximation of the mixed partial in equation (4.65). Here we use an approach based on the Taylor series. If we expand the solution, $u(x,y,t)$, from the base point, (x,y), to

Table 4.3 Numerical Output from Programs 4.3a to 4.3d. *Continued next pages.*

```
RUN NO. -    1  Mickens, R. E., Trans SCS, pp 209-227, 1991

INITIAL T -  0.000D+00

  FINAL T -  0.500D+01

  PRINT T -  0.100D+01

NUMBER OF DIFFERENTIAL EQUATIONS -  11

INTEGRATION ALGORITHM - RKF45

MAXIMUM INTEGRATION ERROR -  0.100D-05


T =  0.00

        X     U NUM     U ANAL     ERROR      U*UXX      -U*UX        UT
    0.000   2.00000   2.00000   0.00000    1.99981   -3.99995  -2.00014
    0.100   2.20517   2.20517   0.00000    2.43711   -4.64227  -2.20516
    0.200   2.42140   2.42140   0.00000    2.95750   -5.37890  -2.42140
    0.300   2.64986   2.64986   0.00000    3.57693   -6.22681  -2.64988
    0.400   2.89182   2.89182   0.00000    4.31409   -7.20594  -2.89185
    0.500   3.14872   3.14872   0.00000    5.19136   -8.34011  -3.14875
    0.600   3.42212   3.42212   0.00000    6.23550   -9.65765  -3.42215
    0.700   3.71375   3.71375   0.00000    7.47857  -11.19237  -3.71380
    0.800   4.02554   4.02554   0.00000    8.95900  -12.98459  -4.02559
    0.900   4.35960   4.35960   0.00000   10.72296  -15.08255  -4.35958
    1.000   4.71828   4.71828   0.00000   12.82483  -17.54368  -4.71886


T =  1.00
        X     U NUM     U ANAL     ERROR      U*UXX      -U*UX        UT
    0.000   0.99995   1.00000  -0.00005    0.49995   -0.99994  -0.50002
    0.100   1.10259   1.10259   0.00000    0.60927   -1.16057  -0.55129
    0.200   1.21070   1.21070   0.00000    0.73937   -1.34472  -0.60535
    0.300   1.32493   1.32493   0.00000    0.89424   -1.55670  -0.66246
    0.400   1.44591   1.44591   0.00000    1.07853   -1.80148  -0.72296
    0.500   1.57436   1.57436   0.00000    1.29785   -2.08503  -0.78718
    0.600   1.71106   1.71106   0.00000    1.55888   -2.41441  -0.85553
    0.700   1.85688   1.85688   0.00000    1.86966   -2.79809  -0.92844
    0.800   2.01277   2.01277   0.00000    2.23976   -3.24615  -1.00639
    0.900   2.17980   2.17980   0.00000    2.68074   -3.77064  -1.08990
    1.000   2.35872   2.35914  -0.00042    3.20563   -4.38514  -1.17972
        .
        .
        .

T =  2.00
        .
        .
        .
```

Table 4.3 *Continued.*

```
T =  3.00
       .
       .
       .

T =  4.00
       .
       .
       .

T =  5.00

         X      U NUM      U ANAL      ERROR       U*UXX       -U*UX         UT
     0.000    0.33326    0.33333   -0.00008    0.05555   -0.11108   -0.05555
     0.100    0.36753    0.36753    0.00000    0.06770   -0.12895   -0.06126
     0.200    0.40357    0.40357    0.00000    0.08215   -0.14941   -0.06726
     0.300    0.44164    0.44164    0.00000    0.09936   -0.17297   -0.07360
     0.400    0.48197    0.48197    0.00000    0.11983   -0.20016   -0.08033
     0.500    0.52479    0.52479    0.00000    0.14421   -0.23167   -0.08746
     0.600    0.57035    0.57035    0.00000    0.17320   -0.26827   -0.09507
     0.700    0.61896    0.61896    0.00000    0.20776   -0.31090   -0.10314
     0.800    0.67092    0.67092    0.00000    0.24885   -0.36068   -0.11184
     0.900    0.72660    0.72660    0.00000    0.29785   -0.41896   -0.12111
     1.000    0.78568    0.78638   -0.00070    0.35609   -0.48689   -0.13092


RUN NO. -    2  Mickens, R. E., Trans SCS, pp 209-227, 1991

INITIAL T -  0.000D+00

  FINAL T -  0.500D+01

  PRINT T -  0.100D+01

NUMBER OF DIFFERENTIAL EQUATIONS -  11

INTEGRATION ALGORITHM - RKF45

MAXIMUM INTEGRATION ERROR -  0.100D-05


T =  0.00
         X      U NUM      U ANAL      ERROR       U*UXX       -U*UX         UT
     0.000    2.00000    2.00000    0.00000    1.99998   -4.00000   -2.00002
     0.100    2.20517    2.20517    0.00000    2.43711   -4.64227   -2.20516
     0.200    2.42140    2.42140    0.00000    2.95750   -5.37890   -2.42140
     0.300    2.64986    2.64986    0.00000    3.57693   -6.22681   -2.64988
     0.400    2.89182    2.89182    0.00000    4.31409   -7.20594   -2.89185
     0.500    3.14872    3.14872    0.00000    5.19136   -8.34011   -3.14875
     0.600    3.42212    3.42212    0.00000    6.23550   -9.65765   -3.42215
     0.700    3.71375    3.71375    0.00000    7.47857  -11.19237   -3.71380
```

Table 4.3 *Continued.*

```
     0.800   4.02554   4.02554   0.00000   8.95900 -12.98459  -4.02559
     0.900   4.35960   4.35960   0.00000  10.72296 -15.08255  -4.35958
     1.000   4.71828   4.71828   0.00000  12.82555 -17.54390  -4.71836


 T =  1.00
         X     U NUM    U ANAL     ERROR     U*UXX     -U*UX        UT
     0.000   1.00000   1.00000   0.00000   0.49999  -1.00000  -0.50000
     0.100   1.10258   1.10259   0.00000   0.60927  -1.16056  -0.55129
     0.200   1.21070   1.21070   0.00000   0.73937  -1.34472  -0.60535
     0.300   1.32492   1.32493   0.00000   0.89423  -1.55670  -0.66247
     0.400   1.44591   1.44591   0.00000   1.07852  -1.80148  -0.72296
     0.500   1.57436   1.57436   0.00000   1.29784  -2.08502  -0.78718
     0.600   1.71105   1.71106   0.00000   1.55888  -2.41441  -0.85553
     0.700   1.85687   1.85688   0.00000   1.86965  -2.79808  -0.92844
     0.800   2.01277   2.01277   0.00000   2.23975  -3.24614  -1.00639
     0.900   2.17980   2.17980   0.00000   2.68073  -3.77063  -1.08990
     1.000   2.35914   2.35914   0.00000   3.20640  -4.38597  -1.17957
         .
         .
         .

 T =  2.00
         .
         .
         .

 T =  3.00
         .
         .
         .

 T =  4.00
         .
         .
         .

 T =  5.00

         X     U NUM    U ANAL     ERROR     U*UXX     -U*UX        UT
     0.000   0.33333   0.33333   0.00000   0.05555  -0.11111  -0.05556
     0.100   0.36752   0.36753   0.00000   0.06770  -0.12895  -0.06125
     0.200   0.40356   0.40357   0.00000   0.08215  -0.14941  -0.06726
     0.300   0.44164   0.44164   0.00000   0.09936  -0.17297  -0.07361
     0.400   0.48197   0.48197   0.00000   0.11984  -0.20016  -0.08033
     0.500   0.52478   0.52479   0.00000   0.14420  -0.23167  -0.08746
     0.600   0.57035   0.57035   0.00000   0.17321  -0.26827  -0.09506
     0.700   0.61895   0.61896   0.00000   0.20774  -0.31090  -0.10316
     0.800   0.67092   0.67092   0.00000   0.24886  -0.36068  -0.11182
     0.900   0.72660   0.72660   0.00000   0.29786  -0.41896  -0.12110
     1.000   0.78638   0.78638   0.00000   0.35626  -0.48733  -0.13106
```

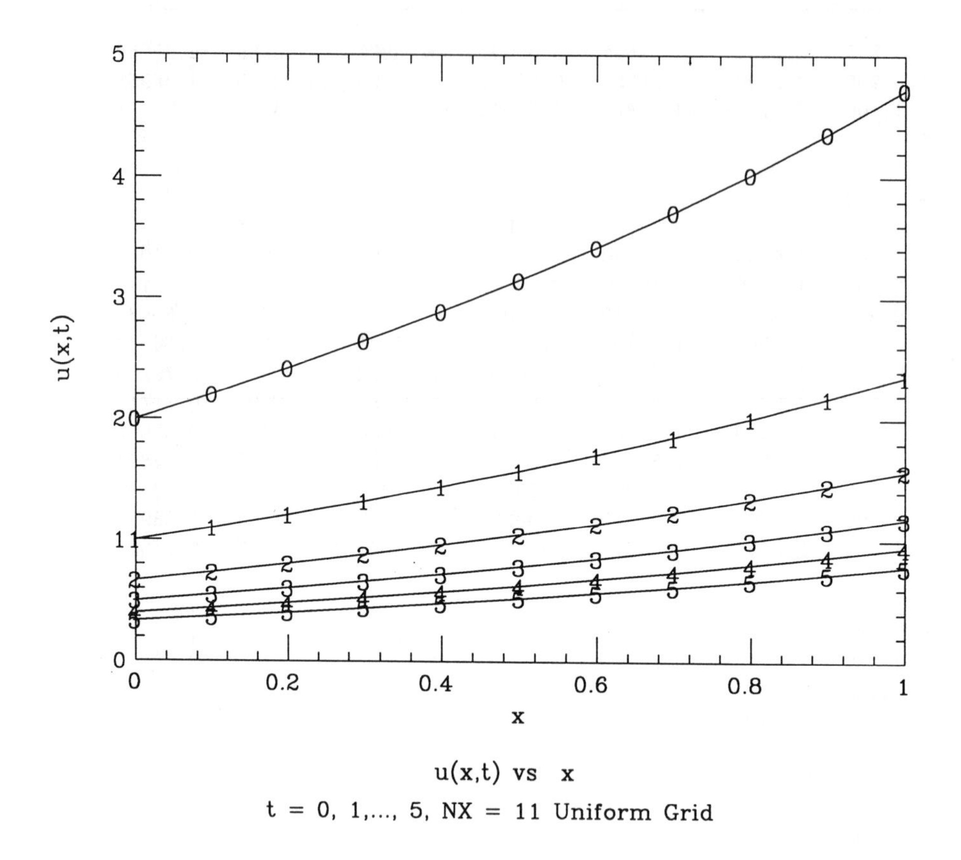

Figure 4.5 Numerical (Points) and Analytical (Solid Curves) Solutions of Equation (4.63) from Programs 4.3a to 4.3d (NORUN=2).

$(x_i + \Delta x, y_i + \Delta y)$, the Taylor series up to and including the second-order terms in Δx and Δy is

$$u(x_i + \Delta x, y_i + \Delta y) = u(x_i, y_i) + u_x(x_i, y_i)\Delta x + u_y(x_i, y_i)\Delta y + u_{xx}(x_i, y_i)\frac{\Delta x^2}{2!} + 2u_{xy}(x_i, y_i)\frac{\Delta x \Delta y}{2!} + u_{yy}(x_i, y_i)\frac{\Delta y^2}{2!} \quad (4.72)$$

Similarly, an expansion around (x, y), to $(x_i - \Delta x, y_i - \Delta y)$ is

$$u(x_i - \Delta x, y_i - \Delta y) = u(x_i, y_i) - u_x(x_i, y_i)\Delta x - u_y(x_i, y_i)\Delta y + u_{xx}(x_i, y_i)\frac{\Delta x^2}{2!} + 2u_{xy}(x_i, y_i)\frac{\Delta x \Delta y}{2!} + u_{yy}(x_i, y_i)\frac{\Delta y^2}{2!} \quad (4.73)$$

```
      SUBROUTINE INITAL
      IMPLICIT DOUBLE PRECISION (A-H,O-Z)
      PARAMETER (NX=11,NY=11)
      COMMON/T/          T,      NSTOP,      NORUN
     1      /Y/  U(NX,NY)
     2      /F/ UT(NX,NY)
     3      /S/ UX(NX,NY), UY(NX,NY),UXX(NX,NY),UYY(NX,NY),UXY(NX,NY),
     4                X(NX),      Y(NY),         DX,         DY
     5      /I/        IP
C...
C... GRID SPACING IN X AND Y
      DX=1.0D0/DFLOAT(NX-1)
      DY=1.0D0/DFLOAT(NY-1)
C...
C... SPATIAL GRID AND INITIAL CONDITION
      DO 1 I=1,NX
      DO 1 J=1,NY
C...
C...    GRID IN X
        X(I)=DX*DFLOAT(I-1)
C...
C...    GRID IN Y
        Y(J)=DY*DFLOAT(J-1)
C...
C...    INITIAL CONDITION IS THE EXACT SOLUTION AT T = 0
        U(I,J)=EXACT(X(I),Y(J),T)
C...
C...    THE MIXED DERIVATIVE IS ZEROED INITIALLY SINCE IT IS NOT
C...    COMPUTED ON THE BOUNDARIES (X = 0 AND 1, Y = 0 AND 1) IN
C...    SUBROUTINE DERV
        UXY(I,J)=0.0D0
1     CONTINUE
C...
C... INITIAL DERIVATIVES
      CALL DERV
      IP=0
      RETURN
      END
```

Program 4.4a Subroutine INITAL for the Solution of Equation (4.65).

We can now solve equations (4.72) and (4.73) for the mixed partial derivative, $u_{xy}(x_i, y_i)$, in terms of $u_{xx}(x_i, y_i)$ and $u_{yy}(x_i, y_i)$

$$u_{xy}(x_i, y_i) = \frac{u(x_i + \Delta x, y_i + \Delta y) - 2u(x_i, y_i) + u(x_i - \Delta x, y_i - \Delta y)}{2\Delta x \Delta y} - u_{xx}(x_i, y_i)\frac{(\Delta x/\Delta y)}{2} - u_{yy}(x_i, y_i)\frac{(\Delta y/\Delta x)}{2} \tag{4.74}$$

Equation (4.74) is then used to calculate $\partial^2 u/\partial x \partial y$ $(= u_{xy}(x_i, y_i))$ in equation (4.65).

Subroutine INITAL is listed in Program 4.4a for the method of lines solution of equation (4.65) on an 11×11 grid. We can note the following points about subroutine INITAL:

1. The partial derivatives in equation (4.65) are in arrays in the usual COMMON blocks:

```
      PARAMETER (NX=11,NY=11)
      COMMON/T/          T,       NSTOP,       NORUN
     1      /Y/  U(NX,NY)
     2      /F/ UT(NX,NY)
     3      /S/ UX(NX,NY), UY(NX,NY),UXX(NX,NY),UYY(NX,NY),UXY(NX,NY),
     4               X(NX),     Y(NY),          DX,          DY
     5      /I/         IP
```

in particular, the mixed partial, $\partial^2 u/\partial x \partial y$ is in array UXY(NX,NY).

2. The grids in x and y, stored in arrays X and Y, and the initial condition of equation (4.67) are then defined in DO loop 1:

```
C...
C...  SPATIAL GRID AND INITIAL CONDITION
      DO 1 I=1,NX
      DO 1 J=1,NY
C...
C...     GRID IN X
         X(I)=DX*DFLOAT(I-1)
C...
C...     GRID IN Y
         Y(J)=DY*DFLOAT(J-1)
C...
C...     INITIAL CONDITION IS THE EXACT SOLUTION AT T = 0
         U(I,J)=EXACT(X(I),Y(J),T)
1     CONTINUE
```

The analytical solution of equation (4.66) is implemented in function EXACT, which defines initial condition (4.67) at $t = 0$:

```
      DOUBLE-PRECISION FUNCTION EXACT(X,Y,T)
C...
C...  FUNCTION EXACT COMPUTES THE EXACT SOLUTION
C...
C...     U(X,Y,T)=EXP(X)*EXP(Y)*EXP(-T)
C...
      IMPLICIT DOUBLE PRECISION (A-H,O-Z)
      EXACT=DEXP(X)*DEXP(Y)*DEXP(-T)
      RETURN
      END
```

Subroutine DERV is listed in Program 4.4b. We can note the following points about subroutine DERV:

1. The boundary conditions at $x = 0$, $y = 0$, $x = 1$, and $y = 1$ are implemented through the temporal derivatives $\partial u(0, y, t)/\partial t$, $\partial u(x, 0, t)/\partial t$, $\partial u(1, y, t)/\partial t$, and $\partial u(x, 1, t)/\partial t$, respectively. These temporal derivatives are computed from the

```
      SUBROUTINE DERV
      IMPLICIT DOUBLE PRECISION (A-H,O-Z)
      PARAMETER (NX=11,NY=11)
      COMMON/T/          T,      NSTOP,      NORUN
     1      /Y/  U(NX,NY)
     2      /F/ UT(NX,NY)
     3      /S/ UX(NX,NY), UY(NX,NY),UXX(NX,NY),UYY(NX,NY),UXY(NX,NY),
     4              X(NX),     Y(NY),         DX,         DY
     5      /I/         IP
C...
C...  ARRAYS FOR THE INTERMEDIATE ONE-DIMENSIONAL SOLUTION AND DERI-
C...  VATIVES
      DIMENSION US(NX), USX(NX), USY(NY), USXX(NX), USYY(NY)
C...
C...  BC AT X = 0
      DO 1 J=1,NY
         UT(1,J)=-EXACT(X(1),Y(J),T)
1     CONTINUE
C...
C...  BC AT Y = 0
      DO 2 I=1,NX
         UT(I,1)=-EXACT(X(I),Y(1),T)
2     CONTINUE
C...
C...  BC AT X = 1
      DO 3 J=1,NY
         UT(NX,J)=-EXACT(X(NX),Y(J),T)
3     CONTINUE
C...
C...  BC AT Y = 1
      DO 4 I=1,NX
         UT(I,NY)=-EXACT(X(I),Y(NY),T)
4     CONTINUE
C...
C...  UXX
C...
C...     0 LE Y LE 1
         DO 5 J=1,NY
C...
C...     0 LE X LE 1
         DO 6 I=1,NX
C...
C...     TRANSFER U(I,J) TO US(I)
6        US(I)=U(I,J)
C...
C...     USXX
         NL=1
         NU=1
         CALL DSS044(0.0D0,1.0D0,NX,US,USX,USXX,NL,NU)
C...
C...     TRANSFER USXX TO UXX
```

Program 4.4b Subroutine DERV for the Solution of Equation (4.65). *Continued next page.*

```
      DO 7 I=1,NX
7     UXX(I,J)=USXX(I)
5   CONTINUE
C...
C... UYY
C...
C...   0 LE X LE 1
       DO 8 I=1,NX
C...
C...   0 LE Y LE 1
       DO 9 J=1,NY
C...
C...   TRANSFER U(I,J) TO US(J)
9      US(J)=U(I,J)
C...
C...   USYY
       CALL DSS044(0.0D0,1.0D0,NY,US,USY,USYY,NL,NU)
C...
C...   TRANSFER USYY TO UYY
       DO 10 J=1,NY
10     UYY(I,J)=USYY(J)
8      CONTINUE
C...
C... UXY
     DO 11 I=2,NX-1
     DO 11 J=2,NY-1
       UXY(I,J)=
    1    (U(I+1,J+1)-2.0D0*U(I,J)+U(I-1,J-1))/(2.0D0*DX*DY)
    2    -UXX(I,J)*(DX/DY)/2.0D0-UYY(I,J)*(DY/DX)/2.0D0
11   CONTINUE
C...
C... PDE
     DO 12 I=2,NX-1
     DO 12 J=2,NY-1
       UT(I,J)=UXX(I,J)-3.0D0*UXY(I,J)+UYY(I,J)
12   CONTINUE
     RETURN
     END
```

Program 4.4b *Continued.*

analytical solution of equation (4.66), through calls to function EXACT in DO loops 1 to 4, respectively:

```
C...
C... BC AT X = 0
     DO 1 J=1,NY
       UT(1,J)=-EXACT(X(1),Y(J),T)
1    CONTINUE
C...
C... BC AT Y = 0
     DO 2 I=1,NX
       UT(I,1)=-EXACT(X(I),Y(1),T)
2    CONTINUE
```

```
C...
C...  BC AT X = 1
      DO 3 J=1,NY
         UT(NX,J)=-EXACT(X(NX),Y(J),T)
3     CONTINUE
C...
C...  BC AT Y = 1
      DO 4 I=1,NX
         UT(I,NY)=-EXACT(X(I),Y(NY),T)
4     CONTINUE
```

2. The second derivative $\partial^2 u/\partial x^2$ in equation (4.65) is computed by a call to subroutine DSS044. This requires the transfer of the solution, $u(x, y, t)$, and the derivative, $\partial^2 u/\partial x^2$, between one- and two-dimensional arrays since the numerical method of lines solution is computed using two-dimensional arrays, but DSS044 can accept only one-dimensional arrays. The procedure is straightforward, and demonstrates how multidimensional PDE problems can be addressed with one-dimensional spatial differentiation subroutines:

```
C...
C...  UXX
C...
C...     0 LE Y LE 1
         DO 5 J=1,NY
C...
C...     0 LE X LE 1
         DO 6 I=1,NX
C...
C...     TRANSFER U(I,J) TO US(I)
6        US(I)=U(I,J)
C...
C...     USXX
         NL=1
         NU=1
         CALL DSS044(0.0D0,1.0D0,NX,US,USX,USXX,NL,NU)
C...
C...     TRANSFER USXX TO UXX
         DO 7 I=1,NX
7        UXX(I,J)=USXX(I)
5     CONTINUE
```

The two-dimensional arrays for u and $\partial^2 u/\partial x^2$ are U and UXX, respectively, and the one-dimensional arrays for u, $\partial u/\partial x$, and $\partial^2 u/\partial x^2$ are US, USX, and USXX, respectively. DO loop 5 steps through a series of values of y, from $y = 0$ to $y = 1$. For each y, DO loops 6 and 7 step through a series of values of x from $x = 0$ to $x = 1$. Note that DO loops 6 and 7 are the inner loops that provide differentiation with respect to x (for the calculation of $\partial^2 u/\partial x^2$) by the call to DSS044; that is, US is differentiated to USXX by DSS044, where the variation within US and USXX is with respect to x.

3. The procedure is then reversed for the calculation of $\partial^2 u/\partial y^2$ in equation (4.65):

```
C...
C...  UYY
```

```
C...
C...      0 LE X LE 1
          DO 8 I=1,NX
C...
C...      0 LE Y LE 1
          DO 9 J=1,NY
C...
C...      TRANSFER U(I,J) TO US(J)
9         US(J)=U(I,J)
C...
C...      USYY
          CALL DSS044(0.0D0,1.0D0,NY,US,USY,USYY,NL,NU)
C...
C...      TRANSFER USYY TO UYY
          DO 10 J=1,NY
10        UYY(I,J)=USYY(J)
8         CONTINUE
```

Now DO loop 8 steps through a series of values of x from $x = 0$ to 1. For each value of x, DO loops 9 and 10 step through a series of values of y; thus the call to DSS044 gives differentiation with respect to y, that is, the calculation of $\partial^2 u/\partial y^2$.

4. Once $\partial^2 u/\partial x^2$ and $\partial^2 u/\partial y^2$ are calculated on the interior of the unit square, $\partial^2 u/\partial x \partial y$ is calculated from equation (4.74):

```
C...
C...  UXY
      DO 11 I=2,NX-1
      DO 11 J=2,NY-1
         UXY(I,J)=
     1      (U(I+1,J+1)-2.0D0*U(I,J)+U(I-1,J-1))/(2.0D0*DX*DY)
     2      -UXX(I,J)*(DX/DY)/2.0D0-UYY(I,J)*(DY/DX)/2.0D0
11    CONTINUE
```

5. Equation (4.65) is then programmed in DO loop 12:

```
C...
C...  PDE
      DO 12 I=2,NX-1
      DO 12 J=2,NY-1
         UT(I,J)=UXX(I,J)-3.0D0*UXY(I,J)+UYY(I,J)
12    CONTINUE
```

Note that the programming in DO loop 12 is for the interior of the unit square. Since the temporal derivatives on the boundaries of the unit square were previously computed by DO loops 1 to 4, all $11 \times 11 = 121$ temporal derivatives are computed at the conclusion of DO loop 12. In other words, the 121 derivatives can now be sent to the ODE integrator through COMMON/F/, and the 121 solutions will be returned from the ODE integrator through COMMON/Y/.

Subroutine PRINT is listed in Program 4.4c. We can note the following points about subroutine PRINT:

```
      SUBROUTINE PRINT(NI,NO)
      IMPLICIT DOUBLE PRECISION (A-H,O-Z)
      PARAMETER (NX=11,NY=11)
      COMMON/T/          T,      NSTOP,      NORUN
     1      /Y/  U(NX,NY)
     2      /F/ UT(NX,NY)
     3      /S/ UX(NX,NY), UY(NX,NY),UXX(NX,NY),UYY(NX,NY),UXY(NX,NY),
     4              X(NX),     Y(NY),         DX,         DY
     5      /I/         IP
C...
C...  DIMENSION THE ARRAYS FOR PRINTING AND PLOTTING
      DIMENSION TP(51), UP(2,51), DIFF(51), UE(NX,NY)
C...
C...  MONITOR THE SOLUTION
      WRITE(*,*)IP,T
C...
C...  PRINT THE SOLUTION EVERY TENTH CALL TO SUBROUTINE PRINT
      IP=IP+1
      IF(((IP-1)/10*10).EQ.(IP-1))THEN
C...
C...  PRINT THE NUMERICAL U(X,Y,T)
      WRITE(NO,1)T,((U(I,J),I=1,NX,2),J=1,NY,2)
1     FORMAT(' T = ',F6.3,//,'  U(X,Y,T)',/,
     1 15X,'X = 0',4X,'X = 0.2',3X,'X = 0.4',3X,'X = 0.6',
     2            3X,'X = 0.8',3X,'X = 1.0',/,
     33X,' Y = 0 ',6F10.3,/,3X,'Y = 0.2',6F10.3,/,3X,'Y = 0.4',6F10.3,/,
     43X,'Y = 0.6',6F10.3,/,3X,'Y = 0.8',6F10.3,/,3X,'Y = 1.0',6F10.3,/)
C...
C...  COMPUTE THE EXACT U(X,Y,T)
      DO 2 J=1,NY
      DO 2 I=1,NX
         UE(I,J)=EXACT(X(I),Y(J),T)
2     CONTINUE
C...
C...  PRINT THE EXACT U(X,Y,T)
      WRITE(NO,3)((UE(I,J),I=1,NX,2),J=1,NY,2)
3     FORMAT(//,'  UE(X,Y,T)',/,
     1 15X,'X = 0',4X,'X = 0.2',3X,'X = 0.4',3X,'X = 0.6',
     2            3X,'X = 0.8',3X,'X = 1.0',/,
     33X,' Y = 0 ',6F10.3,/,3X,'Y = 0.2',6F10.3,/,3X,'Y = 0.4',6F10.3,/,
     43X,'Y = 0.6',6F10.3,/,3X,'Y = 0.8',6F10.3,/,3X,'Y = 1.0',6F10.3,/)
C...
C...  PRINT THE NUMERICAL UT(X,Y,T)
      WRITE(NO,4)((UT(I,J),I=1,NX,2),J=1,NY,2)
4     FORMAT(//,'  UT(X,Y,T)',/,
     1 15X,'X = 0',4X,'X = 0.2',3X,'X = 0.4',3X,'X = 0.6',
     2            3X,'X = 0.8',3X,'X = 1.0',/,
     33X,' Y = 0 ',6F10.3,/,3X,'Y = 0.2',6F10.3,/,3X,'Y = 0.4',6F10.3,/,
     43X,'Y = 0.6',6F10.3,/,3X,'Y = 0.8',6F10.3,/,3X,'Y = 1.0',6F10.3,/)
C...
```

Program 4.4c Subroutine PRINT for Printing the Solution of Equation (4.65) with All of the Partial Derivatives. *Continued next page.*

```
C...  PRINT THE NUMERICAL UXX(X,Y,T)
      WRITE(NO,5)((UXX(I,J),I=1,NX,2),J=1,NY,2)
5     FORMAT(//, ' UXX(X,Y,T)',/,
     1 15X,'X = 0',4X,'X = 0.2',3X,'X = 0.4',3X,'X = 0.6',
     2            3X,'X = 0.8',3X,'X = 1.0',/,
     33X,' Y = 0 ',6F10.3,/,3X,'Y = 0.2',6F10.3,/,3X,'Y = 0.4',6F10.3,/,
     43X,'Y = 0.6',6F10.3,/,3X,'Y = 0.8',6F10.3,/,3X,'Y = 1.0',6F10.3,/)
C...
C...  PRINT THE NUMERICAL UXY IN THE INTERIOR AND THE ANALYTICAL UXY ON
C...  THE BOUNDARIES.  THE LATTER ARE FIRST COMPUTED IN THE FOLLOWING
C...  SERIES OF FOUR DO LOOPS
      DO 6 J=1,NY
6        UXY(1,J) =EXACT(X(1) ,Y(J),T)
      DO 7 J=1,NY
7        UXY(NX,J)=EXACT(X(NX),Y(J),T)
      DO 8 I=1,NX
8        UXY(I,1) =EXACT(X(I) ,Y(1),T)
      DO 9 I=1,NX
9        UXY(I,NY)=EXACT(X(I),Y(NY),T)
      WRITE(NO,10)((UXY(I,J),I=1,NX,2),J=1,NY,2)
10    FORMAT(//, ' UXY(X,Y,T)',/,
     1 15X,'X = 0',4X,'X = 0.2',3X,'X = 0.4',3X,'X = 0.6',
     2            3X,'X = 0.8',3X,'X = 1.0',/,
     33X,' Y = 0 ',6F10.3,/,3X,'Y = 0.2',6F10.3,/,3X,'Y = 0.4',6F10.3,/,
     43X,'Y = 0.6',6F10.3,/,3X,'Y = 0.8',6F10.3,/,3X,'Y = 1.0',6F10.3,/)
C...
C...  PRINT THE NUMERICAL UYY
      WRITE(NO,11)((UYY(I,J),I=1,NX,2),J=1,NY,2)
11    FORMAT(//, ' UYY(X,Y,T)',/,
     1 15X,'X = 0',4X,'X = 0.2',3X,'X = 0.4',3X,'X = 0.6',
     2            3X,'X = 0.8',3X,'X = 1.0',/,
     33X,' Y = 0 ',6F10.3,/,3X,'Y = 0.2',6F10.3,/,3X,'Y = 0.4',6F10.3,/,
     43X,'Y = 0.6',6F10.3,/,3X,'Y = 0.8',6F10.3,/,3X,'Y = 1.0',6F10.3,/)
      END IF
C...
C...  STORE THE NUMERICAL AND ANALYTICAL U(1/2,1/2,T) AND THEIR
C...  DIFFERENCE FOR PRINTING
      TP(IP)=T
      UP(1,IP)=U(6,6)
      UP(2,IP)=EXACT(X(6),Y(6),T)
      DIFF(IP)=(UP(1,IP)-UP(2,IP))/UP(2,IP)*100.0D0
C...
C...  PRINT THE NUMERICAL AND ANALYTICAL U(1/2,1/2,T) AND THEIR
C...  DIFFERENCE VS T
      IF(IP.NE.51)RETURN
      WRITE(NO,12)(TP(I),UP(1,I),UP(2,I),DIFF(I),I=1,51)
12    FORMAT(//,10X,
     1 '   T   ',6X,'U(1/2,1/2,T)',' U(1/2,1/2,T)','   O/O DIFF  ',/,
     2          21X,'     NUM     ','      ANAL    '              ,/,
     3 (10X,F6.2,2F15.4,F12.2))
      RETURN
      END
```

Program 4.4c *Continued.*

1. All of the partial derivatives in equation (4.65) are printed every 10th call to subroutine PRINT through the counter IP used in the IF statement

```
C...
C...  PRINT THE SOLUTION EVERY TENTH CALL TO SUBROUTINE PRINT
      IP=IP+1
      IF(((IP-1)/10*10).EQ.(IP-1))THEN
```

2. The numerical solution in array U is first printed over the unit square:

```
C...
C...  PRINT THE NUMERICAL U(X,Y,T)
      WRITE(NO,1)T,((U(I,J),I=1,NX,2),J=1,NY,2)
1     FORMAT(' T = ',F6.3,//,'   U(X,Y,T)',/,
     1 15X,'X = 0',4X,'X = 0.2',3X,'X = 0.4',3X,'X = 0.6',
     2             3X,'X = 0.8',3X,'X = 1.0',/,
     33X,' Y = 0 ',6F10.3,/,3X,'Y = 0.2',6F10.3,/,3X,'Y = 0.4',6F10.3,/,
     43X,'Y = 0.6',6F10.3,/,3X,'Y = 0.8',6F10.3,/,3X,'Y = 1.0',6F10.3,/)
```

3. The analytical solution is then put in array UE and printed:

```
C...
C...  COMPUTE THE EXACT U(X,Y,T)
      DO 2 J=1,NY
      DO 2 I=1,NX
         UE(I,J)=EXACT(X(I),Y(J),T)
2     CONTINUE
C...
C...  PRINT THE EXACT U(X,Y,T)
      WRITE(NO,3)((UE(I,J),I=1,NX,2),J=1,NY,2)
3     FORMAT(//,'  UE(X,Y,T)',/,
     1 15X,'X = 0',4X,'X = 0.2',3X,'X = 0.4',3X,'X = 0.6',
     2             3X,'X = 0.8',3X,'X = 1.0',/,
     33X,' Y = 0 ',6F10.3,/,3X,'Y = 0.2',6F10.3,/,3X,'Y = 0.4',6F10.3,/,
     43X,'Y = 0.6',6F10.3,/,3X,'Y = 0.8',6F10.3,/,3X,'Y = 1.0',6F10.3,/)
```

This process of detailed printing continues for the numerical partial derivatives $\partial u/\partial t$, $\partial^2 u/\partial x^2$, $\partial^2 u/\partial x \partial y$, and $\partial^2 u/\partial y^2$. In the case of the mixed partial, $\partial^2 u/\partial x \partial y$, the analytical derivative is computed and printed on the boundaries of the unit square since equation (4.74) does not provide the boundary values.

4. The centerpoint numerical and analytical solutions are then stored and printed vs. t:

```
C...
C...  STORE THE NUMERICAL AND ANALYTICAL U(1/2,1/2,T) AND THEIR
C...  DIFFERENCE FOR PRINTING
      TP(IP)=T
      UP(1,IP)=U(6,6)
      UP(2,IP)=EXACT(X(6),Y(6),T)
      DIFF(IP)=(UP(1,IP)-UP(2,IP))/UP(2,IP)*100.0D0
C...
C...  PRINT THE NUMERICAL AND ANALYTICAL U(1/2,1/2,T) AND THEIR
C...  DIFFERENCE VS T
```

```
TWO-DIMENSIONAL PDE WITH MIXED PARTIAL DERIVATIVE
0.         0.5        0.01
  121                  0.000001
END OF RUNS
```

Program 4.4d Data File for Equation (4.65).

```
      IF(IP.NE.51)RETURN
      WRITE(NO,12)(TP(I),UP(1,I),UP(2,I),DIFF(I),I=1,51)
12    FORMAT(//,10X,
     1 '   T   ',6X,'U(1/2,1/2,T)',' U(1/2,1/2,T)','    0/0 DIFF  ',/,
     2         21X,'     NUM      ','        ANAL      '           ,/,
     3 (10X,F6.2,2F15.4,F12.2))
```

The data file read by the main program is listed in Program 4.4d. 121 ODEs are specified in accordance with the 11 $\times$ 11 point spatial grid. The integration is by RKF45, called by a main program which is not listed, but is very similar to Program 2.2a.

Abbreviated numerical output and graphical output from Programs 4.4a to d are given in Table 4.4. We can note the following points about the output in Table 4.4:

1. To evaluate the numerical solution, we should first observe that the analytical solution and the various partial derivatives have the same values at a given point in the unit square, except for the derivative $\partial u/\partial t$ which has a sign reversal. This property follows directly from the analytical solution, equation (4.66). That is,

$$u = -\frac{\partial u}{\partial t} = \frac{\partial^2 u}{\partial x^2} = \frac{\partial^2 u}{\partial x \partial y} = \frac{\partial^2 u}{\partial y^2} = e^x e^y e^{-t}$$

   Also, the solution and its various derivatives have the same values for the same values of $x+y$ (or more generally, for the same values of $x+y-t$), which follows immediately from the analytical solution

$$u = e^x e^y e^{-t} = e^{x+y-t}$$

   Thus, this high degree of symmetry in the analytical solution permits a rapid evaluation of the numerical solution by inspection.

2. This symmetry is apparent in the analytical solution at $t = 0$, which is also the numerical solution at $t = 0$, since the two solutions are the same according to initial condition (4.67) programmed in subroutine INITAL:

UE(X,Y,T)

	X = 0	X = 0.2	X = 0.4	X = 0.6	X = 0.8	X = 1.0
Y = 0	1.000	1.221	1.492	1.822	2.226	2.718
Y = 0.2	1.221	1.492	1.822	2.226	2.718	3.320
Y = 0.4	1.492	1.822	2.226	2.718	3.320	4.055
Y = 0.6	1.822	2.226	2.718	3.320	4.055	4.953
Y = 0.8	2.226	2.718	3.320	4.055	4.953	6.050
Y = 1.0	2.718	3.320	4.055	4.953	6.050	7.389

   For example, for $x+y = 0.8$, $u(x,y,0) = e^{0.8} = 2.226$. Also for $x+y = 1$, $e^1 = 2.718$ (note the diagonals running through the analytical solution with these values).

Table 4.4 Numerical and Graphical Output from Programs 4.4a to d. *Continued next pages.*

RUN NO. - 1 TWO-DIMENSIONAL PDE WITH MIXED PARTIAL DERIVATIVE

INITIAL T - 0.000D+00

FINAL T - 0.500D+00

PRINT T - 0.100D-01

NUMBER OF DIFFERENTIAL EQUATIONS - 121

INTEGRATION ALGORITHM - RKF45

MAXIMUM INTEGRATION ERROR - 0.100D-05

T = 0.000

U(X,Y,T)

	X = 0	X = 0.2	X = 0.4	X = 0.6	X = 0.8	X = 1.0
Y = 0	1.000	1.221	1.492	1.822	2.226	2.718
Y = 0.2	1.221	1.492	1.822	2.226	2.718	3.320
Y = 0.4	1.492	1.822	2.226	2.718	3.320	4.055
Y = 0.6	1.822	2.226	2.718	3.320	4.055	4.953
Y = 0.8	2.226	2.718	3.320	4.055	4.953	6.050
Y = 1.0	2.718	3.320	4.055	4.953	6.050	7.389

UE(X,Y,T)

	X = 0	X = 0.2	X = 0.4	X = 0.6	X = 0.8	X = 1.0
Y = 0	1.000	1.221	1.492	1.822	2.226	2.718
Y = 0.2	1.221	1.492	1.822	2.226	2.718	3.320
Y = 0.4	1.492	1.822	2.226	2.718	3.320	4.055
Y = 0.6	1.822	2.226	2.718	3.320	4.055	4.953
Y = 0.8	2.226	2.718	3.320	4.055	4.953	6.050
Y = 1.0	2.718	3.320	4.055	4.953	6.050	7.389

UT(X,Y,T)

	X = 0	X = 0.2	X = 0.4	X = 0.6	X = 0.8	X = 1.0
Y = 0	-1.000	-1.221	-1.492	-1.822	-2.226	-2.718
Y = 0.2	-1.221	-1.522	-1.859	-2.270	-2.773	-3.320
Y = 0.4	-1.492	-1.859	-2.270	-2.773	-3.387	-4.055
Y = 0.6	-1.822	-2.270	-2.773	-3.387	-4.136	-4.953
Y = 0.8	-2.226	-2.773	-3.387	-4.136	-5.052	-6.050
Y = 1.0	-2.718	-3.320	-4.055	-4.953	-6.050	-7.389

UXX(X,Y,T)

	X = 0	X = 0.2	X = 0.4	X = 0.6	X = 0.8	X = 1.0
Y = 0	1.000	1.221	1.492	1.822	2.226	2.718
Y = 0.2	1.221	1.492	1.822	2.226	2.718	3.320
Y = 0.4	1.492	1.822	2.226	2.718	3.320	4.055

Table 4.4 *Continued.*

Y = 0.6	1.822	2.226	2.718	3.320	4.055	4.953
Y = 0.8	2.225	2.718	3.320	4.055	4.953	6.049
Y = 1.0	2.718	3.320	4.055	4.953	6.050	7.389

UXY(X,Y,T)

	X = 0	X = 0.2	X = 0.4	X = 0.6	X = 0.8	X = 1.0
Y = 0	1.000	1.221	1.492	1.822	2.226	2.718
Y = 0.2	1.221	1.502	1.834	2.240	2.736	3.320
Y = 0.4	1.492	1.834	2.240	2.736	3.342	4.055
Y = 0.6	1.822	2.240	2.736	3.342	4.082	4.953
Y = 0.8	2.226	2.736	3.342	4.082	4.986	6.050
Y = 1.0	2.718	3.320	4.055	4.953	6.050	7.389

UYY(X,Y,T)

	X = 0	X = 0.2	X = 0.4	X = 0.6	X = 0.8	X = 1.0
Y = 0	1.000	1.221	1.492	1.822	2.225	2.718
Y = 0.2	1.221	1.492	1.822	2.226	2.718	3.320
Y = 0.4	1.492	1.822	2.226	2.718	3.320	4.055
Y = 0.6	1.822	2.226	2.718	3.320	4.055	4.953
Y = 0.8	2.226	2.718	3.320	4.055	4.953	6.050
Y = 1.0	2.718	3.320	4.055	4.953	6.049	7.389

T = 0.100

U(X,Y,T)

	X = 0	X = 0.2	X = 0.4	X = 0.6	X = 0.8	X = 1.0
Y = 0	0.905	1.105	1.350	1.649	2.014	2.460
Y = 0.2	1.105	1.350	1.648	2.012	2.457	3.004
Y = 0.4	1.350	1.648	2.011	2.456	3.001	3.669
Y = 0.6	1.649	2.012	2.456	3.000	3.667	4.482
Y = 0.8	2.014	2.457	3.001	3.667	4.480	5.474
Y = 1.0	2.460	3.004	3.669	4.482	5.474	6.686

UE(X,Y,T)

	X = 0	X = 0.2	X = 0.4	X = 0.6	X = 0.8	X = 1.0
Y = 0	0.905	1.105	1.350	1.649	2.014	2.460
Y = 0.2	1.105	1.350	1.649	2.014	2.460	3.004
Y = 0.4	1.350	1.649	2.014	2.460	3.004	3.669
Y = 0.6	1.649	2.014	2.460	3.004	3.669	4.482
Y = 0.8	2.014	2.460	3.004	3.669	4.482	5.474
Y = 1.0	2.460	3.004	3.669	4.482	5.474	6.686

UT(X,Y,T)

	X = 0	X = 0.2	X = 0.4	X = 0.6	X = 0.8	X = 1.0
Y = 0	-0.905	-1.105	-1.350	-1.649	-2.014	-2.460
Y = 0.2	-1.105	-1.347	-1.648	-2.022	-2.470	-3.004
Y = 0.4	-1.350	-1.648	-2.026	-2.481	-3.016	-3.669

Table 4.4 *Continued.*

Y = 0.6	-1.649	-2.022	-2.481	-3.021	-3.672	-4.482
Y = 0.8	-2.014	-2.470	-3.016	-3.672	-4.479	-5.474
Y = 1.0	-2.460	-3.004	-3.669	-4.482	-5.474	-6.686

UXX(X,Y,T)

	X = 0	X = 0.2	X = 0.4	X = 0.6	X = 0.8	X = 1.0
Y = 0	0.905	1.105	1.350	1.649	2.014	2.459
Y = 0.2	1.108	1.338	1.635	2.026	2.538	3.118
Y = 0.4	1.338	1.625	2.021	2.514	3.076	3.666
Y = 0.6	1.593	2.017	2.504	3.057	3.695	4.478
Y = 0.8	2.017	2.510	3.038	3.686	4.483	5.482
Y = 1.0	2.459	3.004	3.669	4.482	5.474	6.685

UXY(X,Y,T)

	X = 0	X = 0.2	X = 0.4	X = 0.6	X = 0.8	X = 1.0
Y = 0	0.905	1.105	1.350	1.649	2.014	2.460
Y = 0.2	1.105	1.341	1.636	2.022	2.506	3.004
Y = 0.4	1.350	1.636	2.023	2.500	3.044	3.669
Y = 0.6	1.649	2.022	2.500	3.045	3.684	4.482
Y = 0.8	2.014	2.506	3.044	3.684	4.482	5.474
Y = 1.0	2.460	3.004	3.669	4.482	5.474	6.686

UYY(X,Y,T)

	X = 0	X = 0.2	X = 0.4	X = 0.6	X = 0.8	X = 1.0
Y = 0	0.905	1.108	1.338	1.593	2.017	2.459
Y = 0.2	1.105	1.338	1.625	2.017	2.510	3.004
Y = 0.4	1.350	1.635	2.021	2.504	3.038	3.669
Y = 0.6	1.649	2.026	2.514	3.057	3.686	4.482
Y = 0.8	2.014	2.538	3.076	3.695	4.483	5.474
Y = 1.0	2.459	3.118	3.666	4.478	5.482	6.685

.
.
.

T = 0.200

.
.
.

T = 0.300

.
.
.

T = 0.400

.
.
.

T = 0.500

Table 4.4 Continued.

U(X,Y,T)

	X = 0	X = 0.2	X = 0.4	X = 0.6	X = 0.8	X = 1.0
Y = 0	0.607	0.741	0.905	1.105	1.350	1.649
Y = 0.2	0.741	0.905	1.104	1.348	1.646	2.014
Y = 0.4	0.905	1.104	1.347	1.645	2.011	2.460
Y = 0.6	1.105	1.348	1.645	2.010	2.458	3.004
Y = 0.8	1.350	1.646	2.011	2.458	3.003	3.669
Y = 1.0	1.649	2.014	2.460	3.004	3.669	4.482

UE(X,Y,T)

	X = 0	X = 0.2	X = 0.4	X = 0.6	X = 0.8	X = 1.0
Y = 0	0.607	0.741	0.905	1.105	1.350	1.649
Y = 0.2	0.741	0.905	1.105	1.350	1.649	2.014
Y = 0.4	0.905	1.105	1.350	1.649	2.014	2.460
Y = 0.6	1.105	1.350	1.649	2.014	2.460	3.004
Y = 0.8	1.350	1.649	2.014	2.460	3.004	3.669
Y = 1.0	1.649	2.014	2.460	3.004	3.669	4.482

UT(X,Y,T)

	X = 0	X = 0.2	X = 0.4	X = 0.6	X = 0.8	X = 1.0
Y = 0	-0.607	-0.741	-0.905	-1.105	-1.350	-1.649
Y = 0.2	-0.741	-0.905	-1.105	-1.348	-1.646	-2.014
Y = 0.4	-0.905	-1.105	-1.347	-1.645	-2.011	-2.460
Y = 0.6	-1.105	-1.348	-1.645	-2.010	-2.458	-3.004
Y = 0.8	-1.350	-1.646	-2.011	-2.458	-3.003	-3.669
Y = 1.0	-1.649	-2.014	-2.460	-3.004	-3.669	-4.482

UXX(X,Y,T)

	X = 0	X = 0.2	X = 0.4	X = 0.6	X = 0.8	X = 1.0
Y = 0	0.606	0.741	0.905	1.105	1.350	1.649
Y = 0.2	0.754	0.887	1.083	1.371	1.728	2.078
Y = 0.4	0.901	1.064	1.363	1.722	2.066	2.421
Y = 0.6	1.038	1.347	1.712	2.066	2.454	2.998
Y = 0.8	1.336	1.704	2.055	2.462	2.992	3.685
Y = 1.0	1.649	2.014	2.460	3.004	3.669	4.481

UXY(X,Y,T)

	X = 0	X = 0.2	X = 0.4	X = 0.6	X = 0.8	X = 1.0
Y = 0	0.607	0.741	0.905	1.105	1.350	1.649
Y = 0.2	0.741	0.893	1.084	1.355	1.693	2.014
Y = 0.4	0.905	1.084	1.358	1.693	2.044	2.460
Y = 0.6	1.105	1.355	1.693	2.048	2.458	3.004
Y = 0.8	1.350	1.693	2.044	2.458	2.996	3.669
Y = 1.0	1.649	2.014	2.460	3.004	3.669	4.482

UYY(X,Y,T)

	X = 0	X = 0.2	X = 0.4	X = 0.6	X = 0.8	X = 1.0
Y = 0	0.606	0.754	0.901	1.038	1.336	1.649

Table 4.4 *Continued.*

Y = 0.2	0.741	0.887	1.064	1.347	1.704	2.014
Y = 0.4	0.905	1.083	1.363	1.712	2.055	2.460
Y = 0.6	1.105	1.371	1.722	2.066	2.462	3.004
Y = 0.8	1.350	1.728	2.066	2.454	2.992	3.669
Y = 1.0	1.649	2.078	2.421	2.998	3.685	4.481

T	U(1/2,1/2,T) NUM	U(1/2,1/2,T) ANAL	0/0 DIFF
0.00	2.7183	2.7183	0.00
0.01	2.6907	2.6912	-0.02
0.02	2.6634	2.6645	-0.04
0.03	2.6363	2.6379	-0.06
0.04	2.6096	2.6117	-0.08
0.05	2.5831	2.5857	-0.10
0.06	2.5570	2.5600	-0.12
0.07	2.5312	2.5345	-0.13
0.08	2.5056	2.5093	-0.15
0.09	2.4804	2.4843	-0.16
0.10	2.4554	2.4596	-0.17
	.	.	
	.	.	
	.	.	
0.40	1.8173	1.8221	-0.27
0.41	1.7992	1.8040	-0.27
0.42	1.7813	1.7860	-0.27
0.43	1.7636	1.7683	-0.27
0.44	1.7460	1.7507	-0.27
0.45	1.7286	1.7333	-0.27
0.46	1.7114	1.7160	-0.27
0.47	1.6944	1.6989	-0.27
0.48	1.6775	1.6820	-0.27
0.49	1.6608	1.6653	-0.27
0.50	1.6443	1.6487	-0.27

3. Even at $t = 0$, the numerical partial derivatives are in error since they are computed by the finite difference approximations in subroutine DSS044. For example, for $\partial u(x, y, 0)/\partial t$, we have

UT(X,Y,T)

	X = 0	X = 0.2	X = 0.4	X = 0.6	X = 0.8	X = 1.0
Y = 0	-1.000	-1.221	-1.492	-1.822	-2.226	-2.718
Y = 0.2	-1.221	-1.522	-1.859	-2.270	-2.773	-3.320
Y = 0.4	-1.492	-1.859	-2.270	-2.773	-3.387	-4.055
Y = 0.6	-1.822	-2.270	-2.773	-3.387	-4.136	-4.953
Y = 0.8	-2.226	-2.773	-3.387	-4.136	-5.052	-6.050
Y = 1.0	-2.718	-3.320	-4.055	-4.953	-6.050	-7.389

Note the substantial departures of the $x + y = 1$ interior values, which should be $e^1 = 2.718$. Similar conclusions apply to the other interior values. For $\partial^2 u(x,y,0)/\partial x^2$, we have

UXX(X,Y,T)

	X = 0	X = 0.2	X = 0.4	X = 0.6	X = 0.8	X = 1.0
Y = 0	1.000	1.221	1.492	1.822	2.226	2.718
Y = 0.2	1.221	1.492	1.822	2.226	2.718	3.320
Y = 0.4	1.492	1.822	2.226	2.718	3.320	4.055
Y = 0.6	1.822	2.226	2.718	3.320	4.055	4.953
Y = 0.8	2.225	2.718	3.320	4.055	4.953	6.049
Y = 1.0	2.718	3.320	4.055	4.953	6.050	7.389

which are in substantially better agreement with the exact values than those for $\partial u(x,y,0)/\partial t$. The discrepancies in $\partial u(x,y,0)/\partial t$ can be explained by the fact that they are computed according to the PDE, equation (4.65), as the algebraic sum of three derivatives, each of which is in error (the programming is in `DO` loop 12 of subroutine `DERV` in Program 4.4b). Fortunately, substantial errors in temporal derivatives, e.g., $\partial u/\partial t$, are often diminished through the temporal integration process; thus the errors in the numerical solution are not as great as the errors in the temporal derivatives would suggest.

4. The preceding discussion in (3) is for $t = 0$, before the temporal integration process started. This integration can in turn introduce errors (although a tight error tolerance of 0.000001 was used for the temporal integration by `RKF45`, as indicated in the data file of Program 4.4d). To gain some idea of the accuracy in the solution for $t > 0$, we consider the output at $t = 0.5$. First, the exact solution is

UE(X,Y,T)

	X = 0	X = 0.2	X = 0.4	X = 0.6	X = 0.8	X = 1.0
Y = 0	0.607	0.741	0.905	1.105	1.350	1.649
Y = 0.2	0.741	0.905	1.105	1.350	1.649	2.014
Y = 0.4	0.905	1.105	1.350	1.649	2.014	2.460
Y = 0.6	1.105	1.350	1.649	2.014	2.460	3.004
Y = 0.8	1.350	1.649	2.014	2.460	3.004	3.669
Y = 1.0	1.649	2.014	2.460	3.004	3.669	4.482

These numbers, of course, follow from the analytical solution, equation (4.66), at $t = 0.5$. For example, for $x + y - 1 = 0.8 - 0.5$, $e^{x+y-t} = e^{0.8-0.5} = e^{0.3} = 1.350$. The corresponding numerical solution is

U(X,Y,T)

	X = 0	X = 0.2	X = 0.4	X = 0.6	X = 0.8	X = 1.0
Y = 0	0.607	0.741	0.905	1.105	1.350	1.649
Y = 0.2	0.741	0.905	1.104	1.348	1.646	2.014
Y = 0.4	0.905	1.104	1.347	1.645	2.011	2.460
Y = 0.6	1.105	1.348	1.645	2.010	2.458	3.004
Y = 0.8	1.350	1.646	2.011	2.458	3.003	3.669
Y = 1.0	1.649	2.014	2.460	3.004	3.669	4.482

The numerical solution $u(x = 0.4, y = 0.4, t = 0.5) = 1.347$ has a relative error of $(1.347 - 1.350)/1.350 = -0.00222$ (-0.222%). The partial derivatives have varying degrees of error. For example, the numerical $\partial u(x, y, 0.5)/\partial t$ is

UT(X,Y,T)

	X = 0	X = 0.2	X = 0.4	X = 0.6	X = 0.8	X = 1.0
Y = 0	-0.607	-0.741	-0.905	-1.105	-1.350	-1.649
Y = 0.2	-0.741	-0.905	-1.105	-1.348	-1.646	-2.014
Y = 0.4	-0.905	-1.105	-1.347	-1.645	-2.011	-2.460
Y = 0.6	-1.105	-1.348	-1.645	-2.010	-2.458	-3.004
Y = 0.8	-1.350	-1.646	-2.011	-2.458	-3.003	-3.669
Y = 1.0	-1.649	-2.014	-2.460	-3.004	-3.669	-4.482

which is close to the accuracy of the numerical $u(x, y, 0.5)$, but $\partial^2 u/\partial x^2$ is in substantially greater error:

UXX(X,Y,T)

	X = 0	X = 0.2	X = 0.4	X = 0.6	X = 0.8	X = 1.0
Y = 0	0.606	0.741	0.905	1.105	1.350	1.649
Y = 0.2	0.754	0.887	1.083	1.371	1.728	2.078
Y = 0.4	0.901	1.064	1.363	1.722	2.066	2.421
Y = 0.6	1.038	1.347	1.712	2.066	2.454	2.998
Y = 0.8	1.336	1.704	2.055	2.462	2.992	3.685
Y = 1.0	1.649	2.014	2.460	3.004	3.669	4.481

The numerical solution $\partial^2 u(x = 0.6, y = 0.2, t = 0.5)/\partial x^2 = 1.371$ has a relative error of $(1.371 - 1.350)/1.350 = 0.0155(1.55\%)$. Fortunately, as mentioned previously, the solution computed by integration of the partial derivatives generally has better accuracy than the derivatives themselves, i.e., integration is a smoothing process.

5. To gain an overall perspective of the accuracy of the numerical solution, the numerical and analytical solutions at $x = 0.5$, $y = 0.5$ are tabulated as a function of t. A sample of the output from Table 4.4

T	U(1/2,1/2,T) NUM	U(1/2,1/2,T) ANAL	O/O DIFF
0.00	2.7183	2.7183	0.00
0.01	2.6907	2.6912	-0.02
0.02	2.6634	2.6645	-0.04
0.03	2.6363	2.6379	-0.06
0.04	2.6096	2.6117	-0.08
0.05	2.5831	2.5857	-0.10
	.	.	
	.	.	
	.	.	
0.45	1.7286	1.7333	-0.27
0.46	1.7114	1.7160	-0.27
0.47	1.6944	1.6989	-0.27
0.48	1.6775	1.6820	-0.27
0.49	1.6608	1.6653	-0.27
0.50	1.6443	1.6487	-0.27

indicates that the numerical solution at the centerpoint is less than 0.3% in error; in fact, the maximum error over $0 \le t \le 0.5$ is -0.27%.

This example of a PDE with a mixed partial derivative presents one approach to a method of lines solution through the approximation of equation (4.74) (which was derived from the Taylor series). Another possibility would be to construct a direct finite difference approximation of $\partial^2 u/\partial x \partial y$ (rather than approximate $\partial^2 u/\partial x \partial y$ in terms of $\partial^2 u/\partial x^2$ and $\partial^2 u/\partial y^2$ as in equation (4.74)) with the boundary conditions included in the approximation. Another approach that we tried and found did not work is to compute $\partial^2 u/\partial x \partial y$ by two successive first-order differentiations, the first with respect to x and the second with respect to y (using subroutine `DSS004`, for example). This apparently gave an unstable system of ODEs, since initially $t = 0$, the various derivatives had acceptable accuracy, but the solution soon became unstable during the integration of the ODEs in t. We explored this approach only briefly; perhaps the use of an implicit ODE integrator would correct the problem.

This discussion indicates that the method of lines is an open-ended procedure with generally more than one possible approach. Some experimentation and reformulation of the approximations is not at all uncommon when developing the solution to a new problem. Also, of course, there is the fundamental question of when do we accept the output as having reasonable accuracy? This is the most difficult question to answer in the numerical integration of PDEs. Clearly every effort should be made to evaluate the solution by performing a quantitative error analysis. Physical and mathematical insights also generally play an important role; certainly a solution cannot be accepted that does not make sense physically or mathematically.

4.4 PDE Solution with a DAE Solver

We now consider a PDE with *variable coefficients* (functions of the independent variables that multiply the partial derivatives) and several nonlinearities [Skeel and Berzins (1990)]

$$uu_t = (1/x^2)(x^2 uu_x)_x + 5u^2 + 4xuu_x \tag{4.75}$$

Clearly, we are putting the subscript notation for PDEs to good use in this problem (i.e., this complex PDE can be stated succinctly with subscript notation).

The initial and boundary conditions for equation (4.75) are

$$u(x,0) = e^{(1-x^2)} \tag{4.76}$$

$$u_x(0,t) = 0, \qquad u(1,t) = e^{-t} \qquad (4.77)(4.78)$$

The analytical solution to this problem is surprisingly simple,

$$u(x,t) = e^{(1-x^2-t)} \tag{4.79}$$

The spatial domain for this problem is $0 \le x \le 1$ (note that boundary conditions (4.77) and (4.78) are defined at $x = 0$ and $x = 1$). A problem developes with equation (4.75) at $x = 0$. If we apply the chain rule to the first right-hand term of equation (4.75) we have

$$(1/x^2)(x^2 uu_x)_x = uu_{xx} + {u_x}^2 + (2/x)uu_x \tag{4.80}$$

The third right-hand term of equation (4.80), $(2/x)uu_x$, is indeterminate at $x = 0$ since $u_x(0,t) = 0$ from boundary condition (4.77). If we apply l'Hôspital's rule to this third term

$$\lim_{x \to 0}(2/x)uu_x = 2(uu_{xx} + u_x^2) \tag{4.81}$$

Equation (4.81) is then used at $x = 0$ in the subsequent coding.

```
      SUBROUTINE INITAL
      IMPLICIT DOUBLE PRECISION (A-H,O-Z)
      PARAMETER (NX=161)
      COMMON/T/          T,      NSTOP,      NORUN
     1      /Y/    U(NX)
     2      /F/   UT(NX)
     3      /S/   UX(NX),   UXX(NX)
     4      /C/    X(NX),          N,       NORD,          IP
C...
C...  SELECT THE ORDER OF THE SECOND DERIVATIVE FORMULAS
      NORD=2
C...  NORD=4
C...
C...  INITIAL CONDITION
      IF(NORD.EQ.2)CALL INTAL1
      IF(NORD.EQ.4)CALL INTAL2
      RETURN
      END
```

Program 4.5a Subroutine INITAL for Initial Condition (4.76).

We now consider a method of lines solution to equations (4.75) to (4.78) over a spatial grid for which the number of points is varied through five values, $n = 11$, 21, 41, 81, and 161. Also, we compute the spatial derivatives in equation (4.75) by second- and fourth-order finite difference approximations (e.g., in subroutines DSS042 and DSS044, respectively). In this way, we can investigate the variation in the errors of the numerical solutions with the grid spacing and the order of the approximations for the spatial derivatives.

Subroutine INITAL is listed in Program 4.5a. INITAL first defines the order of the finite differences used to approximate the spatial derivatives in equation (4.75), i.e., NORD=2 and 4 for second and fourth-order approximations, respectively. Then INITAL calls two other initialization routines, INTAL1 and INTAL2, depending on the order selected (NORD).

INTAL1 and INTAL2 are listed in Program 4.5b. INTAL1 and INTAL2 (which are actually identical) first define the number of spatial grid points for five runs, NORUN=1 to 5, then implement initial condition (4.76) (as expected, the same way for the second- and fourth-order finite differences by writing the initial value of array U into COMMON/Y/).

This *hierarchical approach* to programming (i.e., INTAL1 and INTAL2 called by INITAL) is quite useful for relatively complex PDE problems. All that is basically required in using the numerical method of lines format developed in preceding problems is subroutines INITAL, DERV, and PRINT. However, these three subroutines can in turn call other (subordinate) subroutines in order to divide the coding into more manageable parts. The only requirement to keep in mind is the use of the three basic COMMON blocks, /T/, /Y/, and /F/, to ensure communication with the ODE integrator. In the case of INTAL1 and INTAL2, COMMON/Y/ is required to provide the initial value of u (array U) to the ODE integrator. However, other COMMON blocks can be included, and in fact, a good programming procedure is to make up one comprehensive COMMON block, then duplicate it in all of the subroutines for the PDE problem.

Subroutine DERV and subordinate routines DERV1 and DERV2 are listed in Program 4.5c. Again, DERV calls either DERV1 or DERV2, depending on the order of the finite difference approximation, NORD (=2 for DERV1 and =4 for DERV2). We can note

```
      SUBROUTINE INTAL1
      IMPLICIT DOUBLE PRECISION (A-H,O-Z)
      PARAMETER (NX=161)
      COMMON/T/          T,      NSTOP,      NORUN
     1      /Y/     U(NX)
     2      /F/    UT(NX)
     3      /S/    UX(NX),   UXX(NX)
     4      /C/     X(NX),          N,       NORD,          IP
C...
C...  SET THE NUMBER OF GRID POINTS
      IF(NORUN.EQ.1)N=11
      IF(NORUN.EQ.2)N=21
      IF(NORUN.EQ.3)N=41
      IF(NORUN.EQ.4)N=81
      IF(NORUN.EQ.5)N=161
C...
C...  INITIAL CONDITION
      DO 1 I=1,N
         X(I)=DFLOAT(I-1)/DFLOAT(N-1)
         U(I)=DEXP(1.0D0-X(I)**2)
1     CONTINUE
C...
C...  INITIAL DERIVATIVES
      CALL DERV
      IP=0
      RETURN
      END

      SUBROUTINE INTAL2
      IMPLICIT DOUBLE PRECISION (A-H,O-Z)
      PARAMETER (NX=161)
      COMMON/T/          T,      NSTOP,      NORUN
     1      /Y/     U(NX)
     2      /F/    UT(NX)
     3      /S/    UX(NX),  UXX(NX)
     4      /C/     X(NX),          N,       NORD,          IP
C...
C...  SET THE NUMBER OF GRID POINTS
      IF(NORUN.EQ.1)N=11
      IF(NORUN.EQ.2)N=21
      IF(NORUN.EQ.3)N=41
      IF(NORUN.EQ.4)N=81
      IF(NORUN.EQ.5)N=161
C...
C...  INITIAL CONDITION
      DO 1 I=1,N
         X(I)=DFLOAT(I-1)/DFLOAT(N-1)
         U(I)=DEXP(1.0D0-X(I)**2)
1     CONTINUE
C...
```

Program 4.5b Subroutines INTAL1 and INTAL2 Called by Subroutine INITAL. *Continued next page.*

```
C...  INITIAL DERIVATIVES
      CALL DERV
      IP=0
      RETURN
      END
```

Program 4.5b *Continued.*

the following points about subroutine DERV1:

1. Boundary condition (4.78) is first programmed to be consistent with the analytical solution, equation (4.79)

```
C...
C...  FIRST DERIVATIVE UX INCLUDING THE DIRICHLET BOUNDARY CONDITION
C...  AT X = 1
      U (N)= DEXP(-T)
      UT(N)=-DEXP(-T)
      CALL DSS002(0.D0,1.D0,N,U,UX)
```

2. Subroutine DSS002 is then called to compute the derivative u_x. DSS002 has three-point centered, (second-order correct) finite difference approximations and is structured the same way as DSS004.

3. Boundary conditions (4.77) and (4.78) are next specified (NL=2, NU=1) and subroutine DSS042 is then called to compute the derivative u_{xx}. DSS042 has three-point centered, (second-order correct) finite difference approximations and is structured the same way as DSS044.

```
C...
C...  SECOND DERIVATIVE INCLUDING THE SYMMETRY (NEUMANN) BOUNDARY
C...  CONDITION AT X =0
      NL=2
      NU=1
      UX(1)=0.D0
      CALL DSS042(0.D0,1.D0,N,U,UX,UXX,NL,NU)
```

4. Finally, equation (4.75) is implemented in DO loop 1. For I = 1, X = 0 and equation (4.81) is used to avoid the indeterminate form in equation (4.75). In other words, equation (4.75) for $x \neq 0$ is

$$\begin{aligned} uu_t &= (1/x^2)(x^2 uu_x)_x + 5u^2 + 4xuu_x \\ &= uu_{xx} + {u_x}^2 + (2/x)uu_x + 5u^2 + 4xuu_x \end{aligned} \tag{4.82}$$

and for $x = 0$, equation (4.75) is (from equation (4.81))

$$\begin{aligned} uu_t &= uu_{xx} + u_x^2 + 2(uu_{xx} + u_x^2) + 5u^2 + 4xuu_x \\ &= 3(uu_{xx} + {u_x}^2) + 5u^2 + 4xuu_x \end{aligned} \tag{4.83}$$

```
      SUBROUTINE DERV
      IMPLICIT DOUBLE PRECISION (A-H,O-Z)
      PARAMETER (NX=161)
      COMMON/T/          T,      NSTOP,      NORUN
     1      /Y/    U(NX)
     2      /F/   UT(NX)
     3      /S/   UX(NX),   UXX(NX)
     4      /C/    X(NX),          N,       NORD,           IP
C...
C...  SELECT THE SECOND-DERIVATIVE FORMULAS
      IF(NORD.EQ.2)CALL DERV1
      IF(NORD.EQ.4)CALL DERV2
      RETURN
      END

      SUBROUTINE DERV1
      IMPLICIT DOUBLE PRECISION (A-H,O-Z)
      PARAMETER (NX=161)
      COMMON/T/          T,      NSTOP,      NORUN
     1      /Y/    U(NX)
     2      /F/   UT(NX)
     3      /S/   UX(NX),   UXX(NX)
     4      /C/    X(NX),          N,       NORD,           IP
C...
C...  FIRST DERIVATIVE UX INCLUDING THE DIRICHLET BOUNDARY CONDITION
C...  AT X = 1
      U (N)= DEXP(-T)
      UT(N)=-DEXP(-T)
      CALL DSS002(0.D0,1.D0,N,U,UX)
C...
C...  SECOND DERIVATIVE INCLUDING THE SYMMETRY (NEUMANN) BOUNDARY
C...  CONDITION AT X=0
      NL=2
      NU=1
      UX(1)=0.D0
      CALL DSS042(0.D0,1.D0,N,U,UX,UXX,NL,NU)
C...
C...  PDE
      DO 1 I=1,N-1
C...     X = 0
         IF(I.EQ.1)THEN
            UT(1)=(1.D0/U(1))*3.D0*(U(1)*UXX(1)+UX(1)**2)+5.D0*U(1)
C...
C...     0 LT X LT 1
         ELSE
            UT(I)=(1.D0/U(I))*(U(I)*UXX(I)+UX(I)**2+2.D0/X(I)*U(I)*UX(I)
     1           +5.D0*U(I)**2+4.D0*X(I)*U(I)*UX(I))
         END IF
1     CONTINUE
      RETURN
      END
```

Program 4.5c Subroutines DERV, DERV1, and DERV2 for Equation (4.75). *Continued next page.*

```
      SUBROUTINE DERV2
      IMPLICIT DOUBLE PRECISION (A-H,O-Z)
      PARAMETER (NX=161)
      COMMON/T/          T,      NSTOP,      NORUN
     1      /Y/     U(NX)
     2      /F/    UT(NX)
     3      /S/    UX(NX),    UXX(NX)
     4      /C/     X(NX),          N,       NORD,          IP
C...
C...  FIRST DERIVATIVE UX INCLUDING THE DIRICHLET BOUNDARY CONDITION
C...  AT X = 1
      U (N)= DEXP(-T)
      UT(N)=-DEXP(-T)
      CALL DSS004(0.D0,1.D0,N,U,UX)
C...
C...  SECOND DERIVATIVE INCLUDING THE SYMMETRY (NEUMANN) BOUNDARY
C...  CONDITION AT X = 0
      NL=2
      NU=1
      UX(1)=0.D0
      CALL DSS044(0.D0,1.D0,N,U,UX,UXX,NL,NU)
C...
C...  PDE
      DO 1 I=1,N-1
C...     X = 0
         IF(I.EQ.1)THEN
            UT(1)=(1.D0/U(1))*3.D0*(U(1)*UXX(1)+UX(1)**2)+5.D0*U(1)
C...
C...     0 LT X LT 1
         ELSE
            UT(I)=(1.D0/U(I))*(U(I)*UXX(I)+UX(I)**2+2.D0/X(I)*U(I)*UX(I)
     1              +5.D0*U(I)**2+4.D0*X(I)*U(I)*UX(I))
         END IF
1     CONTINUE
      RETURN
      END
```

Program 4.5c *Continued.*

Equations (4.82) and (4.83), solved explicitly for u_t, are coded in DO loop 1 of DERV1

```
C...
C...  PDE
      DO 1 I=1,N-1
C...     X = 0
         IF(I.EQ.1)THEN
            UT(1)=(1.D0/U(1))*3.D0*(U(1)*UXX(1)+UX(1)**2)+5.D0*U(1)
C...
C...     0 LT X LT 1
         ELSE
            UT(I)=(1.D0/U(I))*(U(I)*UXX(I)+UX(I)**2+2.D0/X(I)*U(I)*UX(I)
     1              +5.D0*U(I)**2+4.D0*X(I)*U(I)*UX(I))
         END IF
1     CONTINUE
```

The only difference between DERV1 and DERV2 are the calls to subroutines DSS004 and DSS044 in place of the calls to DSS002 and DSS042, thus fourth-order approximations are used in DERV2 for equation (4.75).

5. The coding in subroutines DERV1 and DERV2 again illustrates the modest and straightforward programming requirements for the method of lines solution of nonlinear PDEs like equation (4.75).

Subroutine PRINT calls subroutines PRINT1 and PRINT2 for NORD=2 and 4, respectively, as indicated in Program 4.5d. We can note the following points about PRINT1:

1. The analytical solution, UA, and the difference between the analytical and numerical solutions, ERR, are computed over the spatial grid (the last point, I = N, is not included since the exact solution is used for boundary condition (4.78)). Then the maximum error, ERRMAX, over the grid is retained for subsequent printing.

```
C...
C...  COMPUTE THE EXACT SOLUTION, THE DIFFERENCE BETWEEN THE NUMERICAL
C...  AND ANALYTICAL SOLUTIONS, AND SAVE THE LARGEST NUMERICAL ERROR
      ERRMAX=0.DO
      DO 1 I=1,N-1
         UA=DEXP(1.0DO-X(I)**2-T)
         ERR=DABS(UA-U(I))
         IF(ERRMAX.LT.ERR)ERRMAX=ERR
1     CONTINUE
```

2. A label is then printed for the output via FORMAT 3 and the values of t (T), the number of spatial grid points (N), and the maximum error over the spatial grid (ERRMAX) are printed after the first call to PRINT1 (for $t > 0$).

```
C...
C...  LABEL AND PRINT THE LARGEST ERROR (NOTE - AT T = 0, THE ERROR
C...  IS ZERO)
      IP=IP+1
      IF(IP.EQ.1)THEN
C...
C...     PRINT A HEADING AT T = 0
         WRITE(NO,3)
3        FORMAT(//,8X,' SOLUTION BY SECOND-ORDER APPROXIMATIONS',//,
     1   19X,'T',4X,'N',9X,'ERRMAX')
         ERRPLT=ERRMAX
      ELSE IF(IP.NE.1)THEN
C...
C...     PRINT ERRMAX VS T
         WRITE(NO,2)T,N,ERRMAX
2        FORMAT(10X,F10.3,I5,D15.4)
         IF(ERRMAX.GT.ERRPLT)ERRPLT=ERRMAX
      END IF
```

3. The $\log_{10}$ of the maximum error for the entire run, ERRPLT, is stored with the

```
      SUBROUTINE PRINT(NI,NO)
      IMPLICIT DOUBLE PRECISION (A-H,O-Z)
      PARAMETER (NX=161)
      COMMON/T/          T,      NSTOP,      NORUN
     1      /Y/    U(NX)
     2      /F/   UT(NX)
     3      /S/   UX(NX),   UXX(NX)
     4      /C/    X(NX),           N,       NORD,          IP
C...
C...  SELECT THE OUTPUT ROUTINE
      IF(NORD.EQ.2)CALL PRINT1(NI,NO)
      IF(NORD.EQ.4)CALL PRINT2(NI,NO)
      RETURN
      END

      SUBROUTINE PRINT1(NI,NO)
      IMPLICIT DOUBLE PRECISION (A-H,O-Z)
      PARAMETER (NX=161)
      COMMON/T/          T,      NSTOP,      NORUN
     1      /Y/    U(NX)
     2      /F/   UT(NX)
     3      /S/   UX(NX),   UXX(NX)
     4      /C/    X(NX),           N,       NORD,          IP
C...
C...  ARRAYS FOR PLOTTING
      DIMENSION SLOGN(11), SLOGE(11)
C...
C...  MONITOR THE CALCULATION
      WRITE(*,*)NORD,NORUN,N
C...
C...  COMPUTE THE EXACT SOLUTION, THE DIFFERENCE BETWEEN THE NUMERICAL
C...  AND ANALYTICAL SOLUTIONS, AND SAVE THE LARGEST NUMERICAL ERROR
      ERRMAX=0.D0
      DO 1 I=1,N-1
         UA=DEXP(1.0D0-X(I)**2-T)
         ERR=DABS(UA-U(I))
         IF(ERRMAX.LT.ERR)ERRMAX=ERR
1     CONTINUE
C...
C...  LABEL AND PRINT THE LARGEST ERROR (NOTE - AT T = 0, THE ERROR
C...  IS ZERO)
      IP=IP+1
      IF(IP.EQ.1)THEN
C...
C...     PRINT A HEADING AT T = 0
         WRITE(NO,3)
3        FORMAT(//,8X,' SOLUTION BY SECOND ORDER APPROXIMATIONS',//,
     1   19X,'T',4X,'N',9X,'ERRMAX')
         ERRPLT=ERRMAX
      ELSE IF(IP.NE.1)THEN
C...
C...     PRINT ERRMAX VS T
         WRITE(NO,2)T,N,ERRMAX
```

Program 4.5d Subroutines PRINT, PRINT1, and PRINT2 for Equation (4.75). *Continued next pages.*

```
2        FORMAT(10X,F10.3,I5,D15.4)
         IF(ERRMAX.GT.ERRPLT)ERRPLT=ERRMAX
      END IF
C...
C...  STORE LOG10(ERRPLT) VS LOG10(N) FOR SUBSEQUENT PLOTTING
      IF(IP.EQ.11)THEN
         SLOGN(NORUN)=DLOG10(DFLOAT(N))
         SLOGE(NORUN)=DLOG10(ERRPLT)
C...
C...  AT THE END OF THE FIFTH RUN (WITH N = 161), PLOT LOG10(ERR)
C...  VS LOG10(N)
      IF(NORUN.EQ.5)THEN
         WRITE(NO,4)(SLOGN(I),SLOGE(I),I=1,NORUN)
4        FORMAT(//,22X,' DATA FOR PLOT',//,
     1          18X,'LOG10(N)',6X,'LOG10(ERR)',/,
     2          (10X,2F15.4))
         CALL SPLOTS(1,NORUN,SLOGN,SLOGE)
C...
C...     COMPUTE AND PRINT ESTIMATED ORDER
         SLOPE=(SLOGE(1)-SLOGE(NORUN-1))/(SLOGN(1)-SLOGN(NORUN-1))
         WRITE(NO,5)SLOPE
5        FORMAT(//,' LOG10(MAXERR) VS LOG10(N), APPARENT ORDER = ',
     1              F6.2)
      END IF
      END IF
      RETURN
      END

      SUBROUTINE PRINT2(NI,NO)
      IMPLICIT DOUBLE PRECISION (A-H,O-Z)
      PARAMETER (NX=161)
      COMMON/T/          T,      NSTOP,      NORUN
     1      /Y/     U(NX)
     2      /F/    UT(NX)
     3      /S/    UX(NX),    UXX(NX)
     4      /C/     X(NX),          N,       NORD,          IP
C...
C...  ARRAYS FOR PLOTTING
      DIMENSION SLOGN(11), SLOGE(11)
C...
C...  MONITOR THE CALCULATION
      WRITE(*,*)NORD,NORUN,N
C...
C...  COMPUTE THE EXACT SOLUTION, THE DIFFERENCE BETWEEN THE NUMERICAL
C...  AND ANALYTICAL SOLUTIONS, AND SAVE THE LARGEST NUMERICAL ERROR
      ERRMAX=0.D0
      DO 1 I=1,N-1
         UA=DEXP(1.0D0-X(I)**2-T)
         ERR=DABS(UA-U(I))
         IF(ERRMAX.LT.ERR)ERRMAX=ERR
1     CONTINUE
C...
```

Program 4.5d *Continued.*

```
C...  LABEL AND PRINT THE LARGEST ERROR (NOTE - AT T = 0, THE ERROR
C...  IS ZERO)
      IP=IP+1
      IF(IP.EQ.1)THEN
C...
C...     PRINT A HEADING AT T = 0
         WRITE(NO,3)
3        FORMAT(//,8X,' SOLUTION BY FOURTH ORDER APPROXIMATIONS',//,
     1   19X,'T',4X,'N',9X,'ERRMAX')
         ERRPLT=ERRMAX
      ELSE IF(IP.NE.1)THEN
C...
C...     PRINT ERRMAX VS T
         WRITE(NO,2)T,N,ERRMAX
2        FORMAT(10X,F10.3,I5,D15.4)
         IF(ERRMAX.GT.ERRPLT)ERRPLT=ERRMAX
      END IF
C...
C...  STORE LOG10(ERRPLT) VS LOG10(N) FOR SUBSEQUENT PLOTTING
      IF(IP.EQ.11)THEN
         SLOGN(NORUN)=DLOG10(DFLOAT(N))
         SLOGE(NORUN)=DLOG10(ERRPLT)
C...
C...  AT THE END OF THE FIFTH RUN (WITH N = 161), PLOT LOG10(ERR)
C...  VS LOG10(N)
      IF(NORUN.EQ.5)THEN
         WRITE(NO,4)(SLOGN(I),SLOGE(I),I=1,NORUN)
4        FORMAT(//,22X,' DATA FOR PLOT',//,
     1         18X,'LOG10(N)',6X,'LOG10(ERR)',/,
     2         (10X,2F15.4))
         CALL SPLOTS(1,NORUN,SLOGN,SLOGE)
         WRITE(NO,5)
5        FORMAT(//,' LOG10(MAXERR) VS LOG10(N)')
      END IF
      END IF
      RETURN
      END
```

Program 4.5d *Continued.*

$\log_{10}$ of the number of grid points, N, for printing and plotting at the end of the fifth run

```
C...
C...  STORE LOG10(ERRPLT) VS LOG10(N) FOR SUBSEQUENT PLOTTING
      IF(IP.EQ.11)THEN
         SLOGN(NORUN)=DLOG10(DFLOAT(N))
         SLOGE(NORUN)=DLOG10(ERRPLT)
```

4. At the end of the fifth run, $\log_{10}$ (ERRPLT) is printed and plotted vs. $\log_{10}$ (N). After the call to SPLOTS, the slope of the $\log_{10} - \log_{10}$ plot is computed from the points for $n = 11$, 21, 41, and 81. The last point for $n = 161$ was not included since it appears to depart from a straight line through the other four points;

```
EQS. (4.75) TO (4.78), N = 11
0.          1.0         0.1
   11                         0.00001
EQS. (4.75) TO (4.78), N = 21
0.          1.0         0.1
   21                         0.00001
EQS. (4.75) TO (4.78), N = 41
0.          1.0         0.1
   41                         0.00001
EQS. (4.75) TO (4.78), N = 81
0.          1.0         0.1
   81                         0.00001
EQS. (4.75) TO (4.78), N = 161
0.          1.0         0.1
  161                         0.00001
END OF RUNS
```

Program 4.5e Data File for Programs 4.5a to d.

possible reasons for this departure are considered during the discussion of the output.

```
C...
C...  AT THE END OF THE FIFTH RUN (WITH N = 161), PLOT LOG10(ERR)
C...  VS LOG10(N)
      IF(NORUN.EQ.5)THEN
         WRITE(NO,4)(SLOGN(I),SLOGE(I),I=1,NORUN)
4        FORMAT(//,22X,' DATA FOR PLOT',//,
     1          18X,'LOG10(N)',6X,'LOG10(ERR)',/,
     2          (10X,2F15.4))
         CALL SPLOTS(1,NORUN,SLOGN,SLOGE)
C...
C...     COMPUTE AND PRINT ESTIMATED ORDER
         SLOPE=(SLOGE(1)-SLOGE(NORUN-1))/(SLOGN(1)-SLOGN(NORUN-1))
         WRITE(NO,5)SLOPE
5        FORMAT(//,' LOG10(MAXERR) VS LOG10(N), APPARENT ORDER = ',
     1             F6.2)
      END IF
      END IF
```

5. Subroutine PRINT2 is essentially identical to PRINT1. The only difference is the deletion of the slope calculation after the call to SPLOTS since, as we will see in considering the output for NORD=4 (the fourth-order approximation of equation (4.75)), the five points plotted by SPLOTS do not form a straight line.

The data file read by the main program, which calls a *DAE solver*, DASSL, for the ODE integration, is given in Program 4.5e. As expected, five runs are specified with 11, 21, 41, 81, and 161 ODEs.

The output from Programs 4.5a to e for NORD=2 is given in Table 4.5a. We can note the following points about the output in Table 4.5a:

1. The maximum error in the numerical solution drops from 0.1385D-01 for $n = 11$ to 0.9091D-04 for $n = 161$. From equation (4.79), we see that the solution varies

Table 4.5 (a) Output from Programs 4.5a to e for Second-Order Finite Difference Approximation of Equations (4.75) to (4.78). *Continued next pages.*

```
RUN NO. -    1  EQS. (4.75) TO (4.78), N = 11

INITIAL T -  0.000D+00

  FINAL T -  0.100D+01

  PRINT T -  0.100D+00

NUMBER OF DIFFERENTIAL EQUATIONS -  11

MAXIMUM INTEGRATION ERROR -  0.100D-04


        SOLUTION BY SECOND-ORDER APPROXIMATIONS

                 T     N          ERRMAX
             0.100    11      0.8760D-02
             0.200    11      0.1243D-01
             0.300    11      0.1375D-01
             0.400    11      0.1385D-01
             0.500    11      0.1332D-01
             0.600    11      0.1249D-01
             0.700    11      0.1155D-01
             0.800    11      0.1059D-01
             0.900    11      0.9669D-02
             1.000    11      0.8796D-02


COMPUTATIONAL STATISTICS

LAST STEP SIZE                           0.810D-01

LAST ORDER OF THE METHOD                         3

TOTAL NUMBER OF STEPS TAKEN                     39

NUMBER OF FUNCTION EVALUATIONS                  54

NUMBER OF JACOBIAN EVALUATIONS                  12

RUN NO. -    2  EQS. (4.75) TO (4.78), N = 21

INITIAL T -  0.000D+00

  FINAL T -  0.100D+01

  PRINT T -  0.100D+00

NUMBER OF DIFFERENTIAL EQUATIONS -  21

MAXIMUM INTEGRATION ERROR -  0.100D-04
```

Table 4.5 (a) *Continued.*

```
        SOLUTION BY SECOND-ORDER APPROXIMATIONS

              T      N          ERRMAX
            0.100   21      0.2268D-02
            0.200   21      0.3122D-02
            0.300   21      0.3446D-02
            0.400   21      0.3478D-02
            0.500   21      0.3322D-02
            0.600   21      0.3118D-02
            0.700   21      0.2881D-02
            0.800   21      0.2635D-02
            0.900   21      0.2405D-02
            1.000   21      0.2185D-02


COMPUTATIONAL STATISTICS

LAST STEP SIZE                               0.562D-01

LAST ORDER OF THE METHOD                             3

TOTAL NUMBER OF STEPS TAKEN                         38

NUMBER OF FUNCTION EVALUATIONS                      51

NUMBER OF JACOBIAN EVALUATIONS                      11

RUN NO. -    3  EQS. (4.75) TO (4.78), N = 41

INITIAL T -  0.000D+00

  FINAL T -  0.100D+01

  PRINT T -  0.100D+00

NUMBER OF DIFFERENTIAL EQUATIONS -  41

MAXIMUM INTEGRATION ERROR -  0.100D-04

        SOLUTION BY SECOND-ORDER APPROXIMATIONS

              T      N          ERRMAX
            0.100   41      0.6104D-03
            0.200   41      0.7686D-03
            0.300   41      0.8563D-03
            0.400   41      0.8662D-03
            0.500   41      0.8256D-03
            0.600   41      0.7787D-03
            0.700   41      0.7174D-03
            0.800   41      0.6593D-03
            0.900   41      0.6002D-03
            1.000   41      0.5473D-03
```

Table 4.5 (a) *Continued.*

```
COMPUTATIONAL STATISTICS

LAST STEP SIZE                              0.493D-01

LAST ORDER OF THE METHOD                            3

TOTAL NUMBER OF STEPS TAKEN                        40

NUMBER OF FUNCTION EVALUATIONS                     55

NUMBER OF JACOBIAN EVALUATIONS                     11


RUN NO. -   4  EQS. (4.75) TO (4.78), N = 81

INITIAL T -  0.000D+00

  FINAL T -  0.100D+01

  PRINT T -  0.100D+00

NUMBER OF DIFFERENTIAL EQUATIONS -  81

MAXIMUM INTEGRATION ERROR -  0.100D-04


        SOLUTION BY SECOND-ORDER APPROXIMATIONS

                  T     N          ERRMAX
              0.100    81      0.1949D-03
              0.200    81      0.1791D-03
              0.300    81      0.2076D-03
              0.400    81      0.2138D-03
              0.500    81      0.2001D-03
              0.600    81      0.1934D-03
              0.700    81      0.1771D-03
              0.800    81      0.1651D-03
              0.900    81      0.1508D-03
              1.000    81      0.1340D-03


COMPUTATIONAL STATISTICS

LAST STEP SIZE                              0.490D-01

LAST ORDER OF THE METHOD                            4

TOTAL NUMBER OF STEPS TAKEN                        40

NUMBER OF FUNCTION EVALUATIONS                     55

NUMBER OF JACOBIAN EVALUATIONS                     11
```

Table 4.5 (a) *Continued.*

```
RUN NO. -   5  EQS. (4.75) TO (4.78), N = 161

INITIAL T -  0.000D+00

  FINAL T -  0.100D+01

  PRINT T -  0.100D+00

NUMBER OF DIFFERENTIAL EQUATIONS - 161

MAXIMUM INTEGRATION ERROR -  0.100D-04


         SOLUTION BY SECOND-ORDER APPROXIMATIONS

                    T      N          ERRMAX
                  0.100   161       0.9091D-04
                  0.200   161       0.3155D-04
                  0.300   161       0.4540D-04
                  0.400   161       0.5066D-04
                  0.500   161       0.4376D-04
                  0.600   161       0.4706D-04
                  0.700   161       0.4201D-04
                  0.800   161       0.4142D-04
                  0.900   161       0.3804D-04
                  1.000   161       0.3162D-04

                         DATA FOR PLOT

                  LOG10(N)         LOG10(ERR)
                   1.0414           -1.8585
                   1.3222           -2.4586
                   1.6128           -3.0624
                   1.9085           -3.6700
                   2.2068           -4.0414
```

from $e^1 = 2.718$ (for $x = t = 0$) to $e^{-1} = 1/2.781$ for $x = t = 1$, thus the error is about 1 part in 10^4 for $n = 161$.

2. The plot of $\log_{10}(error)$ vs. $\log_{10}(n)$ is approximately linear for n = 11, 21, 41, and 81; the slope of the line is −2.09. This might seem reasonable since second-order approximations are used in subroutines DSS002 and DSS042. Remember, however, that these are second-order approximations for the first and second derivatives, u_x and u_{xx}, in equation (4.75), and not for the solution, u. In general, we cannot expect the solution accuracy to be of the same order as the approximations used for the spatial derivatives in the PDE(s).

3. The computational statistics for DASSL, a DAE solver based on the BDFs (equation (2.160)), indicate a modest computational effort. For example, for the fifth

Table 4.5 (a) *Continued.*

```
             ..1....1....1....1....1....1....1....1....1....1....1..
  -0.186D+01+  1                                                   +I
            -                                                      -I
            -                                                      -I
            -                                                      -I
            -           1                                          -I
  -0.259D+01+                                                      +I
            -                                                      -I
            -                                                      -I
            -                        1                             -I
            -                                                      -I
  -0.331D+01+                                                      +I
            -                                                      -I
            -                                    1                 -I
            -                                                      -I
            -                                                      -I
  -0.404D+01+                                                    1 +I
             ..1....1....1....1....1....1....1....1....1....1....1..
          0.104D+01  0.13D+01  0.15D+01  0.17D+01  0.20D+01  0.22D+01


LOG10(MAXERR) VS LOG10(N), APPARENT ORDER =  -2.09

COMPUTATIONAL STATISTICS

LAST STEP SIZE                            0.490D-01

LAST ORDER OF THE METHOD                          4

TOTAL NUMBER OF STEPS TAKEN                      40

NUMBER OF FUNCTION EVALUATIONS                   55

NUMBER OF JACOBIAN EVALUATIONS                   11
```

run (for $n = 161$), the statistics are

```
COMPUTATIONAL STATISTICS

LAST STEP SIZE                            0.490D-01

LAST ORDER OF THE METHOD                          4

TOTAL NUMBER OF STEPS TAKEN                      40

NUMBER OF FUNCTION EVALUATIONS                   55

NUMBER OF JACOBIAN EVALUATIONS                   11
```

The last integration step, `0.490D-01`, is almost half of the output interval, 0.1, and only 40 steps were required to compute the complete solution to $t = 1$. The higher order BDF methods, such as for the last order of 4 (q in equation (2.160)), provided computational efficiency so that only 55 function evaluations were required, including those for the 11 Jacobian evaluations. Thus, DASSL performed well on this problem as an ODE integrator. Subsequently we briefly consider the features and use of DASSL.

The output from Programs 4.5a to e for the fourth-order approximation of the spatial derivatives in equation (4.75) (NORD=4 in subroutine INITAL of Program 4.5a) is given in Table 4.5b. We can note the following points about the output in Table 4.5b:

1. The errors in the numerical solution for small numbers of grid points are substantially smaller for the fourth-order approximations than for the second-order approximations. These smaller errors are indicated in Table 4.5c which lists the maximum error for $0 \leq t \leq 1$ for varying numbers of spatial grid points. Thus the fourth-order error for $n = 21$, 0.3187D-04, is smaller than the second-order error for $n = 161$, 0.9091D-04. In other words, the fourth-order approximations produced a solution of comparable accuracy to the second-order approximations with about 1/8 as many grid points. This increased accuracy required a modest increase in computation at each of the 21 grid points, i.e., five values of the solution, u, are used in a weighted sum to calculate u_x (in subroutine DSS004 in subroutine DERV2 of Program 4.5c) and u_{xx} (in subroutine DSS044) for the fourth-order approximations compared with three points in a weighted sum for the second-order approximations (in subroutines DSS002 and DSS042 in subroutine DERV1 of Program 4.5c). However, this additional calculation at each grid point was more than offset by the reduced number of grid points for the fourth-order approximations which achieved comparable accuracy with the second-order approximations. In other words, the higher order approximations are substantially more efficient.

2. The error in each case reached a lower bound of approximately 0.5E-04 to 0.2E-04, beyond which, the addition of grid points apparently would produce little improvement in the accuracy of the solution. This lower limit might be due to the accuracy in the time integration, which was defined for subroutine DASSL as 0.00001 in the data of Table 4.5e, and by roundoff effects. In any case, the plot in Table 4.5b indicates that the lower limit was reached essentially with $n = 21$, and little improvement in the fourth-order solution resulted from the use of more grid points. This also meant that a line to estimate the order of the error could not be drawn through the points in this plot.

3. As with the second-order approximations, the computation required by DASSL was quite modest. Note in particular in Table 4.5b that the last integration step for $n = 11$ was $0.112D+00$, which exceeds the output interval of 0.1 (see the data of Program 4.5e). This is an indication of the particular BDF implementation in DASSL (as well as LSODE) that permits the integration to proceed past the output point, with interpolation to give the numerical solution at the desired output values of the independent variable, t.

Finally, we briefly describe DASSL, a BDF solver for the DAE problem considered previously in Section 1.1.3 of Chapter 1 (and repeated here for discussion)

$$\bar{f}(\bar{y}, d\bar{y}/dt, t) = \bar{0} \tag{4.84}$$

$$\bar{f}(\bar{y}(t_0), d\bar{y}(t_0)/dt, t_0) = \bar{0} \tag{4.85}$$

We have used a special case of equation (4.84) in the method of lines solution of equa-

Table 4.5 (b) Output from Programs 4.5a to d for Fourth-Order Finite Difference Approximation of Equations (4.75) to (4.78). *Continued next pages.*

```
RUN NO. -   1  EQS. (4.75) TO (4.78), N = 11

INITIAL T -  0.000D+00

  FINAL T -  0.100D+01

  PRINT T -  0.100D+00

NUMBER OF DIFFERENTIAL EQUATIONS -  11

MAXIMUM INTEGRATION ERROR -  0.100D-04


        SOLUTION BY FOURTH-ORDER APPROXIMATIONS

                  T     N          ERRMAX
              0.100    11      0.9049D-04
              0.200    11      0.1532D-03
              0.300    11      0.1876D-03
              0.400    11      0.1964D-03
              0.500    11      0.1910D-03
              0.600    11      0.1767D-03
              0.700    11      0.1547D-03
              0.800    11      0.1347D-03
              0.900    11      0.1259D-03
              1.000    11      0.1162D-03


COMPUTATIONAL STATISTICS

LAST STEP SIZE                           0.112D+00

LAST ORDER OF THE METHOD                         4

TOTAL NUMBER OF STEPS TAKEN                     30

NUMBER OF FUNCTION EVALUATIONS                  41

NUMBER OF JACOBIAN EVALUATIONS                  12


RUN NO. -   2  EQS. (4.75) TO (4.78), N = 21

INITIAL T -  0.000D+00

  FINAL T -  0.100D+01

  PRINT T -  0.100D+00

NUMBER OF DIFFERENTIAL EQUATIONS -  21

MAXIMUM INTEGRATION ERROR -  0.100D-04
```

Table 4.5 (b) *Continued.*

```
        SOLUTION BY FOURTH-ORDER APPROXIMATIONS

              T      N          ERRMAX
            0.100   21       0.5598D-05
            0.200   21       0.9247D-05
            0.300   21       0.2523D-04
            0.400   21       0.3069D-04
            0.500   21       0.3187D-04
            0.600   21       0.2648D-04
            0.700   21       0.1436D-04
            0.800   21       0.6215D-05
            0.900   21       0.1938D-04
            1.000   21       0.1068D-04


COMPUTATIONAL STATISTICS

LAST STEP SIZE                           0.101D+00

LAST ORDER OF THE METHOD                         4

TOTAL NUMBER OF STEPS TAKEN                     29

NUMBER OF FUNCTION EVALUATIONS                  38

NUMBER OF JACOBIAN EVALUATIONS                  12


RUN NO. -    3  EQS. (4.75) TO (4.78), N = 41

INITIAL T -  0.000D+00

  FINAL T -  0.100D+01

  PRINT T -  0.100D+00

NUMBER OF DIFFERENTIAL EQUATIONS -  41

MAXIMUM INTEGRATION ERROR -  0.100D-04


        SOLUTION BY FOURTH-ORDER APPROXIMATIONS

              T      N          ERRMAX
            0.100   41       0.1271D-04
            0.200   41       0.3036D-05
            0.300   41       0.1345D-04
            0.400   41       0.1914D-04
            0.500   41       0.2102D-04
            0.600   41       0.1631D-04
            0.700   41       0.4900D-05
            0.800   41       0.4977D-05
```

Table 4.5 (b) *Continued.*

```
             0.900   41      0.1743D-04
             1.000   41      0.3554D-05


COMPUTATIONAL STATISTICS

LAST STEP SIZE                          0.100D+00

LAST ORDER OF THE METHOD                        4

TOTAL NUMBER OF STEPS TAKEN                    29

NUMBER OF FUNCTION EVALUATIONS                 37

NUMBER OF JACOBIAN EVALUATIONS                 12


RUN NO. -   4  EQS. (4.75) TO (4.78), N = 81

INITIAL T -  0.000D+00

  FINAL T -  0.100D+01

  PRINT T -  0.100D+00

NUMBER OF DIFFERENTIAL EQUATIONS -  81

MAXIMUM INTEGRATION ERROR -  0.100D-04


        SOLUTION BY FOURTH-ORDER APPROXIMATIONS

                 T    N          ERRMAX
             0.100   81      0.1329D-04
             0.200   81      0.4167D-05
             0.300   81      0.1271D-04
             0.400   81      0.1863D-04
             0.500   81      0.2059D-04
             0.600   81      0.1593D-04
             0.700   81      0.4435D-05
             0.800   81      0.5771D-05
             0.900   81      0.2124D-04
             1.000   81      0.3232D-05


COMPUTATIONAL STATISTICS

LAST STEP SIZE                          0.100D+00

LAST ORDER OF THE METHOD                        4

TOTAL NUMBER OF STEPS TAKEN                    29
```

Table 4.5 (b) *Continued.*

```
NUMBER OF FUNCTION EVALUATIONS                    37

NUMBER OF JACOBIAN EVALUATIONS                    12


RUN NO. -   5  EQS. (4.75) TO (4.78), N = 161

INITIAL T -  0.000D+00

  FINAL T -  0.100D+01

  PRINT T -  0.100D+00

NUMBER OF DIFFERENTIAL EQUATIONS - 161

MAXIMUM INTEGRATION ERROR -  0.100D-04


        SOLUTION BY FOURTH-ORDER APPROXIMATIONS

                      T    N          ERRMAX
                  0.100  161       0.1336D-04
                  0.200  161       0.4446D-05
                  0.300  161       0.1268D-04
                  0.400  161       0.1874D-04
                  0.500  161       0.2071D-04
                  0.600  161       0.1604D-04
                  0.700  161       0.4473D-05
                  0.800  161       0.6037D-05
                  0.900  161       0.2374D-04
                  1.000  161       0.3348D-05


                           DATA FOR PLOT

                  LOG10(N)       LOG10(ERR)
                   1.0414         -3.7068
                   1.3222         -4.4967
                   1.6128         -4.6775
                   1.9085         -4.6728
                   2.2068         -4.6245
```

tion (4.75) in which each component of the vector of functions, $\bar{f}$, defines a single-time derivative explicitly, that is, the elements of array UT in COMMON/F/ of programs 4.5a to d; these functions are programmed at the end of subroutines DERV1 and DERV2 (in DO loop 1) of Program 4.5c.

The main program and subordinate routines to call subroutine DASSL are listed in Program 4.5f. Program 4.5f is similar in most respects to Program 2.8b which calls

Table 4.5 (b) *Continued.*

```
             ..1....1....1....1....1....1....1....1....1....1....1..
  -0.371D+01+  1                                                    +I
            -                                                       -I
            -                                                       -I
            -                                                       -I
            -                                                       -I
  -0.403D+01+                                                       +I
            -                                                       -I
            -                                                       -I
            -                                                       -I
            -                                                       -I
  -0.435D+01+                                                       +I
            -                                                       -I
            -            1                                          -I
            -                                                       -I
            -                                        1           1  -I
  -0.468D+01+                            1                          +I
             ..1....1....1....1....1....1....1....1....1....1....1..
          0.104D+01  0.13D+01  0.15D+01  0.17D+01  0.20D+01  0.22D+01


LOG10(MAXERR) VS LOG10(N)


COMPUTATIONAL STATISTICS

LAST STEP SIZE                           0.100D+00

LAST ORDER OF THE METHOD                         4

TOTAL NUMBER OF STEPS TAKEN                     29

NUMBER OF FUNCTION EVALUATIONS                  37

NUMBER OF JACOBIAN EVALUATIONS                  12
```

Table 4.5 (c) Maximum Errors in the Numerical Solutions for Second- and Fourth-Order Approximations of Equation (4.75).

n	Second order (from Table 4.5a)	Fourth order (from Table 4.5b)
11	0.1385D-01	0.1964D-03
21	0.3478D-02	0.3187D-04
41	0.8662D-03	0.2102D-04
81	0.2138D-03	0.2059D-04
161	0.5066D-04	0.2374D-04

LSODE. The following points should be noted about Program 4.5f:

1. The absolute dimensioning is for 450 ODEs as indicated by the following COMMON block:

```
C...
C...  THE FOLLOWING CODING IS FOR 450 ODES.  IF MORE ODES ARE TO
C...  BE INTEGRATED, ALL OF THE 450'S SHOULD BE CHANGED TO THE
C...  REQUIRED NUMBER
      IMPLICIT DOUBLE PRECISION (A-H,O-Z)
      COMMON/T/          T,      NSTOP,       NORUN
     1      /Y/     Y(450)
     2      /F/     F(450)
```

   Again, the ODE dependent variable vector is in COMMON/Y/ and the temporal derivative vector is in COMMON/F/. DASSL also requires a real work array and an integer work array with sizes defined by the number of ODEs; in Program 4.5f, the absolute dimensioning for the real work array, WORK, is 5000 and for the integer work array, IWORK, is 475. Additionally, an array, INFO, sized to 15 is required

```
C...
C...  ABSOLUTE DIMENSIONING OF THE ARRAYS REQUIRED BY DASSL
      DIMENSION YV(450), YPRIME(450), DELTA(450),
     1      RWORK(5000)
      INTEGER IWORK(475),    INFO(15)
```

   The sizes of these arrays are defined by formulas given in the documentation of DASSL (an extensive set of comments at the beginning of the code; an example calculation of the size of the real work array for LSODE is given in the comments of Program 2.8b).

2. Two external subroutines are called by DASSL, as defined by the EXTERNAL statement:

```
C...
C...  EXTERNAL THE RESIDUAL AND DAE JACOBIAN MATRIX ROUTINES
C...  CALLED BY DASSL
      EXTERNAL RES, JAC
```

   RES defines the DAE function vector, $\bar{f}$, in equation (4.84). JAC defines the DAE Jacobian matrix if an option of DASSL is selected which requires the user to define the Jacobian matrix. Generally, this option is not selected because for large DAE problems, programming the Jacobian matrix is impractical; rather, an option is selected to have DASSL compute a numerical Jacobian by finite differences. However, a dummy subroutine JAC is provided to satisfy the loader (this requirement varies with the requirements of the loader for different computers).

3. A set of data is then read and tested for the characters END OF RUNS, as before with the main programs that called RKF45 and LSODE

```
C...
C...  BEGIN A RUN
1     NORUN=NORUN+1
C...
```

```
      PROGRAM DASSLT
C...
C...  PROGRAM DASSLT CALLS: (1) SUBROUTINE INITAL TO DEFINE THE ODE
C...  INITAL CONDITIONS, (2) SUBROUTINE DASSL TO INTEGRATE THE ODES,
C...  AND (3) SUBROUTINE PRINT TO PRINT THE SOLUTION.
C...
C...  THE FOLLOWING CODING IS FOR 450 ODES.  IF MORE ODES ARE TO BE
C...  INTEGRATED, ALL OF THE 450'S SHOULD BE CHANGED TO THE REQUIRED
C...  NUMBER
      IMPLICIT DOUBLE PRECISION (A-H,O-Z)
      COMMON/T/          T,     NSTOP,      NORUN
     1      /Y/    Y(450)
     2      /F/    F(450)
C...
C...  COMMON AREA TO PROVIDE THE INPUT/OUTPUT UNIT NUMBERS TO OTHER
C...  SUBROUTINES
      COMMON/IO/       NI,         NO
C...
C...  THE NUMBER OF DIFFERENTIAL EQUATIONS IS AVAILABLE THROUGH COMMON
C...  /N/ FOR USE IN SUBROUTINE RES
      COMMON/N/      NEQN
C...
C...  ABSOLUTE DIMENSIONING OF THE ARRAYS REQUIRED BY DASSL
      DIMENSION YV(450), YPRIME(450), DELTA(450),
     1      RWORK(5000)
      INTEGER IWORK(475),   INFO(15)
C...
C...  EXTERNAL THE DERIVATIVE AND ODE JACOBIAN MATRIX ROUTINES CALLED
C...  BY DASSL
      EXTERNAL RES, JAC
C...
C...  ARRAY FOR THE TITLE (FIRST LINE OF DATA), CHARACTERS  END OF RUNS
      CHARACTER TITLE(20)*4, ENDRUN(3)*4
C...
C...  DEFINE THE CHARACTERS  END OF RUNS
      DATA ENDRUN/'END ','OF R','UNS '/
C...
C...  DEFINE THE INPUT/OUTPUT UNIT NUMBERS
      NI=5
      NO=6
C...
C...  OPEN INPUT AND OUTPUT FILES
      OPEN(NI,FILE='DATA'  ,STATUS='OLD')
      OPEN(NO,FILE='OUTPUT',STATUS='NEW')
C...
C...  INITIALIZE THE RUN COUNTER
      NORUN=0
C...
C...  BEGIN A RUN
1     NORUN=NORUN+1
C...
```

Program 4.5f Main Program DASSLT and Subordinate Routines to Call Subroutine DASSL. *Continued on next pages.*

```
C...  INITIALIZE THE RUN TERMINATION VARIABLE
      NSTOP=0
C...
C...  READ THE FIRST LINE OF DATA
      READ(NI,1000,END=999)(TITLE(I),I=1,20)
C...
C...  TEST FOR  END OF RUNS  IN THE DATA
      DO 2 I=1,3
      IF(TITLE(I).NE.ENDRUN(I))GO TO 3
2     CONTINUE
C...
C...  AN END OF RUNS HAS BEEN READ, SO TERMINATE EXECUTION
999   STOP
C...
C...  READ THE SECOND LINE OF DATA
3     READ(NI,1001,END=999)T0,TF,TP
C...
C...  READ THE THIRD LINE OF DATA
      READ(NI,1002,END=999)NEQN,ERROR
C...
C...  PRINT A DATA SUMMARY
      WRITE(NO,1003)NORUN,(TITLE(I),I=1,20),
     1                    T0,TF,TP,
     2                    NEQN,ERROR
C...
C...  INITIALIZE TIME
      T=T0
C...
C...  SET THE INITIAL CONDITIONS
      CALL INITAL
C...
C...  SET THE INITIAL DERIVATIVES (FOR POSSIBLE PRINTING)
      CALL DERV
C...
C...  PRINT THE INITIAL CONDITIONS
      CALL PRINT(NI,NO)
C...
C...  SET THE INITIAL CONDITIONS FOR SUBROUTINE DASSL
      TV=T0
      DO 5 I=1,NEQN
      YV(I)=Y(I)
      YPRIME(I)=F(I)
5     CONTINUE
C...
C...  SET THE PARAMETERS FOR SUBROUTINE DASSL
      RTOL=ERROR
      ATOL=ERROR
      LRW=5000
      LIW=475
      DO 7 I=1,15
      INFO(I)=0
```

Program 4.5f *Continued.*

```
7     CONTINUE
      TOUT=TV+TP
C...
C...  SPECIFY THAT THE DASSL BANDED ODE JACOBIAN MATRIX OPTION IS TO
C...  BE USED
      ML=3
      MU=3
      IWORK(1)=ML
      IWORK(2)=MU
      INFO(6)=1
C...
C...  CALL SUBROUTINE DASSL TO COVER ONE PRINT INTERVAL
4     CALL DASSL(RES,NEQN,TV,YV,YPRIME,TOUT,INFO,RTOL,ATOL,IDID,
     1           RWORK,LRW,IWORK,LIW,RPAR,IPAR,JAC)
C...
C...  PRINT THE SOLUTION AT THE NEXT PRINT POINT
      T=TV
      DO 6 I=1,NEQN
      Y(I)=YV(I)
6     CONTINUE
      CALL DERV
      CALL PRINT(NI,NO)
C...
C...  TEST FOR AN ERROR CONDITION
      IF(IDID.LT.0)THEN
C...
C...     PRINT A MESSAGE INDICATING AN ERROR CONDITION
         WRITE(NO,1005)RWORK(7),IWORK(8),IWORK(11),IWORK(12),IWORK(13)
         WRITE(NO,1004)IDID
C...
C...     GO ON TO THE NEXT RUN
         GO TO 1
      END IF
C...
C...  CHECK FOR A RUN TERMINATION
      IF(NSTOP.NE.0)GO TO 1
C...
C...  CHECK FOR THE END OF THE RUN
      TOUT=TV+TP
      IF(TV.LT.(TF-0.5D0*TP))GO TO 4
C...
C...  THE CURRENT RUN IS COMPLETE, SO PRINT THE COMPUTATIONAL STA-
C...  TISTICS FOR DASSL AND GO ON TO THE NEXT RUN
      WRITE(NO,1005)RWORK(7),IWORK(8),IWORK(11),IWORK(12),IWORK(13)
      GO TO 1
C...
C...  ****************************************************************
C...
C...  FORMATS
C...
1000  FORMAT(20A4)
1001  FORMAT(3D10.0)
```

Program 4.5f *Continued.*

```
1002  FORMAT(I5,20X,D10.0)
1003  FORMAT(1H1,
     1 ' RUN NO. - ',I3,2X,20A4,//,
     2 ' INITIAL T - ',D10.3,//,
     3 '   FINAL T - ',D10.3,//,
     4 '   PRINT T - ',D10.3,//,
     5 ' NUMBER OF DIFFERENTIAL EQUATIONS - ',I3,//,
     6 ' MAXIMUM INTEGRATION ERROR - ',D10.3,//,
     7 1H1)
1004  FORMAT(1H ,//,' IDID = ',I3,//,
     1 ' INDICATING AN INTEGRATION ERROR, SO THE CURRENT RUN'     ,/,
     2 ' IS TERMINATED.  PLEASE REFER TO THE DOCUMENTATION FOR'   ,/,
     3 ' SUBROUTINE',//,25X,'DASSL',//,
     4 ' FOR AN EXPLANATION OF THESE ERROR INDICATORS'            )
1005  FORMAT(1H ,//,' COMPUTATIONAL STATISTICS'                    ,//,
     1 ' LAST STEP SIZE                        ',          D10.3,//,
     2 ' LAST ORDER OF THE METHOD              ',            I10,//,
     3 ' TOTAL NUMBER OF STEPS TAKEN           ',            I10,//,
     4 ' NUMBER OF FUNCTION EVALUATIONS        ',            I10,//,
     5 ' NUMBER OF JACOBIAN EVALUATIONS        ',            I10,/)
      END

      SUBROUTINE RES(TV,YV,YPRIME,DELTA,IRES,RPAR,IPAR)
C...
C...  SUBROUTINE RES IS AN INTERFACE ROUTINE BETWEEN SUBROUTINES DASSL
C...  AND DERV
C...
C...  COMMON AREA
      IMPLICIT DOUBLE PRECISION (A-H,O-Z)
      COMMON/T/          T,      NSTOP,      NORUN
     1      /Y/      Y(1)
     2      /F/      F(1)
C...
C...  THE NUMBER OF DIFFERENTIAL EQUATIONS IS AVAILABLE THROUGH COMMON
C...  /N/
      COMMON/N/      NEQN
C...
C...  ABSOLUTE DIMENSION THE DEPENDENT, DERIVATIVE, AND RESIDUAL ARRAYS
      DIMENSION YV(450), YPRIME(450), DELTA(450)
C...
C...  TRANSFER THE INDEPENDENT VARIABLE, DEPENDENT VARIABLE VECTOR
C...  FOR USE IN SUBROUTINE DERV
      T=TV
      DO 1 I=1,NEQN
      Y(I)=YV(I)
1     CONTINUE
C...
C...  EVALUATE THE DERIVATIVE VECTOR
      CALL DERV
C...
C...  TRANSFER THE DERIVATIVE VECTOR FOR USE BY SUBROUTINE DASSL
```

Program 4.5f *Continued.*

```
      DO 2 I=1,NEQN
      DELTA(I)=YPRIME(I)-F(I)
2     CONTINUE
      RETURN
      END

      SUBROUTINE JAC(T,Y,YPRIME,PD,CJ,RPAR,IPAR)
C...
C...  SUBROUTINE JAC IS CALLED ONLY IF AN OPTION OF DASSL IS SELECTED
C...  FOR WHICH THE USER MUST SUPPLY THE ANALYTICAL JACOBIAN MATRIX OF
C...  THE ODE SYSTEM.  THE PROGRAMMING OF THE ANALYTICAL JACOBIAN IN
C...  SUBROUTINE JAC IS DESCRIBED IN DETAIL IN THE DOCUMENTATION
C...  COMMENTS AT THE BEGINNING OF SUBROUTINE DASSL.
C...
      RETURN
      END
```

Program 4.5f *Continued.*

```
C...  INITIALIZE THE RUN TERMINATION VARIABLE
      NSTOP=0
C...
C...  READ THE FIRST LINE OF DATA
      READ(NI,1000,END=999)(TITLE(I),I=1,20)
C...
C...  TEST FOR  END OF RUNS  IN THE DATA
      DO 2 I=1,3
      IF(TITLE(I).NE.ENDRUN(I))GO TO 3
2     CONTINUE
C...
C...  AN END OF RUNS HAS BEEN READ, SO TERMINATE EXECUTION
999   STOP
C...
C...  READ THE SECOND LINE OF DATA
3     READ(NI,1001,END=999)TO,TF,TP
C...
C...  READ THE THIRD LINE OF DATA
      READ(NI,1002,END=999)NEQN,ERROR
C...
C...  PRINT A DATA SUMMARY
      WRITE(NO,1003)NORUN,(TITLE(I),I=1,20),
     1                    TO,TF,TP,
     2                    NEQN,ERROR
```

After the data are read they are printed in a data summary.

4. The initial conditions for the DAE system, according to equation (4.85), are then defined by a call to INITAL, followed by a call to DERV to compute the initial derivatives; a call to PRINT then prints the initial conditions.

```
C...
C...  INITIALIZE TIME
      T=TO
```

```
C...
C...  SET THE INITIAL CONDITIONS
      CALL INITAL
C...
C...  SET THE INITIAL DERIVATIVES
      CALL DERV
C...
C...  PRINT THE INITIAL CONDITIONS
      CALL PRINT(NI,NO)
C...
C...  SET THE INITIAL CONDITIONS FOR SUBROUTINE DASSL
      TV=TO
      DO 5 I=1,NEQN
      YV(I)=Y(I)
      YPRIME(I)=F(I)
5     CONTINUE
```

Finally, the initial conditions for the dependent variable vector (in array `Y` from `COMMON/Y/`) and the derivative vector (in array `F` from `COMMON/F/`) are put into arrays `YV` and `YPRIME` used by `DASSL` to start the solution. Note that the use of the initial dependent and derivative vectors, $\bar{y}(t_0)$ and $d\bar{y}(t_0)/dt$, at t_0, follows from equation (4.85).

5. A set of parameters is defined before the first call to `DASSL`:

```
C...
C...  SET THE PARAMETERS FOR SUBROUTINE DASSL
      RTOL=ERROR
      ATOL=ERROR
      LRW=5000
      LIW=475
      DO 7 I=1,15
      INFO(I)=0
7     CONTINUE
      TOUT=TV+TP
```

Briefly these parameters are: (a) a relative error tolerance (RTOL) and an absolute error tolerance (ATOL) applied to the dependent variables of magnitude `ERROR` which is read from the data file, Program 4.5e, as `1.0D-05`; (b) the lengths of the real and integer work arrays, `LRW=5000`, `LIW=475` (these lengths may be over-specified, and the only disadvantage is the use of additional memory); (c) the array to specify options when `DASSL` is called is initialized to zero, `INFO(I)=0`; and (d) the output time, `TOUT`, which equals the initial time (or current time in subsequent calls to `DASSL`), `TV`, advanced by the print or output interval, `TP`.

6. Option 6 for a banded Jacobian matrix of the DAE system is then selected (`INFO(6)=1`). The upper and lower half bandwidths, `ML` and `MU`, are each set to 3, then stored in the first and second elements of the integer work array, `IWORK(1)` and `IWORK(2)`, for use by `DASSL`:

```
C...
C...  SPECIFY THAT THE DASSL BANDED JACOBIAN MATRIX OPTION IS TO
C...  BE USED
      ML=3
      MU=3
```

```
      IWORK(1)=ML
      IWORK(2)=MU
      INFO(6)=1
```

This total bandwidth of ML + MU + 1 = 7 is adequate to accommodate all of the nonzero elements of the Jacobian matrix since the five-point approximations of subroutines DSS004 and DSS044 called in subroutine DERV2 of Program 4.5c would produce only five diagonals along the main diagonal of the Jacobian matrix (a pentadiagonal system). Alternatively, the Jacobian matrix could be taken as full by using the default INFO(6)=0. This would be wasteful of computer resources (storage and computation time) since the Jacobian matrix consists largely of zero elements (due to the pentadiagonal structure), particularly for the larger numbers of grid points, e.g., $n = 161$.

7. The call to DASSL moves the solution forward from TV to TOUT:

```
C...
C...  CALL SUBROUTINE DASSL TO COVER ONE PRINT INTERVAL
4     CALL DASSL(RES,NEQN,TV,YV,YPRIME,TOUT,INFO,RTOL,ATOL,IDID,
     1           RWORK,LRW,IWORK,LIW,RPAR,IPAR,JAC)
```

8. The solution at the next output point, returned by DASSL in array YV, is then printed along with the updated derivative vector produced by a call to DERV:

```
C...
C...  PRINT THE SOLUTION AT THE NEXT PRINT POINT
      T=TV
      DO 6 I=1,NEQN
      Y(I)=YV(I)
6     CONTINUE
      CALL DERV
      CALL PRINT(NI,NO)
```

9. An error condition is reported by DASSL as a nonzero value of IDID, for which an error message is printed and the next run is initiated. If an error condition is not reported by DASSL, a test for the end of the current run is made, and if the solution is not complete, DASSL is called again (by a return to statement 4). If the run is complete, the computational statistics reported by DASSL are printed before initiating the next run:

```
C...
C...  TEST FOR AN ERROR CONDITION
      IF(IDID.LT.0)THEN
C...
C...     PRINT A MESSAGE INDICATING AN ERROR CONDITION
         WRITE(NO,1005)RWORK(7),IWORK(8),IWORK(11),IWORK(12),IWORK(13)
         WRITE(NO,1004)IDID
C...
C...     GO ON TO THE NEXT RUN
         GO TO 1
      END IF
C...
C...  CHECK FOR A RUN TERMINATION
      IF(NSTOP.NE.0)GO TO 1
```

```
C...
C...   CHECK FOR THE END OF THE RUN
       TOUT=TV+TP
       IF(TV.LT.(TF-0.5D0*TP))GO TO 4
C...
C...   THE CURRENT RUN IS COMPLETE, SO PRINT THE COMPUTATIONAL
C...   STATISTICS FOR DASSL AND GO ON TO THE NEXT RUN
       WRITE(NO,1005)RWORK(7),IWORK(8),IWORK(11),IWORK(12),IWORK(13)
       GO TO 1
```

10. Subroutine RES is called by DASSL as its first argument, and again, it is an EXTERNAL. RES defines the residuals of the DAE system, equation (4.84). The principal inputs to RES are the independent variable, t (TV), the dependent variable vector, $\bar{y}$ (YV), and the derivative vector, $d\bar{y}/dt$ (YPRIME). The function (or residual) vector, $\bar{f}(\bar{y}, d\bar{y}/dt, t)$, of equation (4.84) is computed in RES and returned to the calling program in array DELTA (of course, the purpose of DASSL is to compute the dependent variable vector, $\bar{y}$, to make this residual vector essentially a *zero vector* according to equation (4.84)). Integer IRES and arrays RPAR and IPAR are somewhat incidental. IRES provides a means for terminating execution of DASSL if it is set to a nonzero value in RES (this might be used, for example, if any of the elements of $\bar{y}$ assume physically unrealistic values). Arrays RPAR and IPAR provide a means for bringing real and integer problem parameters into RES for programming the residual vector (in DELTA). For the case of an explicit system of ODEs, which is what we integrate for the method of lines approximation of equation (4.84), the residual vector, $\Delta(\bar{y}, t)$, is of the form

$$\Delta(\bar{y}, t) = d\bar{y}/dt - \bar{f}(\bar{y}, t) \tag{4.86}$$

where $\bar{f}(\bar{y}, t)$ (not the $\bar{f}$ of equation (4.84)) is computed by the call to DERV in RES and $d\bar{y}/dt$ is in array YPRIME (an input to RES). Thus, the programming of this residual in RES is

```
C...
C...   COMPUTE THE RESIDUAL VECTOR FOR USE BY SUBROUTINE DASSL
       DO 2 I=1,NEQN
       DELTA(I)=YPRIME(I)-F(I)
2      CONTINUE
```

11. DASSL has an option for the user to provide an *analytical Jacobian* for the DAE system of equation (4.84) which is selected by setting INFO(5)=1. Since INFO(5) was previously set to zero (in DO loop 7 before the call to DASSL), a *numerical Jacobian* is computed by finite differences; therefore subroutine JAC for an analytical Jacobian is not called by DASSL, and a dummy JAC is provided only to satisfy the loader.

In summary, we are using DASSL as an explicit ODE integrator for the method of lines solution of equation (4.75); it is, however, capable of solving a more general set of DAEs (for which the ODEs are a special case) defined by equation (4.84). Only a few basic features of DASSL have been described through this example application. DASSL has many options which are described in detail in the prologue of comments at the beginning of DASSL [Brenan et al. (1989); Schiesser (1991)].

4.5 Multiregion PDEs.

We now consider the method of lines solution of a system of PDEs defined over different spatial regions or domains. Such problems arise frequently in scientific and engineering applications, like heat transfer or diffusion in two or more phases. Diffusion between two phases is illustrated in Figure 4.6. Within each region, we assume the concentration of a substance is defined by the one-dimensional Fourier's second law (also called Fick's second law) in Cartesian coordinates

$$\frac{\partial C_a}{\partial t} = D_a \frac{\partial^2 C_a}{\partial x^2}, \qquad \frac{\partial C_b}{\partial t} = D_b \frac{\partial^2 C_b}{\partial x^2} \tag{4.87)(4.88}$$

where C_a, C_b = concentrations of the diffusing substance in the two regions ($kg\text{-}mol/m^3$); x = position within the two regions (m); t = time (s); and D_a, D_b = diffusivities in the two regions (m^2/s). Diffusion can take place across the interface between the two regions (at $x = 0$) that must be described as boundary conditions. Two types of boundary conditions are generally used for the description of the conditions at the interface:

Case I

$$C_a(0,t) = mC_b(0,t) \tag{4.89}$$

$$D_a \frac{\partial C_a(0,t)}{\partial x} = D_b \frac{\partial C_b(0,t)}{\partial x} \tag{4.90}$$

Case II

$$D_a \frac{\partial C_a(0,t)}{\partial x} = h_i(C_b(0,t) - C_a(0,t)) \tag{4.91}$$

$$D_a \frac{\partial C_a(0,t)}{\partial x} = D_b \frac{\partial C_b(0,t)}{\partial x} \tag{4.92}$$

For Case I, equation (4.89) states that the two phases (regions) are in equilibrium at the interface, and the proportionality constant, m, is called an *equilibrium constant*. For thermal systems, C_a and C_b are temperatures and generally $m = 1$ which simply states that the two phases are at the same temperature at the interface. For mass transfer systems, in which diffusion takes place due to concentration differences or gradients, generally $m \neq 1$.

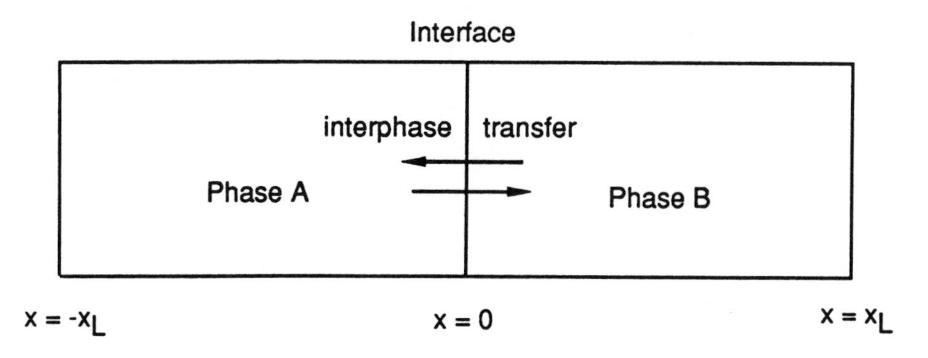

Figure 4.6 Diffusion in Two Regions.

Equation (4.90) states that the flux across the interface is continuous. In other words, the rate at which heat or mass crosses the interface from one phase equals the rate to the other phase; thus there is no accumulation of heat or mass at the interface.

For Case II, equation (4.91) states that the flux across the interface in phase A, $D_a \partial C_a(0,t)/\partial x$, is proportional to the difference between the concentrations in the two phases at the interface, $h_i(C_b(0,t) - C_a(0,t))$. h_i is called a heat or mass transfer coefficient or a thermal contact resistance. Again, equation (4.92) ensures continuity of the flux across the interface.

We now consider a method of lines implementation of Case II (a formulation for Case I follows directly from the Case II analysis). A 121-point spatial grid is defined over the interval $-60 \le x \le 60$, as depicted in Figure 4.6 with $x_L = -60$, $x_U = 60$. For the interval $-60 \le x < 0$, C_a is defined according to equation (4.87). Similarly, for $0 < x \le 60$, C_b is defined according to equation (4.88). We use the initial conditions

$$C_a(x,0) = C_{a0}, C_b(x,0) = C_{b0} \qquad (4.93)(4.94)$$

with $C_{a0} = 0$, $C_{b0} = 1$ so initially, a unit step in concentration occurs at $x = 0$. Subroutine `INITAL` to define the grid in x and implement initial conditions (4.93) and (4.94) is listed in Program 4.6a. We can note the following points about `INITAL`:

1. The `COMMON` area has two arrays for $C_a(x,t)$ and $C_b(x,t)$, `CA(NX)` and `CB(NX)`, with the associated spatial and temporal derivatives. Since `NX=61`, there are 122 ODEs with derivatives in `COMMON/F/`.

2. The spatial grid has unit spacing as defined by the code

```
C...
C...  HALF LENGTH
      XL=60.D0
C...
C...  GRID INCREMENT
      DX=XL/DFLOAT(NX-1)
C...
C...  GRID POINTS
      DO 1 I=1,NX
         XA(I)=-XL+DFLOAT(I-1)*DX
         XB(I)=    DFLOAT(I-1)*DX
1     CONTINUE
```

3. Initial conditions (4.93) and (4.94) are then defined on this grid

```
C...
C...  INITIAL CONDITION
      CA0=0.D0
      CB0=1.0D0
      DO 2 I=1,NX
         CA(I)=CA0
         CB(I)=CB0
2     CONTINUE
```

4. The diffusivities, D_a and D_b, and transfer coefficient, h_i (in equation (4.91)), are

```
      SUBROUTINE INITAL
      IMPLICIT DOUBLE PRECISION (A-H,O-Z)
      PARAMETER(NX=61)
      COMMON/T/          T,      NSTOP,     NORUN
     1      /Y/   CA(NX),    CB(NX)
     2      /F/  CAT(NX),   CBT(NX)
     3      /S/  CAX(NX),  CAXX(NX),  CBX(NX),  CBXX(NX),
     4      /R/   XA(NX),    XB(NX),       XL,         DX,
     5                DA,         DB,      CA0,        CB0,
     6                 H
     7      /I/       IP
C...
C...  HALF LENGTH
      XL=60.D0
C...
C...  GRID INCREMENT
      DX=XL/DFLOAT(NX-1)
C...
C...  GRID POINTS
      DO 1 I=1,NX
         XA(I)=-XL+DFLOAT(I-1)*DX
         XB(I)=    DFLOAT(I-1)*DX
1     CONTINUE
C...
C...  DIFFUSIVITIES
      DA=1.0D0
      DB=1.0D0
C...
C...  TRANSFER COEFFICIENT
      H=0.1D0
C...
C...  INITIAL CONDITION
      CA0=0.D0
      CB0=1.0D0
      DO 2 I=1,NX
         CA(I)=CA0
         CB(I)=CB0
2     CONTINUE
C...
C...  INITIALIZE COUNTER FOR PRINTED AND PLOTTED SOLUTION
      IP=0
      RETURN
      END
```

Program 4.6a Subroutine INITAL for Initial Conditions (4.93) and (4.94).

given numerical values and passed through COMMON to DERV and PRINT

```
C...
C...  DIFFUSIVITIES
      DA=1.0D0
      DB=1.0D0
C...
C...  TRANSFER COEFFICIENT
      H=0.1D0
```

Since equations (4.87) and (4.88) are second order in x, they each require two boundary conditions. Thus, in addition to boundary conditions (4.91) and (4.92) we use as boundary conditions

$$C_a(-60,t) = C_{a0}, \qquad C_b(60,t) = C_{b0} \tag{4.95)(4.96}$$

where C_{a0} and C_{b0} are the initial concentrations of equations (4.93) and (4.94). In using equations (4.95) and (4.96), we are assuming that the boundary concentrations at $x = -60$ and $x = 60$ do not depart from their initial values. In other words, we assume for the time over which the solution is computed, the diffusion between the two phases has not penetrated to the outer boundaries. This assumption can be checked by obsering how far the concentration profiles resulting from diffusion have penetrated each phase.

Subroutine DERV for equations (4.87), (4.88), (4.91), (4.92), (4.95), and (4.96) is listed in Program 4.6b. We can note the following points about DERV:

1. Boundary condition (4.95) is first implemented at the left boundary, which corresponds to grid point 1 in array CA:

```
C...
C...  BOUNDARY CONDITION AT X = -XL
      NL=1
      CA(1)=CAO
```

2. Boundary condition (4.91) is then implemented at the interface:

```
C...
C...  BOUNDARY CONDITION AT X = 0 FOR CA
      NU=2
      CAX(NX)=-(H/DA)*(CA(NX)-CB(1))
```

   Note that this is a Neumann boundary condition because it defines $\partial C_a(0,t)/\partial x$ according to equation (4.91).

3. Both boundary conditions for equation (4.87) are now defined so the second derivative, $\partial^2 C_a(x,t)/\partial x^2$, in equation (4.87) can be computed by a call to DSS044:

```
C...
C...  DERIVATIVE CA
C...                 XX
      CALL DSS044(-XL,0.DO,NX,CA,CAX,CAXX,NL,NU)
```

4. The boundary conditions for $C_b(x,t)$ are next implemented, equations (4.92) and (4.96):

```
C...
C...  BOUNDARY AT X = 0 FOR CB
      NL=2
      CBX(1)=(DA/DB)*CAX(NX)
C...
C...  BOUNDARY CONDITION AT X = XL
      NU=1
      CB(NX)=CBO
```

```
      SUBROUTINE DERV
      IMPLICIT DOUBLE PRECISION (A-H,O-Z)
      PARAMETER(NX=61)
      COMMON/T/          T,      NSTOP,     NORUN
     1      /Y/   CA(NX),     CB(NX)
     2      /F/  CAT(NX),    CBT(NX)
     3      /S/  CAX(NX),  CAXX(NX),   CBX(NX),   CBXX(NX),
     4      /R/   XA(NX),     XB(NX),        XL,          DX,
     5                DA,         DB,       CA0,         CB0,
     6                 H
     7      /I/       IP
C...
C...  BOUNDARY CONDITION AT X = -XL
      NL=1
      CA(1)=CA0
C...
C...  BOUNDARY CONDITION AT X = 0 FOR CA
      NU=2
      CAX(NX)=-(H/DA)*(CA(NX)-CB(1))
C...
C...  DERIVATIVE CA
C...               XX
      CALL DSS044(-XL,0.D0,NX,CA,CAX,CAXX,NL,NU)
C...
C...  BOUNDARY AT X = 0 FOR CB
      NL=2
      CBX(1)=(DA/DB)*CAX(NX)
C...
C...  BOUNDARY CONDITION AT X = XL
      NU=1
      CB(NX)=CB0
C...
C...  DERIVATIVE CB
C...               XX
      CALL DSS044(0.D0,XL,NX,CB,CBX,CBXX,NL,NU)
C...
C...  PDE FOR CA
      DO 1 I=1,NX
         CAT(I)=DA*CAXX(I)
1     CONTINUE
C...
C...  PDE FOR CA
      DO 2 I=1,NX
         CBT(I)=DB*CBXX(I)
2     CONTINUE
      RETURN
      END
```

Program 4.6b Subroutine DERV for Equations (4.87), (4.88), (4.91), (4.92), (4.95), and (4.96).

5. Both boundary conditions for equation (4.88) are defined so the second derivative, $\partial^2 C_b(x,t)/\partial x^2$, in equation (4.88) can be computed by a call to DSS044:

```
C...
C...  DERIVATIVE CB
C...                     XX
      CALL DSS044(0.D0,XL,NX,CB,CBX,CBXX,NL,NU)
```

6. With both spatial derivatives in equations (4.87) and (4.88) now computed, the corresponding derivatives in time can be computed to complete subroutine DERV

```
C...
C...  PDE FOR CA
      DO 1 I=1,NX
         CAT(I)=DA*CAXX(I)
1     CONTINUE
C...
C...  PDE FOR CA
      DO 2 I=1,NX
         CBT(I)=DB*CBXX(I)
2     CONTINUE
```

DO loops 1 and 2 define 122 ODEs. Also, the outer boundary ODEs are included in these DO loops, rather than using explicit programming such as CAT(1)=0.0D0, CBT(NX)=0.0D0 for equations (4.95) and (4.96). This approach works because the correct boundary values are set according to equations (4.95) and (4.96) each time DERV is called and the time integration does not move $C_a(-60,t)$ and $C_b(60,t)$ away from the values prescribed by boundary conditions (4.95) and (4.96) for more than one time integration step, after which they are reset again to the correct values by the next call to DERV. Of course, CAT(1)=0.0D0, CBT(NX)=0.0D0 could be added after DO loop 2.

Subroutine PRINT is listed in Program 4.6c. We can note the following points about PRINT:

1. A heading for the numerical solution at $x = -60$, -30, 0, 30, and 60 is printed in the first call to PRINT (at $t = 0$):

```
C...
C...  PRINT A HEADING FOR THE SOLUTION
      IP=IP+1
      IF(IP.EQ.1)THEN
         WRITE(NO,1)
1        FORMAT(7X,'t'
     1   ,1X,'Ca(-60,t)',1X,'Ca(-30,t)',1X,'  Ca(0,t)'
     2   ,1X,'  Cb(0,t)',1X,' Cb(30,t)',1X,' Cb(60,t)')
      END IF
```

2. The numerical solution is then printed and written to a file for plotting by a call to PLOT1:

```
C...
C...  PRINT THE NUMERICAL SOLUTION
      NIN=30
```

```
      WRITE(NO,3)T,(CA(I),I=1,NX,NIN),(CB(I),I=1,NX,NIN)
3     FORMAT(F8.3,6F10.5,/)
C...
C...  PREPARE A FILE FOR PLOTTING
      CALL PLOT1
```

Again, we do not list subroutine PLOT1 because it is similar to previous plotting routines, e.g., Program 3.5d.

The data file for Programs 4.6a to c is given in Program 4.6d. In summary, subroutine PRINT is called 101 times, and 122 ODEs are integrated with an error tolerance of 0.00001 (by RKF45).

The numerical solution from Programs 4.6a to d is partially listed in Table 4.6 and plotted in Figure 4.7. We can note the following points about the solution in Table 4.6 and Figure 4.7:

1. A finite jump occurs in the concentration profiles at the interface (at $x = 0$), which is a result of boundary condition (4.91). In other words, a difference in the concentrations, $(C_b(0,t) - C_a(0,t))$, is required to provide a flux across the interface, which is $D_a \partial C_a(0,t)/\partial x$ or $h_i(C_b(0,t) - C_a(0,t))$. This flux must occur,

```
      SUBROUTINE PRINT(NI,NO)
      IMPLICIT DOUBLE PRECISION (A-H,O-Z)
      PARAMETER(NX=61)
      COMMON/T/          T,     NSTOP,    NORUN
     1      /Y/   CA(NX),    CB(NX)
     2      /F/  CAT(NX),   CBT(NX)
     3      /S/  CAX(NX),  CAXX(NX),  CBX(NX),  CBXX(NX),
     4      /R/   XA(NX),    XB(NX),       XL,        DX,
     5                DA,        DB,      CA0,       CB0,
     6                 H
     7      /I/       IP
C...
C...  PRINT A HEADING FOR THE SOLUTION
      IP=IP+1
      IF(IP.EQ.1)THEN
         WRITE(NO,1)
1        FORMAT(7X,'t'
     1   ,1X,'Ca(-60,t)',1X,'Ca(-30,t)',1X,'  Ca(0,t)'
     2   ,1X,'  Cb(0,t)',1X,' Cb(30,t)',1X,' Cb(60,t)')
      END IF
C...
C...  PRINT THE NUMERICAL SOLUTION
      NIN=30
      WRITE(NO,3)T,(CA(I),I=1,NX,NIN),(CB(I),I=1,NX,NIN)
3     FORMAT(F8.3,6F10.5,/)
C...
C...  PREPARE A FILE FOR PLOTTING
      CALL PLOT1
      RETURN
      END
```

Program 4.6c Subroutine PRINT for Printing and Plotting the Numerical Solution to Equations (4.87), (4.88), (4.91), (4.92), (4.95), and (4.96).

```
TWO-REGION DIFFUSION
0.         100.0     1.0
  122                      0.00001
END OF RUNS
```

Program 4.6d Data File for Programs 4.6a to c.

Table 4.6 Abbreviated Numerical Output from Programs 4.6a to d.

```
RUN NO. -    1  TWO-REGION DIFFUSION

INITIAL T -  0.000D+00

  FINAL T -  0.100D+03

  PRINT T -  0.100D+01

NUMBER OF DIFFERENTIAL EQUATIONS - 122

INTEGRATION ALGORITHM - RKF45

MAXIMUM INTEGRATION ERROR -  0.100D-04

      t Ca(-60,t) Ca(-30,t)   Ca(0,t)   Cb(0,t)  Cb(30,t)  Cb(60,t)
  0.000   0.00000   0.00000   0.00000   1.00000   1.00000   1.00000

  1.000   0.00000   0.00000   0.09581   0.90419   1.00000   1.00000

  2.000   0.00000   0.00000   0.12643   0.87357   1.00000   1.00000

  3.000   0.00000   0.00000   0.14790   0.85210   1.00000   1.00000

  4.000   0.00000   0.00000   0.16448   0.83552   1.00000   1.00000

  5.000   0.00000   0.00000   0.17803   0.82197   1.00000   1.00000
                .                                    .
                .                                    .
                .                                    .

 95.000   0.00000   0.00747   0.36955   0.63045   0.99253   1.00000

 96.000   0.00000   0.00772   0.37011   0.62989   0.99228   1.00000

 97.000   0.00000   0.00798   0.37067   0.62933   0.99202   1.00000

 98.000   0.00000   0.00824   0.37122   0.62878   0.99176   1.00000

 99.000   0.00000   0.00850   0.37177   0.62823   0.99150   1.00000

100.000   0.00000   0.00876   0.37231   0.62769   0.99124   1.00000
```

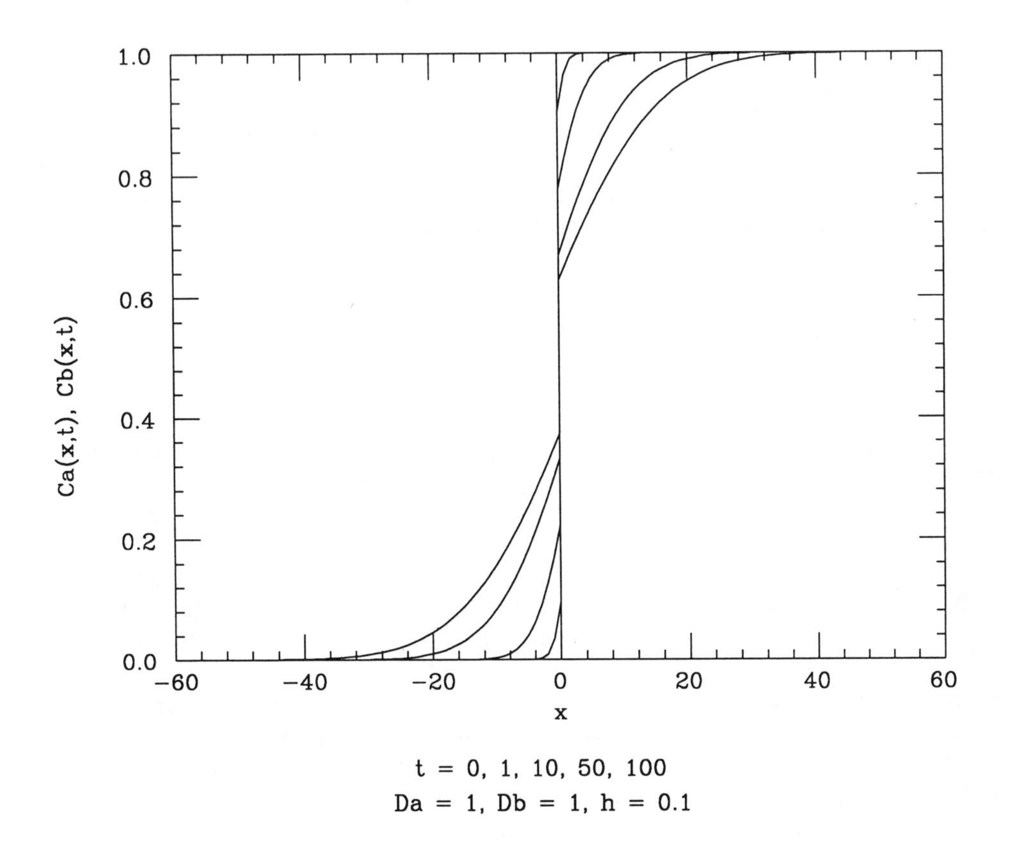

Figure 4.7 Graphical Output from Programs 4.6a to d.

of course, because of the difference in concentrations in the two phases, starting with initial conditions (4.93) and (4.94).

2. Figure 4.7 indicates the assumption of negligible penetration of the concentration profiles to the outer boundaries, as reflected in boundary conditions (4.95) and (4.96), is justified. This might not be the case in general, though. For example, the problem time might be increased beyond $t = 100$ (see Program 6.4d) which would allow a longer time for diffusion to the outer boundaries, or the rate of diffusion might be increased by increasing the diffusivities D_a and D_b. If conditions are selected so the assumption of negligible penetration of the concentration profiles to the boundaries is not valid, some other boundary conditions might be required. For example, no flux conditions might be applied at the

outer boundaries if they are impermeable; that is, in place of equations (4.95) and (4.96) we use as boundary conditions

$$\frac{\partial C_a(-60,t)}{\partial x} = \frac{\partial C_b(60,t)}{\partial x} = 0 \qquad (4.97)(4.98)$$

We conclude this example by noting that the solution in Table 4.6 and Figure 4.7 agrees with the solution of Liu and Aguirre (1990) which was calculated by the method of lines, but with a somewhat different formulation. Through the preceding example, we have demonstrated some general techniques for multiregion PDE problems. The two essential new features are: (a) the definition of a grid for each region and (b) the use of conditions at the interface(s) of the regions.

4.6 *Bandwidth Reduction in the Method of Lines*

We consider again the method of lines analysis of equations (3.11) to (3.15). We observed that the method of lines solution of these equations produced a Jacobian map with a main diagonal band and several outlying diagonals; see, for example, the maps in Table 3.1a. This Jacobian structure is typical for simultaneous PDEs.

If we use an implicit ODE integrator, like `LSODE` or `DASSL`, the structure of the Jacobian matrix is an important consideration. Recall again that the Jacobian matrix reflects the structure of the linear algebraic equations that must be solved iteratively in Newton's method to obtain the solution of the nonlinear BDF equations. We can, of course, always use the full Jacobian matrix, which will include the outlying diagonals of a PDE Jacobian matrix, but this leads to the well-known n^3 operations count to solve the linear equations in the Newton iteration where n is the number of ODEs; as n increases, n^3 increases very rapidly, and generally we try to avoid this unfavorable operations count for large sets of ODEs resulting from the method of lines solution of PDEs. For example, in the solution of equation (4.75), we used a 161-point grid, and if we called `DASSL` with the full matrix option (which is the default), each solution of the linear algebraic equations in the Newton iteration would have required a number of arithmetic operations proportional to $161^3 = 4,173,281$. However, we called `DASSL` with a bandwidth of seven; thus the operations count was actually proportional to $(161)(7^2) = 7889$, and therefore the operations count was reduced by $7889/4,173,281 = 0.00189$ or approximately a 500-fold reduction in the number of operations (and recall that the linear algebraic equations are solved many times during the solution of the ODE system).

In general, if an $n \times n$ linear algebraic system has a bandwidth m, the number of arithmetic operations required to solve the banded system is proportional to nm^2, and clearly, making m as small as possible is very advantageous (as the preceding example demonstrated, even though $n = 161$ is a problem of modest size). We were able to use a bandwidth of seven in the case of equation (4.75) because we were computing the solution to only one PDE, and when the spatial derivatives were replaced with five-point approximations (in subroutines `DSS004` and `DSS044`), a pentadiagonal ODE system resulted (with all of the diagonals clustered around the main diagonal).

In the case of simultaneous PDEs, however, like equations (3.11) and (3.12), the Jacobian matrix has outlying diagonals due to the coupling between the PDEs, and therefore m is of the order of n (the bandwidth is not much smaller than the full matrix dimension). In other words, we cannot exploit the structure of the matrix to arrive at the much smaller operations count of nm^2 rather than n^3 since m is not much smaller than n. This conclusion, however, presupposes that we formulate the method of lines solutions of the simultaneous PDEs as we did previously (in the case of equations (3.11) to (3.15), for example). The essential feature of this formulation is the use of COMMON/Y/ and /F/ of the form (from Program 3.1c)

```
      COMMON/T/          T,      NSTOP,      NORUN
     +      /Y/   U1(NX),     U2(NX),        U1A
     +      /F/  U1T(NX),    U2T(NX),       U1AT
```

In other words, the ODEs in COMMON/Y/ and their associated temporal derivatives in COMMON/F/ are ordered according to the PDEs, e.g., U1 and U1T for equation (3.11), then U2 and U2T for equation (3.12). However, an alternative ordering is by grid point, e.g., U1(1),U2(1),U1(2),U2(2),...,U1(NX),U2(NX) which will clearly change the structure of the Jacobian matrix (recall that the map of the Jacobian matrix has the index of the dependent variables across the top, and the index of the ODEs down the left side as they are ordered in COMMON/Y/ and /F/, respectively; clearly the dependent variables and their temporal derivatives have been reordered in COMMON/Y/ and/F/).

We now investigate the effect of the grid-point ordering (rather than the previous PDE ordering) on the structure of the Jacobian matrix. To do this, we return to Programs 3.1a to f for boundary condition function (3.32). Recall that subroutines INIT1, DERV1, and PRINT1 were the three basic subroutines called by INITAL, DERV, and PRINT for boundary condition (3.32); DERV1 in turn called three spatial differentiators, DSS012, DSS018, and DSS020. We now add INIT2, DERV2, and PRINT2 for the solution of equations (3.11) to (3.15) by grid-point ordering with spatial differentiation by DSS012, DSS018, and DSS020. Thus, we have a total of six runs, (PDE ordering + grid-point ordering)(DSS012 + DSS018 + DSS020) = (2)(3) = 6.

Subroutines INITAL, DERV, and PRINT are first listed in Program 4.7a. Note that in each subroutine, NORUN=1 to 3 calls the previous subroutines INIT1, DERV1, and PRINT1 of Programs 3.1b to d (for PDE ordering), and NORUN=4 to 6 calls subroutines INIT2, DERV2, and PRINT2 for grid-point ordering; these latter subroutines, i.e., the "2" series, will now be considered.

INIT2 is similar to INIT1; the only essential changes are a reorganization of the ODE COMMON area and the addition of DO loop 2 to implement the initial conditions, equations (3.13) and (3.15), in grid-point ordering. We can note the following points about INIT2:

1. COMMON/Y/ has been modified so it now contains one continuous dependent variable array, U, of length 2*NX+1=2*21+1=43. Also, as expected, COMMON/F/ contains one temporal derivative vector, UT, of length of 2*NX+1

```
      PARAMETER (NX=21)
      COMMON/T/          T,      NSTOP,      NORUN
     +      /Y/ U(2*NX+1)
     +      /F/UT(2*NX+1)
     +     /Y1/    U1(NX),     U2(NX),        U1A
     +     /F1/   U1T(NX),    U2T(NX),       U1AT
     +      /S/   U1X(NX)
     +      /C/         C1,         C2,          V,          X,
     +                EXP1,        ARG
     +      /I/       NPDE
```

Arrays U1(NX) and U2(NX) have been retained, but in COMMON/Y1/, arrays U1T(NX) and U2T(NX) have been put in COMMON/F1/.

```
      SUBROUTINE INITAL
      IMPLICIT DOUBLE PRECISION (A-H,O-Z)
      COMMON/T/          T,     NSTOP,     NORUN
C...
C...  V(0,T) = 1 - EXP(-C2*T)
C...
C...  EQUATION ORDERING
      IF((NORUN-1)*(NORUN-2)*(NORUN-3).EQ.0)CALL INIT1
C...
C...  GRID-POINT ORDERING
      IF((NORUN-4)*(NORUN-5)*(NORUN-6).EQ.0)CALL INIT2
      RETURN
      END

      SUBROUTINE DERV
      IMPLICIT DOUBLE PRECISION (A-H,O-Z)
      COMMON/T/          T,     NSTOP,     NORUN
C...
C...  V(0,T) = 1 - EXP(-C2*T)
C...
C...  EQUATION ORDERING
      IF((NORUN-1)*(NORUN-2)*(NORUN-3).EQ.0)CALL DERV1
C...
C...  GRID-POINT ORDERING
      IF((NORUN-4)*(NORUN-5)*(NORUN-6).EQ.0)CALL DERV2
      RETURN
      END

      SUBROUTINE PRINT(NI,NO)
      IMPLICIT DOUBLE PRECISION (A-H,O-Z)
      COMMON/T/          T,     NSTOP,     NORUN
C...
C...  V(0,T) = 1 - EXP(-C2*T)
C...
C...  EQUATION ORDERING
      IF((NORUN-1)*(NORUN-2)*(NORUN-3).EQ.0)CALL PRINT1(NI,NO)
C...
C...  GRID-POINT ORDERING
      IF((NORUN-4)*(NORUN-5)*(NORUN-6).EQ.0)CALL PRINT2(NI,NO)
      RETURN
      END
```

Program 4.7a Subroutines INITAL, DERV, and PRINT for Equations (3.11) to (3.15) with PDE and Grid-Point Ordering.

2. DO loop 2 has an additional integer counter, J, that is incremented by NPDE, the number of PDEs, which in this case is two for equations (3.11) and (3.12).

```
C...
C...  TRANSFER TO ARRAY U
      J=0
      DO 2 I=1,NX
         J=J+NPDE
```

```
      SUBROUTINE INIT2
      IMPLICIT DOUBLE PRECISION (A-H,O-Z)
      PARAMETER (NX=21)
      COMMON/T/          T,      NSTOP,      NORUN
     +      /Y/ U(2*NX+1)
     +      /F/UT(2*NX+1)
     +     /Y1/    U1(NX),    U2(NX),        U1A
     +     /F1/   U1T(NX),   U2T(NX),       U1AT
     +      /S/   U1X(NX)
     +      /C/         C1,         C2,          V,          X,
     +                EXP1,        ARG
     +      /I/       NPDE
C...
C...  SET THE PROBLEM PARAMETERS
         C1=0.560D+00
         C2=0.100D+00
          V=2.031D+00
          X=1.0D+00
C...
C...  NUMBER OF PDES
      NPDE=2
C...
C...  PRECOMPUTE SOME QUANTITIES USED IN THE ANALYTICAL SOLUTION
      EXP1=DEXP(-C1/V*X)
C...
C...  INITIAL CONDITIONS
      DO 1 I=1,NX
         U1(I)=0.0D+00
         U2(I)=0.0D+00
1     CONTINUE
      U1A=0.0D+00
C...
C...  TRANSFER TO ARRAY U
      J=0
      DO 2 I=1,NX
         J=J+NPDE
         U(J-1)=U1(I)
         U(J  )=U2(I)
2     CONTINUE
      U(2*NX+1)=U1A
C...
C...  INITIAL DERIVATIVES
      CALL DERV2
      RETURN
      END
```

Program 4.7b Subroutine INIT2 for Grid-Point Ordering of Initial Conditions (3.13) and (3.15).

```
         U(J-1)=U1(I)
         U(J  )=U2(I)
2     CONTINUE
      U(2*NX+1)=U1A
```

For the first pass through DO loop 2, I=1, and J=2. Then U(J-1)=U(1)=U1(1) and

U(J)=U(2)=U2(1). For the second pass through DO loop 2, I=2, J=4, and U(J-1)=U(3)= U1(2), U(J)=U(4)=U2(2). Thus, U(1), U(2), U(3), U(4),... contain U1(1), U2(1), U1(2), U2(2),... as required by the grid-point ordering. In general, to convert the initialization routine from PDE ordering to grid-point ordering, all that is required is the minor reorganization of the COMMON, and the addition of a DO loop at the end of the routine. In other words, the programming of the initial conditions can be done in the familiar PDE-ordering format, and the grid-point ordering is then implemented by the final DO loop.

DERV2 is similar to DERV1; the only changes are again, the rearrangement of the COMMON, the addition of a DO loop at the beginning for the transfer of array U to arrays U1 and U2, and the addition of a DO loop at the end for the transfer of arrays U1T and U2T to array UT, as indicated in Program 4.7c. We can note the following points about DERV2:

1. The COMMON/Y/ and /F/ are reorganized as in subroutine INIT2 or Program 4.7b.
2. DO loop 2 transfers array U to arrays U1 and U2, i.e., it performs the reverse operations of DO loop 2 in INIT2. This direction of transfer, from U to U1 and U2, is required because U comes into DERV2 through COMMON/Y/ from the ODE integrator, but U1 and U2 are required in the programming of the PDEs, equations (3.11) and (3.12).

```
C...
C...  TRANSFER TO ARRAYS U1 AND U2
      J=0
      DO 2 I=1,NX
         J=J+NPDE
         U1(I)=U(J-1)
         U2(I)=U(J)
2     CONTINUE
      U1A=U(2*NX+1)
```

3. DO loop 3 transfers arrays U1T and U2T to array UT. This direction of transfer, from U1T and U2T to UT, is required because UT is sent from DERV2 to the ODE integrator through COMMON/F/, but U1T and U2T are computed in the programming of the PDEs, equations (3.11) and (3.12).

In summary, all that is required in programming the grid-point ordering in the ODE derivative routine is to modify the routine in three places: (a) the COMMON/Y/ and /F/ areas have single arrays for the dependent variable and derivative vectors, respectively; (b) a DO loop is placed at the beginning of the routine to transfer the dependent variable vector in COMMON/Y/ to the arrays of dependent variables used in the programming of the PDEs; and (c) a DO loop is placed at the end of the routine to transfer the derivative vectors computed in the routine to the single derivative array in COMMON/F/.

Subroutine PRINT2 is the same as subroutine PRINT1 except a call to DERV2 is included to transfer the single, dependent variable array in COMMON/Y/ to the PDE dependent variable arrays for printing, e.g., U to U1 and U2. Because this is the operation done by the first DO loop added to DERV2 (DO loop 2 in Program 4.7c), this transfer need not be programmed in PRINT2; rather, the transfer can be effected by merely calling DERV2. The remainder of the programming is the same as for Programs 3.1a to f (i.e., subroutine MAP, function BESSI0, and the data file), so it is not repeated here.

The output from Programs 4.7a to d consists of the numerical solution to equations (3.11) and (3.12) (for boundary condition (3.32)). This output is substantial; thus

```
      SUBROUTINE DERV2
      IMPLICIT DOUBLE PRECISION (A-H,O-Z)
      PARAMETER (NX=21)
      COMMON/T/           T,      NSTOP,      NORUN
     +      /Y/ U(2*NX+1)
     +      /F/UT(2*NX+1)
     +     /Y1/    U1(NX),    U2(NX),       U1A
     +     /F1/   U1T(NX),   U2T(NX),      U1AT
     +      /S/   U1X(NX)
     +      /C/          C1,         C2,          V,          X,
     +                 EXP1,        ARG
     +      /I/        NPDE
C...
C...  TRANSFER TO ARRAYS U1 AND U2
      J=0
      DO 2 I=1,NX
         J=J+NPDE
         U1(I)=U(J-1)
         U2(I)=U(J)
2     CONTINUE
      U1A=U(2*NX+1)
C...
C...  BOUNDARY CONDITION
       U1(1)=1.0D+00-DEXP(-C2*T)
      U1T(1)=0.0D+00
C...
C...  U1X
C...
C...     TWO-POINT UPWIND
         IF(NORUN.EQ.4)CALL DSS012(0.D+0,X,NX,U1,U1X,1.0D+00)
C...
C...     FOUR-POINT BIASED UPWIND
         IF(NORUN.EQ.5)CALL DSS018(0.D+0,X,NX,U1,U1X,1.0D+00)
C...
C...     FIVE-POINT BIASED UPWIND
         IF(NORUN.EQ.6)CALL DSS020(0.D+0,X,NX,U1,U1X,1.0D+00)
C...
C...  PDES
      DO 1 I=1,NX
         IF(I.NE.1)THEN
            U1T(I)=C1*(U2(I)-U1(I))-V*U1X(I)
         END IF
            U2T(I)=C2*(U1(I)-U2(I))
1     CONTINUE
C...
C...  TEST FOR THE ELAPSED TIME
C...  T - X/V LT 0
      IF((T-X/V).LT.0.0D+00)THEN
       U1AT=0.0D+00
C...
C...  T - X/V GE 0
```

Program 4.7c Subroutine DERV2 for Grid-Point Ordering of PDEs (3.11) and (3.12). *Continued next page.*

```
      ELSE
         ARG=2.0D+00*DSQRT((C1*C2*X/V)*(T-X/V))
        U1AT=EXP1*C2*DEXP(-C2*(T-X/V))*BESSI0(ARG)
      END IF
C...
C...  TRANSFER TO ARRAY UT
      J=0
      DO 3 I=1,NX
         J=J+NPDE
         UT(J-1)=U1T(I)
         UT(J  )=U2T(I)
3     CONTINUE
      UT(2*NX+1)=U1AT
      RETURN
      END
```

Program 4.7c *Continued.*

we now discuss only selected portions. First, we consider the output for subroutine DSS012 (two-point upwind approximation of the spatial derivative in equation (3.11)), which is produced in runs 1 and 4.

We can note the following points about the output in Table 4.7a:

1. Whereas the Jacobian matrix has two outlying diagonals when PDE ordering is used (run 1), the Jacobian matrix has only a single, main diagonal band when grid-point ordering is used (run 4). To illustrate this point, the fifth row of each map (for the fifth ODE in the system of 42) is listed below.

```
DSS012, PDE ordering

       5      66                    4

DSS012, grid-point ordering

       5      6 64
```

   Thus, the bandwidth has been reduced from 23 (the horizontal distance from the 66 to the 4 in the fifth equation with PDE ordering) to four (the width of 6 64 for the fifth equation with grid-point ordering). Thus, the number of operations required for these two cases is in the ratio $23^2/4^2 = 33.1$ (since, as we noted earlier, the operations count for a banded matrix is proportional to the square of the bandwidth); clearly bandwidth reduction is worthwhile if an implicit, banded ODE solver is applied to a system of simultaneous PDEs. Also, we should keep in mind that this is a small problem (42 ODEs), and the advantage of bandwidth reduction will generally increase with increasing problem size.

2. The change from PDE ordering to grid-point ordering does not change the solu-

```
      SUBROUTINE PRINT2(NI,NO)
      IMPLICIT DOUBLE PRECISION (A-H,O-Z)
      PARAMETER (NX=21)
      COMMON/T/          T,      NSTOP,      NORUN
     +      /Y/ U(2*NX+1)
     +      /F/UT(2*NX+1)
     +     /Y1/    U1(NX),    U2(NX),        U1A
     +     /F1/   U1T(NX),   U2T(NX),       U1AT
     +      /S/   U1X(NX)
     +      /C/        C1,         C2,          V,         X,
     +               EXP1,        ARG
     +      /I/      NPDE
C...
C...  INITIALIZE A COUNTER FOR THE PRINTING AND PLOTTING
      DATA IP/0/
C...
C...  TRANSFER TO ARRAY U (BY A CALL TO DERV)
      CALL DERV
      IP=IP+1
      IF(IP.EQ.1)THEN
C...
C...  MAP THE ODE JACOBIAN MATRIX AT THE BEGINNING OF THE SOLUTION
      CALL MAP
C...
C...  PRINT A HEADING
      WRITE(NO,1)
1     FORMAT(//,14X,'T',8X,'U1(X,T)',3X,'U1(X,T) ANAL')
      END IF
C...
C...  PRINT THE SOLUTION
      WRITE(NO,2)T,U1(NX),U1A
2     FORMAT(F15.2,2F15.4)
C...
C...  WRITE NORUN, IP AND T ON THE SCREEN TO MONITOR THE PROGRESS OF
C...  THE CALCULATIONS
      WRITE(*,*)' NORUN = ',NORUN,' IP = ',IP,' T = ',T
C...
C...  MAP THE ODE JACOBIAN MATRIX AT THE END OF THE SOLUTION
      IF(IP.LT.11)RETURN
      CALL MAP
C...
C...  RESET THE INTEGER COUNTER FOR THE NEXT RUN
      IP=0
      RETURN
      END
```

Program 4.7d Subroutine PRINT2 to Print and Map the Jacobian Matrix of Equations (3.11) and (3.12) with Grid-Point Ordering of the ODEs.

Table 4.7 (a) Abbreviated Output of Programs 4.7a to d from DSS012. *Continued next pages.*

```
RUN NO. -    1  BATEMAN TEST PROBLEM - DSS012 - EQUATION ORDERING

INITIAL T -  0.000D+00

  FINAL T -  0.100D+02

  PRINT T -  0.100D+01

NUMBER OF DIFFERENTIAL EQUATIONS -  43

MAXIMUM INTEGRATION ERROR -  0.100D-05


DEPENDENT VARIABLE COLUMN INDEX J (FOR YJ) IS PRINTED HORIZONTALLY

DERIVATIVE ROW INDEX I (FOR DYI/DT = FI(Y1,Y2,...,YJ,...,YN) IS PRINTED
VERTICALLY

JACOBIAN MATRIX ELEMENT IN THE MAP WITH INDICES I,J IS FOR PFI/PYJ
WHERE P DENOTES A PARTIAL DERIVATIVE


                111111111122222222223333333333444
       123456789012345678901234567890123456789012
   1
   2    6                    4
   3    66                    4
   4     66                    4
   5      66                    4
   6       66                    4
   7        66                    4
   8         66                    4
   9          66                    4
  10           66                    4
  11            66                    4
  12             66                    4
  13              66                    4
  14               66                    4
  15                66                    4
  16                 66                    4
  17                  66                    4
  18                   66                    4
  19                    66                    4
  20                     66                    4
  21                      66                    4
  22                            3
  23    3                    3
  24     3                    3
  25      3                    3
  26       3                    3
  27        3                    3
  28         3                    3
```

Table 4.7 (a) *Continued.*

```
     29          3                    3
     30           3                    3
     31            3                    3
     32             3                    3
     33              3                    3
     34               3                    3
     35                3                    3
     36                 3                    3
     37                  3                    3
     38                   3                    3
     39                    3                    3
     40                     3                    3
     41                      3                    3
     42                       3                    3


          T        U1(X,T)   U1(X,T) ANAL
        0.00        0.0000         0.0000
        1.00        0.0383         0.0379
        2.00        0.1089         0.1085
        3.00        0.1745         0.1741
        4.00        0.2355         0.2350
        5.00        0.2921         0.2916
        6.00        0.3447         0.3442
        7.00        0.3934         0.3931
        8.00        0.4387         0.4384
        9.00        0.4807         0.4804
       10.00        0.5197         0.5194
                      .
                      .
                      .


RUN NO. -   4  BATEMAN TEST PROBLEM - DSS012 - GRID-POINT ORDERING

INITIAL T -  0.000D+00

  FINAL T -  0.100D+02

  PRINT T -  0.100D+01

NUMBER OF DIFFERENTIAL EQUATIONS -  43

MAXIMUM INTEGRATION ERROR -  0.100D-05


DEPENDENT VARIABLE COLUMN INDEX J (FOR YJ) IS PRINTED HORIZONTALLY

DERIVATIVE ROW INDEX I (FOR DYI/DT = FI(Y1,Y2,...,YJ,...,YN) IS PRINTED
VERTICALLY

JACOBIAN MATRIX ELEMENT IN THE MAP WITH INDICES I,J IS FOR PFI/PYJ
WHERE P DENOTES A PARTIAL DERIVATIVE
```

Table 4.7 (a) *Continued.*

```
                111111111122222222223333333333444
       123456789012345678901234567890123456789012
   1
   2    3
   3     64
   4     33
   5     6 64
   6       33
   7       6 64
   8         33
   9         6 64
  10           33
  11           6 64
  12             33
  13             6 64
  14               33
  15               6 64
  16                 33
  17                 6 64
  18                   33
  19                   6 64
  20                     33
  21                     6 64
  22                       33
  23                       6 64
  24                         33
  25                         6 64
  26                           33
  27                           6 64
  28                             33
  29                             6 64
  30                               33
  31                               6 64
  32                                 33
  33                                 6 64
  34                                   33
  35                                   6 64
  36                                     33
  37                                     6 64
  38                                       33
  39                                       6 64
  40                                         33
  41                                         6 64
  42                                           33
```

T	U1(X,T)	U1(X,T) ANAL
0.00	0.0000	0.0000
1.00	0.0383	0.0379
2.00	0.1089	0.1085
3.00	0.1745	0.1741

Table 4.7 (a) *Continued.*

```
   4.00        0.2355        0.2350
   5.00        0.2921        0.2916
   6.00        0.3447        0.3442
   7.00        0.3934        0.3931
   8.00        0.4387        0.4384
   9.00        0.4807        0.4804
  10.00        0.5197        0.5194
```

tion, which can be confirmed by comparing the two solutions in Table 4.7a. For example the solutions at $t = 10$ are

```
DSS012, PDE ordering

       10.00        0.5197        0.5194

DSS012, grid point ordering

       10.00        0.5197        0.5194
```

Finally, we consider the output for subroutine DSS020 (five-point biased upwind approximation of the spatial derivative in equation (3.11)), which is produced in runs 3 and 6. We can note the following points about the output in Table 4.7b:

1. The Jacobian matrix again has two outlying diagonals when PDE ordering is used (run 3), but the Jacobian matrix has only a single, main diagonal band when grid-point ordering is used (run 6). The fifth row of each map (for the fifth ODE in the system of 42) is listed below.

```
DSS020, PDE ordering

     5   56666                  4

DSS020, grid-point ordering

     5    6 446 5
```

The effect of the five-point biased upwind approximation of the spatial derivative in equation (3.11) is clear (the main diagonal band is pentadiagonal). The bandwidth has been reduced from 25 (the horizontal distance from the 56666 to the 4 in the fifth equation with PDE ordering) to seven (the width of 6 446 5 for the fifth equation with grid-point ordering). Thus, the number of operations required for these two cases is in the ratio $25^2/7^2 = 12.8$, which is substantially less than for DSS012 (but recall that DSS020 gives considerably better approximations of the PDE spatial derivative than DSS012, as reflected in the better agreement between the numerical and analytical solutions of equations (3.11) to (3.15) listed in Tables 4.7a and b). Also, we should again keep in mind that this is a small problem (42 ODEs), and the advantage of bandwidth reduction will generally increase with increasing problem size.

Table 4.7 (b) Abbreviated Output of Programs 4.7a to d from DSS020. *Continued next pages.*

```
RUN NO. -   3  BATEMAN TEST PROBLEM - DSS020 - EQUATION ORDERING

INITIAL T -  0.000D+00

  FINAL T -  0.100D+02

  PRINT T -  0.100D+01

NUMBER OF DIFFERENTIAL EQUATIONS -  43

MAXIMUM INTEGRATION ERROR -  0.100D-05


DEPENDENT VARIABLE COLUMN INDEX J (FOR YJ) IS PRINTED HORIZONTALLY

DERIVATIVE ROW INDEX I (FOR DYI/DT = FI(Y1,Y2,...,YJ,...,YN) IS PRINTED
VERTICALLY

JACOBIAN MATRIX ELEMENT IN THE MAP WITH INDICES I,J IS FOR PFI/PYJ
WHERE P DENOTES A PARTIAL DERIVATIVE


                  1111111111222222222233333333334444
         1234567890123456789012345678901234567890123
     1
     2    6665                 4
     3    6465                  4
     4    6666                   4
     5    56666                   4
     6     56666                   4
     7      56666                   4
     8       56666                   4
     9        56666                   4
    10         56666                   4
    11          56666                   4
    12           56666                   4
    13            56666                   4
    14             56666                   4
    15              56666                   4
    16               56666                   4
    17                56666                   4
    18                 56666                   4
    19                  56666                   4
    20                   56666                   4
    21                   66776                    4
    22                        3
    23    3                    3
    24     3                    3
    25      3                    3
    26       3                    3
    27        3                    3
    28         3                    3
```

Table 4.7 (b) *Continued.*

```
     29          3                   3
     30           3                   3
     31            3                   3
     32             3                   3
     33              3                   3
     34               3                   3
     35                3                   3
     36                 3                   3
     37                  3                   3
     38                   3                   3
     39                    3                   3
     40                     3                   3
     41                      3                   3
     42                       3                   3


            T        U1(X,T)   U1(X,T) ANAL
          0.00        0.0000         0.0000
          1.00        0.0378         0.0378
          2.00        0.1084         0.1084
          3.00        0.1740         0.1740
          4.00        0.2349         0.2349
          5.00        0.2916         0.2916
          6.00        0.3442         0.3442
          7.00        0.3930         0.3930
          8.00        0.4383         0.4383
          9.00        0.4803         0.4803
         10.00        0.5193         0.5193
                        .
                        .
                        .

RUN NO. -   6  BATEMAN TEST PROBLEM - DSS020 - GRID POINT ORDERING

INITIAL T -  0.000D+00

  FINAL T -  0.100D+02

  PRINT T -  0.100D+01

NUMBER OF DIFFERENTIAL EQUATIONS -  43

MAXIMUM INTEGRATION ERROR -  0.100D-05


DEPENDENT VARIABLE COLUMN INDEX J (FOR YJ) IS PRINTED HORIZONTALLY

DERIVATIVE ROW INDEX I (FOR DYI/DT = FI(Y1,Y2,...,YJ,...,YN) IS PRINTED
VERTICALLY

JACOBIAN MATRIX ELEMENT IN THE MAP WITH INDICES I,J IS FOR PFI/PYJ
WHERE P DENOTES A PARTIAL DERIVATIVE
```

Table 4.7 (b) *Continued.*

```
                111111111122222222223333333333444
       123456789012345678901234567890123456789012
  1
  2     3
  3      646 6 5
  4      33
  5      6 446 5
  6        33
  7      6 6 646
  8          33
  9      5 6 6 646
 10            33
 11        5 6 6 646
 12              33
 13          5 6 6 646
 14                33
 15            5 6 6 646
 16                  33
 17              5 6 6 646
 18                    33
 19                5 6 6 646
 20                      33
 21                  5 6 6 646
 22                        33
 23                    5 6 6 646
 24                          33
 25                      5 6 6 646
 26                            33
 27                        5 6 6 646
 28                              33
 29                          5 6 6 646
 30                                33
 31                            5 6 6 646
 32                                  33
 33                              5 6 6 646
 34                                    33
 35                                5 6 6 646
 36                                      33
 37                                  5 6 6 646
 38                                        33
 39                                    5 6 6 646
 40                                          33
 41                                    6 6 7 7 64
 42                                            33

          T        U1(X,T)   U1(X,T) ANAL
       0.00         0.0000         0.0000
       1.00         0.0378         0.0378
       2.00         0.1084         0.1084
       3.00         0.1740         0.1740
       4.00         0.2349         0.2349
```

Table 4.7 (b) *Continued.*

```
       5.00          0.2916          0.2916
       6.00          0.3442          0.3442
       7.00          0.3930          0.3930
       8.00          0.4383          0.4383
       9.00          0.4803          0.4803
      10.00          0.5193          0.5193
```

2. As before, the change from PDE ordering to grid-point ordering does not change the solution, which can be confirmed by comparing the two solutions in Table 4.7b. For example the solutions at $t = 10$ are

```
DSS020, PDE ordering

          10.00          0.5193          0.5193

DSS020, grid point ordering

          10.00          0.5193          0.5193
```

These results suggest that bandwidth reduction should be considered for applications of simultaneous PDEs, when the approximating ODEs are to be integrated with an implicit banded solver. The reordering by grid point is straightforward and does not affect the solution, but it can lead to substantial reductions in the computation for the ODE solution. In fact, bandwidth reduction is used routinely as an option in large finite element codes. We, in fact, investigated the use of bandwidth reduction algorithms which are used in finite element codes to automate this process (rather than having the user rearrange the `COMMON` and program the `DO` loops as in Programs 4.7b and c). Our conclusion is that these algorithms do not perform any better than the procedure we have described. We did not, however, determine if this conclusion is valid for large, multidimensional PDE problems.

4.7 Weighted Residual Methods

To this point, we have considered essentially only finite difference methods for approximating the spatial (boundary value) derivatives in PDEs. An essential requirement in this approach is the assessment of the error in the numerical solution which results from using the finite difference approximations. This error analysis is usually done in a rather indirect way, for example, by increasing the number of grid points, which is termed *h-refinement*, or by changing the order of the approximation, which is termed *p-refinement*, and observing the effect on the numerical solution. We have used both approaches in computing the solutions to the preceding problems in PDEs. For example, the number of grid points was varied in the solution of equation (4.75) through the values $n = 11$, 21, 41, 81, and 161, and we observed convergence to the analytical solution (see Table 4.5a). Also, we varied the order of the approximation of the spatial derivative in equation (3.11) by using subroutines `DSS012` (first order), `DSS018` (fourth order), and `DSS020` (fifth order), and observed the convergence to the analytical solution.

If the process of h- and/or p-refinement is continued until we do not observe any further change in the numerical solution beyond a certain number of figures, we then assume that the solution is accurate to that number of figures. These procedures are based on the assumption that h- and p-refinement converge to the correct solution. Also, the process of changing the number of grid points and the order of the approximation can be automated within the programming. We did this only to the extent of having the programs execute a series of runs in which either the number of grid points or the order of the approximation was changed from run to run, so we could observe the changes in the solutions for a series of runs.

An alternate approach is to operate more directly on the error resulting from the approximations in the numerical methods applied to a PDE problem system. For example, the method of weighted residuals, as the name implies, is a procedure for minimizing the residuals resulting from approximate numerical solutions of the problem system equations. We now illustrate the method of weighted residuals through a one-dimensional parabolic problem, Fourier's or Fick's second law

$$\frac{\partial u}{\partial t} = D\frac{\partial^2 u}{\partial x^2} \tag{4.99}$$

Equation (4.99) is just equation (4.87) with the usual convention of using u for the dependent variable. Since equation (4.99) is second order in x and first order in t, we require two boundary conditions and one initial condition which we take as

$$u(x_0, t) = 0, \qquad \frac{\partial u(x_L, t)}{\partial t} = 0 \tag{4.100)(4.101}$$

$$u(x, 0) = 1 \tag{4.102}$$

where x_0 and x_L are the boundary values of x.

We now assume a solution to equations (4.99) to (4.102) of the form

$$u(x,t) \approx \sum_{i=1}^{N} c_i(t)\phi_i(x) \tag{4.103}$$

Equation (4.103) is called a *separated solution* because it is a sum of a product of two functions, $c_i(t)$ which is a function of t only, and $\phi_i(x)$ which is a function of x only; in other words, x and t are separated between these two functions. This particular form for the trial solution is often used because of the ease with which it can be manipulated mathematically, but it is not the only possibility. In fact, any trial solution which the analyst considers a good candidate can be used; that is, it will likely produce a final solution of reasonable accuracy with reasonable effort.

One of the major advantages of a separated solution is the ease with which the derivatives in the PDE(s) can be evaluated. For example, we have from equation (4.103)

$$u_t(x,t) \approx \sum_{i=1}^{N} c_i'(t)\phi_i(x) \tag{4.104}$$

$$u_{xx}(x,t) \approx \sum_{i=1}^{N} c_i(t)\phi_i''(x) \tag{4.105}$$

where ($'$) denotes differentiation (there is no confusion about this differentiation since $c_i(t)$ and $\phi_i(x)$ are functions of only one variable). If equations (4.104) and (4.105) are substituted in equation (4.99) we have

$$\sum_{i=1}^{N} c_i'(t)\phi_i(x) \approx D\sum_{i=1}^{N} c_i(t)\phi_i''(x) \tag{4.106}$$

Equation (4.103) will probably, at best, be only an approximate solution, and therefore equation (4.99) will not be satisfied exactly. Thus we arrange equation (4.106) as

$$\sum_{i=1}^{N} c_i'(t)\phi_i(x) - D\sum_{i=1}^{N} c_i(t)\phi_i''(x) = R(x,t) \tag{4.107}$$

where $R(x,t)$ is a residual that ideally is zero for all x and t (if we could find the exact solution to equation (4.99)). Because equation (4.103) (or (4.106)) will only be approximately correct, we attempt to minimize $R(x,t)$ in equation (4.107) in some fashion. Generally, this is done by forming the integral

$$\int_{x_0}^{x_L} w(x)R(x,t)\,dx = 0 \tag{4.108}$$

where $w(x)$ is a *weighting function* selected by the analyst. Thus the use of equation (4.108) is called the *method of weighted residuals* (as suggested by the integrand in equation (4.108)). In other words, we are making the residual, $R(x,t)$, orthogonal to the weighting function, $w(t)$.

In order to obtain $c_i'(t)$, $i = 1, 2, \ldots, N$, we multiply equation (4.107) by $\phi_j(x)$ and integrate from x_0 to x_L

$$\int_{x_0}^{x_L} \phi_j(x)\sum_{i=1}^{N} c_i'(t)\phi_i(x)\,dx - D\int_{x_0}^{x_L} \phi_j(x)\sum_{i=1}^{N} c_i(t)\phi_i''(x)\,dx = \int_{x_0}^{x_L} \phi_j(x)R(x,t)\,dx = 0$$

or

$$\sum_{i=1}^{N} c_i'(t)\int_{x_0}^{x_L} \phi_j(x)\phi_i(x)\,dx - D\sum_{i=1}^{N} c_i(t)\int_{x_0}^{x_L} \phi_j(x)\phi_i''(x)\,dx = \int_{x_0}^{x_L} \phi_j(x)R(x,t)\,dx = 0 \tag{4.109}$$

where we have interchanged the order of integration and summation. The right-hand integral is just a statement of *Galerkin's method*; that is, we make the residual, $R(x,t)$, orthogonal to the *basis functions*, $\phi_i(x,t)$.

In order to evaluate the integrals of equation (4.109), we must select the basis (*shape*) functions, $\phi_i(x)$. Many different basis functions have been used in the weighted residual method, and we consider only one possibility. Specifically, we select

$$\phi_i(x) = \sin(\lambda_i x), \quad \lambda_i = \pi(i - 1/2), \qquad i = 1, 2, \ldots \tag{4.110}$$

and we have also taken $x_0 = 0$, $x_L = 1$ (only for convenience; the interval in x can remain general). This choice of the $\phi_i(x)$ was made because they satisfy boundary conditions (4.100) and (4.101), and as we will see, they also facilitate the evaluation of the $c_i(t)$ in equation (4.103).

Substitution of the $\phi_i(x)$ of equation (4.110) in equation (4.109) gives

$$\sum_{i=1}^{N} c_i'(t) \int_0^1 \sin(\lambda_j x)\sin(\lambda_i x)\,dx + D\sum_{i=1}^{N} c_i(t) \int_0^1 \lambda_i^2 \sin(\lambda_j x)\sin(\lambda_i x)\,dx$$

$$= \int_0^1 \sin(\lambda_j x)R(x,t)\,dx = 0 \qquad (4.111)$$

In order for the first line of equation (4.111) to equal zero, we require

$$\sum_{i=1}^{N} \{c_i'(t) + D\lambda_i^2 c_i(t)\} \int_0^1 \sin(\lambda_j x)\sin(\lambda_i x)\,dx = 0 \qquad (4.112)$$

The integral in equation (4.112) has the value

$$\int_0^1 \sin(\lambda_j x)\sin(\lambda_i x)\,dx = 0, \qquad i \neq j \qquad (4.113)$$

due to the orthogonality of the sine functions. Thus, only one term in the sum of equation (4.112) remains for $i = j$, and we have from equation (4.112)

$$\{c_i'(t) + D\lambda_i^2 c_i(t)\} \int_0^1 \sin^2(\lambda_i x) = 0 \qquad (4.114)$$

Because the integral is clearly not zero (note that the integrand is positive), we require that

$$c_i'(t) + D\lambda_i^2 c_i(t) = 0, \qquad i = 1, 2, \ldots \qquad (4.115)$$

Thus, we have reduced the original PDE problem, equation (4.99), to a system of approximating ODEs, equations (4.115). This corresponds closely to what we have done previously in the numerical method of lines, and in fact, this development starting with equation (4.103) is an application of the *analytical method of lines* (it is also known as *separation of variables*). A central requirement in the analytical method of lines is the analytical solution of the approximating ODEs. In this case, we have as the solution to equations (4.115)

$$c_i(t) = a_i e^{-D(\lambda_i)^2 t} \qquad (4.116)$$

where a_i, $i = 1, 2, \ldots$ are constants to be determined.

The solution to this point is (from equations (4.103), (4.110), and (4.116))

$$u(x,t) \approx \sum_{i=1}^{N} a_i e^{-D\lambda_i{}^2 t} \sin(\lambda_i x) \qquad (4.117)$$

$u(x,t)$ of equation (4.117) satisfies the original PDE, equation (4.99) (note that it makes the residual, $R(x,t)$, in equation (4.107) zero for all x and t). It also satisfies the boundary conditions, equations (4.100) and (4.101). Therefore, only the initial condition (4.102) remains for consideration. Equation (4.117) for $t = 0$ becomes

$$u(x,0) = 1 \approx \sum_{i=1}^{N} a_i \sin(\lambda_i x) \qquad (4.118)$$

Equation (4.118) defines a celebrated problem. One indication that it may require some unusual methods of analysis to obtain the coefficients a_i is the fact that the left-hand side is a constant, 1, but the right-hand side is a function of x, and therefore presumably will change value with changing x. The resolution of this apparent dilemma is attributed to Fourier, and the right-hand side of equation (4.118) is a *Fourier sine series* if we take $N = \infty$

$$u(x,0) = 1 = \sum_{i=1}^{\infty} a_i \sin(\lambda_i x) \tag{4.119}$$

The problem then is to determine the Fourier coefficients, a_i, $i = 1, 2, \ldots$. If equation (4.119) is multiplied by $\sin(\lambda_j x)$

$$\int_0^1 1 \sin(\lambda_j x)\, dx = \int_0^1 \sin(\lambda_j x) \sum_{i=0}^{\infty} a_n \sin(\lambda_i x)\, dx \tag{4.120}$$

Interchanging the orders of integration and summation in the right-hand side of equation (4.120), and using the orthogonality property of the sine functions, equation (4.113) (so only the term in the series corresponding to $i = j$ remains)

$$\int_0^1 \sin^2(\lambda_i x)\, dx = \int_0^1 (1/2)\,\{1 - \cos(2\lambda_i x)\}\, dx$$

$$= (1/2)\left\{x - \left(\frac{1}{2\lambda_i}\right)\sin(2\lambda_i x)\right\}\Bigg|_0^1 = 1/2 \tag{4.121}$$

The left-hand side integral of equation (4.120) is (with $\lambda_i = \lambda_j$)

$$\int_0^1 1 \sin(\lambda_i x)\, dx = -\left(\frac{1}{\lambda_i}\right)\cos(\lambda_i x)\Big|_0^1 = \left(\frac{1}{\lambda_i}\right) \tag{4.122}$$

Thus, from equations (4.120), (4.121), and (4.122),

$$a_i = \left(\frac{2}{\lambda_i}\right), \qquad i = 1, 2, \ldots \tag{4.123}$$

which are the *Fourier sine coeffcients* for 1 with λ_i given by equation (4.110). The final solution to equations (4.99) to (4.102) is therefore from equations (4.117) and (4.123)

$$u(x,t) = \sum_{i=1}^{\infty} (2/\lambda_i) e^{-D\lambda_i{}^2 t} \sin(\lambda_i x), \qquad \lambda_i = \pi(i - 1/2), \qquad i = 1, 2, \ldots \tag{4.124}$$

Equation (4.124) is an example of the application of the analytical method of lines (and also, separation of variables). We selected a PDE problem, equations (4.99) to (4.102), for which this method of solution is straightforward; in general, it is not this easy. Some indication of the difficulties that might develop is given by (a) the requirement to evaluate the integrals resulting from equation (4.108) (the definition of the weighted residual method); (b) the requirement to solve the approximating ODEs, in this case, equations (4.115); and (c) the requirement to impose the initial condition for the approximating ODEs, in this case, equation (4.102). Often these requirements are essentially insurmountable if attempted analytically, and typically the strategy is to switch to a numerical procedure. In other words, the weighted residual methods we are considering are semi-analytical in the sense that analytical methods are carried as far as practical, and then numerical methods are used to complete the solution.

For example, the integrals from equation (4.108) can be evaluated by *numerical quadrature*. The approximating ODEs can be integrated numerically, and the initial conditions for these ODEs can be applied as part of the numerical integration. The ease with which these various numerical steps can be executed is determined by:

(a) The selection of the weighting functions, $w(x)$, in equation (4.108). For example, if delta functions are used, which makes the integration in equation (4.108) particularly easy, the resulting weighted residual method is called *collocation*. If the basis functions are used as the weighting functions, the resulting weighted residual method is called Galerkin's method.

(b) The selection of the basis functions, $\phi_i(x)$, in equation (4.103). These are often selected to identically satisfy the boundary conditions, as was done in the case of equation (4.110). Another possibility is to select a set of weighting functions which have only local support. That is, they are nonzero over only a relatively small interval in x and are zero everywhere else. This is the basis of the finite element method; the finite elements are the basis functions which are nonzero over relatively small intervals. Such a choice of basis functions facilitates the spatial integrations of equation (4.108), the integration of the approximating ODEs, and the imposition of the ODE initial conditions.

We will consider some of these aspects of the method of weighted residuals in the form of the finite element method in a subsequent example. For now, we return to equations (4.99) to (4.102), and perform a numerical method of lines solution for comparison with the analytical method of lines solution of equation (4.124). Equations (4.99) to (4.102) are of modest complexity compared to some of the PDE problems we have considered previously, and therefore their method of lines solution is straightforward. Here we consider two approaches, one by the repeated use of `DSS004` to compute $\partial^2 u/\partial x^2$, and a second in which this derivative is computed directly from using `DSS044`.

Subroutine `INITAL` is a straightforward implementation of equation (4.102) on a 21-point grid, as indicated in Program 4.8a. The length X_L is set to one, and then initial condition (4.102) is programmed in `DO` loop 1.

`DERV` calls `DERV1` for the first case (solution by `DSS004`) and `DERV2` for the second case (solution by `DSS044`), as indicated in Program 4.8b. In `DERV1` and `DERV2`, boundary conditions (4.100) and (4.101) are first programmed. Then the temporal derivatives of equation (4.99) are computed in `DO` loop 1.

Subroutine `PRINT` is listed in Program 4.8c. Subroutine `PRINT` calls function `UANAL` for the calculation of the analytical solution. Then the numerical and analytical solutions are printed for comparison. Note that 200 terms are used in the series solution of equation (4.124) in function `UANAL`.

The data file is listed in Program 4.8d. The integration of the 21 ODEs is by `RKF45` to an accuracy of 0.00001.

Abbreviated output from Programs 4.8a to d is listed in Table 4.8. We can note the following points about the output in Table 4.8:

1. Generally, the solutions agree to within four figures. In fact, the only time when this is not the case is for $t = 0$, when the analytical solution, equation (4.124), is

```
      SUBROUTINE INITAL
      IMPLICIT DOUBLE PRECISION (A-H,O-Z)
      PARAMETER (NX=21)
      COMMON/T/        T,  NSTOP,  NORUN
     1      /Y/  U(NX)
     2      /F/  UT(NX)
     3      /S/  UX(NX),UXX(NX)
     4      /C/      XL,      IP
C...
C...  TOTAL LENGTH
      XL=1.0D0
C...
C...  INITIAL CONDITION
      DO 1 I=1,NX
         U(I)=1.0D0
1     CONTINUE
      RETURN
      END
```

Program 4.8a Subroutine `INITAL` for Initial Condition (4.102).

in error because the series converges slowly. That is, for $t = 0$, equation (4.124) is

$$u(x,0) = \sum_{i=1}^{\infty}(2/\lambda_i)\sin(\lambda_i x), \quad \lambda_i = \pi(i-1/2), \quad i = 1,2,\ldots \tag{4.125}$$

and the individual terms in the series decrease as $(2/\lambda_i)$. However, for $t > 0$, the series converges more rapidly because of the exponential term

$$e^{-D\lambda_i{}^2 t}$$

which decreases rapidly with increasing i. Also, the analytical solution might be a source of error at $x = 0$, $t > 0$ since a *discontinuity (unit jump)* between initial (4.102) and boundary condition (4.100) occurs at $t = 0$. This discontinuity can produce oscillation in the analytical solution which is called the *Gibbs' phenomenon*. This oscillation will occur no matter how many terms are used in the series solution of equation (4.124).

2. The successive or stagewise differentiation to compute $\partial^2 u/\partial x^2$ in equation (4.99) by two calls to `DSS004` (in subroutine `DERV1`) produces a Jacobian matrix with a bandwidth approximately twice that when $\partial^2 u/\partial x^2$ is calculated directly by a call to `DSS044`. This bandwidth expansion is one of the disadvantages of stagewise differentiation. The other disadvantage is that the procedure may lead to inaccurate solutions for PDEs which are strongly first-order hyperbolic such as Burgers' equation (4.41) for small μ/ρ (i.e., PDEs for which convection is substantially greater than diffusion). We suggest some caution and careful evaluation of the solution in this case (although, clearly stagewise differentiation worked well when applied to Burgers' equation with successive first-order differentiations by subroutine `DSS038` in Program 4.2c). We can recommend the use of stagewise differentiation for PDEs with a significant parabolic (and also elliptic) characteristic. Stagewise differentiation is particularly useful for nonlinear, second-order (in space) PDEs; for example, the second-order (parabolic) term $\partial(D(u)\partial u/\partial x)/\partial x$

```
      SUBROUTINE DERV
      IMPLICIT DOUBLE PRECISION (A-H,O-Z)
      PARAMETER (NX=21)
      COMMON/T/      T,  NSTOP,  NORUN
C...
C...  STAGEWISE DIFFERENTIATION
      IF(NORUN.EQ.1)CALL DERV1
C...
C...  DIRECT DIFFERENTIATION
      IF(NORUN.EQ.2)CALL DERV2
      RETURN
      END

      SUBROUTINE DERV1
      IMPLICIT DOUBLE PRECISION (A-H,O-Z)
      PARAMETER (NX=21)
      COMMON/T/      T,  NSTOP,  NORUN
     1      /Y/  U(NX)
     2      /F/ UT(NX)
     3      /S/ UX(NX),UXX(NX)
     4      /C/     XL,     IP
C...
C...  BOUNDARY CONDITION AT X = 0
      U(1)=0.0D0
C...
C...  UX
      CALL DSS004(0.0D0,XL,NX,U,UX)
C...
C...  BOUNDARY CONDITION AT X = XL
      UX(NX)=0.0D0
C...
C...  UXX
      CALL DSS004(0.0D0,XL,NX,UX,UXX)
C...
C...  PDE
      DO 1 I=1,NX
         UT(I)=UXX(I)
1     CONTINUE
      RETURN
      END

      SUBROUTINE DERV2
      IMPLICIT DOUBLE PRECISION (A-H,O-Z)
      PARAMETER (NX=21)
      COMMON/T/      T,  NSTOP,  NORUN
     1      /Y/  U(NX)
     2      /F/ UT(NX)
     3      /S/ UX(NX),UXX(NX)
     4      /C/     XL,     IP
C...
C...  BOUNDARY CONDITION AT X = 0
```

Program 4.8b Subroutines DERV, DERV1, and DERV2 for the Solution of Equations (4.99), (4.100), and (4.101). *Continued next page.*

```
      NL=1
      U(1)=0.0D0
C...
C...  BOUNDARY CONDITION AT X = XL
      NU=2
      UX(NX)=0.0D0
C...
C...  UXX
      CALL DSS044(0.0D0,XL,NX,U,UX,UXX,NL,NU)
C...
C...  PDE
      DO 1 I=1,NX
         UT(I)=UXX(I)
1     CONTINUE
      RETURN
      END
```

Program 4.8b *Continued.*

can be computed by a call to DSS004 to obtain $\partial u/\partial x$ from u. Then the product $D(u)\partial u/\partial x$ can be formed. Finally, a second call to DSS004 applied to this product gives $\partial(D(u)\partial u/\partial x)/\partial x$. The alternative would be to program this nonlinear term explicitly as was done in equation (4.54), for example.

We conclude this example (equations (4.99) to (4.102)) by considering a technique to facilitate the use of the analytical method of lines. If a *finite sine transform* of $f(x)$ is defined as

$$\bar{f}(\lambda_i) = \int_0^{x_L} f(x)\sin(\lambda_i x)\,dx \tag{4.126}$$

then the *inverse transform* is

$$f(x) = (2/x_L)\sum_{i=1}^{\infty}\bar{f}(\lambda_i)\sin(\lambda_i x) \tag{4.127}$$

where $\lambda_i = \pi(i-1/2)/x_L$, $i = 1, 2, \ldots$. We recognize equation (4.127) as just a Fourier sine series with Fourier coefficients $\bar{f}(\lambda_i)$ given by equation (4.126).

We can now follow the usual procedure of tabulating a *table of transforms* for various functions of x:

$f(x)$	$\bar{f}(\lambda_i)$
$af(x)$	$a\bar{f}(\lambda_i)$
$af_1(x) + bf_2(x)$	$a\bar{f}_1(\lambda_i) + b\bar{f}_2(\lambda_i)$
1	$1/\lambda_i$
$d^2f(x)/dx^2$	$(-1)^{i+1}df(x_L)/dx + \lambda_i f(0) - \lambda_i^2\bar{f}(\lambda_i)$

where a and b are constants. These *transform pairs* follow directly from the definition

```
      SUBROUTINE PRINT(NI,NO)
      IMPLICIT DOUBLE PRECISION (A-H,O-Z)
      PARAMETER (NX=21)
      COMMON/T/      T,  NSTOP,  NORUN
     1      /Y/  U(NX)
     2      /F/ UT(NX)
     3      /S/ UX(NX),UXX(NX)
     4      /C/      XL,      IP
      DIMENSION  XA(6),  UA(6)
C...
C...  INITIALIZE A COUNTER FOR THE PRINTING AND PLOTTING
      DATA IP/0/
      IP=IP+1
      IF(IP.EQ.1)THEN
C...
C...  MAP THE ODE JACOBIAN MATRIX AT THE BEGINNING OF THE SOLUTION
      CALL MAP
      END IF
C...
C...  MONITOR THE CALCULATIONS
      WRITE(*,*)NORUN,T
C...
C...  COMPUTE THE ANALYTICAL SOLUTION AT X = 0, 0.2*XL, 0.4*XL, 0.6*XL,
C...  0.8*XL, XL
      XA(1)=0.D0
      XA(2)=0.2D0*XL
      XA(3)=0.4D0*XL
      XA(4)=0.6D0*XL
      XA(5)=0.8D0*XL
      XA(6)=XL
      DO 1 I=1,6
         UA(I)=UANAL(XA(I),T,XL)
1     CONTINUE
C...
C...  PRINT THE NUMERICAL AND ANALYTICAL SOLUTIONS AT X = 0, 0.2*XL,
C...  0.4*XL, 0.6*XL, 0.8*XL, XL
      IF(T.EQ.0.)WRITE(NO,4)
4     FORMAT(10X,'T',8X,'X=0',3X,'X=0.2*XL',3X,'X=0.4*XL',
     1                        3X,'X=0.6*XL',3X,'X=0.8*XL',
     2                        7X,'X=XL')
      WRITE(NO,2)T,(U(I),I=1,NX,4)
2     FORMAT(F11.2,/,'  NUMERICAL',6F11.4)
      WRITE(NO,3)(UA(I),I=1,6)
3     FORMAT(' ANALYTICAL',6F11.4,//)
C...
C...  MAP THE ODE JACOBIAN MATRIX AT THE END OF THE SOLUTION
      IF(IP.LT.11)RETURN
      CALL MAP
C...
C...  RESET THE INTEGER COUNTER FOR THE NEXT RUN
```

Program 4.8c Subroutine PRINT and Function UANAL to Print the Numerical and Analytical Solutions to the Solution of Equations (4.99), and (4.100) to (4.102). *Continued next page.*

```
      IP=0
      RETURN
      END

      DOUBLE PRECISION FUNCTION UANAL(X,T,XL)
      IMPLICIT DOUBLE PRECISION (A-H,O-Z)
C...
C...  SUM 200 TERMS IN THE SERIES SOLUTION.  THIS LARGE NUMBER WAS
C...  SELECTED BECAUSE OF THE SLOW CONVERGENCE OF THE SERIES, I.E.,
C...  THE FOURIER COEFFICIENTS DECREASE AS 1/I DUE TO THE DISCONTINUITY
C...  AT X = 0
      SUM=0.0D0
      DO 1 I=1,200
C...
C...  EIGENVALUES
      EN=(DFLOAT(I)-0.5D0)*3.1415927D0/XL
C...
C...  FOURIER COEFFICIENTS
      CN=2.0D0/(EN*XL)
C...
C...  ARGUMENT OF THE EXPONENTIAL
      ARG=-(EN**2)*T
C...
C...  BRANCH TO AVOID AN UNDERFLOW OF THE EXPONENTIAL FUNCTION
      IF(ARG.LT.-200.0D0)THEN
         EX=0.
      ELSE
         EX=DEXP(ARG)
      END IF
C...
C...  NEXT TERM IN THE SERIES
      TERM=CN*EX*DSIN(EN*X)
C...
C...  SUM THE SERIES
      SUM=SUM+TERM
1     CONTINUE
C...
C...  200 TERMS HAVE BEEN SUMMED AS THE ANALYTICAL SOLUTION
      UANAL=SUM
      RETURN
      END
```

Program 4.8c *Continued.*

of the transform, equation (4.126). To illustrate this, we now compute the transform of d^2f/dx^2. Applying the definition of the transform, equation (4.126), to d^2f/dx^2, we have

$$\int_0^{x_L} d^2f/dx^2 \sin(\lambda_i x)\, dx \tag{4.128}$$

```
PARABOLIC PDE WITH INCONSISTENT IC AND BC, DSS004
0.        1.0       0.1
   21                    0.00001
PARABOLIC PDE WITH INCONSISTENT IC AND BC, DSS044
0.        1.0       0.1
   21                    0.00001
END OF RUNS
```

Program 4.8d Data for the Solution of Equations (4.99), and (4.100) to (4.102).

Integration by parts applied to integral (4.128) gives

$$\int_0^{x_L} d^2f/dx^2 \sin(\lambda_i x)\,dx = df/dx \sin(\lambda_i x)\Big|_0^{x_L} - \lambda_i \int_0^{x_L} df/dx \cos(\lambda_i x)\,dx$$

$$= (-1)^{i+1} df(x_L)/dx - \lambda_i f \cos(\lambda_i x)\Big|_0^{x_L}$$

$$- \lambda_i^2 \int_0^{x_L} f \sin(\lambda_i x)\,dx$$

$$= (-1)^{i+1} df(x_L)/dx + \lambda_i f(0) - \lambda_i^2 \bar{f}(\lambda_i) \tag{4.129}$$

in which the last right-hand side of equation (4.129) follows from equation (4.126). The transform of d^2f/dx^2 from equation (4.129) suggests that transform pairs (4.126) and (4.127) can be applied to second-order PDEs with a Dirichlet boundary condition at $x = 0$ (due to the term $\lambda_i f(0)$) and a Neumann boundary condition at $x = x_L$ (due to the term $(-1)^{i+1} df(x_L)/dx$).

For example, the sine transform of equation (4.126) can be applied to equations (4.99) to (4.102). First we multiply equation (4.99) by the *kernel* of the transform, $\sin(\lambda_i x)$, and integrate from $x = 0$ to $x = x_L$:

$$\int_0^{x_L} \partial u/\partial t \sin(\lambda_i x)\,dx = D \int_0^{x_L} \partial^2 u/\partial x^2 \sin(\lambda_i x)\,dx \tag{4.130}$$

If we interchange the order of differentiation and integration in the left-hand side integral of equation (4.130), we obtain

$$\partial/\partial t \int_0^{x_L} u \sin(\lambda_i x)\,dx = d\bar{u}/dt \tag{4.131}$$

The right-hand side of equation (4.130) is from equations (4.100), (4.101), and (4.129)

$$D \int_0^{x_L} \partial^2 u/\partial x^2 \sin(\lambda_i x)\,dx = -\lambda_i^2 \bar{u}(\lambda_i) \tag{4.132}$$

Equations (4.131) and (4.132) then combine to give

$$d\bar{u}(\lambda_i, t)/dt = -\lambda_i^2 \bar{u}(\lambda_i, t) \tag{4.133}$$

Table 4.8 Comparison of the Numerical and Analytical Solutions to Equations (4.99) to (4.102). *Continued next pages.*

```
RUN NO. -    1  PARABOLIC PDE WITH INCONSISTENT IC AND BC, DSS004

INITIAL T -  0.000D+00

  FINAL T -  0.100D+01

  PRINT T -  0.100D+00

NUMBER OF DIFFERENTIAL EQUATIONS -  21

MAXIMUM INTEGRATION ERROR -  0.100D-04


DEPENDENT VARIABLE COLUMN INDEX J (FOR YJ) IS PRINTED HORIZONTALLY

DERIVATIVE ROW INDEX I (FOR DYI/DT = FI(Y1,Y2,...,YJ,...,YN) IS
PRINTED VERTICALLY

JACOBIAN MATRIX ELEMENT IN THE MAP WITH INDICES I,J IS FOR PFI/PYJ
WHERE P DENOTES A PARTIAL DERIVATIVE


                   111111111122
          123456789012345678901
      1    888775
      2    767765
      3    777765
      4    7676765
      5    67676765
      6    567676765
      7     567676765
      8      567676765
      9       567676765
     10        567676765
     11         567676765
     12          567676765
     13           567676765
     14            567676765
     15             567676765
     16              567676765
     17               567676765
     18                56767676
     19                 5677776
     20                 5677777
     21                 5778877


         T       X=0   X=0.2*XL   X=0.4*XL   X=0.6*XL   X=0.8*XL      X=XL
      0.00
 NUMERICAL    0.0000     1.0000     1.0000     1.0000     1.0000    1.0000
ANALYTICAL    0.0000     0.9949     0.9973     0.9980     0.9983    0.9984
```

Table 4.8 *Continued.*

0.10						
NUMERICAL	0.0000	0.3452	0.6286	0.8185	0.9191	0.9493
ANALYTICAL	0.0000	0.3452	0.6286	0.8185	0.9191	0.9493
0.20						
NUMERICAL	0.0000	0.2442	0.4616	0.6304	0.7363	0.7723
ANALYTICAL	0.0000	0.2442	0.4616	0.6304	0.7363	0.7723
0.30						
NUMERICAL	0.0000	0.1881	0.3575	0.4915	0.5773	0.6068
ANALYTICAL	0.0000	0.1881	0.3575	0.4915	0.5773	0.6068
0.40						
NUMERICAL	0.0000	0.1467	0.2790	0.3839	0.4513	0.4745
ANALYTICAL	0.0000	0.1467	0.2790	0.3839	0.4513	0.4745
0.50						
NUMERICAL	0.0000	0.1146	0.2179	0.3000	0.3526	0.3708
ANALYTICAL	0.0000	0.1146	0.2179	0.3000	0.3526	0.3708
0.60						
NUMERICAL	0.0000	0.0895	0.1703	0.2344	0.2755	0.2897
ANALYTICAL	0.0000	0.0895	0.1703	0.2344	0.2755	0.2897
0.70						
NUMERICAL	0.0000	0.0699	0.1331	0.1831	0.2153	0.2264
ANALYTICAL	0.0000	0.0700	0.1331	0.1831	0.2153	0.2264
0.80						
NUMERICAL	0.0000	0.0547	0.1040	0.1431	0.1682	0.1769
ANALYTICAL	0.0000	0.0547	0.1040	0.1431	0.1682	0.1769
0.90						
NUMERICAL	0.0000	0.0427	0.0812	0.1118	0.1314	0.1382
ANALYTICAL	0.0000	0.0427	0.0812	0.1118	0.1314	0.1382
1.00						
NUMERICAL	0.0000	0.0334	0.0635	0.0874	0.1027	0.1080
ANALYTICAL	0.0000	0.0334	0.0635	0.0874	0.1027	0.1080

Table 4.8 *Continued.*

```
RUN NO. -   2  PARABOLIC PDE WITH INCONSISTENT IC AND BC, DSS044

INITIAL T -  0.000D+00

  FINAL T -  0.100D+01

  PRINT T -  0.100D+00

NUMBER OF DIFFERENTIAL EQUATIONS -  21

MAXIMUM INTEGRATION ERROR -  0.100D-04


DEPENDENT VARIABLE COLUMN INDEX J (FOR YJ) IS PRINTED HORIZONTALLY

DERIVATIVE ROW INDEX I (FOR DYI/DT = FI(Y1,Y2,...,YJ,...,YN) IS PRINTED
VERTICALLY

JACOBIAN MATRIX ELEMENT IN THE MAP WITH INDICES I,J IS FOR PFI/PYJ
WHERE P DENOTES A PARTIAL DERIVATIVE


                 111111111122
        123456789012345678901
    1    88887
    2    77776
    3    7776
    4    67776
    5     67776
    6      67776
    7       67776
    8        67776
    9         67776
   10          67776
   11           67776
   12            67776
   13             67776
   14              67776
   15               67776
   16                67776
   17                 67776
   18                  67776
   19                   67776
   20                  677777
   21                   67888

          T        X=0   X=0.2*XL   X=0.4*XL   X=0.6*XL   X=0.8*XL         X=XL
       0.00
  NUMERICAL     0.0000     1.0000     1.0000     1.0000     1.0000       1.0000
 ANALYTICAL     0.0000     0.9949     0.9973     0.9980     0.9983       0.9984
```

Table 4.8 *Continued.*

0.10						
NUMERICAL	0.0000	0.3452	0.6286	0.8185	0.9191	0.9493
ANALYTICAL	0.0000	0.3452	0.6286	0.8185	0.9191	0.9493
0.20						
NUMERICAL	0.0000	0.2442	0.4616	0.6304	0.7363	0.7723
ANALYTICAL	0.0000	0.2442	0.4616	0.6304	0.7363	0.7723
0.30						
NUMERICAL	0.0000	0.1881	0.3575	0.4915	0.5773	0.6068
ANALYTICAL	0.0000	0.1881	0.3575	0.4915	0.5773	0.6068
0.40						
NUMERICAL	0.0000	0.1467	0.2790	0.3839	0.4513	0.4745
ANALYTICAL	0.0000	0.1467	0.2790	0.3839	0.4513	0.4745
0.50						
NUMERICAL	0.0000	0.1146	0.2179	0.3000	0.3526	0.3708
ANALYTICAL	0.0000	0.1146	0.2179	0.3000	0.3526	0.3708
0.60						
NUMERICAL	0.0000	0.0895	0.1703	0.2344	0.2755	0.2897
ANALYTICAL	0.0000	0.0895	0.1703	0.2344	0.2755	0.2897
0.70						
NUMERICAL	0.0000	0.0700	0.1331	0.1831	0.2153	0.2264
ANALYTICAL	0.0000	0.0700	0.1331	0.1831	0.2153	0.2264
0.80						
NUMERICAL	0.0000	0.0547	0.1040	0.1431	0.1682	0.1769
ANALYTICAL	0.0000	0.0547	0.1040	0.1431	0.1682	0.1769
0.90						
NUMERICAL	0.0000	0.0427	0.0812	0.1118	0.1314	0.1382
ANALYTICAL	0.0000	0.0427	0.0812	0.1118	0.1314	0.1382
1.00						
NUMERICAL	0.0000	0.0334	0.0635	0.0874	0.1027	0.1080
ANALYTICAL	0.0000	0.0334	0.0635	0.0874	0.1027	0.1080

Finally, equation (4.102) transforms to

$$\bar{u}(\lambda_i, 0) = \int_0^{x_L} u(x,0)\sin(\lambda_i x) = \int_0^{x_L} 1\sin(\lambda_i x) = 1/\lambda_i \tag{4.134}$$

Again, we have reduced a problem in PDEs, equation (4.99) to (4.102), to a problem in ODEs, equations (4.133) and (4.134) via the analytical method of lines implemented by the sine transform of equation (4.126).

The solution to equation (4.133) subject to initial condition (4.134) is easily obtained as

$$\bar{u}(\lambda_i, t) = (1/\lambda_i)e^{-\lambda_i{}^2 t} \tag{4.135}$$

The inversion of $u(\lambda_i, t)$ from equation (4.135) merely requires substitution in the inverse transform of equation (4.127)

$$\begin{aligned} u(x,t) &= (2/x_L)\sum_{i=1}^{\infty} \bar{u}(\lambda_i, t)\sin(\lambda_i x) \\ &= (2/x_L)\sum_{i=1}^{\infty}(1/\lambda_i)e^{-\lambda_i{}^2 t}\sin(\lambda_i x) \end{aligned} \tag{4.136}$$

Equation (4.136) is the previous solution, equation (4.124), with $x_L = 1$. This last step illustrates one of the positive features of finite integral transforms like the sine transform of equation (4.126) (they are *finite transforms* because the limits on the defining integral are finite); the inversion can always be accomplished easily since it merely involves substitution in an infinite series.

The previous procedure of defining a finite integral transform, then deriving its inverse as an infinite series, suggests a generalization leading to four transforms, two with the sine kernel function $\sin(\lambda_i x)$ (we have already considered one, the transform of equation (4.126) with $\lambda_i = (i - 1/2)\pi/x_l$) and two for the cosine kernel function $\cos(\lambda_i x)$; for each of the two kernels, we have $\lambda_i = i\pi/x_L$ or $\lambda_i = (i - 1/2)\pi/x_L$, thus there are a total of four possible transforms. We now derive the corresponding four transforms for the second derivative, $d^2 f(x)/dx^2$, in terms of the transform of $f(x)$, which we again denote as $\bar{f}(\lambda_i)$. Of course, other transform pairs can be derived as needed in the solution of second-order PDEs.

For the sine function, the transform of the second derivative, $d^2 f/dx^2$, is

$$\begin{aligned} \int_0^{x_L} d^2f/dx^2 \sin(\lambda_i x)\,dx &= df/dx \sin(\lambda_i x)\Big|_0^{x_L} - \lambda_i \int_0^{x_L} df/dx \cos(\lambda_i x)\,dx \\ &= df(x_L)/dx \sin(\lambda_i x_L) - \lambda_i f\cos(\lambda_i x)\Big|_0^{x_L} - \lambda_i^2 \int_0^{x_L} f \sin(\lambda_i x)\,dx \\ &= df(x_L)/dx \sin(\lambda_i x_L) - \lambda_i f(x_L)\cos(\lambda_i x_L) \\ &\quad + \lambda_i f(0) - \lambda_i{}^2 \bar{f}(\lambda_i) \end{aligned} \tag{4.137}$$

Case I: For $\lambda_i = i\pi/x_L$, equation (4.137) gives

$$\int_0^{x_L} d^2f/dx^2 \sin(\lambda_i x)\,dx = (-1)^{i+1}\lambda_i f(x_L) + \lambda_i f(0) - \lambda_i{}^2 \bar{f}(\lambda_i) \tag{4.138}$$

which would be used if $f(0)$ and $f(x_L)$ are given as boundary conditions, i.e., Dirichlet boundary conditions at $x = 0$ and $x = x_L$.

Case II: For $\lambda_i = (i - 1/2)\pi/x_L$, equation (4.137) gives

$$\int_0^{x_L} d^2f/dx^2 \sin(\lambda_i x)\,dx = (-1)^{i+1} df(x_L)/dx + \lambda_i f(0) - \lambda_i^2 \bar{f}(\lambda_i) \tag{4.139}$$

which would be used if $f(0)$ and $df(x_L)/dx$ are given as boundary conditions, i.e., a Dirichlet boundary condition at $x = 0$ and a Neumann boundary condition at $x = x_L$. Equation (4.139) is just equation (4.129).

Similarly, for a cosine transform of the second derivative, d^2f/dx^2, we have

$$\begin{aligned}\int_0^{x_L} d^2f/dx^2 \cos(\lambda_i x)\,dx &= df/dx \cos(\lambda_i x)\Big|_0^{x_L} + \lambda_i \int_0^{x_L} df/dx \sin(\lambda_i x)\,dx \\ &= df(x_L)/dx \cos(\lambda_i x_L) - df(0)/dx + \lambda_i f \sin(\lambda_i x)\Big|_0^{x_L} \\ &\quad - \lambda_i^2 \int_0^{x_L} f \cos(\lambda_i x)\,dx \\ &= df(x_L)/dx \cos(\lambda_i x_L) - df(0)/dx + \lambda_i f(x_L) \sin(\lambda_i x_L) \\ &\quad - \lambda_i^2 \bar{f}(\lambda_i)\end{aligned} \tag{4.140}$$

Case III: For $\lambda_i = i\pi/x_L$, equation (4.140) gives

$$\int_0^{x_L} d^2f/dx^2 \cos(\lambda_i x)\,dx = (-1)^i df(x_L)/dx - df(0)/dx - \lambda_i^2 \bar{f}(\lambda_i) \tag{4.141}$$

which would be used if $df(0)/dx$ and $df(x_L)/dx$ are given as boundary conditions, i.e., Neumann boundary conditions at $x = 0$ and $x = x_L$.

Case IV: For $\lambda_i = (i - 1/2)\pi/x_L$, equation (4.137) gives

$$\int_0^{x_L} d^2f/dx^2 \cos(\lambda_i x)\,dx = -df(0)/dx + (-1)^{i+1} \lambda_i f(x_L) - \lambda_i^2 \bar{f}(\lambda_i) \tag{4.142}$$

which would be used if $df(0)/dx$ and $f(x_L)$ are given as boundary conditions, i.e., a Neumann boundary condition at $x = 0$ and a Dirichlet boundary condition at $x = x_L$.

For Case III, the inverse transform would be

$$f(x) = (1/x_L)\bar{f}(\lambda_0) + (2/x_L)\sum_{i=1}^{\infty} \bar{f}(\lambda_i)\cos(\lambda_i x) \tag{4.143}$$

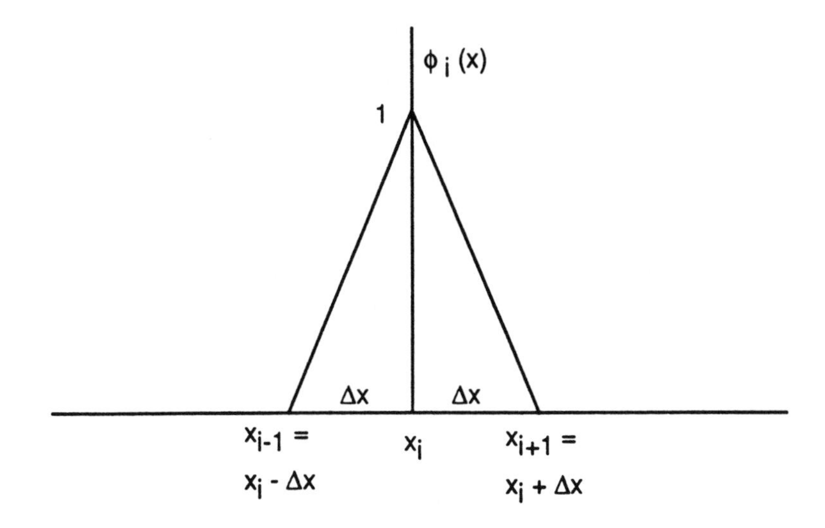

Figure 4.8 (a) The Linear Finite Element.

The transforms of equations (4.138), (4.139), (4.141), and (4.142) facilitate the separation of variables or analytical method of lines solution of linear PDEs which are second order in space. Also, many other finite integral transforms are possible through the use of various orthogonal functions, e.g., *Hankel transforms* based on Bessel functions. We will use the Case III transform in a subsequent example to verify the numerical solution of equation (4.99) by finite differences, finite elements, and finite volumes.

4.8 *The Finite Element Method*

In the preceding section, we considered the series solution of one-dimensional PDEs with basis functions $\phi_i(x)$ (see equation (4.103), for example). The series solutions that were developed had the sine and cosine as basis functions. We now consider another type of basis function, the *finite element*. Many different finite elements have been used in the approximate solution of PDEs, and we consider just one possibility, the *linear finite element* or *hat* or *chapeau function*, illustrated in Figures 4.8a and b.

We can note the following properties of the linear finite element in Figure 4.8a:

1. $\phi_i(x)$ is centered at $x = x_i$ and consists of two linear segments that intersect the x axis at $x_i \pm \Delta x$ ($= x_{i+1}$ and x_{i-1}, respectively); also $\phi_i(x_i) = 1$, $\phi_i(x_i - \Delta x) = \phi_i(x_i + \Delta x) = 0$. In other words, the function $\phi_i(x)$ is nonzero only in the interval $x_i - \Delta x \le x \le x_i + \Delta x$. Δx is the basic increment of the linear finite element (which we assume to be uniform for all x, although this uniformity is not required). These linear finite elements make a weighted residual analysis relatively easy because they are easily differentiated and integrated.

2. Since $\phi_i(x)$ is linear, its first derivative, $d\phi_i(x)/dx$, is piecewise constant (see Figure 4.8b).

With these properties in mind, we now substitute the linear finite element function into the basic series solution, equation (4.103). We can then substitute this series solution into the problem system PDE equation(s), and thus develop the equation

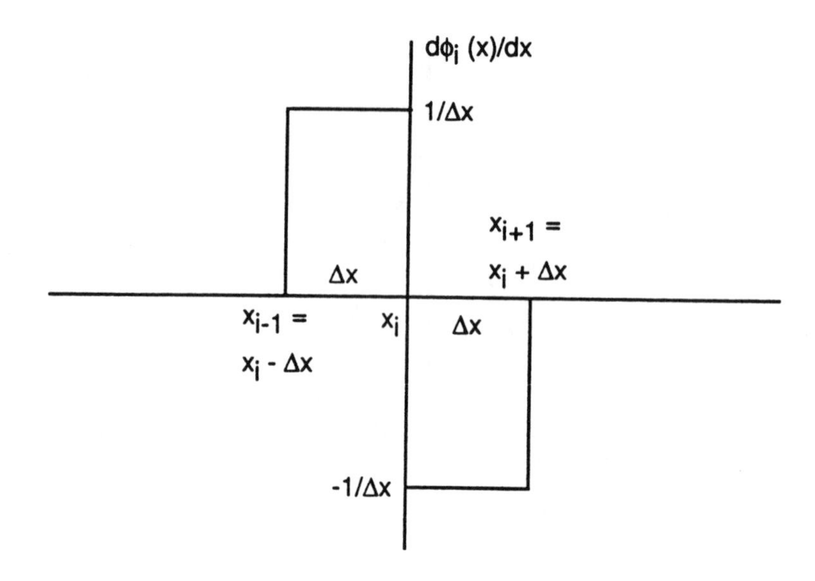

Figure 4.8 (b) The First Derivative of the Linear Finite Element.

residuals. First, however, we restate the PDE problem with some minor changes in notation which will facilitate the weighted residual analysis. The PDE, equation (4.99), in subscript notation is

$$u_t = Du_{xx} \tag{4.144}$$

with the initial condition

$$u(x,0) = g(x) \tag{4.145}$$

Also, we now consider the Neumann boundary conditions (note the subscript in x)

$$u_x(x_1,t) = h_1(t), \qquad u_x(x_N,t) = h_N(t) \qquad (4.146)(4.147)$$

The left and right boundary values of x are denoted as x_1 and x_N, respectively, which have been selected to give a closer correspondence with the finite element grid. Equations (4.144) to (4.147) are then the PDE problem system (and again, we take $x_1 = 0$ and $x_N = 1$ to facilitate the analysis, but this is not required).

For a specific problem, we take as the initial condition function

$$u(x,0) = g(x) = \begin{cases} 0, & 0 \le x < 0.5 \\ 1, & 0.5 < x \le 1 \end{cases} \tag{4.148}$$

and the boundary condition functions

$$u_x(0,t) = h_1(t) = 0, \qquad u_x(1,t) = h_N(t) = 0 \qquad (4.149)(4.150)$$

The analytical solution to equations (4.144), (4.148), (4.149), and (4.150), which will serve as a standard for the numerical (finite element) solution, can be derived using the Case III cosine transform discussed previously. The finite cosine transform of the derivative u_{xx} in equation (4.144) is from equation (4.141)

$$\int_0^1 u_{xx}(x,t)\cos(\lambda_i x)\,dx = (-1)^i u_x(1,t) - u_x(0,t) - \lambda_i^2 \bar{u}(\lambda_i,t), \;\; \lambda_i = i\pi \tag{4.151}$$

or, from boundary conditions (4.149) and (4.150)

$$\int_0^1 u_{xx}(x,t)\cos(\lambda_i x)\,dx = -\lambda_i^2 \bar{u}(\lambda_i,t), \;\; \lambda_i = i\pi \tag{4.152}$$

Therefore, the transform of equation (4.144) is

$$d\bar{u}(\lambda_i,t)/dt = -D\lambda_i^2 \bar{u}(\lambda_i,t) \tag{4.153}$$

Initial condition (4.148) transforms to

$$\bar{u}(\lambda_i,0) = \int_0^1 u(x,0)\cos(\lambda_i x)\,dx = \int_{0.5}^1 1\cos(\lambda_i x)\,dx = (1/\lambda_i)\sin(\lambda_i x)\Big|_{0.5}^1$$

$$= -(1/\lambda_i)\sin(\lambda_i 0.5) \tag{4.154}$$

Note that $\bar{u}(\lambda_i,0)$ given by equation (4.154) has the successive nonzero values $-1/\pi$, $1/(3\pi)$, $-1/(5\pi)$, $1/(7\pi), \ldots$. The solution of equation (4.153), subject to initial condition (4.154), is

$$\bar{u}(\lambda_i,t) = (-1/\lambda_i)\sin(\lambda_i 0.5)e^{-D\lambda_i^2 t} \tag{4.155}$$

The inverse transform is given by equation (4.143)

$$u(x,t) = \bar{u}(\lambda_0,t) + 2\sum_{i=1}^{\infty} \bar{u}(\lambda_i,t)\cos(\lambda_i x) \tag{4.156}$$

Before we can use equation (4.156), we must consider $\bar{u}(\lambda_0,t)$. Since $\lambda_0 = 0\pi = 0$, we have from equation (4.153)

$$d\bar{u}(\lambda_0,t)/dt = 0 \tag{4.157}$$

with the initial condition from equation (4.154)

$$\bar{u}(\lambda_0,0) = \int_0^1 u(x,0)\,dx = \int_{0.5}^1 1\,dx = x\Big|_{0.5}^1 = 0.5 \tag{4.158}$$

Thus, from equations (4.157) and (4.158) we have

$$\bar{u}(\lambda_0,t) = 0.5 \tag{4.159}$$

Finally, we can substitute equations (4.155) and (4.159) in equation (4.156)

$$u(x,t) = 0.5 + 2\sum_{i=1}^{\infty}(-1/\lambda_i)\sin(\lambda_i 0.5)e^{-D\lambda_i^2 t}\cos(\lambda_i x) \tag{4.160}$$

which is the analytical solution to be compared with the numerical solution computed by the finite element method.

We can also (a) compute the total energy of the system, which should remain constant because of the *insulated boundary conditions*, equations (4.149) and (4.150), and (b) observe the approach to the steady-state solution $u(x,\infty) = 0.5$ (which, of course, follows from equation (4.160)). The total energy of the solution, E, is

$$E = \int_0^1 H(u)\,dx \tag{4.161}$$

where $H(u)$ is the *enthalpy per unit volume* which is a function of temperature u, and can be computed as

$$H(u) = \int_0^u \rho(u)C_p(u)\,du \tag{4.162}$$

where ρ and C_p are the density and constant pressure specific heat, respectively. For equation (4.144), $D = k/(\rho C_p) = 1$, therefore we take $\rho = 1$ and $C_p = 1$ in equation (4.162). Thus, $H(u) = u$, which, when substituted in equation (4.161), gives

$$E(t) = \int_0^1 u(x,t)\,dx \tag{4.163}$$

The integral in equation (4.163) is evaluated numerically by the *trapezoidal rule* in subroutine `PRINT`, and should have the constant value $E(t) = 0.5$.

We have now gone through the preliminaries of formulating the PDE problem, equation (4.144) and (4.148) to (4.150), the analytical solution, equation (4.160), and an energy balance for the numerical solution, equation (4.163). We can now proceed to the formulation of the finite element solution and the associated programming. We start by returning to equation (4.109) (renumbered here)

$$\sum_{i=1}^{N} c_i'(t)\int_{x_1}^{x_N}\phi_j(x)\phi_i(x)\,dx - D\sum_{i=1}^{N} c_i(t)\int_{x_1}^{x_N}\phi_j(x)\phi_i''(x)\,dx = \int_{x_1}^{x_N}\phi_j(x)R(x,t)\,dx = 0 \tag{4.164}$$

Equation (4.164) is a statement of Galerkin's method in which we make the residual, $R(x,t)$, orthogonal to the basis functions, $\phi_i(x)$. As before, in order to evaluate the integrals of equation (4.164), we must select the basis (shape) functions, $\phi_i(x)$. Instead of using the sine functions of equation (4.110), we now use the linear finite elements of Figure 4.8a.

We must now consider three cases for each of the two left-hand integrals in equation (4.164). For the first integral (with reference to Figure 4.8a)

1. $\displaystyle\int_{x_1}^{x_N}\phi_j(x)\phi_i(x)\,dx$

(1.1) $i = j$

$$\int_{x_1}^{x_N}\phi_i(x)\phi_i(x)\,dx = \int_{x_{i-1}}^{x_i}\left\{\frac{x - x_{i-1}}{x_i - x_{i-1}}\right\}^2 dx + \int_{x_i}^{x_{i+1}}\left\{\frac{x_{i+1} - x}{x_{i+1} - x_i}\right\}^2 dx$$

$$= \frac{1}{3\Delta x^2}(x - x_{i-1})^3\Big|_{x_{i-1}}^{x_i} - \frac{1}{3\Delta x^2}(x_{i+1} - x)^3\Big|_{x_i}^{x_{i+1}}$$

$$= (2/3)\Delta x \tag{4.165}$$

(1.2) $i = j + 1$ (or $j = i - 1$)

$$\int_{x_1}^{x_N} \phi_i(x)\phi_{i+1}(x)\,dx = \int_{x_i}^{x_{i+1}} \left\{\frac{x_{i+1} - x}{x_{i+1} - x_i}\right\}\left\{\frac{x - x_i}{x_{i+1} - x_i}\right\} dx$$

$$= -\int_{x_i}^{x_{i+1}} \left\{\frac{x - x_{i+1}}{x_{i+1} - x_i}\right\}\left\{\frac{x - x_i}{x_{i+1} - x_i}\right\} dx$$

Using integration by parts we have

$$-\int_{x_i}^{x_{i+1}} \left\{\frac{x - x_{i+1}}{x_{i+1} - x_i}\right\}\left\{\frac{x - x_i}{x_{i+1} - x_i}\right\} dx$$

$$= \frac{-1}{\Delta x^2}\left\{(x - x_{i+1})(1/2)(x - x_i)^2\Big|_{x_i}^{x_{i+1}} - (1/2)\int_{x_i}^{x_{i+1}} (x - x_i)^2\,dx\right\}$$

$$= \frac{1}{\Delta x^2}\left\{(1/6)(x - x_i)^3\Big|_{x_i}^{x_{i+1}}\right\} = (1/6)\Delta x$$

so

$$\int_{x_1}^{x_N} \phi_i(x)\phi_{i+1}(x)\,dx = (1/6)\Delta x \tag{4.166}$$

(1.3) $i = j - 1$ (or $j = i + 1$)

$$\int_{x_1}^{x_N} \phi_i(x)\phi_{i-1}(x)\,dx = \int_{x_{i-1}}^{x_i} \left\{\frac{x - x_{i-1}}{x_i - x_{i-1}}\right\}\left\{\frac{x_i - x}{x_i - x_{i-1}}\right\} dx$$

$$= -\int_{x_{i-1}}^{x_i} \left\{\frac{x - x_{i-1}}{x_i - x_{i-1}}\right\}\left\{\frac{x - x_i}{x_i - x_{i-1}}\right\} dx$$

$$= (1/6)\Delta x \tag{4.167}$$

which follows from the preceding case with $i + 1 \to i$.

Note that in the evaluation of these three integrals, the limits x_1 to x_N could be replaced with limits corresponding to integration over at most two elements, e.g., x_{i-1} to x_{i+1} in integral (1.1). This was possible because the linear finite elements of Figure 4.8a, $\phi_i(x)$, have the property of *local support*, i.e., for any given x_i, only the three functions

ϕ_{i-1}, ϕ_i, and ϕ_{i+1} are nonzero in the interval $x_i \pm \Delta x$. Combining equations (4.165) to (4.167), we have for the first term on the left-hand side of equation (4.164)

$$\sum_{i=1}^{N} c_i'(t) \int_{x_1}^{x_N} \phi_j(x)\phi_i(x)\,dx = (1/6)c_{i-1}' + (4/6)c_i' + (1/6)c_{i+1}' \tag{4.168}$$

For the second left-hand term of equation (4.164)

$$-D\sum_{i=1}^{N} c_i(t) \int_{x_1}^{x_N} \phi_j(x)\phi_i''(x)\,dx$$

we consider the intergal in three steps (with reference to Figures 4.8a and b):

2. $$\int_{x_1}^{x_N} \phi_j(x)\phi_i''(x)\,dx = \phi_j(x)\phi_i'(x)\Big|_{x_1}^{x_N} - \int_{x_1}^{x_N} \phi_j'(x)\phi_i'(x)\,dx$$

(2.1) i and $j \neq 1$ or N (for which $\phi_j(x)\phi_i'(x)\Big|_{x_1}^{x_N}$ in the preceding equation is zero)

$$c_i \int_{x_{i-1}}^{x_{i+1}} \phi_i'(x)\phi_i'(x)\,dx + c_{i-1} \int_{x_{i-1}}^{x_i} \phi_i'(x)\phi_{i-1}'(x)\,dx + c_{i+1} \int_{x_i}^{x_{i+1}} \phi_i'(x)\phi_{i+1}'(x)\,dx$$

$$= c_i \left(\int_{x_i}^{x_{i+1}} \phi_i'(x)\phi_i'(x)\,dx + \int_{x_{i-1}}^{x_i} \phi_i'(x)\phi_i'(x)\,dx \right)$$

$$+ c_{i-1} \int_{x_{i-1}}^{x_i} \phi_i'(x)\phi_{i-1}'(x)\,dx + c_{i+1} \int_{x_i}^{x_{i+1}} \phi_i'(x)\phi_{i+1}'(x)\,dx$$

$$= c_i \left(\int_{x_i}^{x_{i+1}} (-1/\Delta x)^2\,dx + \int_{x_{i-1}}^{x_i} (1/\Delta x)^2\,dx \right)$$

$$+ c_{i-1} \int_{x_{i-1}}^{x_i} (1/\Delta x)(-1/\Delta x)\,dx$$

$$+ c_{i+1} \int_{x_i}^{x_{i+1}} (-1/\Delta x)(1/\Delta x)\,dx$$

$$= c_i \left((-1/\Delta x)^2 x\Big|_{x_i}^{x_{i+1}} + (1/\Delta x)^2 x\Big|_{x_{i-1}}^{x_i} \right)$$

$$- c_{i-1}(1/\Delta x)^2 x|_{x_{i-1}}^{x_i} - c_{i+1}(1/\Delta x)^2 x|_{x_i}^{x_{i+1}} \}$$

$$= c_i\left((1/\Delta x) + (1/\Delta x)\right) - c_{i-1}(1/\Delta x) - c_{i+1}(1/\Delta x) \tag{4.169}$$

Then, combining equations (4.164), (4.168), and (4.169), we have as the approximation of equation (4.144) at the interior grid points (for $i \neq 1$ or N)

$$\begin{aligned}(\Delta x/6)c'_{i+1} + (4\Delta x/6)c'_i + (\Delta x/6)c'_{i-1} \\ = -D\left\{c_i\left((1/\Delta x) + (1/\Delta x)\right) - c_{i-1}(1/\Delta x) - c_{i+1}(1/\Delta x)\right\}\end{aligned}$$

or

$$(1/6)c'_{i+1} + (4/6)c'_i + (1/6)c'_{i-1} = D\left\{\frac{c_{i+1} - 2c_i + c_{i-1}}{\Delta x^2}\right\} \tag{4.170}$$

Equation (4.166) is the (local) Galerkin approximation of equation (4.144), based on the linear finite element of Figure 4.8a. We note in particular that the right-hand side is just the well-known, second-order central difference approximation for the second derivative u_{xx}. However, the left-hand side is a weighted sum of three time derivatives.

We must now include in our analysis boundary conditions (4.146) and (4.147). To do this, we first consider the residual of equation (4.144)

$$\int_{x_1}^{x_N} w(x)R(x,t)\,dx = \int_{x_1}^{x_N} w(x)\left\{u_t - Du_{xx}\right\}dx = \int_{x_1}^{x_N} \phi_i(x)\left\{u_t - Du_{xx}\right\}dx = 0 \tag{4.171}$$

where we have again applied the Galerkin method, i.e., $w(x) = \phi_i(x)$. We then apply equation (4.171) at the boundaries, $x = x_1$ and $x = x_N$ (along with equations (4.146) and (4.147)).

(2.2) $i = 1$

Considering first the integral with u_t in equation (4.171) we have

$$\begin{aligned}\int_{x_1}^{x_2} \phi_1(x)u_t\,dx &= c'_1(t)\int_{x_1}^{x_2} \phi_1(x)\phi_1(x)\,dx + c'_2(t)\int_{x_1}^{x_2} \phi_1(x)\phi_2(x)\,dx \\ &= (2\Delta x/6)c'_1 + (\Delta x/6)c'_2\end{aligned} \tag{4.172}$$

where we have made use of equations (4.165) (with integration over Δx rather than $2\Delta x$) and (4.166). Considering next the integral with u_{xx} in equation (4.171), we have

$$\begin{aligned}-D\int_{x_1}^{x_N} \phi_i(x)u_{xx}\,dx &= -D\left\{\phi_i(x)u_x\Big|_{x_1}^{x_N} - \int_{x_1}^{x_N} \phi'_i(x)u_x\,dx\right\} \\ &= -D\left\{-u_x(x_1,t) - \int_{x_1}^{x_N} \phi'_i(x)\sum_{i=1}^{N} c_i(t)\phi'_i(x)\,dx\right\} \\ &= -D\left\{-u_x(x_1,t) - c_1\int_{x_1}^{x_2} \phi'_1(x)\phi'_1(x)\,dx - c_2\int_{x_1}^{x_2} \phi'_1(x)\phi'_2(x)\,dx\right\} \\ &= -D\left\{-u_x(x_1,t) - c_1\int_{x_1}^{x_2} (-1/\Delta x)^2\,dx - c_2\int_{x_1}^{x_2} (-1/\Delta x)(1/\Delta x)\,dx\right\} \\ &= -D\left\{-u_x(x_1,t) - c_1(-1/\Delta x)^2 x\Big|_{x_1}^{x_2} + c_2(1/\Delta x)^2 x\Big|_{x_1}^{x_2}\right\} \\ &= -D\left\{-u_x(x_1,t) - c_1(1/\Delta x) + c_2(1/\Delta x)\right\}\end{aligned} \tag{4.173}$$

Substitution of equations (4.172) and (4.173) in (4.171) gives

$$(2\Delta x/6)c_1' + (\Delta x/6)c_2' = D\{-u_x(x_1,t) - c_1(1/\Delta x) + c_2(1/\Delta x)\}$$

or

$$(2/6)c_1' + (1/6)c_2' = -Dh_1(t)/\Delta x + D\left\{\frac{c_2 - c_1}{\Delta x^2}\right\} \tag{4.174}$$

where we have used boundary condition (4.146).

(2.3) $i = N$

Considering first the integral with u_t in equation (4.171) we have

$$\int_{x_{N-1}}^{x_N} \phi_N(x)u_t\,dx = c_N'(t)\int_{x_{N-1}}^{x_N} \phi_N(x)\phi_N(x)\,dx + c_{N-1}'(t)\int_{x_{N-1}}^{x_N} \phi_N(x)\phi_{N-1}(x)\,dx$$

$$= (2\Delta x/6)c_N' + (\Delta x/6)c_{N-1}' \tag{4.175}$$

where we have made use of equations (4.165) and (4.167) (with integration over Δx rather than $2\Delta x$ for the former).

Considering next the integral with u_{xx} in equation (4.171) we have

$$-D\int_{x_1}^{x_N} \phi_i(x)u_{xx}\,dx$$

$$= -D\left\{\phi_i(x)u_x\Big|_{x_1}^{x_N} - \int_{x_1}^{x_N} \phi_i'(x)u_x\,dx\right\}$$

$$= -D\left\{u_x(x_N,t) - \int_{x_1}^{x_N} \phi_i'(x)\sum_{i=1}^{N} c_i(t)\phi_i'(x)\,dx\right\}$$

$$= -D\left\{u_x(x_N,t) - c_N\int_{x_{N-1}}^{x_N} \phi_N'(x)\phi_N'(x)\,dx - c_{N-1}\int_{x_{N-1}}^{x_N} \phi_N'(x)\phi_{N-1}'(x)\,dx\right\}$$

$$= -D\left\{u_x(x_N,t) - c_N\int_{x_{N-1}}^{x_N} (1/\Delta x)^2\,dx - c_{N-1}\int_{x_{N-1}}^{x_N} (1/\Delta x)(-1/\Delta x)\,dx\right\}$$

$$= -D\left\{u_x(x_N,t) - c_N(1/\Delta x)^2 x\Big|_{x_{N-1}}^{x_N} + c_{N-1}(1/\Delta x)^2 x\Big|_{x_{N-1}}^{x_N}\right\}$$

$$= -D\{u_x(x_N,t) - c_N(1/\Delta x) + c_{N-1}(1/\Delta x)\} \tag{4.176}$$

Substitution of equations (4.175) and (4.176) in (4.171) gives

$$(2\Delta x/6)c_N' + (\Delta x/6)c_{N-1}' = D\{u_x(x_N,t) - c_N(1/\Delta x) + c_{N-1}(1/\Delta x)\}$$

or

$$(2/6)c_N' + (1/6)c_{N-1}' = Dh_N(t)/\Delta x + D\left\{\frac{c_{N-1} - c_N}{\Delta x^2}\right\} \tag{4.177}$$

where we have used boundary condition (4.147).

The complete set of approximating ODEs for equations (4.144) to (4.147) is therefore

$$(2/6)c_1' + (1/6)c_2' = -Dh_1(t)/\Delta x + D\left\{\frac{c_2 - c_1}{\Delta x^2}\right\}$$

$$(1/6)c_1' + (4/6)c_2' + (1/6)c_3' = D\left\{\frac{c_3 - 2c_2 + c_1}{\Delta x^2}\right\}$$

$$\vdots \quad \vdots$$

$$(1/6)c_{i-1}' + (4/6)c_i' + (1/6)c_{i+1}' = D\left\{\frac{c_{i+1} - 2c_i + c_{i-1}}{\Delta x^2}\right\} \tag{4.178}$$

$$\vdots \quad \vdots$$

$$(1/6)c_{N-2}' + (4/6)c_{N-1}' + (1/6)c_N' = D\left\{\frac{c_N - 2c_{N-1} + c_{N-2}}{\Delta x^2}\right\}$$

$$(1/6)c_{N-1}' + (2/6)c_N' = Dh_N(t)/\Delta x + D\left\{\frac{c_{N-1} - c_N}{\Delta x^2}\right\}$$

This is a system of N semi-implicit (tridiagonal) ODEs which can be integrated to obtain $c_1(t), c_2(t), \ldots, c_{N-1}(t), c_N(t)$ subject to the initial conditions (from equation (4.145))

$$c_1(0) = g(x_1), \quad c_2(0) = g(x_2), \ldots, \quad c_{N-1}(0) = g(x_{N-1}), \quad c_N(0) = g(x_N) \tag{4.179}$$

Then, the computed coefficients, $c_1(t)$ to $c_N(t)$ can be substituted in equation (4.103) to give the required solution (these last steps of numerical integration to obtain $c_1(t)$ to $c_N(t)$, followed by substitution in the series solution, equation (4.103), again illustrates the semi-analytical (or semi-numerical?) characteristic of this finite element method).

Because equations (4.179) are semi-implicit ODEs of the form of equations (1.12), with two or three derivatives appearing in the left-hand side of each equation, we need to either use an ODE integrator which can accommodate these semi-implicit ODEs, e.g., `LSODI` [Byrne and Hindmarsh (1987)], `DASSL` [Brenan et al. (1989)], or we must uncouple the ODEs to make them explicit (with one derivative in each equation), then use an integrator for explicit ODEs, e.g., `RKF45` or `LSODE`. In the subsequent programming, we use the latter approach.

Subroutine `INITAL` for initial condition (4.148) defined on a 51-point grid is listed in Program 4.9a. We can note the following points about subroutine `INITAL`:

1. The `COMMON` area has the usual sections, i.e., `/T/`, `/Y/`, and `/F/` plus section `/XG/` for the spatial grid, and `/FE/` for the additional requirements of the finite element formulation. In particular, arrays `AL(N)`, `BM(N)`, and `CU(N)` contain the coefficients

```
      SUBROUTINE INITAL
      IMPLICIT DOUBLE PRECISION (A-H,O-Z)
      PARAMETER (N=51)
      COMMON/T/      T,  NSTOP,  NORUN
     1      /Y/   U(N)
     2      /F/  UT(N)
     3     /XG/      L,     DX,   X(N)
     4      /I/     IP
     5     /FE/  AL(N),  BM(N),  CU(N), BRHS(N)
C...
C...  TYPE SELECTED VARIABLES AS DOUBLE PRECISION
      DOUBLE PRECISION L
C...
C...  LENGTH
      L=1.0D0
C...
C...  GRID SPACING
      DX=L/DFLOAT(N-1)
C...
C...  SET UP THE COEFFICIENT MATRIX IN BAND STORAGE MODE. THIS IS
C...  DONE ONLY ONCE SINCE THE COEFFICIENT MATRIX IS CONSTANT
C...
C...  LOWER DIAGONAL
      DO 2 I=1,N
         IF(I.EQ.1)THEN
            AL(1)=0.0D0
         ELSE
     +   IF(I.EQ.N)THEN
            AL(N)=1.0D0/6.0D0
         ELSE
            AL(I)=1.0D0/6.0D0
         END IF
2     CONTINUE
C...
C...  MAIN DIAGONAL
      DO 3 I=1,N
         IF(I.EQ.1)THEN
            BM(1)=2.0D0/6.0D0
         ELSE
     +   IF(I.EQ.N)THEN
            BM(N)=2.0D0/6.0D0
         ELSE
            BM(I)=4.0D0/6.0D0
         END IF
3     CONTINUE
C...
C...  UPPER DIAGONAL
      DO 4 I=1,N
         IF(I.EQ.1)THEN
            CU(1)=1.0D0/6.0D0
         ELSE
     +   IF(I.EQ.N)THEN
            CU(N)=0.0D0
```

Program 4.9a Subroutine INITAL for Initial Condition (4.148). *Continued next page.*

```
          ELSE
             CU(I)=1.0D0/6.0D0
          END IF
4       CONTINUE
C...
C...    INITIAL CONDITION
        DO 1 I=1,N
          X(I)=DFLOAT(I-1)*DX
          IF(I.LT.(N+1)/2)U(I)=0.0D0
          IF(I.GT.(N+1)/2)U(I)=1.0D0
          IF(I.EQ.(N+1)/2)U(I)=0.5D0
1       CONTINUE
C...
C...    INITIAL DERIVATIVES
        CALL DERV
        IP=0
        RETURN
        END
```

Program 4.9a *Continued.*

in the lower, main, and upper diagonal of the left-hand side of equations (4.178). For example, BM(N) has the elements 2/6, 4/6, 4/6, ..., 4/6, 2/6.

```
      COMMON/T/        T,  NSTOP,  NORUN
     1      /Y/    U(N)
     2      /F/   UT(N)
     3     /XG/       L,      DX,    X(N)
     4      /I/      IP
     5     /FE/  AL(N),  BM(N),  CU(N), BRHS(N)
```

2. After the unit length of the grid and the grid spacing are defined:

```
C...
C...  LENGTH
      L=1.0D0
C...
C...  GRID SPACING
      DX=L/DFLOAT(N-1)
```

the tridiagonal coefficients of equation (4.178) are stored in arrays AL(N), BM(N), and CU(N):

```
C...
C...  SET UP THE COEFFICIENT MATRIX IN BAND STORAGE MODE. THIS IS
C...  DONE ONLY ONCE SINCE THE COEFFICIENT MATRIX IS CONSTANT
C...
C...  LOWER DIAGONAL
      DO 2 I=1,N
         IF(I.EQ.1)THEN
            AL(1)=0.0D0
         ELSE
```

```
     +    IF(I.EQ.N)THEN
             AL(N)=1.0D0/6.0D0
          ELSE
             AL(I)=1.0D0/6.0D0
          END IF
2      CONTINUE
C...
C...   MAIN DIAGONAL
       DO 3 I=1,N
          IF(I.EQ.1)THEN
             BM(1)=2.0D0/6.0D0
          ELSE
     +    IF(I.EQ.N)THEN
             BM(N)=2.0D0/6.0D0
          ELSE
             BM(I)=4.0D0/6.0D0
          END IF
3      CONTINUE
C...
C...   UPPER DIAGONAL
       DO 4 I=1,N
          IF(I.EQ.1)THEN
             CU(1)=1.0D0/6.0D0
          ELSE
     +    IF(I.EQ.N)THEN
             CU(N)=0.0D0
          ELSE
             CU(I)=1.0D0/6.0D0
          END IF
4      CONTINUE
```

The coding in DO loops 2, 3, and 4 follows directly from the left-hand side of equation (4.178). Also, since these coefficients are constant, they are defined once in INITAL, then passed to other subroutines through COMMON/FE/. The first element of AL (AL(1)), and the last element of CU (CU(N)) are zeroed, but they are not used in the subsequent calculations (they are not part of the tridiagonal system of equations (4.178)).

3. Initial condition (4.148) is then programmed in DO loop 1. Only a numerical approximation is possible in the neighborhood of the finite jump, and an average value is used at $x = 1/2$, i.e., $u(1/2, t) = 0.5$:

```
C...
C...   INITIAL CONDITION
       DO 1 I=1,N
          X(I)=DFLOAT(I-1)*DX
          IF(I.LT.(N+1)/2)U(I)=0.0D0
          IF(I.GT.(N+1)/2)U(I)=1.0D0
          IF(I.EQ.(N+1)/2)U(I)=0.5D0
1      CONTINUE
```

Note that this programming in DO loop 1 is also an implementation of initial condition (4.179), which is the method of lines representation of initial condition (4.145).

```
      SUBROUTINE DERV
      IMPLICIT DOUBLE PRECISION (A-H,O-Z)
      PARAMETER (N=51)
      COMMON/T/       T,  NSTOP,  NORUN
     1      /Y/   U(N)
     2      /F/  UT(N)
     3     /XG/       L,      DX,    X(N)
     4      /I/      IP
     5     /FE/  AL(N),  BM(N),  CU(N), BRHS(N)
C...
C...  TYPE-SELECTED VARIABLES AS DOUBLE PRECISION
      DOUBLE PRECISION L
C...
C...  LEFT BOUNDARY CONDITION
      H1=0.0D0
C...
C...  RIGHT BOUNDARY CONDITION
      HN=0.0D0
C...
C...  RIGHT HAND SIDE VECTOR (WITH D = 1)
      DO 1 I=1,N
         IF(I.EQ.1)THEN
            BRHS(I)=-H1/DX+(U(I+1)-U(I))/DX**2
         ELSE
     +   IF(I.EQ.N)THEN
            BRHS(I)= HN/DX+(U(I-1)-U(I))/DX**2
         ELSE
            BRHS(I)=(U(I+1)-2.0D0*U(I)+U(I-1))/DX**2
         END IF
1     CONTINUE
C...
C...  SOLVE THE LINEAR ALGEBRAIC EQUATIONS BY SUBROUTINE TRIDAG,
C...  WHICH RETURNS THE DERIVATIVE VECTOR UT IN COMMON/F/
      CALL TRIDAG(AL,BM,CU,BRHS,UT,N)
      RETURN
      END
```

Program 4.9b Subroutine DERV for PDE (4.144) and Boundary Conditions (4.149) and (4.150).

Subroutine DERV for PDE (4.144) and boundary conditions (4.149) and (4.150) is listed in Program 4.9b. We can note the following points about DERV:

1. The general boundary condition functions, $h_1(t)$ and $h_N(t)$, in equations (4.146) and (4.147) are zeroed according to the special cases of boundary conditions (4.149) and (4.150), i.e., the insulated boundary conditions with zero heat flux at the boundaries at $x = 0$ and $x = 1$

```
C...
C...  LEFT BOUNDARY CONDITION
      H1=0.0D0
C...
C...  RIGHT BOUNDARY CONDITION
      HN=0.0D0
```

2. The right-hand vector of Equations (4.178) is then computed:

```
C...
C...  RIGHT-HAND SIDE VECTOR (WITH D = 1)
      DO 1 I=1,N
         IF(I.EQ.1)THEN
            BRHS(I)=-H1/DX+(U(I+1)-U(I))/DX**2
         ELSE
     +   IF(I.EQ.N)THEN
            BRHS(I)= HN/DX+(U(I-1)-U(I))/DX**2
         ELSE
            BRHS(I)=(U(I+1)-2.0D0*U(I)+U(I-1))/DX**2
         END IF
1     CONTINUE
```

For example, the right-hand side of the first ODE in equations (4.178)

$$(2/6)c_1' + (1/6)c_2' = -Dh_1(t)/\Delta x + D\left\{\frac{c_2 - c_1}{\Delta x^2}\right\}$$

is programmed (with $D = 1$) as

```
         IF(I.EQ.1)THEN
            BRHS(I)=-H1/DX+(U(I+1)-U(I))/DX**2
         ELSE
```

Similarly, the right-hand side of the last ODE in equations (4.178)

$$(1/6)c_{N-1}' + (2/6)c_N' = Dh_N(t)/\Delta x + D\left\{\frac{c_{N-1} - c_N}{\Delta x^2}\right\}$$

is programmed as

```
     +   IF(I.EQ.N)THEN
            BRHS(I)= HN/DX+(U(I-1)-U(I))/DX**2
         ELSE
```

The right-hand sides of ODEs $i = 2$ to $i = N - 1$ in equations (4.178)

$$(1/6)c_{i-1}' + (4/6)c_i' + (1/6)c_{i+1}' = D\left\{\frac{c_{i+1} - 2c_i + c_{i-1}}{\Delta x^2}\right\}$$

are programmed as

```
         ELSE
            BRHS(I)=(U(I+1)-2.0D0*U(I)+U(I-1))/DX**2
         END IF
```

3. At this point, we have a system of tridiagonal, semi-implicit ODEs, equations (4.178), because the left-hand sides of these ODEs have already been programmed in subroutine `INITAL` of Program 4.9a. However, we cannot send this system of tridiagonal ODEs to an explicit ODE integrator because such an integrator requires that each ODE have only one derivative (while the system of equations (4.178) has two or three derivatives in each ODE). We call a tridiagonal

solver, TRIDAG [Press et al. (1986)], which uncouples the ODEs, i.e., computes the temporal derivative vector explicitly:

```
C...
C...  SOLVE THE LINEAR ALGEBRAIC EQUATIONS BY SUBROUTINE TRIDAG,
C...  WHICH RETURNS THE DERIVATIVE VECTOR UT IN COMMON/F/
      CALL TRIDAG(AL,BM,CU,BRHS,UT,N)
```

TRIDAG has as inputs arrays AL, BM, CU, BRHS, and N, as expected, to solve the Nth order tridiagonal ODE system, equations (4.178). The solution, which is the derivative vector, $du_1/dt, du_2/dt, \ldots du_i/dt, \ldots du_{N-1}/dt, du_N/dt$ of equations (4.178), is returned in array UT. Note that UT is also in COMMON/F/; thus it is integrated by the explicit ODE integrator, LSODE in this case, to produce the solution vector in array U (in COMMON/Y/), which can then be used in the next call to DERV to compute the RHS vector in array BRHS.

This completes the programming of the tridiagonal system of ODEs, equations (4.178). Subroutine PRINT prints the numerical solution computed by the integration of the 51 ODEs programmed in DERV. Additionally, the energy of the solution is computed according to equation (4.163), as indicated in Program 4.9c.

We can note the following points about subroutine PRINT:

1. The tridiagonal coefficient matrtix of equations (4.178) is printed during the first call to PRINT to confirm that it was defined correctly:

```
C...
C...  PRINT THE COEFFICIENT MATRIX FOR VERIFICATION
      IP=IP+1
      IF(IP.EQ.1)THEN
         WRITE(NO,13)
13       FORMAT(/,' COEFFICIENT MATRIX',/)
         DO 14 I=1,N
            WRITE(NO,15)I,AL(I),BM(I),CU(I)
15          FORMAT(I5,3F12.4)
14       CONTINUE
         WRITE(NO,16)
16       FORMAT(//)
      END IF
```

2. The energy of the system given by equation (4.163) is then computed by the trapezoidal rule:

```
C...
C...  COMPUTE THE TOTAL ENERGY (RHO = CP = 1)
      CALL DERV
      ENERGY=0.0D0
      DO 1 I=2,N
         ENERGY=ENERGY+0.5D0*(U(I)+U(I-1))*DX
1     CONTINUE
```

3. The analytical (series) solution of equation (4.144), equation (4.160), is computed by a call to function SERIES (listed in Program 4.9d), and the difference between

```
      SUBROUTINE PRINT(NI,NO)
      IMPLICIT DOUBLE PRECISION (A-H,O-Z)
      PARAMETER (N=51)
      COMMON/T/      T,  NSTOP,  NORUN
     1      /Y/   U(N)
     2      /F/  UT(N)
     3     /XG/      L,      DX,    X(N)
     4      /I/      IP
     5     /FE/  AL(N),  BM(N),  CU(N), BRHS(N)
C...
C...  TYPE-SELECTED VARIABLES AS DOUBLE PRECISION
      DOUBLE PRECISION L
C...
C...  DIMENSION ARRAYS FOR THE ANALYTICAL SOLUTION
      DIMENSION UE(N), DIFF(N)
C...
C...  PRINT THE COEFFICIENT MATRIX FOR VERIFICATION
      IP=IP+1
      IF(IP.EQ.1)THEN
         WRITE(NO,13)
13       FORMAT(/,' COEFFICIENT MATRIX',/)
         DO 14 I=1,N
            WRITE(NO,15)I,AL(I),BM(I),CU(I)
15          FORMAT(I5,3F12.4)
14       CONTINUE
         WRITE(NO,16)
16       FORMAT(//)
      END IF
C...
C...  COMPUTE THE TOTAL ENERGY (RHO = CP = 1)
      CALL DERV
      ENERGY=0.0D0
      DO 1 I=2,N
         ENERGY=ENERGY+0.5D0*(U(I)+U(I-1))*DX
1     CONTINUE
C...
C...  CALCULATE THE EXACT SOLUTION AND THE DIFFERENCE BETWEEN THE
C...  NUMERICAL AND EXACT SOLUTIONS
      DO 3 I=1,N,10
         UE(I)=SERIES(I)
         DIFF(I)=U(I)-UE(I)
3     CONTINUE
C...
C...  PRINT THE NUMERICAL AND EXACT SOLUTIONS
      WRITE(NO,2)T,ENERGY,(U(I),I=1,N,10),
     1                     (UE(I),I=1,N,10),
     2                   (DIFF(I),I=1,N,10)
```

Program 4.9c Subroutine PRINT for Equations (4.144), (4.148), (4.149) and (4.150). *Continued next page.*

```
2     FORMAT('   T = ',F6.2,'  ENERGY = ',F9.5,
     1          /,14X,'   X=0',5X,'X=0.2',5X,'X=0.4',
     2              5X,'X=0.6',5X,'X=0.8',5X,'X=1.0D0'/,
     3                        '    U(X,T)',6F10.6,/,
     4                        '   UE(X,T)',6F10.6,/,
     5                        ' DIFF(X,T)',6F10.6,/)
      RETURN
      END
```

Program 4.9c *Continued.*

the numerical and the analytical solutions is then computed at every 10th point along the grid:

```
C...
C...  CALCULATE THE EXACT SOLUTION AND THE DIFFERENCE BETWEEN THE
C...  NUMERICAL AND EXACT SOLUTIONS
      DO 3 I=1,N,10
         UE(I)=SERIES(I)
         DIFF(I)=U(I)-UE(I)
3     CONTINUE
```

4. The numerical and analytical solutions, and their difference, are finally printed via Format 2:

```
C...
C...  PRINT THE NUMERICAL AND EXACT SOLUTIONS
      WRITE(NO,2)T,ENERGY,(U(I),I=1,N,10),
     1                    (UE(I),I=1,N,10),
     2                  (DIFF(I),I=1,N,10)
2     FORMAT('   T = ',F6.2,'  ENERGY = ',F9.5,
     1          /,14X,'   X=0',5X,'X=0.2',5X,'X=0.4',
     2              5X,'X=0.6',5X,'X=0.8',5X,'X=1.0'/,
     3                        '    U(X,T)',6F10.6,/,
     4                        '   UE(X,T)',6F10.6,/,
     5                        ' DIFF(X,T)',6F10.6,/)
```

Function SERIES to evaluate the series solution of equation (4.160) is listed in Program 4.9d. We can note the following points about function SERIES:

1. The sum and the alternating sign $(-1)^i$ are initialized before DO loop 1:

```
C...
C...  EVALUATE SERIES SOLUTION
      SUM =1.0D0
      SIGN=1.0D0
```

2. The series of equation (4.160) is summed in DO loop 1. For the ith pass through the DO loop, the eigenvalue, $\lambda_i = i\pi$, is first computed as EI; the exponent (with

```
      DOUBLE PRECISION FUNCTION SERIES(IG)
C...
C...  FUNCTION SERIES EVALUATES THE ANALYTICAL SOLUTION
C...
      IMPLICIT DOUBLE PRECISION (A-H,O-Z)
      PARAMETER (N=51)
      COMMON/T/       T,  NSTOP,  NORUN
     1      /Y/    U(N)
     2      /F/   UT(N)
     3     /XG/       L,      DX,    X(N)
     4      /I/      IP
C...
C...  TYPE SELECTED VARIABLES AS DOUBLE PRECISION
      DOUBLE PRECISION L
C...
C...  DEFINE MAXIMUM NUMBER OF TERMS IN THE SERIES (THIS IS SET TO A
C...  HIGH VALUE TO ACHIEVE CONVERGENCE TO THREE FIGURES FOR T = 0;
C...  BEYOND T = 0, THE SERIES CONVERGES RAPIDLY)
      PARAMETER (IS=1000)
C...
C...  PI
      PI=4.0D0*DATAN(1.0D0)
C...
C...  EVALUATE SERIES SOLUTION
      SUM=0.25D0
      SIGN=1.0D0
      DO 1 I=1,IS,2
         EI=DFLOAT(I)*PI
         EXI=-(EI**2)*T
         SIGN=-1.0D0*SIGN
         CI=SIGN/EI
         SUM=SUM+CI*DEXP(EXI)*DCOS(EI*X(IG))
C...
C...     AVOID UNDERFLOW OF THE EXP FUNCTION
         IF(EXI.LT.-100.0D0)GO TO 2
1     CONTINUE
C...
C...  SUMMATION OF SERIES IS COMPLETE
2     SERIES=2.0D0*SUM
      END
```

Program 4.9d Function SERIES for the Series Solution of Equations (4.144) and (4.148) to (4.150).

$D = 1$), $-D\lambda_i^2 t$, and the Fourier coefficients, $((-1)^i/\lambda_i)$, in equation (4.160) are next computed as EXI andCI, respectively:

```
EI=DFLOAT(I)*PI
EXI=-(EI**2)*T
SIGN=-1.0D0*SIGN
CI=SIGN/EI
```

3. The ith term in the series of equation (4.160), $((-1)^i/\lambda_i)e^{-D\lambda_i^2 t}\cos(\lambda_i x)$, is then

```
FINITE ELEMENT SOLUTION
0.         1.0       0.1
   51                     0.0000001
END OF RUNS
```

Program 4.9e Data files for Programs 4.9a to d.

computed and added to SUM as the next term in the series:

```
SUM=SUM+CI*DEXP(EXI)*DCOS(EI*X(IG))
```

In order to avoid an underflow of the EXP function, a test is also made on the exponent, $-D\lambda_i^2 t$. If this exponent has a value smaller (more negative) than -100, the subsequent terms are considered to be negligibly small, and the summation in equation (4.160) is terminated:

```
C...
C...      AVOID UNDERFLOW OF THE EXP FUNCTION
          IF(EXI.LT.-100.0D0)GO TO 2
```

4. After either IS ($=1000$) terms are summed or the summation is terminated by the preceding test of the exponent, $-D\lambda_i^2 t$, the final series is evaluated according to equation (4.160):

```
C...
C...   SUMMATION OF SERIES IS COMPLETE
2      SERIES=2.0D0*SUM
```

The large number of terms (IS = 1000) was programmed because of the slow convergence of the series in equation (4.160), particularly for $t = 0$. This slow convergence is due to the Fourier coefficient, $(-1)^i/\lambda_i$, which is proportional to $1/i$. This is expected since in general, the order of the coefficient, that is p in $1/\lambda_i^p$, equals the order of the $(p-1)$st derivative of the solution which is discontinuous. For the present case, the lowest order derivative which is discontinuous is of order $(p-1) = (1-1) = 0$ since the solution itself is discontinuous as a result of initial condition (4.148).

The data file for Programs 4.9a through d is listed in Program 4.9e. 51 ODEs ($N = 51$ in equations (4.178)) are integrated by LSODE with an accuracy of 0.0000001. The main program is not listed because it is similar to earlier main programs that call LSODE as the ODE integrator, e.g., Program 2.8b.

Abbreviated output from subroutine PRINT of Program 4.9c is listed in Table 4.9a. We can note the following points about the output in Table 4.9a:

1. The tridiagonal coefficient matrix of equations (4.178) is confirmed in the output

```
 COEFFICIENT MATRIX

    1      0.0000      0.3333      0.1667
    2      0.1667      0.6667      0.1667
```

Table 4.9 (a) Numerical Output from Programs 4.9a to e. *Continued next page.*

```
RUN NO. -    1  FINITE ELEMENT SOLUTION

INITIAL T -   0.000D+00

  FINAL T -   0.100D+01

  PRINT T -   0.100D+00

NUMBER OF DIFFERENTIAL EQUATIONS -  51

MAXIMUM INTEGRATION ERROR -   0.100D-06


COEFFICIENT MATRIX

    1       0.0000       0.3333       0.1667
    2       0.1667       0.6667       0.1667
    3       0.1667       0.6667       0.1667
    4       0.1667       0.6667       0.1667
    5       0.1667       0.6667       0.1667
              .                         .
              .                         .
              .                         .

   45       0.1667       0.6667       0.1667
   46       0.1667       0.6667       0.1667
   47       0.1667       0.6667       0.1667
   48       0.1667       0.6667       0.1667
   49       0.1667       0.6667       0.1667
   50       0.1667       0.6667       0.1667
   51       0.1667       0.3333       0.0000



   T =    0.00   ENERGY =    0.50000
                   X=0       X=0.2      X=0.4      X=0.6      X=0.8      X=1.0
   U(X,T)  0.000000  0.000000  0.000000  1.000000  1.000000  1.000000
  UE(X,T)  0.000318  0.000393  0.001030  0.998970  0.999607  0.999682
DIFF(X,T) -0.000318 -0.000393 -0.001030  0.001030  0.000393  0.000318

   T =    0.10   ENERGY =    0.50000
                   X=0       X=0.2      X=0.4      X=0.6      X=0.8      X=1.0
   U(X,T)  0.262911  0.308159  0.426703  0.573297  0.691841  0.737089
  UE(X,T)  0.262756  0.308033  0.426655  0.573345  0.691967  0.737244
DIFF(X,T)  0.000154  0.000126  0.000049 -0.000049 -0.000126 -0.000154

   T =    0.20   ENERGY =    0.50000
                   X=0       X=0.2      X=0.4      X=0.6      X=0.8      X=1.0
   U(X,T)  0.411653  0.428526  0.472699  0.527301  0.571474  0.588347
  UE(X,T)  0.411566  0.428456  0.472673  0.527327  0.571544  0.588434
DIFF(X,T)  0.000086  0.000070  0.000027 -0.000027 -0.000070 -0.000086
```

Table 4.9 (a) *Continued.*

```
  T =   0.30  ENERGY =   0.50000
                X=0       X=0.2      X=0.4      X=0.6      X=0.8      X=1.0
  U(X,T)  0.467083  0.473369  0.489828  0.510172  0.526631  0.532917
 UE(X,T)  0.467040  0.473335  0.489815  0.510185  0.526665  0.532960
DIFF(X,T)  0.000043  0.000035  0.000013 -0.000013 -0.000035 -0.000043

  T =   0.40  ENERGY =   0.50000
                X=0       X=0.2      X=0.4      X=0.6      X=0.8      X=1.0
  U(X,T)  0.487735  0.490078  0.496210  0.503790  0.509922  0.512265
 UE(X,T)  0.487716  0.490062  0.496204  0.503796  0.509938  0.512284
DIFF(X,T)  0.000020  0.000016  0.000006 -0.000006 -0.000016 -0.000020

  T =   0.50  ENERGY =   0.50000
                X=0       X=0.2      X=0.4      X=0.6      X=0.8      X=1.0
  U(X,T)  0.495430  0.496303  0.498588  0.501412  0.503697  0.504570
 UE(X,T)  0.495422  0.496296  0.498585  0.501415  0.503704  0.504578
DIFF(X,T)  0.000009  0.000007  0.000003 -0.000003 -0.000007 -0.000009
                          .                            .
                          .                            .
                          .                            .

  T =   1.00  ENERGY =   0.50000
                X=0       X=0.2      X=0.4      X=0.6      X=0.8      X=1.0
  U(X,T)  0.499967  0.499973  0.499990  0.500010  0.500027  0.500033
 UE(X,T)  0.499967  0.499973  0.499990  0.500010  0.500027  0.500033
DIFF(X,T)  0.000000  0.000000  0.000000  0.000000  0.000000  0.000000


LSODE COMPUTATIONAL STATISTICS

LAST STEP SIZE                          0.398D-01

LAST ORDER OF THE METHOD                        5

TOTAL NUMBER OF STEPS TAKEN                   222

NUMBER OF FUNCTION EVALUATIONS               1891

NUMBER OF JACOBIAN EVALUATIONS                 32
```

```
                  .                       .
                  .                       .

      50      0.1667      0.6667      0.1667
      51      0.1667      0.3333      0.0000
```

2. The numerical and analytical solutions agree to within three figures, except for $t = 0$ where the analytical solution is actually in error (the initial condition is given by equation (4.148)). The error in the analytical solution at $t = 0$ is again due to the slow convergence of the series of equation (4.160) as discussed after Program 4.9d.

3. The energy computed according to equation (4.163) is accurate to five figures, indicating that the finite element method as formulated for this problem (i.e., equations (4.178)) conserves energy according to boundary conditions (4.159) and (4.160).

4. Subroutine LSODE computed a solution with modest computational effort, as reflected in the computational statistics. This is in contrast with the explicit ODE integrator RKF45 that could not produce a solution with reasonable computational effort; apparently the 51 ODEs of equations (4.178) are stiff, and therefore require an implicit ODE integrator.

As a point of interest, Programs 4.9a to e were executed a second time with only one change; the maximum number of terms in the series solution of equation (4.160) was changed from IS=1000 to IS=1 in Program 4.9d. The resulting output is listed in Table 4.9b. We can note the following points about the output in Table 4.9b:

1. The series solution gives poor accuracy at $t = 0$, which is expected since again, the rate of convergence is proportional to $1/\lambda_i$ (or $1/i$ since $\lambda_i = i\pi$), and we have limited i to 1 by using only one term in the series.

2. However, for $t > 0$, the exponential term $e^{-D\lambda_i^2 t}$ in equation (4.160) causes rapid convergence of the series solution. Thus, for $t = 0.1$, the numerical and analytical solutions agree to three figures. The agreement improves with increasing t because of the more rapid convergence of the series with increasing t. In other words, for $t > 0$, the exact solution is defined accurately using only one eigenvalue, $\lambda_1 = \pi$.

This concludes the discussion of the finite element method, which clearly is only the briefest introduction for one interpretation of the method. The finite element method is widely used in science and engineering, particularly for the solution of multidimensional PDEs on irregular domains, and a large literature has developed for this method. We have indicated here only a few of the basic concepts. In summary,

1. The finite element method was formulated within the Galerkin method defined by equation (4.180)

$$\int_{x_1}^{x_N} \phi_j(x) R(x, t)\, dx = 0 \tag{4.180}$$

   The basis functions, $\phi_j(x)$, were selected as the linear finite elements of Figure 4.8a and were made orthogonal to the residual, $R(x, t)$, according to equation (4.180). Many other basis functions have been used in the finite element method, e.g., splines and other polynomials. In any case, the integrals defined by equation (4.180) must be evaluated, which is a major part of the the effort in using the finite element method; examples of these integrals are given by equations (4.165) to (4.167), (4.169), (4.172), (4.173), (4.175), and (4.176). Once these integrals are evaluated, they can often be used in later applications. For example, the integrals of equations (4.165) to (4.167) are generally useful for PDEs with first-order derivatives in t. However, new integrals are generally required for new PDEs, and they may be so complicated as to require numerical quadrature; this is also the case if the basis functions are relatively complicated.

2. Generally, the finite element method might evolve from application of the method

Table 4.9 (b) Output from Programs 4.9a to e with One Term in the Series Solution (4.160) Programmed in Function SERIES. *Continued next page.*

```
RUN NO. -    1  FINITE ELEMENT SOLUTION

INITIAL T -  0.000D+00

  FINAL T -  0.100D+01

  PRINT T -  0.100D+00

NUMBER OF DIFFERENTIAL EQUATIONS -  51

MAXIMUM INTEGRATION ERROR -  0.100D-06


COEFFICIENT MATRIX

    1        0.0000       0.3333       0.1667
    2        0.1667       0.6667       0.1667
               .                         .
               .                         .
               .                         .

   50        0.1667       0.6667       0.1667
   51        0.1667       0.3333       0.0000



  T =    0.00  ENERGY =    0.50000
                  X=0       X=0.2      X=0.4      X=0.6      X=0.8      X=1.0
  U(X,T)  0.000000  0.000000  0.000000  1.000000  1.000000  1.000000
 UE(X,T) -0.136620 -0.015036  0.303274  0.696726  1.015036  1.136620
DIFF(X,T)  0.136620  0.015036 -0.303274  0.303274 -0.015036 -0.136620

  T =    0.10  ENERGY =    0.50000
                  X=0       X=0.2      X=0.4      X=0.6      X=0.8      X=1.0
  U(X,T)  0.262911  0.308159  0.426703  0.573297  0.691841  0.737089
 UE(X,T)  0.262727  0.308042  0.426679  0.573321  0.691958  0.737273
DIFF(X,T)  0.000184  0.000117  0.000025 -0.000025 -0.000117 -0.000184

  T =    0.20  ENERGY =    0.50000
                  X=0       X=0.2      X=0.4      X=0.6      X=0.8      X=1.0
  U(X,T)  0.411653  0.428526  0.472699  0.527301  0.571474  0.588347
 UE(X,T)  0.411566  0.428456  0.472673  0.527327  0.571544  0.588434
DIFF(X,T)  0.000086  0.000070  0.000027 -0.000027 -0.000070 -0.000086

  T =    0.30  ENERGY =    0.50000
                  X=0       X=0.2      X=0.4      X=0.6      X=0.8      X=1.0
  U(X,T)  0.467083  0.473369  0.489828  0.510172  0.526631  0.532917
 UE(X,T)  0.467040  0.473335  0.489815  0.510185  0.526665  0.532960
DIFF(X,T)  0.000043  0.000035  0.000013 -0.000013 -0.000035 -0.000043
```

Table 4.9 (b) *Continued.*

```
 T =    0.40   ENERGY =     0.50000
                       X=0        X=0.2       X=0.4       X=0.6       X=0.8       X=1.0
   U(X,T)   0.487735  0.490078  0.496210  0.503790  0.509922  0.512265
  UE(X,T)   0.487716  0.490062  0.496204  0.503796  0.509938  0.512284
DIFF(X,T)   0.000020  0.000016  0.000006 -0.000006 -0.000016 -0.000020

 T =    0.50   ENERGY =     0.50000
                       X=0        X=0.2       X=0.4       X=0.6       X=0.8       X=1.0
   U(X,T)   0.495430  0.496303  0.498588  0.501412  0.503697  0.504570
  UE(X,T)   0.495422  0.496296  0.498585  0.501415  0.503704  0.504578
DIFF(X,T)   0.000009  0.000007  0.000003 -0.000003 -0.000007 -0.000009
                                 .                              .
                                 .                              .
                                 .                              .

 T =    1.00   ENERGY =     0.50000
                       X=0        X=0.2       X=0.4       X=0.6       X=0.8       X=1.0
   U(X,T)   0.499967  0.499973  0.499990  0.500010  0.500027  0.500033
  UE(X,T)   0.499967  0.499973  0.499990  0.500010  0.500027  0.500033
DIFF(X,T)   0.000000  0.000000  0.000000  0.000000  0.000000  0.000000


LSODE COMPUTATIONAL STATISTICS

LAST STEP SIZE                          0.398D-01

LAST ORDER OF THE METHOD                        5

TOTAL NUMBER OF STEPS TAKEN                   222

NUMBER OF FUNCTION EVALUATIONS               1891

NUMBER OF JACOBIAN EVALUATIONS                 32
```

of weighted residuals defined by the integral

$$\int_{x_1}^{x_N} w_j(x) R(x,t)\, dx = 0 \tag{4.181}$$

where $w_j(x)$ is a weighting function. For example, if $w_j(x) = \delta(x - x_j)$ (the *Dirac delta function*), the resulting method for approximating the PDEs is called *collocation*. This choice of a weighting function is particularly advantageous since the integrals of equation (4.181) are immediately available

$$\int_{x_1}^{x_N} \delta_j(x - x_j) R(x,t)\, dx = R(x_j, t) = 0 \tag{4.182}$$

that is, we merely set the PDE residuals to zero at a series of N collocation points, $x_j = 1, 2, \ldots N$. This again produces a system of N semi-implicit ODEs with a coefficient matrix that is of order N (rather than tridiagonal as in equation (4.178)).

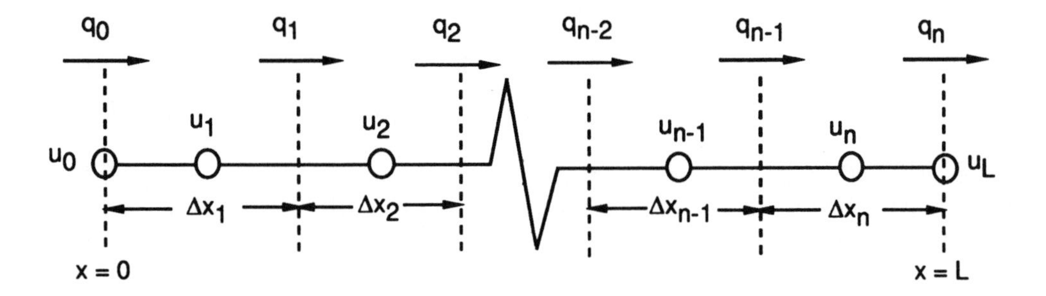

Figure 4.9 Control Volume Discretization of Equations (4.144) to (4.147).

3. The method of weighted residuals therefore generally produces systems of semi-implicit ODEs of the form of equation (1.12). We must therefore have a method available for the integration of semi-implicit ODEs. Special integrators like `LSODI` [Byrne and Hindmarsh (1987)] and `DASSL` [Brenan et al. (1989)] are available for this purpose.

4. Finally, the preceding development of the finite element method was approached through the use of the method of weighted residuals, that is, equation (4.181). Another approach is to use a *variational principle* [Johnson (1990)], which we will not consider here to keep the discussion to a reasonable length.

4.9 The Finite Volume Method

The finite difference and finite element methods considered previously are two examples of general approaches to the discretization of PDEs. In particular, we applied these methods to the discretization of the spatial (boundary value) derivatives of PDEs, then completed the approximation of the PDEs via the numerical method of lines. However, other possibilities for discretization can be considered. We now consider a third widely used approach, the *method of finite volumes*. Briefly, this method is based on the use of basic conservation principles such as the conservation of energy applied to a finite or incremental volume of the system to be analyzed, which is usually called a *control volume* [Patankar (1980), Lick (1989)], as illustrated in Figure 4.9.

To illustrate this approach, we again return to the heat conduction system defined by equations (4.144) to (1.447). A spatial grid in x is represented in terms of a series of control volumes with internal temperatures u_1 and u_N (u, the dependent variable, represents temperature in the subsequent analysis).

An energy balance written for the volume with temperature u_2 gives

$$V_2\rho_2 C_{p2}\frac{du_2}{dt} = A_1 q_1 - A_2 q_2 \tag{4.183}$$

where u_2 = temperature of control volume 2; t = time; V_2 = volume of control volume 2; ρ_2 = density of material in control volume 2; C_{p2} = specific heat of material in control volume 2; q_1 = heat flux due to temperature difference $u_1 - u_2$; A_1 = area across which q_1 flows; q_2 = heat flux due to temperature difference $u_2 - u_3$; and A_2 = area across which q_2 flows.

If we compute heat fluxes according to Fourier's first law, q_1 and q_2 are

$$q_1 = k_{1,2}\left(\frac{u_1 - u_2}{\Delta x_1/2 + \Delta x_2/2}\right),$$

$$q_2 = k_{2,3}\left(\frac{u_2 - u_3}{\Delta x_2/2 + \Delta x_3/2}\right) \tag{4.184)(4.185}$$

where $k_{1,2}$ = thermal conductivity as a function of u_1 and u_2 (to be defined); $k_{2,3}$ = thermal conductivity as a function of u_2 and u_3 (to be defined); and $\Delta x_1, \Delta x_2, \Delta x_3$ = lengths indicated in Figure 4.9.

Then substitution of equations (4.184) and (4.185) in equation (4.183) gives an ODE for the temperature u_2. However, in a computer code, it is often more convenient to compute the fluxes separately, for example, from equations (4.184) and (4.185), then use them in the energy balance, e.g., equation (4.183).

To further develop this approach, we can now consider an *interface temperature*, u_i, at the left face of volume V_2. Then, if we assume continuity in the heat flux q_1 across the left face

$$q_1 = k_{1,2}\left(\frac{u_1 - u_2}{\Delta x_1/2 + \Delta x_2/2}\right) = k_1\left(\frac{u_1 - u_i}{\Delta x_1/2}\right) = k_2\left(\frac{u_i - u_2}{\Delta x_2/2}\right) \tag{4.186}$$

u_i obtained from the last of equations (4.186) is

$$\left\{\left(\frac{k_1}{\Delta x_1/2}\right) + \left(\frac{k_2}{\Delta x_2/2}\right)\right\} u_i = \left(\frac{k_1}{\Delta x_1/2}\right) u_1 + \left(\frac{k_2}{\Delta x_2/2}\right) u_2$$

or

$$u_i = \frac{\left(\frac{k_1}{\Delta x_1/2}\right) u_1 + \left(\frac{k_2}{\Delta x_2/2}\right) u_2}{\left(\frac{k_1}{\Delta x_1/2}\right) + \left(\frac{k_2}{\Delta x_2/2}\right)} \tag{4.187}$$

Then, we can form the temperature difference u_i - u_2 (after cancellation of the common factor of 2 in equation (4.187))

$$u_i - u_2 = \frac{\left(\frac{k_1}{\Delta x_1}\right) u_1 + \left(\frac{k_2}{\Delta x_2}\right) u_2}{\left(\frac{k_1}{\Delta x_1}\right) + \left(\frac{k_2}{\Delta x_2}\right)} - u_2$$

or

$$u_i - u_2 = \frac{\left(\frac{k_1}{\Delta x_1}\right) u_1 + \left(\frac{k_2}{\Delta x_2}\right) u_2 - u_2\left\{\left(\frac{k_1}{\Delta x_1}\right) + \left(\frac{k_2}{\Delta x_2}\right)\right\}}{\left(\frac{k_1}{\Delta x_1}\right) + \left(\frac{k_2}{\Delta x_2}\right)} = \frac{\left(\frac{k_1}{\Delta x_1}\right)(u_1 - u_2)}{\left(\frac{k_1}{\Delta x_1}\right) + \left(\frac{k_2}{\Delta x_2}\right)} \tag{4.188}$$

Then, from equations (4.186) and (4.188),

$$q_1 = k_{1,2}\left(\frac{u_1 - u_2}{\Delta x_1/2 + \Delta x_2/2}\right) = k_2\left(\frac{u_i - u_2}{\Delta x_2/2}\right)$$

$$= \frac{2k_1k_2/(\Delta x_1 \Delta x_2)}{\left(\frac{k_1}{\Delta x_1}\right) + \left(\frac{k_2}{\Delta x_2}\right)}(u_1 - u_2) = \frac{2}{\left(\frac{\Delta x_1}{k_1}\right) + \left(\frac{\Delta x_2}{k_2}\right)}(u_1 - u_2)$$

or from equation (4.184)

$$k_{1,2} = \frac{(\Delta x_1 + \Delta x_2)}{\left(\frac{\Delta x_1}{k_1}\right) + \left(\frac{\Delta x_2}{k_2}\right)} \tag{4.189}$$

which is the result reported by Patankar [(1980) equation (4.9)]. For the special case $\Delta x_1 = \Delta x_2 = \Delta x$, equations (4.189) reduce to

$$k_{1,2} = \frac{2k_1k_2}{k_1 + k_2} \tag{4.190}$$

so that $k_{1,2}$ is the *harmonic mean* of k_1 and k_2.

We can now generalize equations (4.183), (4.184), (4.185), and (4.189) to

$$V_i\rho_iC_{pi}\frac{du_i}{dt} = A_{i-1}q_{i-1} - A_iq_i \tag{4.191}$$

$$q_{i-1} = k_{i-1,i}\left(\frac{u_{i-1} - u_i}{\Delta x_{i-1}/2 + \Delta x_i/2}\right) \tag{4.192}$$

$$q_i = k_{i,i+1}\left(\frac{u_i - u_{i+1}}{\Delta x_i/2 + \Delta x_{i+1}/2}\right) \tag{4.193}$$

$$k_{i-1,i} = \frac{(\Delta x_{i-1} + \Delta x_i)}{\left(\frac{\Delta x_{i-1}}{k_{i-1}}\right) + \left(\frac{\Delta x_i}{k_i}\right)} \tag{4.194}$$

where i is a grid index and does not denote an interface. By writing equations (4.191) to (4.194) in terms of the grid index i, it is possible to vary the properties as a function of u_i, i.e., nonlinear characteristics can be included by programming ρ_i, C_{pi}, $k_{i-1,i}$, and $k_{i,i+1}$ as a function of u_i. Also, the geometric parameters can be varied with i, i.e., Δx_i, V_i, A_{i-1}, and A_i as a function of i, which could be used to implement equation (4.191) in various coordinate systems, e.g., cylindrical or spherical.

We now consider the application of equations (4.191) to (4.194) to equations (4.144), (4.148), (4.149), and (4.150) for the special case $\Delta x_{i-1} = \Delta x_i = \Delta x_{i+1} = \Delta x$ (a uniformly spaced grid), $A_{i-1} = A_i = A$, $V_i = V$ (constant area and volume), $k_{i-1} = k_i = k_{i+1} = k$, $\rho_i = \rho$, and $C_{pi} = C_p$ (constant physical properties) and $D/(\rho C_p) = 1$ (unit thermal diffusivity). Also, since we have considered the solution of equations (4.144) and (4.148) to (4.150) previously by the finite element method (in Section 4.8), and we have the approximations readily available to solve these equations by finite differences, we now combine the three approaches (finite differences, elements, and volumes) in a single program that will facilitate a comparison between them.

Subroutine `INITAL` call subroutines `INIT1`, `INIT2`, and `INIT3` for the three methods of approximations, as indicated in Program 4.10a.

We can note the following points about these subroutines:

1. Subroutine `INITAL` calls `INIT1` for `NORUN=1` and 2, `INIT2` for `NORUN=3`, and `INIT4` for `NORUN=4`. Subroutine `INIT1` is called for two cases: spatial differentiation by `DSS042` and `DSS044`.

2. Subroutine `INIT1` is a straightforward implementation of initial condition (4.148), and, in fact, is the same as `INIT2`, which was discussed previously as Program 4.9a, except that the definition of the tridiagonal coefficients (of equation (4.178)) is not included since the finite difference method leads directly to explicit ODEs (with only one derivative in each ODE).

3. Subroutine `INIT2` was discussed previously in Program 4.9a.

```
      SUBROUTINE INITAL
      IMPLICIT DOUBLE PRECISION (A-H,O-Z)
      COMMON/T/      T,  NSTOP,  NORUN
C...
C...  FINITE DIFFERENCES
      IF((NORUN.EQ.1).OR.(NORUN.EQ.2))CALL INIT1
C...
C...  LINEAR FINITE ELEMENTS
      IF(NORUN.EQ.3)CALL INIT2
C...
C...  FINITE VOLUMES
      IF(NORUN.EQ.4)CALL INIT3
      RETURN
      END

      SUBROUTINE INIT1
C...
C...  SOLUTION BY FINITE DIFFERENCES
C...
      IMPLICIT DOUBLE PRECISION (A-H,O-Z)
      PARAMETER (N=51)
      COMMON/T/      T,  NSTOP,  NORUN
     1      /Y/   U(N)
     2      /F/  UT(N)
     3     /XG/      L,     DX,   X(N)
     4      /I/     IP
C...
C...  TYPE-SELECTED VARIABLES AS DOUBLE PRECISION
      DOUBLE PRECISION L
C...
C...  LENGTH
      L=1.0D0
C...
C...  GRID SPACING
      DX=L/DFLOAT(N-1)
C...
C...  INITIAL CONDITION
      DO 1 I=1,N
         X(I)=DFLOAT(I-1)*DX
         IF(I.LT.(N+1)/2)U(I)=0.0D0
         IF(I.GT.(N+1)/2)U(I)=1.0D0
         IF(I.EQ.(N+1)/2)U(I)=0.5D0
1     CONTINUE
C...
C...  INITIALIZE THE TEMPORAL DERIVATIVES IN COMMON/F/
      CALL DERV
      IP=0
      RETURN
      END

      SUBROUTINE INIT2
```

Program 4.10a Subroutines INITAL, INIT1, INIT2, and INIT3 for the Method of Lines Solution of Equations (4.144), (4.148), (4.149) and (4.150) by Finite Differences, Elements, and Volumes. *Continued next pages.*

```
C...
C...  SOLUTION BY LINEAR FINITE ELEMENTS (HAT FUNCTIONS)
C...
      IMPLICIT DOUBLE PRECISION (A-H,O-Z)
      PARAMETER (N=51)
      COMMON/T/        T,  NSTOP,  NORUN
     1        /Y/   U(N)
     2        /F/  UT(N)
     3       /XG/       L,      DX,   X(N)
     4        /I/      IP
     5       /FE/  AL(N),  BM(N),  CU(N), BRHS(N)
C...
C...  TYPE-SELECTED VARIABLES AS DOUBLE PRECISION
      DOUBLE PRECISION L
C...
C...  LENGTH
      L=1.0D0
C...
C...  GRID SPACING
      DX=L/DFLOAT(N-1)
C...
C...  SET UP THE COEFFICIENT MATRIX IN BAND STORAGE MODE. THIS IS
C...  DONE ONLY ONCE SINCE THE COEFFICIENT MATRIX IS CONSTANT
C...
C...  LOWER DIAGONAL
      DO 2 I=1,N
         IF(I.EQ.1)THEN
            AL(1)=0.0D0
         ELSE
     +   IF(I.EQ.N)THEN
            AL(N)=1.0D0/6.0D0
         ELSE
            AL(I)=1.0D0/6.0D0
         END IF
2     CONTINUE
C...
C...  MAIN DIAGONAL
      DO 3 I=1,N
         IF(I.EQ.1)THEN
            BM(1)=2.0D0/6.0D0
         ELSE
     +   IF(I.EQ.N)THEN
            BM(N)=2.0D0/6.0D0
         ELSE
            BM(I)=4.0D0/6.0D0
         END IF
3     CONTINUE
C...
C...  UPPER DIAGONAL
      DO 4 I=1,N
         IF(I.EQ.1)THEN
            CU(1)=1.0D0/6.0D0
```

Program 4.10a *Continued.*

```
          ELSE
     +    IF(I.EQ.N)THEN
             CU(N)=0.0D0
          ELSE
             CU(I)=1.0D0/6.0D0
          END IF
4     CONTINUE
C...
C...  INITIAL CONDITION
      DO 1 I=1,N
         X(I)=DFLOAT(I-1)*DX
         IF(I.LT.(N+1)/2)U(I)=0.0D0
         IF(I.GT.(N+1)/2)U(I)=1.0D0
         IF(I.EQ.(N+1)/2)U(I)=0.5D0
1     CONTINUE
      CALL DERV
      IP=0
      RETURN
      END

      SUBROUTINE INIT3
C...
C...  SOLUTION BY FINITE VOLUMES
C...
      IMPLICIT DOUBLE PRECISION (A-H,O-Z)
      PARAMETER (N=50)
      COMMON/T/       T,  NSTOP,  NORUN
     1      /Y/   U(N)
     2      /F/  UT(N)
     3     /XG/      L,     DX,    X(N)
     4      /I/     IP
C...
C...  TYPE-SELECTED VARIABLES AS DOUBLE PRECISION
      DOUBLE PRECISION L
C...
C...  LENGTH
      L=1.0D0
C...
C...  GRID SPACING
      DX=L/DFLOAT(N)
C...
C...  INITIAL CONDITION
      DO 1 I=1,N
         X(I)=DFLOAT(I-1)*DX+0.5D0*DX
         IF(I.LE.(N/2))U(I)=0.0D0
         IF(I.GT.(N/2))U(I)=1.0D0
1     CONTINUE
C...
C...  INITIALIZE THE TEMPORAL DERIVATIVES IN COMMON/F/
      CALL DERV
      IP=0
      RETURN
      END
```

Program 4.10a *Continued.*

4. Subroutine INIT3 has a few minor variations on INIT1 and INIT2. First, note that only 50 grid points are used instead of 51. This redefinition of the spatial grid is indicated in Figure 4.9, where the dependent variables $u_1, u_2, \ldots$ are offset by $\Delta x/2$ to the right of the boundary at $x = 0$, and the last value, u_N, is offset $\Delta x/2$ to the left of the boundary at $x = 1$; note this offset of $\Delta x/2$ in computing the grid points in array X(I) in DO loop 1. Also, we do not consider boundary values of u, i.e., $u(0, t)$ and $u(1, t)$, as in the case of finite differences and elements. This slightly different spatial grid facilitates the use of the fluxes q_{i-1} and q_i in equation (4.191), particularly at the boundaries. Specifically, the Neumann (zero flux) boundary conditions, equations (4.149) and (4.150), can be programmed simply as $q_0 = 0$ and $q_N = 0$ (where, now, $N = 50$).

5. As a result of the grid of Figure 4.9, the point of discontinuity of the initial condition (4.148), $x = 0.5$, falls on the interface between two volumes. This situation facilitates the programming of the discontinuity in equation (4.148) in INIT3:

```
C...
C...  INITIAL CONDITION
      DO 1 I=1,N
         X(I)=DFLOAT(I-1)*DX+0.5D0*DX
         IF(I.LE.(N/2))U(I)=0.0D0
         IF(I.GT.(N/2))U(I)=1.0D0
1     CONTINUE
```

Note that the requirement to assign a value to $u(0.5, t)$ is avoided (in contrast with the programming of boundary condition (4.148) in subroutines INIT1 and INIT2).

Subroutines DERV, DERV1, DERV2, and DERV3 are listed in Program 4.10b. We can note the following points about these subroutines:

1. Subroutine DERV calls subroutines DERV1 for NORUN=1 and 2, DERV2 for NORUN=3, and DERV3 for NORUN=4.

2. Subroutine DERV1 is a straightforward finite difference method of lines solution of equation (4.144). In particular, boundary conditions (4.149) and (4.150) are coded as

```
C...
C...  BOUNDARY CONDITION AT X = 0
      UX(1)=0.0D0
      NL=2
C...
C...  BOUNDARY CONDITION AT X = 1
      UX(N)=0.0D0
      NU=2
```

Then either a three- or five-point approximation for u_{xx} in equation (4.144) is computed by DSS042 (NORUN=1) or DSS044 (NORUN=2):

```
C...
C...  U  BY THREE POINT DIFFERENCES
C...   XX
      IF(NORUN.EQ.1)CALL DSS042(0.,L,N,U,UX,UXX,NL,NU)
C...
```

```
      SUBROUTINE DERV
      IMPLICIT DOUBLE PRECISION (A-H,O-Z)

      COMMON/T/      T,  NSTOP,  NORUN
C...
C...  FINITE DIFFERENCES
      IF((NORUN.EQ.1).OR.(NORUN.EQ.2))CALL DERV1
C...
C...  LINEAR FINITE ELEMENTS
      IF(NORUN.EQ.3)CALL DERV2
C...
C...  FINITE VOLUMES
      IF(NORUN.EQ.4)CALL DERV3
      RETURN
      END

      SUBROUTINE DERV1
C...
C...  FINITE DIFFERENCE SOLUTION
C...
      IMPLICIT DOUBLE PRECISION (A-H,O-Z)
      PARAMETER (N=51)
      COMMON/T/      T,  NSTOP,  NORUN
     1      /Y/   U(N)
     2      /F/  UT(N)
     3     /XG/      L,     DX,   X(N)
     4      /I/     IP
C...
C...  TYPE-SELECTED VARIABLES AS DOUBLE PRECISION
      DOUBLE PRECISION L
C...
C...  DEFINE ARRAYS FOR THE SPATIAL DERIVATIVES
      DIMENSION UX(N), UXX(N)
C...
C...  BOUNDARY CONDITION AT X = 0
      UX(1)=0.0D0
      NL=2
C...
C...  BOUNDARY CONDITION AT X = 1
      UX(N)=0.0D0
      NU=2
C...
C...  U  BY THREE POINT DIFFERENCES
C...   XX
      IF(NORUN.EQ.1)CALL DSS042(0.,L,N,U,UX,UXX,NL,NU)
C...
C...  U  BY FIVE POINT DIFFERENCES
C...   XX
      IF(NORUN.EQ.2)CALL DSS044(0.,L,N,U,UX,UXX,NL,NU)
C...
```

Program 4.10b Subroutines DERV, DERV1, DERV2, and DERV3 for the Method of Lines Solution of Equations (4.144), (4.148), (4.149) and (4.150) by Finite Differences, Elements, and Volumes. *Continued next page.*

```
C...  PDE
      DO 1 I=1,N
         UT(I)=UXX(I)
1     CONTINUE
      RETURN
      END

      SUBROUTINE DERV2
C...
C...  FINITE ELEMENT SOLUTION
C...
      IMPLICIT DOUBLE PRECISION (A-H,O-Z)
      PARAMETER (N=51)
      COMMON/T/      T,  NSTOP,  NORUN
     1      /Y/   U(N)
     2      /F/  UT(N)
     3     /XG/      L,      DX,   X(N)
     4      /I/     IP
     5     /FE/  AL(N),  BM(N),  CU(N), BRHS(N)
C...
C...  TYPE-SELECTED VARIABLES AS DOUBLE PRECISION
      DOUBLE PRECISION L
C...
C...  BOUNDARY CONDITION AT X = 0
      H1=0.0D0
C...
C...  BOUNDARY CONDITION AT X = 1
      HN=0.0D0
C...
C...  RIGHT-HAND SIDE VECTOR (WITH D = 1)
      DO 1 I=1,N
         IF(I.EQ.1)THEN
            BRHS(I)=-H1/DX+(U(I+1)-U(I))/DX**2
         ELSE
     +   IF(I.EQ.N)THEN
            BRHS(I)= HN/DX+(U(I-1)-U(I))/DX**2
         ELSE
            BRHS(I)=(U(I+1)-2.0D0*U(I)+U(I-1))/DX**2
         END IF
1     CONTINUE
C...
C...  SOLVE THE LINEAR ALGEBRAIC EQUATIONS BY SUBROUTINE TRIDAG,
C...  WHICH RETURNS THE DERIVATIVE VECTOR UT IN COMMON/F/
      CALL TRIDAG(AL,BM,CU,BRHS,UT,N)
      RETURN
      END

      SUBROUTINE DERV3
C...
C...  FINITE VOLUME SOLUTION
C...
      IMPLICIT DOUBLE PRECISION (A-H,O-Z)
```

Program 4.10b *Continued.*

```
      PARAMETER (N=50)
      COMMON/T/        T,  NSTOP,  NORUN
     1      /Y/    U(N)
     2      /F/   UT(N)
     3     /XG/       L,      DX,    X(N)
     4      /I/      IP
C...
C...  TYPE-SELECTED VARIABLES AS DOUBLE PRECISION
      DOUBLE PRECISION L
C...
C...  DIMENSION ARRAY FOR THE FLUXES
      DIMENSION Q(0:N)
C...
C...  BOUNDARY CONDITION AT X = 0
      Q(0)=0.0D0
C...
C...  BOUNDARY CONDITION AT X = 1
      Q(N)=0.0D0
C...
C...  COMPUTE THE FLUXES (K/(RHO*CP) = 1)
      DO 2 I=1,N-1
         Q(I)=(U(I)-U(I+1))/DX
2     CONTINUE
C...
C...  PDE
      DO 1 I=1,N
         UT(I)=(Q(I-1)-Q(I))/DX
1     CONTINUE
      RETURN
      END
```

Program 4.10b *Continued.*

```
C...  U  BY FIVE POINT DIFFERENCES
C...   XX
      IF(NORUN.EQ.2)CALL DSS044(0.,L,N,U,UX,UXX,NL,NU)
```

PDE (4.144) is finally programmed in `DO` loop 1:

```
C...
C...  PDE
      DO 1 I=1,N
         UT(I)=UXX(I)
1     CONTINUE
```

3. Subroutine `DERV2` is just a repeat of subroutine `DERV` in Program 4.9b for the finite element formulation.

4. Subroutine `DERV3` has a few details specific to the finite volume formulation. The fluxes in equation (4.191) are dimensioned in array `Q`:

```
C...
C...  DIMENSION ARRAY FOR THE FLUXES
      DIMENSION Q(0:N)
```

Note that the subscript for Q starts at 0, for the flux, Q(0), at the left boundary, $x = 0$, and ends at N, for the flux, Q(N), at the right boundary, $x = 1$. The zero flux boundary conditions, equations (4.149) and (4.150), are then programmed as

```
C...
C...  BOUNDARY CONDITION AT X = 0
      Q(0)=0.0D0
C...
C...  BOUNDARY CONDITION AT X = 1
      Q(N)=0.0D0
```

5. The fluxes are next programmed in terms of equations (4.192) and (4.193), with substantial simplification for the case of equal spacing:

```
C...
C...  COMPUTE THE FLUXES (K/(RHO*CP) = 1)
      DO 2 I=1,N-1
         Q(I)=(U(I)-U(I+1))/DX
2     CONTINUE
```

6. The ODEs of equation (4.191) are finally programmed in terms of the fluxes of equations (4.192) and (4.193):

```
C...
C...  PDE
      DO 1 I=1,N
         UT(I)=(Q(I-1)-Q(I))/DX
1     CONTINUE
      RETURN
      END
```

The use of the interface conductivity given by equation (4.194) was not required because the thermal conductivity was assumed to be constant for all of the finite volumes.

The output from Programs 4.10a and b is printed by subroutine PRINT in Program 4.10c, which calls subroutines PRINT1, PRINT2, and PRINT3 for the finite differences, elements, and volumes, respectively.

We can note the following points about these subroutines:

1. Subroutine PRINT1 is very similar to the earlier subroutine PRINT of Program 4.9c, and subroutine PRINT2 is the same as subroutine PRINT of Program 4.9c. Basically, each subroutine computes the energy according to equation (4.161), then prints this energy along with the numerical and analytical solutions and the difference between these two solutions.

2. Subroutine PRINT3 has a few minor differences from PRINT1 and PRINT2. First, the total energy is calculated by summing the energies of the individual volumes:

```
C...
C...  COMPUTE THE TOTAL ENERGY (RHO = CP = 1)
      CALL DERV3
      ENERGY=0.0D0
      DO 1 I=1,N
         ENERGY=ENERGY+U(I)*DX
1     CONTINUE
```

```
      SUBROUTINE PRINT(NI,NO)
      IMPLICIT DOUBLE PRECISION (A-H,O-Z)
      COMMON/T/       T,  NSTOP,  NORUN
C...
C...  FINITE DIFFERENCES
      IF((NORUN.EQ.1).OR.(NORUN.EQ.2))CALL PRINT1(NI,NO)
C...
C...  LINEAR FINITE ELEMENTS
      IF(NORUN.EQ.3)CALL PRINT2(NI,NO)
C...
C...  FINITE VOLUMES
      IF(NORUN.EQ.4)CALL PRINT3(NI,NO)
      RETURN
      END

      SUBROUTINE PRINT1(NI,NO)
C...
C...  FINITE DIFFERENCE SOLUTION
C...
      IMPLICIT DOUBLE PRECISION (A-H,O-Z)
      PARAMETER (N=51)
      COMMON/T/       T,  NSTOP,  NORUN
     1      /Y/   U(N)
     2      /F/  UT(N)
     3     /XG/       L,     DX,    X(N)
     4      /I/      IP
C...
C...  TYPE-SELECTED VARIABLES AS DOUBLE PRECISION
      DOUBLE PRECISION L
C...
C...  DIMENSION ARRAYS FOR THE ANALYTICAL SOLUTION
      DIMENSION UE(N), DIFF(N)
C...
C...  COMPUTE THE TOTAL ENERGY (RHO = CP = 1)
      CALL DERV1
      ENERGY=0.0D0
      DO 1 I=2,N
         ENERGY=ENERGY+0.5D0*(U(I)+U(I-1))*DX
1     CONTINUE
C...
C...  CALCULATE THE EXACT SOLUTION AND THE DIFFERENCE BETWEEN THE
C...  NUMERICAL AND EXACT SOLUTIONS
      DO 3 I=1,N,10
         UE(I)=SERIES(I)
         DIFF(I)=U(I)-UE(I)
3     CONTINUE
C...
C...  PRINT THE NUMERICAL AND EXACT SOLUTIONS AND THE DIFFERENCE
      WRITE(NO,2)T,ENERGY,(U(I),I=1,N,10),
     1                   (UE(I),I=1,N,10),
     2                 (DIFF(I),I=1,N,10)
```

Program 4.10c Subroutines PRINT, PRINT1, PRINT2, and PRINT3 for the Method of Lines Solution of Equations (4.144), (4.148), (4.149) and (4.150) by Finite Differences, Elements, and Volumes. *Continued next page.*

```
2     FORMAT('    T = ',F6.2,'  ENERGY = ',F9.5,
     1            /,14X,'    X=0',5X,'X=0.2',5X,'X=0.4',
     2                5X,'X=0.6',5X,'X=0.8',5X,'X=1.0D0'/,
     3                             '    U(X,T)',6F10.6,/,
     4                             '   UE(X,T)',6F10.6,/,
     5                             ' DIFF(X,T)',6F10.6,/)
      RETURN
      END

      SUBROUTINE PRINT2(NI,NO)
C...
C...  FINITE ELEMENT SOLUTION
C...
      IMPLICIT DOUBLE PRECISION (A-H,O-Z)
      PARAMETER (N=51)
      COMMON/T/      T,  NSTOP,  NORUN
     1      /Y/   U(N)
     2      /F/  UT(N)
     3     /XG/      L,      DX,    X(N)
     4      /I/     IP
     5     /FE/  AL(N),  BM(N),  CU(N), BRHS(N)
C...
C...  TYPE-SELECTED VARIABLES AS DOUBLE PRECISION
      DOUBLE PRECISION L
C...
C...  DIMENSION ARRAYS FOR THE ANALYTICAL SOLUTION
      DIMENSION UE(N), DIFF(N)
C...
C...  PRINT THE COEFFICIENT MATRIX FOR VERIFICATION
      IP=IP+1
      IF(IP.EQ.1)THEN
         WRITE(NO,13)
13       FORMAT(/,' COEFFICIENT MATRIX',/)
         DO 14 I=1,N
            WRITE(NO,15)I,AL(I),BM(I),CU(I)
15          FORMAT(I5,3F12.4)
14       CONTINUE
         WRITE(NO,16)
16       FORMAT(//)
      END IF
C...
C...  COMPUTE THE TOTAL ENERGY (RHO = CP = 1)
      CALL DERV2
      ENERGY=0.0D0
      DO 1 I=2,N
         ENERGY=ENERGY+0.5D0*(U(I)+U(I-1))*DX
1     CONTINUE
C...
C...  CALCULATE THE EXACT SOLUTION AND THE DIFFERENCE BETWEEN THE
C...  NUMERICAL AND EXACT SOLUTIONS
      DO 3 I=1,N,10
         UE(I)=SERIES(I)
         DIFF(I)=U(I)-UE(I)
```

Program 4.10c *Continued.*

```
3     CONTINUE
C...
C...  PRINT THE NUMERICAL AND EXACT SOLUTIONS AND THE DIFFERENCE
      WRITE(NO,2)T,ENERGY,(U(I),I=1,N,10),
     1                    (UE(I),I=1,N,10),
     2                  (DIFF(I),I=1,N,10)
2     FORMAT('     T = ',F6.2,'  ENERGY = ',F9.5,
     1          /,14X,'    X=0',5X,'X=0.2',5X,'X=0.4',
     2              5X,'X=0.6',5X,'X=0.8',5X,'X=1.0D0'/,
     3                        '     U(X,T)',6F10.6,/,
     4                        '    UE(X,T)',6F10.6,/,
     5                        '  DIFF(X,T)',6F10.6,/)
      RETURN
      END

      SUBROUTINE PRINT3(NI,NO)
C...
C...  FINITE VOLUME SOLUTION
C...
      IMPLICIT DOUBLE PRECISION (A-H,O-Z)
      PARAMETER (N=50)
      COMMON/T/       T,  NSTOP,  NORUN
     1      /Y/   U(N)
     2      /F/  UT(N)
     3     /XG/       L,      DX,    X(N)
     4      /I/      IP
C...
C...  TYPE-SELECTED VARIABLES AS DOUBLE PRECISION
      DOUBLE PRECISION L
C...
C...  DIMENSION ARRAYS FOR THE ANALYTICAL SOLUTION
      DIMENSION UE(N), DIFF(N)
C...
C...  COMPUTE THE TOTAL ENERGY (RHO = CP = 1)
      CALL DERV3
      ENERGY=0.0D0
      DO 1 I=1,N
         ENERGY=ENERGY+U(I)*DX
1     CONTINUE
C...
C...  CALCULATE THE EXACT SOLUTION AND THE DIFFERENCE BETWEEN THE
C...  NUMERICAL AND EXACT SOLUTIONS
      DO 3 I=1,21,5
         UE(I)=SERIES(I)
         DIFF(I)=U(I)-UE(I)
3     CONTINUE
C...
C...  PRINT THE NUMERICAL AND EXACT SOLUTIONS AND THE DIFFERENCE
      UAVG=(U(N/2)+U(N/2+1))/2.0D0
      WRITE(NO,2)T,ENERGY,(U(I),I=1,21,5),UAVG,
     1                    (UE(I),I=1,21,5),
     2                  (DIFF(I),I=1,21,5)
```

Program 4.10c *Continued.*

```
2      FORMAT('    T = ',F6.2,'  ENERGY = ',F9.5,
     1          /,14X,'X=0.01',4X,'X=0.11',4X,'X=0.21',
     2            4X,'X=0.31',4X,'X=0.41',4X,'X=0.50'/,
     3                          '     U(X,T)',6F10.6,/,
     4                          '    UE(X,T)',5F10.6,/,
     5                          '  DIFF(X,T)',5F10.6,/)
       RETURN
       END
```

Program 4.10c *Continued.*

3. Next, the analytical solution is computed for the left half of the system (up to and including volume $N/2 = 50/2 = 25$) and the difference between the numerical and analytical solutions is then computed. These grid points were selected for printing the solution so that the value of the interface temperature, $u(0.5, t)$, could also be conveniently included in the output:

```
C...
C...  CALCULATE THE EXACT SOLUTION, AND THE DIFFERENCE BETWEEN THE
C...  NUMERICAL AND EXACT SOLUTIONS
      DO 3 I=1,21,5
         UE(I)=SERIES(I)
         DIFF(I)=U(I)-UE(I)
3     CONTINUE
```

Note that because of the spatial grid defined in `INIT3` and stored in array in `X(N)` (with an offset $\Delta x/2$). the calls to function `SERIES` give the required values of the analytical solution.

4. The interface temperature at $x = 1/2$ is computed as `UAVG` which is the average of the volume temperatures on either side of $x = 1/2$. Then the numerical and analytical solutions and their difference are printed, including the temperature at $x = 1/2$. Note also the values of x corresponding to these solution values are $x = 0.01$ (due to the offset $\Delta x/2 = 1/((50)(2)) = 0.01$), 0.11, 0.21, 0.31, 0.41, and 0.5.

```
C...
C...  PRINT THE NUMERICAL AND EXACT SOLUTIONS AND THE DIFFERENCE
      UAVG=(U(N/2)+U(N/2+1))/2.0D0
      WRITE(NO,2)T,ENERGY,(U(I),I=1,21,5),UAVG,
     1                   (UE(I),I=1,21,5),
     2                 (DIFF(I),I=1,21,5)
2     FORMAT('    T = ',F6.2,'  ENERGY = ',F9.5,
     1          /,14X,'X=0.01',4X,'X=0.11',4X,'X=0.21',
     2            4X,'X=0.31',4X,'X=0.41',4X,'X=0.50'/,
     3                          '     U(X,T)',6F10.6,/,
     4                          '    UE(X,T)',5F10.6,/,
     5                          '  DIFF(X,T)',5F10.6,/)
```

Function `SERIES` is the same as in Program 4.9d, and therefore it is not listed here. The data file for Programs 4.10a to c is listed in Program 4.10d.

```
FINITE DIFFERENCE
0.          1.0         0.1
    51                        0.0000001
FINITE DIFFERENCE
0.          1.0         0.1
    51                        0.0000001
FINITE ELEMENT
0.          1.0         0.1
    51                        0.0000001
FINITE VOLUME
0.          1.0         0.1
    50                        0.0000001
END OF RUNS
```

Program 4.10d Data File for the Method of Lines Solution of Equations (4.144), (4.148), (4.149), and (4.150) by Finite Differences, Elements, and Volumes.

As discussed previously, the finite difference and finite element solutions are based on the integration of 51 ODEs, while the finite volume solution is based on 50 ODEs. The integration is again by subroutine LSODE; the main program that calls LSODE is not listed because it is the same as earlier main programs which call LSODE, e.g., Program 2.8b.

Abbreviated output from Programs 4.10a to d is listed in Table 4.10a. We can note the following details about the output in Table 4.10a:

1. All three methods of approximation: finite differences, elements, and volumes gave comparable accuracy. This is demonstrated in Table 4.10b, where the solutions are compared at $t = 0.1$.

   Also, in each case, the energy computed according to equation (4.163) was constant to five figures.

2. The computational effort by LSODE was modest in all four runs. Again, LSODE was selected because the runs with RKF45 were long, especially in the case of the finite element formulation, indicating possible stiffness in the ODEs.

4.10 A Two-Dimensional Advective Equation

We conclude this chapter with a final example to demonstrate the method of lines solution of a multidimensional PDE, in this case, the advection equation in two dimensions

$$\frac{\partial u}{\partial t} = -v_x \frac{\partial u}{\partial x} - v_y \frac{\partial u}{\partial y} \tag{4.195}$$

where v_x and v_y are given constants (physically, they are velocities in the x- and y-directions, respectively).

The initial and boundary conditions for equation (4.195) are taken as

$$u(x, y, 0) = f(-x/v_x - y/v_y) \tag{4.196}$$

$$u(0, y, t) = f(2t - y/v_y) \tag{4.197}$$

$$u(x, 0, t) = f(2t - x/v_x) \tag{4.198}$$

Table 4.10 (a) Abbreviated Output from Programs 10.4a to d. *Continued next pages.*

```
RUN NO. -    1  FINITE DIFFERENCE

INITIAL T -  0.000D+00

  FINAL T -  0.100D+01

  PRINT T -  0.100D+00

NUMBER OF DIFFERENTIAL EQUATIONS -  51

MAXIMUM INTEGRATION ERROR -  0.100D-06

   T =    0.00  ENERGY =    0.50000
                 X=0      X=0.2      X=0.4      X=0.6      X=0.8      X=1.0D0
   U(X,T)  0.000000  0.000000  0.000000  1.000000  1.000000  1.000000
  UE(X,T)  0.000318  0.000393  0.001030  0.998970  0.999607  0.999682
DIFF(X,T) -0.000318 -0.000393 -0.001030  0.001030  0.000393  0.000318

   T =    0.10  ENERGY =    0.50000
                 X=0      X=0.2      X=0.4      X=0.6      X=0.8      X=1.0D0
   U(X,T)  0.262758  0.308032  0.426654  0.573346  0.691968  0.737242
  UE(X,T)  0.262756  0.308033  0.426655  0.573345  0.691967  0.737244
DIFF(X,T)  0.000002 -0.000001 -0.000001  0.000001  0.000001 -0.000002

   T =    0.20  ENERGY =    0.50000
                 X=0      X=0.2      X=0.4      X=0.6      X=0.8      X=1.0D0
   U(X,T)  0.411540  0.428434  0.472664  0.527336  0.571566  0.588460
  UE(X,T)  0.411566  0.428456  0.472673  0.527327  0.571544  0.588434
DIFF(X,T) -0.000026 -0.000022 -0.000008  0.000008  0.000022  0.000026

   T =    0.30  ENERGY =    0.50000
                 X=0      X=0.2      X=0.4      X=0.6      X=0.8      X=1.0D0
   U(X,T)  0.467020  0.473319  0.489809  0.510191  0.526681  0.532980
  UE(X,T)  0.467040  0.473335  0.489815  0.510185  0.526665  0.532960
DIFF(X,T) -0.000020 -0.000016 -0.000006  0.000006  0.000016  0.000020

   T =    0.40  ENERGY =    0.50000
                 X=0      X=0.2      X=0.4      X=0.6      X=0.8      X=1.0D0
   U(X,T)  0.487704  0.490053  0.496200  0.503800  0.509947  0.512296
  UE(X,T)  0.487716  0.490062  0.496204  0.503796  0.509938  0.512284
DIFF(X.T) -0.000011 -0.000009 -0.000004  0.000004  0.000009  0.000011

   T =    0.50  ENERGY =    0.50000
                 X=0      X=0.2      X=0.4      X=0.6      X=0.8      X=1.0D0
   U(X,T)  0.495416  0.496291  0.498583  0.501417  0.503709  0.504584
  UE(X,T)  0.495422  0.496296  0.498585  0.501415  0.503704  0.504578
DIFF(X,T) -0.000006 -0.000005 -0.000002  0.000002  0.000005  0.000006
                              .                              .
                              .                              .
                              .                              .

   T =    1.00  ENERGY =    0.50000
                 X=0      X=0.2      X=0.4      X=0.6      X=0.8      X=1.0D0
```

Table 4.10 (a) *Continued.*

```
   U(X,T)  0.499967  0.499973  0.499990  0.500010  0.500027  0.500033
  UE(X,T)  0.499967  0.499973  0.499990  0.500010  0.500027  0.500033
DIFF(X,T)  0.000000  0.000000  0.000000  0.000000  0.000000  0.000000

LSODE COMPUTATIONAL STATISTICS

LAST STEP SIZE                          0.230D-01

LAST ORDER OF THE METHOD                        5

TOTAL NUMBER OF STEPS TAKEN                   210

NUMBER OF FUNCTION EVALUATIONS               1581

NUMBER OF JACOBIAN EVALUATIONS                 26


RUN NO. -    2  FINITE DIFFERENCE

INITIAL T -  0.000D+00

  FINAL T -  0.100D+01

  PRINT T -  0.100D+00

NUMBER OF DIFFERENTIAL EQUATIONS -  51

MAXIMUM INTEGRATION ERROR -  0.100D-06

   T =   0.00  ENERGY =   0.50000
                X=0      X=0.2      X=0.4      X=0.6      X=0.8     X=1.0D0
   U(X,T)  0.000000  0.000000  0.000000  1.000000  1.000000  1.000000
  UE(X,T)  0.000318  0.000393  0.001030  0.998970  0.999607  0.999682
DIFF(X,T) -0.000318 -0.000393 -0.001030  0.001030  0.000393  0.000318

   T =   0.10  ENERGY =   0.50000
                X=0      X=0.2      X=0.4      X=0.6      X=0.8     X=1.0D0
   U(X,T)  0.262834  0.308096  0.426679  0.573321  0.691904  0.737166
  UE(X,T)  0.262756  0.308033  0.426655  0.573345  0.691967  0.737244
DIFF(X,T)  0.000078  0.000063  0.000024 -0.000024 -0.000063 -0.000078

   T =   0.20  ENERGY =   0.50000
                X=0      X=0.2      X=0.4      X=0.6      X=0.8     X=1.0D0
   U(X,T)  0.411595  0.428479  0.472681  0.527319  0.571521  0.588405
  UE(X,T)  0.411566  0.428456  0.472673  0.527327  0.571544  0.588434
DIFF(X,T)  0.000029  0.000023  0.000009 -0.000009 -0.000023 -0.000029

   T =   0.30  ENERGY =   0.50000
                X=0      X=0.2      X=0.4      X=0.6      X=0.8     X=1.0D0
   U(X,T)  0.467051  0.473343  0.489818  0.510182  0.526657  0.532949
  UE(X,T)  0.467040  0.473335  0.489815  0.510185  0.526665  0.532960
DIFF(X,T)  0.000011  0.000009  0.000003 -0.000003 -0.000009 -0.000011
```

Table 4.10 (a) *Continued.*

```
   T =   0.40  ENERGY =   0.50000
                X=0      X=0.2     X=0.4     X=0.6     X=0.8     X=1.0D0
   U(X,T)  0.487719  0.490065  0.496205  0.503795  0.509935  0.512281
  UE(X,T)  0.487716  0.490062  0.496204  0.503796  0.509938  0.512284
DIFF(X,T)  0.000004  0.000003  0.000001 -0.000001 -0.000003 -0.000004

   T =   0.50  ENERGY =   0.50000
                X=0      X=0.2     X=0.4     X=0.6     X=0.8     X=1.0D0
   U(X,T)  0.495423  0.496297  0.498586  0.501414  0.503703  0.504577
  UE(X,T)  0.495422  0.496296  0.498585  0.501415  0.503704  0.504578
DIFF(X,T)  0.000001  0.000001  0.000000  0.000000 -0.000001 -0.000001
                        .                             .
                        .                             .
                        .                             .
   T =   1.00  ENERGY =   0.50000
                X=0      X=0.2     X=0.4     X=0.6     X=0.8     X=1.0D0
   U(X,T)  0.499967  0.499973  0.499990  0.500010  0.500027  0.500033
  UE(X,T)  0.499967  0.499973  0.499990  0.500010  0.500027  0.500033
DIFF(X,T)  0.000000  0.000000  0.000000  0.000000  0.000000  0.000000

LSODE COMPUTATIONAL STATISTICS

LAST STEP SIZE                          0.314D-01

LAST ORDER OF THE METHOD                        4

TOTAL NUMBER OF STEPS TAKEN                   215

NUMBER OF FUNCTION EVALUATIONS               1683

NUMBER OF JACOBIAN EVALUATIONS                 28


RUN NO. -   3  FINITE ELEMENT

INITIAL T -  0.000D+00

  FINAL T -  0.100D+01

  PRINT T -  0.100D+00

NUMBER OF DIFFERENTIAL EQUATIONS -  51

MAXIMUM INTEGRATION ERROR -  0.100D-06


COEFFICIENT MATRIX

   1      0.0000       0.3333       0.1667
   2      0.1667       0.6667       0.1667
            .                         .
            .                         .
            .                         .
```

Table 4.10 (a) *Continued.*

```
 50       0.1667      0.6667      0.1667
 51       0.1667      0.3333      0.0000


  T =    0.00  ENERGY =    0.50000
                X=0        X=0.2       X=0.4       X=0.6       X=0.8       X=1.0D0
  U(X,T)  0.000000  0.000000  0.000000  1.000000  1.000000  1.000000
 UE(X,T)  0.000318  0.000393  0.001030  0.998970  0.999607  0.999682
DIFF(X,T) -0.000318 -0.000393 -0.001030  0.001030  0.000393  0.000318

  T =    0.10  ENERGY =    0.50000
                X=0        X=0.2       X=0.4       X=0.6       X=0.8       X=1.0D0
  U(X,T)  0.262911  0.308159  0.426703  0.573297  0.691841  0.737089
 UE(X,T)  0.262756  0.308033  0.426655  0.573345  0.691967  0.737244
DIFF(X,T)  0.000154  0.000126  0.000049 -0.000049 -0.000126 -0.000154

  T =    0.20  ENERGY =    0.50000
                X=0        X=0.2       X=0.4       X=0.6       X=0.8       X=1.0D0
  U(X,T)  0.411653  0.428526  0.472699  0.527301  0.571474  0.588347
 UE(X,T)  0.411566  0.428456  0.472673  0.527327  0.571544  0.588434
DIFF(X,T)  0.000086  0.000070  0.000027 -0.000027 -0.000070 -0.000086

  T =    0.30  ENERGY =    0.50000
                X=0        X=0.2       X=0.4       X=0.6       X=0.8       X=1.0D0
  U(X,T)  0.467083  0.473369  0.489828  0.510172  0.526631  0.532917
 UE(X,T)  0.467040  0.473335  0.489815  0.510185  0.526665  0.532960
DIFF(X,T)  0.000043  0.000035  0.000013 -0.000013 -0.000035 -0.000043

  T =    0.40  ENERGY =    0.50000
                X=0        X=0.2       X=0.4       X=0.6       X=0.8       X=1.0D0
  U(X,T)  0.487735  0.490078  0.496210  0.503790  0.509922  0.512265
 UE(X,T)  0.487716  0.490062  0.496204  0.503796  0.509938  0.512284
DIFF(X,T)  0.000020  0.000016  0.000006 -0.000006 -0.000016 -0.000020

  T =    0.50  ENERGY =    0.50000
                X=0        X=0.2       X=0.4       X=0.6       X=0.8       X=1.0D0
  U(X,T)  0.495430  0.496303  0.498588  0.501412  0.503697  0.504570
 UE(X,T)  0.495422  0.496296  0.498585  0.501415  0.503704  0.504578
DIFF(X,T)  0.000009  0.000007  0.000003 -0.000003 -0.000007 -0.000009
                          .                             .
                          .                             .
                          .                             .

  T =    1.00  ENERGY =    0.50000
                X=0        X=0.2       X=0.4       X=0.6       X=0.8       X=1.0D0
  U(X,T)  0.499967  0.499973  0.499990  0.500010  0.500027  0.500033
 UE(X,T)  0.499967  0.499973  0.499990  0.500010  0.500027  0.500033
DIFF(X,T)  0.000000  0.000000  0.000000  0.000000  0.000000  0.000000

LSODE COMPUTATIONAL STATISTICS

LAST STEP SIZE                        0.398D-01
```

Table 4.10 (a) *Continued.*

```
LAST ORDER OF THE METHOD                        5

TOTAL NUMBER OF STEPS TAKEN                   222

NUMBER OF FUNCTION EVALUATIONS               1891

NUMBER OF JACOBIAN EVALUATIONS                 32


RUN NO. -   4  FINITE VOLUME

INITIAL T -  0.000D+00

  FINAL T -  0.100D+01

  PRINT T -  0.100D+00

NUMBER OF DIFFERENTIAL EQUATIONS -  50

MAXIMUM INTEGRATION ERROR -  0.100D-06

   T =    0.00  ENERGY =    0.50000
          X=0.01    X=0.11    X=0.21    X=0.31    X=0.41    X=0.50
   U(X,T)  0.000000  0.000000  0.000000  0.000000  0.000000  0.500000
  UE(X,T)  0.000318  0.000338  0.000403  0.000566  0.001141
DIFF(X,T) -0.000318 -0.000338 -0.000403 -0.000566 -0.001141

   T =    0.10  ENERGY =    0.50000
          X=0.01    X=0.11    X=0.21    X=0.31    X=0.41    X=0.50
   U(X,T)  0.262758  0.276660  0.312414  0.366538  0.433748  0.500000
  UE(X,T)  0.262873  0.276769  0.312506  0.366604  0.433781
DIFF(X,T) -0.000115 -0.000109 -0.000092 -0.000066 -0.000033

   T =    0.20  ENERGY =    0.50000
          X=0.01    X=0.11    X=0.21    X=0.31    X=0.41    X=0.50
   U(X,T)  0.411538  0.416727  0.430067  0.450252  0.475308  0.500000
  UE(X,T)  0.411610  0.416795  0.430124  0.450293  0.475328
DIFF(X,T) -0.000072 -0.000068 -0.000057 -0.000040 -0.000020

   T =    0.30  ENERGY =    0.50000
          X=0.01    X=0.11    X=0.21    X=0.31    X=0.41    X=0.50
   U(X,T)  0.467019  0.468953  0.473927  0.481453  0.490794  0.500000
  UE(X,T)  0.467056  0.468989  0.473957  0.481474  0.490804
DIFF(X,T) -0.000038 -0.000036 -0.000030 -0.000021 -0.000011

   T =    0.40  ENERGY =    0.50000
          X=0.01    X=0.11    X=0.21    X=0.31    X=0.41    X=0.50
   U(X,T)  0.487703  0.488425  0.490279  0.493085  0.496568  0.500000
  UE(X,T)  0.487722  0.488442  0.490293  0.493095  0.496573
DIFF(X,T) -0.000018 -0.000017 -0.000014 -0.000010 -0.000005

   T =    0.50  ENERGY =    0.50000
          X=0.01    X=0.11    X=0.21    X=0.31    X=0.41    X=0.50
```

Table 4.10 (a) *Continued.*

```
   U(X,T)  0.495416  0.495684  0.496376  0.497422  0.498720  0.500000
  UE(X,T)  0.495424  0.495692  0.496382  0.497427  0.498723
DIFF(X,T) -0.000008 -0.000008 -0.000006 -0.000005 -0.000002
                          .                             .
                          .                             .
                          .                             .

   T =    1.00   ENERGY =    0.50000
               X=0.01    X=0.11    X=0.21    X=0.31    X=0.41    X=0.50
   U(X,T)  0.499967  0.499969  0.499974  0.499981  0.499991  0.500000
  UE(X,T)  0.499967  0.499969  0.499974  0.499981  0.499991
DIFF(X,T)  0.000000  0.000000  0.000000  0.000000  0.000000

LSODE COMPUTATIONAL STATISTICS

LAST STEP SIZE                                0.356D-01

LAST ORDER OF THE METHOD                              5

TOTAL NUMBER OF STEPS TAKEN                         248

NUMBER OF FUNCTION EVALUATIONS                     2108

NUMBER OF JACOBIAN EVALUATIONS                       36
```

The analytical solution to equations (4.195) to (4.198) is

$$u(x, y, t) = f(2t - x/v_x - y/v_y) \tag{4.199}$$

where f is an arbitrary function.

We now compute a numerical solution to equations (4.195) to (4.198) and use equation (4.199) to evaluate the numerical solution. The choice of the function f is a central issue. If f is smooth, a numerical solution is relatively easy to compute; we will demonstrate this by taking f as the exponential function. If f is not smooth, a numerical solution may be difficult to impossible to compute. For example, if f has a finite discontinuity, this discontinuity will propagate in two dimensions as was the case for the advection equation (3.55) in one dimension.

The following method of lines solution is based on an $11 \times 11 = 121$-point grid in x and y (a total of 121 ODEs). Subroutine INITAL is listed in Program 4.11a. We can note the following points about subroutine INITAL:

1. u of equation (4.195) is in array U(M,N) in COMMON/Y/; the temporal derivative, $\partial u/\partial t$, is in array UT(M,N) in COMMON/F/, and the spatial derivatives, $\partial u/\partial x$ and $\partial u/\partial y$, are in arrays UX(M,N) and UY(M,N) in COMMON/S/, respectively:

```
      PARAMETER (M=11, N=11)
      COMMON     /T/         T,       NFIN,      NRUN
     1           /Y/    U(M,N)
     2           /F/   UT(M,N)
     3           /S/   UX(M,N),   UY(M,N)
```

```
     4          /C/        VX,        VY,       X(M),       Y(N)
     5          /B/        XL,        XU,         YL,         YU
```

2. The velocities, v_x and v_y, in equation (4.195) are set next, followed by the spatial dimensions of the x–y domain over which a solution is to be computed:

```
C...
C...  VELOCITIES
      VX=1.0D0
      VY=1.0D0
C...
C...  BOUNDARY VALUES OF X AND Y
      XL=0.D+00
      XU=1.D+00
      YL=0.D+00
      YU=1.D+00
```

Table 4.10 (b) Comparison of the Finite Difference, Element, and Volume Solutions of Table 4.10a.

```
Finite Differences, DSS042

    T =    0.10  ENERGY =    0.50000
                      X=0      X=0.2      X=0.4      X=0.6      X=0.8      X=1.0D0
    U(X,T)  0.262758  0.308032  0.426654  0.573346  0.691968  0.737242
   UE(X,T)  0.262756  0.308033  0.426655  0.573345  0.691967  0.737244
 DIFF(X,T)  0.000002 -0.000001 -0.000001  0.000001  0.000001 -0.000002


Finite Differences, DSS044

    T =    0.10  ENERGY =    0.50000
                      X=0      X=0.2      X=0.4      X=0.6      X=0.8      X=1.0D0
    U(X,T)  0.262834  0.308096  0.426679  0.573321  0.691904  0.737166
   UE(X,T)  0.262756  0.308033  0.426655  0.573345  0.691967  0.737244
 DIFF(X,T)  0.000078  0.000063  0.000024 -0.000024 -0.000063 -0.000078


Finite Elements

    T =    0.10  ENERGY =    0.50000
                      X=0      X=0.2      X=0.4      X=0.6      X=0.8      X=1.0D0
    U(X,T)  0.262911  0.308159  0.426703  0.573297  0.691841  0.737089
   UE(X,T)  0.262756  0.308033  0.426655  0.573345  0.691967  0.737244
 DIFF(X,T)  0.000154  0.000126  0.000049 -0.000049 -0.000126 -0.000154


Finite Volumes

    T =    0.10  ENERGY =    0.50000
                 X=0.01     X=0.11     X=0.21     X=0.31     X=0.41     X=0.50
    U(X,T)  0.262758  0.276660  0.312414  0.366538  0.433748  0.500000
   UE(X,T)  0.262873  0.276769  0.312506  0.366604  0.433781
 DIFF(X,T) -0.000115 -0.000109 -0.000092 -0.000066 -0.000033
```

```
      SUBROUTINE INITAL
      IMPLICIT DOUBLE PRECISION (A-H,O-Z)
      PARAMETER (M=11, N=11)
      COMMON      /T/          T,        NFIN,       NRUN
     1            /Y/     U(M,N)
     2            /F/    UT(M,N)
     3            /S/    UX(M,N),    UY(M,N)
     4            /C/         VX,          VY,       X(M),       Y(N)
     5            /B/         XL,          XU,         YL,         YU
C...
C...  VELOCITIES
      VX=1.0D0
      VY=1.0D0
C...
C...  BOUNDARY VALUES OF X AND Y
      XL=0.D+00
      XU=1.D+00
      YL=0.D+00
      YU=1.D+00
C...
C...  GRID IN X
      DO 1 I=1,M
         X(I)=DFLOAT(I-1)/DFLOAT(M-1)*(XU-XL)
1     CONTINUE
C...
C...  GRID IN Y
      DO 2 J=1,N
         Y(J)=DFLOAT(J-1)/DFLOAT(N-1)*(YU-YL)
2     CONTINUE
C...
C...  INITIAL CONDITION
      DO 3 I=1,M
      DO 3 J=1,N
         ARG=-X(I)/VX-Y(J)/VY
         U(I,J)=F1(ARG)
3     CONTINUE
      RETURN
      END
```

Program 4.11a Subroutine INITAL for Initial Condition (4.196).

Note that the solution will be computed on a unit square.

3. The spatial grid is computed in DO loops 1 and 2 for use in the boundary conditions and the analytical solution, equation (4.199):

```
C...
C...  GRID IN X
      DO 1 I=1,M
         X(I)=DFLOAT(I-1)/DFLOAT(M-1)*(XU-XL)
1     CONTINUE
C...
C...  GRID IN Y
      DO 2 J=1,N
```

```
         Y(J)=DFLOAT(J-1)/DFLOAT(N-1)*(YU-YL)
2     CONTINUE
```

4. Finally, initial condition (4.196) is defined in DO loops 3 from the function F1 which is selected as the exponential function:

```
C...
C...  INITIAL CONDITION
      DO 3 I=1,M
      DO 3 J=1,N
         ARG=-X(I)/VX-Y(J)/VY
         U(I,J)=F1(ARG)
3     CONTINUE
```

Note that the argument of F1 is ARG=-X(I)/VX-Y(J)/VY in accordance with equation (4.196).

Function F1 is listed in Program (4.11b). Again, the use of the (smooth) exponential function facilitates the numerical solution.

Subroutine DERV is listed in Program 4.11c. We can note the following points about DERV:

1. Boundary conditions (4.197) and (4.198) are first implemented (at $x = 0$ or X = XL and $y = 0$ or Y = YL):

```
C...
C...  BOUNDARY CONDITION AT X = XL
      DO 1 J=1,N
         ARG=2.0D0*T-Y(J)/VY
         U(1,J)=F1(ARG)
1     CONTINUE
C...
C...  BOUNDARY CONDITION AT Y = YL
      DO 2 I=1,M
         ARG=2.0D0*T-X(I)/VX
         U(I,1)=F1(ARG)
2     CONTINUE
```

Note that the arguments of F1 are in accordance with boundary conditions (4.197) and (4.198).

```
      DOUBLE PRECISION FUNCTION F1(ARG)
      IMPLICIT DOUBLE PRECISION (A-H,O-Z)
      F1=DEXP(ARG)
      RETURN
      END
```

Program 4.11b Function F1 for the Initial and Boundary Conditions, Equations (4.196) to (4.198), and the Analytical Solution, Equation (4.199).

```
      SUBROUTINE DERV
      IMPLICIT DOUBLE PRECISION (A-H,O-Z)
      PARAMETER (M=11, N=11)
      COMMON      /T/          T,       NFIN,       NRUN
     1            /Y/    U(M,N)
     2            /F/   UT(M,N)
     3            /S/   UX(M,N),   UY(M,N)
     4            /C/         VX,         VY,       X(M),       Y(N)
     5            /B/         XL,         XU,         YL,         YU
C...
C...  BOUNDARY CONDITION AT X = XL
      DO 1 J=1,N
         ARG=2.0D0*T-Y(J)/VY
         U(1,J)=F1(ARG)
1     CONTINUE
C...
C...  BOUNDARY CONDITION AT Y = YL
      DO 2 I=1,M
         ARG=2.0D0*T-X(I)/VX
         U(I,1)=F1(ARG)
2     CONTINUE
C...
C...  UX
      CALL DSS034(XL,XU,M,N,1,U,UX,1.0D+00)
C...
C...  UY
      CALL DSS034(YL,YU,M,N,2,U,UY,1.0D+00)
C...
C...  PDE
      DO 3 I=1,M
      DO 3 J=1,N
         UT(I,J)=-VX*UX(I,J)-VY*UY(I,J)
3     CONTINUE
      RETURN
      END
```

Program 4.11c Subroutine DERV for Equations (4.195), (4.197), and (4.198).

2. The spatial derivatives in equation (4.195), $\partial u/\partial x$ (in UX) and $\partial u/\partial y$ (in UY), are then computed by two calls to DSS034:

```
C...
C...  UX
      CALL DSS034(XL,XU,M,N,1,U,UX,1.0D+00)
C...
C...  UY
      CALL DSS034(YL,YU,M,N,2,U,UY,1.0D+00)
```

DSS034 in general computes a first-order partial derivative over a two-dimensional domain. The arguments of DSS034, using the first call as an example, are:

(2.1) XL and XU: The lower and upper boundary values of the spatial independent variable for which the first-order partial derivative is to be computed; 0 and 1 for x in equation (4.195).

(2.2) `M,N`: The number of grid points in each direction of the two-dimensional domain; 11 and 11 for x and y in equation (4.195).

(2.3) `1` (or 2): The number of the independent spatial variable for which the first-order partial derivative is to be computed; for equation (4.195), $x = 1$, $y = 2$.

(2.4) `U`: The two-dimensional array containing the dependent variable to be differentiated; u in equation (4.195).

(2.5) `UX`: The two-dimensional array containing the computed first-order partial derivative; $\partial u/\partial x$ in equation (4.195).

(2.6) `1.0D0`: A double-precision constant or variable which indicates the direction of flow for a convective (first-order hyperbolic) problem; for equation (4.195), the velocity, v_x, is positive. If the system is diffusive (parabolic), this argument is zero.

Arguments (2.1) to (2.4) and (2.6) are inputs to `DSS034`. Argument (2.5), the computed first-order partial derivative, is returned by `DSS034`.

Subroutine `DSS034` is listed in Program 4.11d. We can note the following points about subroutine `DSS034`:

1. The first-order partial derivative is computed over a two-dimensional domain by a series of calls to the one-dimensional routines `DSS004` (for parabolic problems) and `DSS020` (for hyperbolic problems) considered previously. Thus, the central requirement in `DSS034` is the transfer between one- and two-dimensional arrays.

2. At the beginning of `DSS034`, a branch is made to either the first or second half of the routine depending on which spatial independent variable is the basis for the partial differentiation:

```
C...
C...  GO TO STATEMENT 2 IF THE PARTIAL DERIVATIVE IS TO BE COMPUTED
C...  WITH RESPECT TO THE SECOND INDEPENDENT VARIABLE
      IF(ND.EQ.2)GO TO 2
```

   In the preceding discussion (paragraph (2.3)), `ND=1` for x and `ND=2` for y.

3. If we consider the first half of `DSS034` (for `ND=1`), a transfer from the two-dimensional array containing the dependent variable to a one-dimensional array is made first:

```
C...
C...  TRANSFER THE DEPENDENT VARIABLE IN THE TWO-DIMENSIONAL ARRAY U2D
C...  TO THE ONE-DIMENSIONAL ARRAY U1D SO THAT SUBROUTINES DSS004 AND
C...  DSS020 CAN BE USED TO CALCULATE THE PARTIAL DERIVATIVE
      DO 11 I=1,N1
      U1D(I)=U2D(I,J)
11    CONTINUE
```

   Note that this is done for a series of values of the second independent variable, as set in `DO` loop 10. In other words, we are making use of the basic idea of a partial derivative, that is, a derivative with respect to an independent variable which varies (according to `DO` loop 11), while the other independent variables remain constant (according to `DO` loop 10).

```
      SUBROUTINE DSS034(XL,XU,N1,N2,ND,U2D,UX2D,V)
C...
C...  SUBROUTINE DSS034 COMPUTES A PARTIAL DERIVATIVE OVER A TWO-
C...  DIMENSIONAL DOMAIN USING EITHER FIVE-POINT CENTERED OR FIVE-
C...  POINT BIASED UPWIND APPROXIMATIONS.  IT IS INTENDED PRIMARILY
C...  FOR THE NUMERICAL METHOD OF LINES (NMOL) NUMERICAL INTEGRATION
C...  OF PARTIAL DIFFERENTIAL EQUATIONS (PDES) IN TWO DIMENSIONS.
C...
C...  ARGUMENT LIST
C...
C...     XL        LOWER VALUE OF THE INDEPENDENT VARIABLE FOR WHICH
C...               THE PARTIAL DERIVATIVE IS TO BE COMPUTED (INPUT)
C...
C...     XU        UPPER VALUE OF THE INDEPENDENT VARIABLE FOR WHICH
C...               THE PARTIAL DERIVATIVE IS TO BE COMPUTED (INPUT)
C...
C...     N1        NUMBER OF GRID POINTS FOR THE FIRST INDEPENDENT
C...               VARIABLE (INPUT)
C...
C...     N2        NUMBER OF GRID POINTS FOR THE SECOND INDEPENDENT
C...               VARIABLE (INPUT)
C...
C...     ND        NUMBER OF THE INDEPENDENT VARIABLE FOR WHICH THE
C...               PARTIAL DERIVATIVE IS TO BE COMPUTED (INPUT)
C...
C...     U2D       TWO-DIMENSIONAL ARRAY CONTAINING THE DEPENDENT VARI-
C...               ABLE WHICH IS TO BE DIFFERENTIATED WITH RESPECT TO
C...               INDEPENDENT VARIABLE ND (INPUT)
C...
C...     UX2D      TWO-DIMENSIONAL ARRAY CONTAINING THE PARTIAL DERI-
C...               VATIVE OF THE DEPENDENT VARIABLE WITH RESPECT TO
C...               INDEPENDENT VARIABLE ND (OUTPUT)
C...
C...     V         VARIABLE TO SELECT EITHER THE FIVE-POINT CENTERED
C...               OR FIVE-POINT BIASED UPWIND APPROXIMATION FOR THE
C...               PARTIAL DERIVATIVE.  V EQ 0 CALLS THE FIVE-POINT
C...               CENTERED APPROXIMATION.  V NE 0 CALLS THE FIVE-POINT
C...               BIASED UPWIND APPROXIMATION (INPUT)
C...
C...  TYPE-SELECTED REAL VARIABLES AS DOUBLE PRECISION
      DOUBLE PRECISION   UX1D,   UX2D,    U1D,    U2D,       V,     XL,
     1                   XU
C...
C...  THE FOLLOWING TWO-DIMENSIONAL ARRAYS CONTAIN THE DEPENDENT
C...  VARIABLE (U2D) AND ITS PARTIAL DERIVATIVE (UX2D)
      DIMENSION   U2D(N1,N2), UX2D(N1,N2)
C...
C...  THE FOLLOWING ONE-DIMENSIONAL ARRAYS CONTAIN THE DEPENDENT
C...  VARIABLE (U1D) AND ITS PARTIAL DERIVATIVE (UX1D).  IN EACH
C...  CASE, ONE OF THE INDEPENDENT VARIABLES IS CONSTANT AND THE
C...  OTHER INDEPENDENT VARIABLE VARIES OVER ITS TOTAL INTERVAL.
C...  THESE ARRAYS ARE USED FOR TEMPORARY STORAGE IN CALLING THE
```

Program 4.11d Subroutine DSS034. *Continued next pages.*

```
C...  ONE-DIMENSIONAL ROUTINES DSS004 AND DSS020.
C...
C...  NOTE THAT THE ARRAYS HAVE ABSOLUTE DIMENSIONS AND MAY THERE-
C...  FORE HAVE TO BE INCREASED IN SIZE.  HOWEVER, WITH A SIZE
C...  OF 51, THE TWO-DIMENSIONAL PROBLEM COULD HAVE A GRID OF
C...  51 X 51 POINTS, THEREBY GENERATING AN APPROXIMATING ODE
C...  SYSTEM WITH A MULTIPLE OF 51 X 51 EQUATIONS, DEPENDING ON
C...  THE NUMBER OF SIMULTANEOUS PDES.  THIS IS A VERY LARGE ODE
C...  PROBLEM, AND THEREFORE THE FOLLOWING ABSOLUTE DIMENSIONING
C...  IS CONSIDERED ADEQUATE FOR MOST PROBLEMS.
      DIMENSION      U1D(51),    UX1D(51)
C...
C...  GO TO STATEMENT 2 IF THE PARTIAL DERIVATIVE IS TO BE COMPUTED
C...  WITH RESPECT TO THE SECOND INDEPENDENT VARIABLE
      IF(ND.EQ.2)GO TO 2
C...
C...  ******************************************************************
C...
C...  THE PARTIAL DERIVATIVE IS TO BE COMPUTED WITH RESPECT TO THE
C...  FIRST INDEPENDENT VARIABLE DEFINED OVER AN INTERVAL CONSISTING
C...  OF N1 GRID POINTS.  COMPUTE THE PARTIAL DERIVATIVE AT THE N1 X
C...  N2 GRID POINTS VIA NESTED DO LOOPS 10, 11 AND 12
      DO 10 J=1,N2
C...
C...  TRANSFER THE DEPENDENT VARIABLE IN THE TWO-DIMENSIONAL ARRAY U2D
C...  TO THE ONE-DIMENSIONAL ARRAY U1D SO THAT SUBROUTINES DSS004 AND
C...  DSS020 CAN BE USED TO CALCULATE THE PARTIAL DERIVATIVE
      DO 11 I=1,N1
      U1D(I)=U2D(I,J)
11    CONTINUE
C...
C...  IF V EQ 0, A FIVE-POINT CENTERED APPROXIMATION IS USED FOR THE
C...  PARTIAL DERIVATIVE
      IF(V.EQ.0.D+00)CALL DSS004(XL,XU,N1,U1D,UX1D)
C...
C...  IF V NE 0, A FIVE-POINT BIASED UPWIND APPROXIMATION IS USED FOR
C...  THE PARTIAL DERIVATIVE
      IF(V.NE.0.D+00)CALL DSS020(XL,XU,N1,U1D,UX1D,V)
C...
C...  RETURN THE PARTIAL DERIVATIVE IN THE ONE-DIMENSIONAL ARRAY UX1D
C...  TO THE TWO-DIMENSIONAL ARRAY UX2D
      DO 12 I=1,N1
      UX2D(I,J)=UX1D(I)
12    CONTINUE
C...
C...  THE PARTIAL DERIVATIVE AT A PARTICULAR VALUE OF THE SECOND INDE-
C...  PENDENT VARIABLE HAS BEEN CALCULATED.  REPEAT THE CALCULATION FOR
C...  THE NEXT VALUE OF THE SECOND INDEPENDENT VARIABLE
10    CONTINUE
C...
C...  THE PARTIAL DERIVATIVE HAS BEEN CALCULATED OVER THE ENTIRE N1 X
C...  N2 GRID.  THEREFORE RETURN TO THE CALLING PROGRAM WITH THE PARTIAL
```

Program 4.11d *Continued.*

```
C...  DERIVATIVE IN THE TWO-DIMENSIONAL ARRAY UX2D
      RETURN
C...
C...  ******************************************************************
C...
C...  THE PARTIAL DERIVATIVE IS TO BE COMPUTED WITH RESPECT TO THE
C...  SECOND INDEPENDENT VARIABLE DEFINED OVER AN INTERVAL CONSISTING
C...  OF N2 GRID POINTS.  COMPUTE THE PARTIAL DERIVATIVE AT THE N1 X
C...  N2 GRID POINTS VIA NESTED DO LOOPS 20 AND 21
2     DO 20 I=1,N1
C...
C...  TRANSFER THE DEPENDENT VARIABLE IN THE TWO-DIMENSIONAL ARRAY U2D
C...  TO THE ONE-DIMENSIONAL ARRAY U1D SO THAT SUBROUTINES DSS004 AND
C...  DSS020 CAN BE USED TO CALCULATE THE PARTIAL DERIVATIVE
      DO 21 J=1,N2
      U1D(J)=U2D(I,J)
21    CONTINUE
C...
C...  IF V EQ 0, A FIVE-POINT CENTERED APPROXIMATION IS USED FOR THE
C...  PARTIAL DERIVATIVE
      IF(V.EQ.0.D+00)CALL DSS004(XL,XU,N2,U1D,UX1D)
C...
C...  IF V NE 0, A FIVE-POINT BIASED UPWIND APPROXIMATION IS USED FOR
C...  THE PARTIAL DERIVATIVE
      IF(V.NE.0.D+00)CALL DSS020(XL,XU,N2,U1D,UX1D,V)
C...
C...  RETURN THE PARTIAL DERIVATIVE IN THE ONE-DIMENSIONAL ARRAY UX1D
C...  TO THE TWO-DIMENSIONAL ARRAY UX2D
      DO 22 J=1,N2
      UX2D(I,J)=UX1D(J)
22    CONTINUE
C...
C...  THE PARTIAL DERIVATIVE AT A PARTICULAR VALUE OF THE FIRST INDE-
C...  PENDENT VARIABLE HAS BEEN CALCULATED.  REPEAT THE CALCULATION FOR
C...  THE NEXT VALUE OF THE FIRST INDEPENDENT VARIABLE
20    CONTINUE
C...
C...  THE PARTIAL DERIVATIVE HAS BEEN CALCULATED OVER THE ENTIRE N1 X
C...  N2 GRID.  THEREFORE RETURN TO THE CALLING PROGRAM WITH THE PARTIAL
C...  DERIVATIVE IN THE TWO-DIMENSIONAL ARRAY UX2D
      RETURN
      END
```

Program 4.11d *Continued.*

4. Then either DSS004 for parabolic problems (with $V = 0$) or DSS020 for hyperbolic problems (with $V \neq 0$) is called to compute the partial derivative (in one dimension):

```
C...
C...  IF V EQ 0, A FIVE-POINT CENTERED APPROXIMATION IS USED FOR
C...  THE PARTIAL DERIVATIVE
      IF(V.EQ.0.D+00)CALL DSS004(XL,XU,N1,U1D,UX1D)
C...
C...  IF V NE 0, A FIVE-POINT BIASED UPWIND APPROXIMATION IS USED
```

```
C...  FOR THE PARTIAL DERIVATIVE
      IF(V.NE.0.D+00)CALL DSS020(XL,XU,N1,U1D,UX1D,V)
```

5. Finally the one-dimensional partial derivative is transferred to the two-dimensional array that contains the final result:

```
C...
C...  RETURN THE PARTIAL DERIVATIVE IN THE ONE-DIMENSIONAL ARRAY
C...  UX1D TO THE TWO-DIMENSIONAL ARRAY UX2D
      DO 12 I=1,N1
      UX2D(I,J)=UX1D(I)
12    CONTINUE
```

   This completes outer DO loop 10 for a given (constant) value of the second independent variable. For the next pass through DO loop 10, the second independent variable is advanced to its next value on the two-dimensional spatial grid, and the process of differentiating with respect to the first independent variable is repeated by cycling through DO loops 11 and 12. This process continues for all values of the second independent variable (i.e., DO loop 10 is completed).

6. The code in the second half of DSS034 is the same (for ND=2), except the roles of the two independent variables are interchanged.

Clearly this procedure could be extended to three (or more) dimensions, and this has been done in subroutine DSS036.

Subroutine PRINT is listed in Program 4.11e. We can note the following points about subroutine PRINT:

1. The analytical solution is first computed according to equation (4.199):

```
C...
C...  COMPUTE THE EXACT SOLUTION
      DO 3 J=1,N,2
      DO 3 I=1,M,2
         ARG=2.0D0*T-X(I)/VX-Y(J)/VY
         UE(I,J)=F1(ARG)
3     CONTINUE
```

2. The numerical and analytical solutions are then printed for comparison:

```
C...
C...  PRINT THE NUMERICAL AND EXACT SOLUTIONS
      WRITE(NO,1)T
1     FORMAT(//,5H T = ,F9.2,//,16X,
     1 10H    X=0    ,10H    X=0.2  ,10H    X=0.4   ,
     2 10H    X=0.6  ,10H    X=0.8  ,10H    X=1.0    )
      WRITE(NO,2)((U(I,J),I=1,M,2),(UE(I,J),I=1,M,2),J=1,N,2)
2     FORMAT(6H  Y=0  ,2X,5H  NUM,6F10.4,/,6X,2X,5HEXACT,6F10.4,//,
     1        6H Y=0.2,2X,5H  NUM,6F10.4,/,6X,2X,5HEXACT,6F10.4,//,
     1        6H Y=0.4,2X,5H  NUM,6F10.4,/,6X,2X,5HEXACT,6F10.4,//,
     1        6H Y=0.6,2X,5H  NUM,6F10.4,/,6X,2X,5HEXACT,6F10.4,//,
     1        6H Y=0.8,2X,5H  NUM,6F10.4,/,6X,2X,5HEXACT,6F10.4,//,
     1        6H Y=1.0,2X,5H  NUM,6F10.4,/,6X,2X,5HEXACT,6F10.4,//)
```

```
      SUBROUTINE PRINT(NI,NO)
      IMPLICIT DOUBLE PRECISION (A-H,O-Z)
      PARAMETER (M=11, N=11)
      COMMON      /T/          T,       NFIN,       NRUN
     1            /Y/    U(M,N)
     2            /F/   UT(M,N)
     3            /S/   UX(M,N)
     4            /C/         VX,         VY,       X(M),       Y(N)
     5            /B/         XL,         XU,         YL,         YU
C...
C...  ARRAY FOR THE EXACT SOLUTION
      DIMENSION UE(M,N)
C...
C...  MONITOR SOLUTION ON SCREEN
      WRITE(*,*)T
C...
C...  COMPUTE THE EXACT SOLUTION
      DO 3 J=1,N,2
      DO 3 I=1,M,2
         ARG=2.0D0*T-X(I)/VX-Y(J)/VY
         UE(I,J)=F1(ARG)
3     CONTINUE
C...
C...  PRINT THE NUMERICAL AND EXACT SOLUTIONS
      WRITE(NO,1)T
1     FORMAT(//,5H T = ,F9.2,//,16X,
     1 10H    X=0   ,10H   X=0.2  ,10H   X=0.4   ,
     2 10H  X=0.6   ,10H   X=0.8  ,10H   X=1.0   )
      WRITE(NO,2)((U(I,J),I=1,M,2),(UE(I,J),I=1,M,2),J=1,N,2)
2     FORMAT(6H  Y=0 ,2X,5H  NUM,6F10.4,/,6X,2X,5HEXACT,6F10.4,//,
     1       6H Y=0.2,2X,5H  NUM,6F10.4,/,6X,2X,5HEXACT,6F10.4,//,
     1       6H Y=0.4,2X,5H  NUM,6F10.4,/,6X,2X,5HEXACT,6F10.4,//,
     1       6H Y=0.6,2X,5H  NUM,6F10.4,/,6X,2X,5HEXACT,6F10.4,//,
     1       6H Y=0.8,2X,5H  NUM,6F10.4,/,6X,2X,5HEXACT,6F10.4,//,
     1       6H Y=1.0,2X,5H  NUM,6F10.4,/,6X,2X,5HEXACT,6F10.4,//)
      RETURN
      END
```

Program 4.11e Subroutine PRINT to Print the Numerical Solution of Equations (4.195) to (4.198) and Analytical Solution (4.199).

```
TWO-DIMENSIONAL ADVECTION PDE
0.        1.5       0.1
  121                      0.00001
END OF RUNS
```

Program 4.11f Data for Programs 4.11a to e.

The data for Programs 4.11a to e are listed in Program 4.11f.

The integration of 121 ODEs is done with an error tolerance of 0.00001 by RKF45. The main program that calls RKF45 is similar to earlier main programs and is not listed here (see Program 2.2a).

Table 4.11 Abbreviated Output from Programs 4.11a to f. *Continued next pages.*

```
RUN NO. -    1  TWO-DIMENSIONAL ADVECTION PDE

INITIAL T -   0.000D+00

  FINAL T -   0.150D+01

  PRINT T -   0.100D+00

NUMBER OF DIFFERENTIAL EQUATIONS - 121

INTEGRATION ALGORITHM - RKF45

MAXIMUM INTEGRATION ERROR -   0.100D-04


T =       0.00

                      X=0      X=0.2     X=0.4     X=0.6     X=0.8     X=1.0
 Y=0       NUM     1.0000    0.8187    0.6703    0.5488    0.4493    0.3679
         EXACT     1.0000    0.8187    0.6703    0.5488    0.4493    0.3679

Y=0.2      NUM     0.8187    0.6703    0.5488    0.4493    0.3679    0.3012
         EXACT     0.8187    0.6703    0.5488    0.4493    0.3679    0.3012

Y=0.4      NUM     0.6703    0.5488    0.4493    0.3679    0.3012    0.2466
         EXACT     0.6703    0.5488    0.4493    0.3679    0.3012    0.2466

Y=0.6      NUM     0.5488    0.4493    0.3679    0.3012    0.2466    0.2019
         EXACT     0.5488    0.4493    0.3679    0.3012    0.2466    0.2019

Y=0.8      NUM     0.4493    0.3679    0.3012    0.2466    0.2019    0.1653
         EXACT     0.4493    0.3679    0.3012    0.2466    0.2019    0.1653

Y=1.0      NUM     0.3679    0.3012    0.2466    0.2019    0.1653    0.1353
         EXACT     0.3679    0.3012    0.2466    0.2019    0.1653    0.1353


T =       0.10

                      X=0      X=0.2     X=0.4     X=0.6     X=0.8     X=1.0
 Y=0       NUM     1.2214    1.0000    0.8187    0.6703    0.5488    0.4493
         EXACT     1.2214    1.0000    0.8187    0.6703    0.5488    0.4493

Y=0.2      NUM     1.0000    0.8187    0.6703    0.5488    0.4493    0.3679
         EXACT     1.0000    0.8187    0.6703    0.5488    0.4493    0.3679

Y=0.4      NUM     0.8187    0.6703    0.5488    0.4493    0.3679    0.3012
         EXACT     0.8187    0.6703    0.5488    0.4493    0.3679    0.3012

Y=0.6      NUM     0.6703    0.5488    0.4493    0.3679    0.3012    0.2466
         EXACT     0.6703    0.5488    0.4493    0.3679    0.3012    0.2466
```

Table 4.11 *Continued.*

Y=0.8	NUM	0.5488	0.4493	0.3679	0.3012	0.2466	0.2019
	EXACT	0.5488	0.4493	0.3679	0.3012	0.2466	0.2019
Y=1.0	NUM	0.4493	0.3679	0.3012	0.2466	0.2019	0.1653
	EXACT	0.4493	0.3679	0.3012	0.2466	0.2019	0.1653

T = 0.20

		X=0	X=0.2	X=0.4	X=0.6	X=0.8	X=1.0
Y=0	NUM	1.4918	1.2214	1.0000	0.8187	0.6703	0.5488
	EXACT	1.4918	1.2214	1.0000	0.8187	0.6703	0.5488
Y=0.2	NUM	1.2214	1.0000	0.8187	0.6703	0.5488	0.4493
	EXACT	1.2214	1.0000	0.8187	0.6703	0.5488	0.4493
Y=0.4	NUM	1.0000	0.8187	0.6703	0.5488	0.4493	0.3679
	EXACT	1.0000	0.8187	0.6703	0.5488	0.4493	0.3679
Y=0.6	NUM	0.8187	0.6703	0.5488	0.4493	0.3679	0.3012
	EXACT	0.8187	0.6703	0.5488	0.4493	0.3679	0.3012
Y=0.8	NUM	0.6703	0.5488	0.4493	0.3679	0.3012	0.2466
	EXACT	0.6703	0.5488	0.4493	0.3679	0.3012	0.2466
Y=1.0	NUM	0.5488	0.4493	0.3679	0.3012	0.2466	0.2019
	EXACT	0.5488	0.4493	0.3679	0.3012	0.2466	0.2019

.
.
.

T = 1.00

		X=0	X=0.2	X=0.4	X=0.6	X=0.8	X=1.0
Y=0	NUM	7.3889	6.0495	4.9529	4.0551	3.3200	2.7182
	EXACT	7.3891	6.0496	4.9530	4.0552	3.3201	2.7183
Y=0.2	NUM	6.0495	4.9530	4.0552	3.3201	2.7183	2.2255
	EXACT	6.0496	4.9530	4.0552	3.3201	2.7183	2.2255
Y=0.4	NUM	4.9529	4.0552	3.3201	2.7183	2.2256	1.8221
	EXACT	4.9530	4.0552	3.3201	2.7183	2.2255	1.8221
Y=0.6	NUM	4.0551	3.3201	2.7183	2.2256	1.8221	1.4918
	EXACT	4.0552	3.3201	2.7183	2.2255	1.8221	1.4918
Y=0.8	NUM	3.3200	2.7183	2.2256	1.8221	1.4918	1.2214
	EXACT	3.3201	2.7183	2.2255	1.8221	1.4918	1.2214
Y=1.0	NUM	2.7182	2.2255	1.8221	1.4918	1.2214	1.0000
	EXACT	2.7183	2.2255	1.8221	1.4918	1.2214	1.0000

Table 4.11 *Continued.*

.
.
.

T = 1.50

		X=0	X=0.2	X=0.4	X=0.6	X=0.8	X=1.0
Y=0	NUM	20.0852	16.4441	13.4633	11.0228	9.0247	7.3888
	EXACT	20.0855	16.4446	13.4637	11.0232	9.0250	7.3891
Y=0.2	NUM	16.4441	13.4637	11.0232	9.0250	7.3891	6.0497
	EXACT	16.4446	13.4637	11.0232	9.0250	7.3891	6.0496
Y=0.4	NUM	13.4633	11.0232	9.0251	7.3891	6.0497	4.9531
	EXACT	13.4637	11.0232	9.0250	7.3891	6.0496	4.9530
Y=0.6	NUM	11.0228	9.0250	7.3891	6.0497	4.9531	4.0552
	EXACT	11.0232	9.0250	7.3891	6.0496	4.9530	4.0552
Y=0.8	NUM	9.0247	7.3891	6.0497	4.9531	4.0552	3.3201
	EXACT	9.0250	7.3891	6.0496	4.9530	4.0552	3.3201
Y=1.0	NUM	7.3888	6.0497	4.9531	4.0552	3.3201	2.7183
	EXACT	7.3891	6.0496	4.9530	4.0552	3.3201	2.7183

Abbreviated output from Programs 4.11a to f is given in Table 4.11. We can note the following points about the output in Table 4.11:

1. The numerical and analytical solutions agree to approximately four figures.

2. A traveling exponential results from the convective (first-order hyperbolic) characteristic of equation (4.195), and the imposition of the exponentially increasing boundary conditions (4.197) and (4.198). The traveling exponential also follows from the analytical solution, equation (4.199), which has a value determined only by the argument $\lambda = (2t - x/v_x - y/v_y)$, that is $u(x, y, t) = f(\lambda)$. For example, $u(0,0,0) = f(2(0) - 0/v_x - 0/v_y) = u(0.2, 0, 0.1) = f(2(0.1) - 0.2/v_x - 0/v_y) = u(0, 0.2, 0.1) = f(2(0.1) - 0/v_x - 0.2/v_y) = f(0)$ with $v_x = v_y = 1$; these values are apparent in comparing the solutions at $t = 0$ and 0.1 in Table 4.11. It also follows that $u(0,0,0) = u(0.1, 0.1, 0.2)$, although the latter value is not included in Table 4.11.

3. Again, we emphasize that the agreement between the numerical and analytical solutions is due to the smoothness of the exponential function. A boundary condition function could easily be devised for which the (numerical) solution would be difficult to compute and/or be significantly in error.

We have to this point considered the method of lines solution of PDEs which are first order in the initial value variable, t. We now consider in the next chapter the numerical integration of a series of PDEs which are second order in t (second-order hyperbolic PDEs), and zeroth order in t (boundary value ODEs and elliptic PDEs).

References

Brenan, K. E., S. L. Campbell and L. R. Petzold (1989), *Numerical Solution of Initial-Value Problems in Differential-Algebraic Equations*, North-Holland, New York.

Byrne, G. D. and A. C. Hindmarsh (1987), "Stiff ODE Solvers: A Review of Current and Coming Attractions," *J. Comput. Phys.*, **70**, pp. 1–62.

Forsythe, G. E., M. A. Malcolm and C. B. Moler (1977), *Computer Methods for Mathematical Computations*, Prentice-Hall, Englewood Cliffs, NJ.

Johnson, C. (1990). *Numerical Solution of Partial Differential Equations by the Finite Element Method*, Cambridge University Press, Cambridge, UK.

Lick, W. J. (1989), *Difference Equations from Differential Equations*, Springer-Verlag, Berlin.

Liu, B. and F. J. Aguirre (1990), "An Efficient Method for Handling Time-dependent Boundary Conditions with the DSS/2 Differential Equation Solver," *Simulation*, June, pp. 274–279.

Madsen, Neil K. and Richard F. Sincovec (1976), "Software for Partial Differential Equations," in *Numerical Methods for Differential Systems*, L. Lapidus and W. E. Schiesser (eds.), Academic Press, New York.

Mickens, R. E. (1991), "Nonstandard Finite Difference Schemes for Partial Differential Equations," *Trans. Soc. Computer Simulation*, **8**(2), pp. 109–117.

Patankar, S. V. (1980), *Numerical Heat Transfer and Fluid Flow*, Hemisphere Publishing Corporation, New York.

Press, W. H., B. P. Flannery, S. A. Teukolsy and W. T. Vetterling (1986), *Numerical Recipes*, Cambridge University Press, Cambridge.

Schiesser, W. E. (1991), *The Numerical Method of Lines Integration of Partial Differential Equations*, Academic Press, San Diego.

Silebi, C. A. and W. E. Schiesser (1992), *Dynamic Modeling of Transport Process Systems*, Academic Press, San Diego.

Skeel, R. D. and M. Berzins (1990), "A Method for the Spatial Discretization of Parabolic Equations in One Space Variable," *SIAM J. Sci. Stat. Comput.*, **11**(1), January, pp. 1–32.

chapter five

Partial Differential Equations Second and Zeroth Order in Time

In Chapters 3 and 4 we considered PDEs that are first order in an initial value variable, which we term "time" just to facilitate the discussion. If these PDEs are also first order in the boundary value independent variables, they are generally termed first-order hyperbolic PDEs; the advection equation (3.55) is one example. If the PDEs are first order in the initial value variable and second order in the boundary value variables, they are termed parabolic; Fourier's second law, equation (4.99), is an example. We now consider PDEs which are: (1) second order in the initial value variable, and if they are also second order in the boundary value variables, they are termed second-order hyperbolic, and (2) zeroth order in the initial value independent variable, i.e., the initial value variable does not appear in any partial derivatives, that are termed elliptic.

5.1 PDEs Second Order in Time

To start our discussion of PDEs second order in time, we return to the advection equation (3.55), renumbered here

$$\frac{\partial u}{\partial t} + v\frac{\partial u}{\partial x} = 0 \tag{5.1}$$

Equation (5.1) is self-contained in the sense that it is one equation for one unknown, $u(x, t)$; of course it requires one initial condition (since it is first order in t) and one boundary condition (since it is first order in x). Therefore, in principle, once these two auxiliary conditions are specified, a complete, well-posed problem has been stated that can then be integrated to give $u(x, t)$.

However, equation (5.1) can be differentiated with respect to x to give

$$\frac{\partial^2 u}{\partial t \partial x} + v\frac{\partial^2 u}{\partial x^2} = 0 \tag{5.2}$$

It can also be differentiated with respect to t to give

$$\frac{\partial^2 u}{\partial t^2} + v\frac{\partial^2 u}{\partial x \partial t} = 0 \tag{5.3}$$

If we now assume that the two mixed partial derivatives, $\partial^2 u/\partial t\partial x$ and $\partial^2 u/\partial x\partial t$, are equal, that is, the order of differentiation is immaterial, this mixed partial can be eliminated from equations (5.2) and (5.3) to give

$$\frac{\partial^2 u}{\partial t^2} = v^2\frac{\partial^2 u}{\partial x^2} \tag{5.4}$$

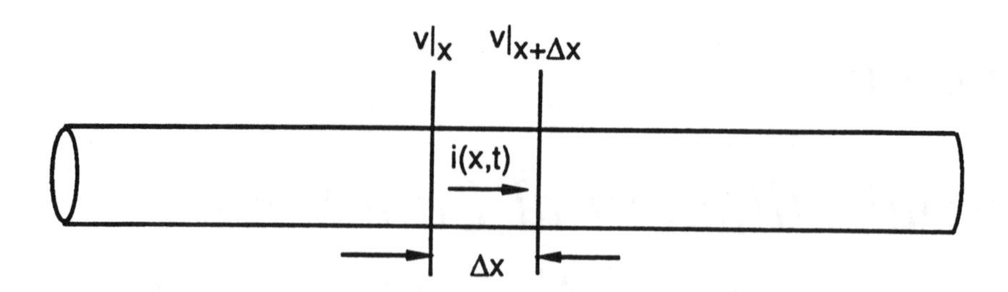

Figure 5.1 Uniform Electric Transmission Line.

Equation (5.4) is the *wave equation*, classified as second-order hyperbolic. However, this excerise really did not accomplish anything useful except possibly to show the relationship between a first-order hyperbolic equation, (5.1), and a second-order hyperbolic equation, (5.4). That is, equation (5.4) is as difficult to integrate, if not more difficult, than equation (5.1) (if for no other reason than it requires two initial conditions and two boundary conditions, which, of course, must be included in a solution).

However, writing two or more first-order hyperbolic PDEs as second-order hyperbolic PDEs can be useful, as we now demonstrate in terms of an example. Consider the uniform electric transmission line illustrated in Figure 5.1 [Myint-U et al. (1987)].

We will derive two PDEs for the current, $I(x,t)$, and voltage, $V(x,t)$, along the line. Starting with a current balance written for an incremental length of the line, we have

$$\Delta x L\frac{\partial I}{\partial t} + \Delta x RI = V|_x - V|_{x+\Delta x} \tag{5.5}$$

where I = line current (*amps*); V = line voltage (*volts*); x = position along the line (*m*); t = time (*s*); L = inductance per unit length of line (*henrys/m*); and R = resistance per unit length of line (*ohms/m*).

Division of equation (5.5) by Δx, followed by the limit $\Delta x \to 0$ gives

$$L\frac{\partial I}{\partial t} + RI = -\frac{\partial V}{\partial x} \tag{5.6}$$

Since equation (5.6) is first-order in space and time, it is first-order hyperbolic.

Also, equation (5.6) has two dependent variables, I and V, thus a second PDE is required. This second equation comes from a voltage balance written on the incremental section of line in Figure 5.1:

$$\Delta x C\frac{\partial V}{\partial t} + \Delta x GV = I|_x - I|_{x+\Delta x} \tag{5.7}$$

where C = capacitance per unit length of line (*farads/m*) and G = leakage conductance per unit length of line (*1/ohms-m*).

Division of equation (5.7) by Δx, followed by the limit $\Delta x \to 0$ gives

$$C\frac{\partial V}{\partial t} + GV = -\frac{\partial I}{\partial x} \tag{5.8}$$

Equations (5.6) and (5.8) are two first-order hyperbolic PDEs for the current, $I(x,t)$, and voltage, $V(x,t)$, along the transmission line of Figure 5.1.

We first consider the method of lines solution of equations (5.6) and (5.8) retained in the format of simultaneous, first-order PDEs. Then we combine equations (5.6)

```
      SUBROUTINE INITAL
      IMPLICIT DOUBLE PRECISION (A-H,O-Z)
      PARAMETER (NX=11)
      COMMON/T/          T,   NSTOP,   NORUN
     1      /Y/   U(NX),   V(NX)
     2      /F/  UT(NX), VT(NX)
     3      /S/  UX(NX), VX(NX)
     4      /P/        DX,       PI
C...
C...  GRID SPACING, PI
      DX=1.0D0/DFLOAT(NX-1)
      PI=4.0D0*DATAN(1.0D0)
C...
C...  INITIAL CONDITIONS
      DO 1 I=1,NX
         U(I)=0.0D0
         V(I)=DCOS((PI/2.0D0)*DX*DFLOAT(I-1))
1     CONTINUE
      RETURN
      END
```

Program 5.1a Subroutine `INITAL` for Initial Conditions (5.11) and (5.12).

and (5.8) into a second-order PDE for solution by the method of lines. For the first approach, we consider the case of a *lossless line*, i.e., $R = G = 0$. Also, to facilitate the programming, the current I is given the variable name U, and we take $L = C = 1$; the final PDEs, written in subscript notation are therefore

$$U_t = -V_x \tag{5.9}$$

$$V_t = -U_x \tag{5.10}$$

Equations (5.9) and (5.10) each require an initial and a boundary condition. We take these as

$$U(x,0) = 0, \qquad V(x,0) = \cos((\pi/2)x) \qquad (5.11)(5.12)$$

$$U(0,t) = 0, \qquad V(1,t) = 0 \qquad (5.13)(5.14)$$

Equations (5.9) to (5.14) are the problem system. An analytical solution can easily be derived as

$$U(x,t) = \sin((\pi/2)x)\sin((\pi/2)t) \tag{5.15}$$

$$V(x,t) = \cos((\pi/2)x)\cos((\pi/2)t) \tag{5.16}$$

Equations (5.15) and (5.16) are programmed in subroutine `PRINT` for comparison with the numerical solution.

Subroutine `INITAL`, listed in Program 5.1a, defines initial conditions (5.11) and (5.12) on an 11-point grid in x.

The `COMMON` area contains $U(x,t)$ and $V(x,t)$ in `/Y/` as arrays `U(NX)` and `V(NX)`, respectively. The corresponding temporal derivatives are in `/F/` as arrays `UT(NX)` and `VT(NX)`, and the spatial derivatives are in `/S/` as arrays `UX(NX)` and `VX(NX)`. Thus, this `COMMON` area closely parallels those of previous examples. Initial conditions (5.11) and (5.12) are then programmed in `DO` loop 1. Note that the interval for x is $0 \le x \le 1$.

```
      SUBROUTINE DERV
      IMPLICIT DOUBLE PRECISION (A-H,O-Z)
      PARAMETER (NX=11)
      COMMON/T/          T,  NSTOP,  NORUN
     1      /Y/   U(NX),  V(NX)
     2      /F/  UT(NX), VT(NX)
     3      /S/  UX(NX), VX(NX)
     4      /P/      DX,     PI
C...
C... BOUNDARY CONDITIONS
        U(1)=0.0D0
       V(NX)=0.0D0
C...
C... UX, VX
C...
C...    SECOND-ORDER DIFFERENCES
        IF(NORUN.EQ.1)THEN
        CALL DSS002(0.0D0,1.0D0,NX,U,UX)
        CALL DSS002(0.0D0,1.0D0,NX,V,VX)
C...
C...    FOURTH-ORDER DIFFERENCES
        ELSE IF(NORUN.EQ.2)THEN
        CALL DSS004(0.0D0,1.0D0,NX,U,UX)
        CALL DSS004(0.0D0,1.0D0,NX,V,VX)
        END IF
C...
C... PDES
      DO 1 I=1,NX
         UT(I)=-VX(I)
         VT(I)=-UX(I)
1     CONTINUE
C...
C... ZERO TEMPORAL DERIVATIVES ACCORDING TO THE BOUNDARY CONDITIONS
        UT(1)=0.0D0
       VT(NX)=0.0D0
      RETURN
      END
```

Program 5.1b Subroutine DERV for Equations (5.10) and (5.11).

Subroutine DERV is listed in Program 5.1b. We can note the following points about DERV:

1. Boundary conditions (5.13) and (5.14) are programmed as

```
C...
C... BOUNDARY CONDITIONS
        U(1)=0.0D0
       V(NX)=0.0D0
```

2. The spatial derivatives, U_x and V_x, in equations (5.9) and (5.10) are then com-

puted by either second-order differences by DSS002 (NORUN=1) or fourth-order differences by DSS004 (NORUN=2):

```
C...
C...      SECOND-ORDER DIFFERENCES
          IF(NORUN.EQ.1)THEN
          CALL DSS002(0.0D0,1.0D0,NX,U,UX)
          CALL DSS002(0.0D0,1.0D0,NX,V,VX)
C...
C...      FOURTH-ORDER DIFFERENCES
          ELSE IF(NORUN.EQ.2)THEN
          CALL DSS004(0.0D0,1.0D0,NX,U,UX)
          CALL DSS004(0.0D0,1.0D0,NX,V,VX)
          END IF
```

3. Finally the 22 ODEs are programmed in DO loop 2 as approximations of PDEs (5.9) and (5.10)

```
C...
C...   PDES
       DO 1 I=1,NX
          UT(I)=-VX(I)
          VT(I)=-UX(I)
1      CONTINUE
C...
C...   ZERO TEMPORAL DERIVATIVES ACCORDING TO THE BOUNDARY CONDITIONS
            UT(1)=0.0D0
          VT(NX)=0.0D0
```

$U_t(0,t)$ and $V_t(1,t)$ are then set to zero to ensure compliance with boundary conditions (5.13) and (5.14).

Subroutine PRINT is listed in Program 5.1c. We can note the following points about subroutine PRINT:

1. The exact solution from equations (5.15) and (5.16) is first computed in DO loop 1:

```
C...
C...   EXACT SOLUTION
       DO 1 I=1,NX,2
          X=DX*DFLOAT(I-1)
          UE(I)=DSIN((PI/2.0D0)*X)*DSIN((PI/2.0D0)*T)
          VE(I)=DCOS((PI/2.0D0)*X)*DCOS((PI/2.0D0)*T)
1      CONTINUE
```

2. The errors in the numerical solution are then computed from the exact solution:

```
C...
C...   ERRORS IN THE NUMERICAL SOLUTION
       DO 4 I=1,NX,2
          EU(I)=U(I)-UE(I)
          EV(I)=V(I)-VE(I)
4      CONTINUE
```

```
      SUBROUTINE PRINT(NI,NO)
      IMPLICIT DOUBLE PRECISION (A-H,O-Z)
      PARAMETER (NX=11)
      COMMON/T/       T,  NSTOP,  NORUN
     1      /Y/   U(NX),  V(NX)
     2      /F/  UT(NX), VT(NX)
     3      /S/  UX(NX), VX(NX)
     4      /P/      DX,     PI
      DIMENSION  UE(NX), VE(NX), EU(NX), EV(NX)
C...
C...  MONITOR THE CALCULATION
      WRITE(*,*)NORUN,T
C...
C...  EXACT SOLUTION
      DO 1 I=1,NX,2
         X=DX*DFLOAT(I-1)
         UE(I)=DSIN((PI/2.0D0)*X)*DSIN((PI/2.0D0)*T)
         VE(I)=DCOS((PI/2.0D0)*X)*DCOS((PI/2.0D0)*T)
1     CONTINUE
C...
C...  ERRORS IN THE NUMERICAL SOLUTION
      DO 4 I=1,NX,2
         EU(I)=U(I)-UE(I)
         EV(I)=V(I)-VE(I)
4     CONTINUE
C...
C...  PRINT THE NUMERICAL SOLUTION AND ERRORS
      WRITE(NO,2)T
2     FORMAT(///,5H T = ,F7.3,/,
     1 5X,'X',9X,'U',8X,'UE',9X,'ERR U',
     2        9X,'V',8X,'VE',9X,'ERR V')
      DO 5 I=1,NX,2
         X=DX*DFLOAT(I-1)
         WRITE(NO,3)X,U(I),UE(I),EU(I),
     1                 V(I),VE(I),EV(I)
3        FORMAT(F6.3,2F10.5,F14.9,
     1                2F10.5,F14.9)
5     CONTINUE
      RETURN
      END
```

Program 5.1c Subroutine PRINT for Printing the Solution to Equations (5.9) to (5.14).

3. Finally, the numerical and exact solutions are printed with the error:

```
C...
C...  PRINT THE NUMERICAL AND EXACT SOLUTIONS AND THE ERRORS
      WRITE(NO,2)T
2     FORMAT(///,5H T = ,F7.3,/,
     1 5X,'X',9X,'U',8X,'UE',9X,'ERR U',
     2        9X,'V',8X,'VE',9X,'ERR V')
      DO 5 I=1,NX,2
         X=DX*DFLOAT(I-1)
         WRITE(NO,3)X,U(I),UE(I),EU(I),
```

```
      1                 V(I),VE(I),EV(I)
   3        FORMAT(F6.3,2F10.5,F14.9,
      1                 2F10.5,F14.9)
   5     CONTINUE
```

The data file for Programs 5.1a to c is listed in Program 5.1d. Two runs with 22 ODEs integrated with an error tolerance of 0.0000001 are specified. The integration was performed by RKF45.

The output from Programs 5.1a to d is listed in Table 5.1. We can note the following points about the output of Table 5.1:

1. The periodic characteristic of the solution is clear, which is a consequence of the $\sin((\pi/2)t)$ and $\cos((\pi/2)t)$ terms in equations (5.15) and (5.16), respectively. This periodic behavior in time is characteristic of hyperbolic systems and contrasts with the exponential time behavior of parabolic PDEs, as illustrated by equation (4.160). This is an important distinction. For example, whereas discontinuities that might arise due to inconsistencies between initial and boundary conditions tend to damp put with increasing time in parabolic problems, such discontinuities are possibly undamped in hyperbolic problems, and therefore can propagate in time indefinitely. An example of this feature of hyperbolic systems is the advection equation (3.55) which can propagate discontinuities as discussed in terms of equations (3.56) to (3.59). Because of these characteristics, we offer the general opinion that hyperbolic problems are inherently more difficult to solve than parabolic (and also elliptic) problems.

2. A comparison of the errors in the numerical solutions computed with the second- and fourth-order approximations in DSS002 and DSS004, respectively, clearly indicates the advantage of using the fourth-order approximations.

In summary, the numerical solutions computed for equations (5.9) to (5.14) are sufficiently accurate for most scientific and engineering applications, particularly the solution from the fourth-order approximations of DSS004. However, this particular problem is relatively easy to solve accurately, as indicated by the smooth analytical solutions of equations (5.15) and (5.16) (which are infinitely differentiable in x and t). More difficult hyperbolic problems can easily be constructed, e.g., equation (3.55), and generally, we can expect more of a challenge in computing the solutions to hyperbolic PDEs.

Next, we consider hyperbolic PDEs second order in time. As we observed initially, second-order hyperbolic PDEs can be derived from first-order hyperbolic PDEs (recall how equation (5.4) followed from equation (5.1)). We now demonstrate again this source of second-order hyperbolic PDEs. Equations (5.6) and (5.8) can be considered

```
TWO FIRST-ORDER HYPERBOLIC PDES - DSS002
0.         4.0        1.0
   22                       0.0000001
TWO FIRST-ORDER HYPERBOLIC PDES - DSS004
0.         4.0        1.0
   22                       0.0000001
END OF RUNS
```

Program 5.1d Data File for Programs 5.1a to c.

Table 5.1 Output from Programs 5.1a to d. *Continued next page.*

```
RUN NO. -    1  TWO FIRST-ORDER HYPERBOLIC PDES - DSS002

INITIAL T -   0.000D+00

  FINAL T -   0.400D+01

  PRINT T -   0.100D+01

NUMBER OF DIFFERENTIAL EQUATIONS -   22

INTEGRATION ALGORITHM - RKF45

MAXIMUM INTEGRATION ERROR -   0.100D-06


T =    0.000
    X         U        UE          ERR U          V        VE          ERR V
0.000   0.00000   0.00000   0.000000000   1.00000   1.00000    0.000000000
0.200   0.00000   0.00000   0.000000000   0.95106   0.95106    0.000000000
0.400   0.00000   0.00000   0.000000000   0.80902   0.80902    0.000000000
0.600   0.00000   0.00000   0.000000000   0.58779   0.58779    0.000000000
0.800   0.00000   0.00000   0.000000000   0.30902   0.30902    0.000000000
1.000   0.00000   0.00000   0.000000000   0.00000   0.00000    0.000000000


T =    1.000
    X         U        UE          ERR U          V        VE          ERR V
0.000   0.00000   0.00000   0.000000000   0.00584   0.00000    0.005839275
0.200   0.30989   0.30902   0.000872997   0.00529   0.00000    0.005293151
0.400   0.58870   0.58779   0.000915749   0.00443   0.00000    0.004429206
0.600   0.80930   0.80902   0.000285517   0.00315   0.00000    0.003150163
0.800   0.95162   0.95106   0.000565818   0.00171   0.00000    0.001708773
1.000   0.99910   1.00000  -0.000897017   0.00000   0.00000    0.000000000


T =    2.000
    X         U        UE          ERR U          V        VE          ERR V
0.000   0.00000   0.00000   0.000000000  -0.99980  -1.00000    0.000203821
0.200   0.00369   0.00000   0.003693374  -0.95090  -0.95106    0.000153662
0.400   0.00646   0.00000   0.006455561  -0.80916  -0.80902   -0.000146921
0.600   0.00849   0.00000   0.008490598  -0.58778  -0.58779    0.000006141
0.800   0.01061   0.00000   0.010607707  -0.30892  -0.30902    0.000098394
1.000   0.00991   0.00000   0.009913405   0.00000   0.00000    0.000000000


T =    3.000
    X         U        UE          ERR U          V        VE          ERR V
0.000   0.00000   0.00000   0.000000000  -0.01716   0.00000   -0.017163329
0.200  -0.30960  -0.30902  -0.000580689  -0.01601   0.00000   -0.016011477
0.400  -0.58841  -0.58779  -0.000620006  -0.01331   0.00000   -0.013311313
0.600  -0.80956  -0.80902  -0.000547903  -0.00955   0.00000   -0.009547907
0.800  -0.95119  -0.95106  -0.000133066  -0.00497   0.00000   -0.004966406
1.000  -0.99981  -1.00000   0.000189491   0.00000   0.00000    0.000000000
```

Table 5.1 *Continued.*

```
T =   4.000
   X          U          UE          ERR U          V         VE         ERR V
0.000   0.00000   0.00000   0.000000000   0.99922   1.00000  -0.000783156
0.200  -0.00697   0.00000  -0.006967274   0.95080   0.95106  -0.000257907
0.400  -0.01302   0.00000  -0.013019440   0.80912   0.80902   0.000105319
0.600  -0.01728   0.00000  -0.017275511   0.58788   0.58779   0.000096445
0.800  -0.02084   0.00000  -0.020844798   0.30877   0.30902  -0.000247367
1.000  -0.02085   0.00000  -0.020848738   0.00000   0.00000   0.000000000


RUN NO. -   2  TWO FIRST-ORDER HYPERBOLIC PDES - DSS004

INITIAL T -  0.000D+00

  FINAL T -  0.400D+01

  PRINT T -  0.100D+01

NUMBER OF DIFFERENTIAL EQUATIONS -  22

INTEGRATION ALGORITHM - RKF45

MAXIMUM INTEGRATION ERROR -  0.100D-06


T =   0.000
   X          U          UE          ERR U          V         VE         ERR V
0.000   0.00000   0.00000   0.000000000   1.00000   1.00000   0.000000000
0.200   0.00000   0.00000   0.000000000   0.95106   0.95106   0.000000000
0.400   0.00000   0.00000   0.000000000   0.80902   0.80902   0.000000000
0.600   0.00000   0.00000   0.000000000   0.58779   0.58779   0.000000000
0.800   0.00000   0.00000   0.000000000   0.30902   0.30902   0.000000000
1.000   0.00000   0.00000   0.000000000   0.00000   0.00000   0.000000000


T =   1.000
   X          U          UE          ERR U          V         VE         ERR V
0.000   0.00000   0.00000   0.000000000   0.00001   0.00000   0.000011729
0.200   0.30901   0.30902  -0.000004958   0.00002   0.00000   0.000017606
0.400   0.58779   0.58779   0.000001843   0.00002   0.00000   0.000023781
0.600   0.80902   0.80902   0.000003475   0.00001   0.00000   0.000012156
0.800   0.95105   0.95106  -0.000003363   0.00001   0.00000   0.000010045
1.000   0.99998   1.00000  -0.000017544   0.00000   0.00000   0.000000000


T =   2.000
   X          U          UE          ERR U          V         VE         ERR V
0.000   0.00000   0.00000   0.000000000  -0.99997  -1.00000   0.000028456
0.200   0.00001   0.00000   0.000013739  -0.95105  -0.95106   0.000005615
0.400   0.00003   0.00000   0.000027219  -0.80902  -0.80902  -0.000005673
0.600   0.00004   0.00000   0.000040447  -0.58779  -0.58779  -0.000002572
0.800   0.00004   0.00000   0.000042422  -0.30901  -0.30902   0.000003402
1.000   0.00003   0.00000   0.000026168   0.00000   0.00000   0.000000000
```

Table 5.1 *Continued.*

T = 3.000

X	U	UE	ERR U	V	VE	ERR V
0.000	0.00000	0.00000	0.000000000	-0.00010	0.00000	-0.000099705
0.200	-0.30901	-0.30902	0.000004030	-0.00007	0.00000	-0.000067759
0.400	-0.58779	-0.58779	-0.000004572	-0.00006	0.00000	-0.000056445
0.600	-0.80902	-0.80902	-0.000003249	-0.00003	0.00000	-0.000034258
0.800	-0.95105	-0.95106	0.000005840	-0.00004	0.00000	-0.000035661
1.000	-0.99999	-1.00000	0.000014705	0.00000	0.00000	0.000000000

T = 4.000

X	U	UE	ERR U	V	VE	ERR V
0.000	0.00000	0.00000	0.000000000	0.99997	1.00000	-0.000028922
0.200	-0.00001	0.00000	-0.000014476	0.95106	0.95106	0.000006186
0.400	-0.00006	0.00000	-0.000056334	0.80901	0.80902	-0.000005694
0.600	-0.00008	0.00000	-0.000079259	0.58780	0.58779	0.000010032
0.800	-0.00009	0.00000	-0.000088198	0.30900	0.30902	-0.000016036
1.000	-0.00007	0.00000	-0.000067817	0.00000	0.00000	0.000000000

as two equations for the two unknowns, $I(x,t)$ and $V(x,t)$. We can then eliminate one of these unknowns between the two equations to arrive at a single equation in either unknown. This can be done in either of two ways.

For example, we can differentiate equation (5.6) with respect to x,

$$L\frac{\partial^2 I}{\partial t \partial x} + R\frac{\partial I}{\partial x} = -\frac{\partial^2 V}{\partial x^2} \tag{5.17}$$

We can also differentiate equation (5.8) with respect to t

$$C\frac{\partial^2 V}{\partial t^2} + G\frac{\partial V}{\partial t} = -\frac{\partial^2 I}{\partial x \partial t} \tag{5.18}$$

If we now assume the two mixed partials in equations (5.17) and (5.18) are equal (i.e., that the order of differentiation is immaterial), we can eliminate the mixed partial

$$C\frac{\partial^2 V}{\partial t^2} + G\frac{\partial V}{\partial t} = (R/L)\frac{\partial I}{\partial x} + (1/L)\frac{\partial^2 V}{\partial x^2} \tag{5.19}$$

Then substitution of $\partial I/\partial x$ from equation (5.8) in equation (5.19) gives

$$C\frac{\partial^2 V}{\partial t^2} + G\frac{\partial V}{\partial t} = (R/L)\left\{-C\frac{\partial V}{\partial t} - GV\right\} + (1/L)\frac{\partial^2 V}{\partial x^2} \tag{5.20}$$

or

$$CL\frac{\partial^2 V}{\partial t^2} + (RC + GL)\frac{\partial V}{\partial t} + RGV = \frac{\partial^2 V}{\partial x^2} \tag{5.21}$$

Equation (5.21) is a second-order hyperbolic PDE in V since it is second order in t and x; another possible classification is hyperbolic (from $\partial^2 V/\partial t^2$) and parabolic (from $\partial V/\partial t$) in t, to distinguish it from hyperbolic and parabolic in space, like equation (4.2). We will develop a method of lines solution of a generalization of equation (5.21) in a subsequent example. Alternatively, we could have eliminated V between equations (5.6) and (5.8) to obtain a second-order PDE in I.

To summarize the geometric classification of PDEs,

Derivative	Type of PDE
$\partial/\partial x$	hyperbolic (with $\partial/\partial t$)
$\partial^2/\partial x^2$	parabolic (with $\partial/\partial t$)
$\partial^2/\partial x^2$	elliptic (without $\partial/\partial t$ or $\partial^2/\partial t^2$)
$\partial/\partial t$	hyperbolic (with $\partial/\partial x$)
$\partial/\partial t$	parabolic (with $\partial^2/\partial x^2$)
$\partial^2/\partial t^2$	hyperbolic (with $\partial^2/\partial x^2$)

where x and t denote boundary value and initial value independent variables, respectively. Combinations are possible, for example, $\partial/\partial t$, $\partial/\partial x$, and $\partial^2/\partial x^2$ would be hyperbolic ($\partial/\partial t$ and $\partial/\partial x$, e.g., equation (3.55)) and parabolic ($\partial/\partial t$ and $\partial^2/\partial x^2$, e.g., equation (4.99)).

A second approach to equation (5.21) is to define a first-order differential operator in t as

$$\frac{\partial}{\partial t} = s \tag{5.22}$$

Substitution of equation (5.22) in equations (5.6) and (5.8) gives

$$LsI + RI = -\frac{\partial V}{\partial x} \tag{5.23}$$

$$CsV + GV = -\frac{\partial I}{\partial x} \tag{5.24}$$

Equations (5.23) and (5.24) can be solved algebraically for I and V, respectively:

$$I = -1/(Ls + R)\frac{\partial V}{\partial x} \tag{5.25}$$

$$V = -1/(Cs + G)\frac{\partial I}{\partial x} \tag{5.26}$$

Substitution of equation (5.25) in equation (5.26) then gives

$$V = 1/\left\{Cs + G)(Ls + R)\right\}\frac{\partial^2 V}{\partial x^2}$$

or

$$CLs^2V + (CR + GL)sV + GRV = \frac{\partial^2 V}{\partial x^2} \tag{5.27}$$

Finally, replacing s by its corresponding derivative in equation (5.27) (with s^2 interpreted as $\partial^2/\partial t^2$), we arrive at equation (5.21).

As a second source of second-order hyperbolic PDEs (in addition to systems of first-order hyperbolic PDEs), we return to Maxwell's equations, (3.74) to (3.77), and related equations (3.78) to (3.81). If equations (3.78) and (3.80) are substituted in equation (3.74)

$$\nabla \times H = \sigma E + \epsilon \partial E/\partial t \tag{5.28}$$

Equation (3.79) can then be substituted in equation (5.28)

$$(1/\mu)\nabla \times B = \sigma E + \epsilon \partial E/\partial t \tag{5.29}$$

If equation (5.29) is differentiated with respect to t,

$$(1/\mu)\nabla \times \partial B/\partial t = \sigma \partial E/\partial t + \epsilon \partial^2 E/\partial t^2 \tag{5.30}$$

where the order of differentiation with respect to t and ∇ (space) have been interchanged in the left-hand side. Then equation (3.75) can be substituted in equation (5.30)

$$-\nabla \times (\nabla \times E) = \mu\sigma \partial E/\partial t + \mu\epsilon \partial^2 E/\partial t^2 \tag{5.31}$$

The identity (equation (3.83))

$$\nabla \times (\nabla \times J) = \nabla(\nabla \bullet J) - \nabla^2 J$$

with $\nabla \bullet J = 0$ (constant charge density in equation (3.81)) and equation (3.80) give

$$\nabla \times (\nabla \times E) = -\nabla^2 E \tag{5.32}$$

Substitution of equation (5.32) in equation (5.31) finally gives

$$\mu\epsilon \partial^2 E/\partial t^2 + \mu\sigma \partial E/\partial t = \nabla^2 E \tag{5.33}$$

Equation (5.33) is the time-dependent Maxwell equation for the electric field E. This same equation also applies to H and J in place of E. For J in place of E, with $\epsilon = 0$ (a good conductor), equation (5.33) reduces to equation (3.86). According to the geometric classifications considered previously, equation (5.33) is both hyperbolic ($\partial^2 E/\partial t^2$ and $\nabla^2 E$) and parabolic ($\partial E/\partial t$ and $\nabla^2 E$).

We now consider the method of lines solution of equation (5.33) for two cases:

Case I: $\epsilon = 0$ (good conductor), $\mu = \sigma = 1$

$$E(x, 0) = 0 \tag{5.34}$$

$$E(0, t) = 1, \qquad \partial E(1, t)/\partial x = 0 \qquad (5.35)(5.36)$$

Case II: $\epsilon = \mu = \sigma = 1$

$$E(x, 0) = \sin(\pi x), \qquad \partial E(x, 0)/\partial t = 0 \qquad (5.37)(5.38)$$

$$\partial E(0, t)/\partial x = 0, \qquad \partial E(1, t)/\partial x = 0 \qquad (5.39)(5.40)$$

Note that for Case I, equation (5.33) is entirely parabolic.

Subroutines INITAL, INIT1 (for initial condition (5.34)), and INIT2 (for initial conditions (5.37) and (5.38)) are listed in Program 5.2a. We can note the following points about Program 5.2a:

1. Subroutine INITAL calls subroutine INIT1 when NORUN=1 and 2 for the Case I problem consisting of equations (5.33) to (5.36), and subroutine INIT2 when NORUN=3 and 4 for the Case II problem consisting of equations (5.33) and (5.37) to (5.40). Each set of two runs is for spatial differentiation via DSS042 and DSS044.

```
      SUBROUTINE INITAL
      IMPLICIT DOUBLE PRECISION (A-H,O-Z)
      PARAMETER (NX=202)
      COMMON/T/      T,  NSTOP,  NORUN
     1      /Y/  U(NX)
     2      /F/ UT(NX)
C...
C...  INITIAL CONDITION(S)
C...
C...     CASE 1
         IF((NORUN.EQ.1).OR.(NORUN.EQ.2))CALL INTAL1
C...
C...     CASE 2
         IF((NORUN.EQ.3).OR.(NORUN.EQ.4))CALL INTAL2
      RETURN
      END

      SUBROUTINE INTAL1
      IMPLICIT DOUBLE PRECISION (A-H,O-Z)
      PARAMETER (NX=51)
      COMMON/T/         T,     NSTOP,     NORUN
     1      /Y/     E(NX)
     2      /F/    ET(NX)
     3      /S/    EX(NX),   EXX(NX)
     4      /C/       EPS,        MU,     SIGMA
     5      /I/         N,        IP
      DOUBLE PRECISION MU
C...
C...  PARAMETERS
      EPS=0.0D0
      MU=1.0D0
      SIGMA=1.0D0
C...
C...  INITIAL CONDITION
      DO 1 I=1,NX
         E(I)=0.0D0
1     CONTINUE
C...
C...  INITIAL DERIVATIVES
      CALL DERV1
      IP=0
      RETURN
      END

      SUBROUTINE INTAL2
      IMPLICIT DOUBLE PRECISION (A-H,O-Z)
      PARAMETER (NX=101)
      COMMON/T/         T,     NSTOP,     NORUN
     1      /Y/    E1(NX),    E2(NX)
     2      /F/   E1T(NX),   E2T(NX)
```

Program 5.2a Subroutines INITAL, INIT1, and INIT2 for Initial Conditions (5.34), (5.37), and (5.38). *Continued next page.*

```
     3      /S/    EX(NX),   EXX(NX)
     4      /C/       EPS,         MU,      SIGMA
     5      /I/         N,         IP
      DOUBLE PRECISION MU
C...
C...  SET THE PARAMETERS
      EPS=1.0D0
      MU=1.0D0
      SIGMA=1.0D0
C...
C...  INITIAL CONDITION
      PI=3.1415927D0
      DO 1 I=1,NX
         X=DFLOAT(I-1)/DFLOAT(NX-1)
         E1(I)=DSIN(PI*X)
1     CONTINUE
C...
C...  INITIAL CONDITION
      DO 2 I=1,NX
         E2(I)=0.0D0
2     CONTINUE
C...
C...  INITIAL DERIVATIVES
      CALL DERV2
      IP=0
      RETURN
      END
```

Program 5.2a *Continued.*

2. For the Case I problem, the COMMON area in INIT1 is for a 51-point grid

```
      PARAMETER (NX=51)
      COMMON/T/          T,     NSTOP,      NORUN
     1      /Y/     E(NX)
     2      /F/    ET(NX)
     3      /S/    EX(NX),   EXX(NX)
     4      /C/       EPS,        MU,      SIGMA
     5      /I/         N,        IP
      DOUBLE PRECISION MU
```

where E in equation (5.33) is in array E(NX).

3. The parameters are set first in INIT1. Of particular importance is $\epsilon = 0$, which takes the second derivative, $\partial^2 E/\partial x^2$, out of equation (5.33), i.e., equation (5.33) is first order in t; thus it is parabolic.

```
C...
C...  PARAMETERS
      EPS=0.0D0
      MU=1.0D0
      SIGMA=1.0D0
```

4. Initial condition (5.34) is then set in DO loop 1 of INIT1:

```
C...
C...  INITIAL CONDITION
      DO 1 I=1,NX
         E(I)=0.0D0
1     CONTINUE
```

5. For Case II, the COMMON area in INIT2 is defined for a 101-point grid. This doubling of the number of grid points over Case I was required in order to get reasonable spatial resolution, i.e., the Case II solution varies more rapidly in space than the Case I solution.

```
      PARAMETER (NX=101)
      COMMON/T/          T,      NSTOP,      NORUN
     1      /Y/    E1(NX),      E2(NX)
     2      /F/   E1T(NX),     E2T(NX)
     3      /S/    EX(NX),     EXX(NX)
     4      /C/        EPS,         MU,      SIGMA
     5      /I/          N,         IP
      DOUBLE PRECISION MU
```

Also, for Case II, $\epsilon = 1$; thus both the first and second temporal derivatives are retained in equation (5.33), and therefore equation (5.33) is hyperbolic parabolic. To accommodate the second derivative in t, we define two state variables, $E_1 = E$ and $E_2 = \partial E/\partial t$ that are in arrays E1(NX) and E2(NX), respectively. Thus, a total of $2 \times 101 = 202$ ODEs are integrated for the Case II solution.

6. The Case II problem parameters are first defined in INIT2:

```
C...
C...  SET THE PARAMETERS
      EPS=1.0D0
      MU=1.0D0
      SIGMA=1.0D0
```

7. Then initial conditions (5.37) and (5.38) are set in DO loops 1 and 2 of INIT2:

```
C...
C...  INITIAL CONDITION
      PI=3.1415927D0
      DO 1 I=1,NX
         X=DFLOAT(I-1)/DFLOAT(NX-1)
         E1(I)=DSIN(PI*X)
1     CONTINUE
C...
C...  INITIAL CONDITION
      DO 2 I=1,NX
         E2(I)=0.0D0
2     CONTINUE
```

This completes the initial conditions for the four runs. Subroutines DERV, DERV1, and

DERV2 are listed in Program 5.2b. We can note the following points about Program 5.2b:

1. Subroutine DERV calls subroutine DERV1 when NORUN=1 and 2 for the Case I problem consisting of equations (5.33) to (5.36), and subroutine DERV2 when NORUN=3 and 4 for the Case II problem consisting of equations (5.33) and (5.37) to (5.40). Each set of two runs is for spatial differentiation via DSS042 and DSS044.

2. The COMMON areas in DERV1 and DERV2 are the same as in INIT1 and INIT2, as expected.

3. In DERV1, boundary conditions (5.35) and (5.36) are first implemented:

```
C...
C...  BOUNDARY CONDITION AT X = 0
       E(1)=1.0D0
      ET(1)=0.0D0
C...
C...  BOUNDARY CONDITION AT X = 1
      EX(NX)=0.0D0
```

4. Next, the derivative $\partial^2 E/\partial x^2$ in equation (5.33) is computed in DERV1 either by second-order differences (NORUN=1 so DSS042 is called), or by fourth-order differences (NORUN=2 so DSS044 is called):

```
C...
C...  UXX
      IF(NORUN.EQ.1)THEN
         CALL DSS042(0.0D0,1.0D0,NX,E,EX,EXX,1,2)
      ELSE IF(NORUN.EQ.2)THEN
         CALL DSS044(0.0D0,1.0D0,NX,E,EX,EXX,1,2)
      END IF
```

Note the combination of Dirichlet and Neumann boundary conditions, as specified by the last two arguments of the differentiation routines, i.e., 1 for $x = 0$ and 2 for $x = 1$.

5. Finally, equation (5.33) with $\epsilon = 0$ is programmed in DO loop 1 of DERV1:

```
C...
C...  PDE
      DO 1 I=2,NX
         ET(I)=EXX(I)
1     CONTINUE
```

6. In DERV2, boundary conditions (5.39) and (5.40) are first implemented:

```
C...
C...  BOUNDARY CONDITION AT X = 0
      EX(1)=0.0D0
C...
C...  BOUNDARY CONDITION AT X = 1
      EX(NX)=0.0D0
```

```
      SUBROUTINE DERV
      IMPLICIT DOUBLE PRECISION (A-H,O-Z)
      PARAMETER (NX=202)
      COMMON/T/      T,  NSTOP,  NORUN
     1      /Y/  U(NX)
     2      /F/ UT(NX)
C...
C...     CASE 1
         IF((NORUN.EQ.1).OR.(NORUN.EQ.2))CALL DERV1
C...
C...     CASE 2
         IF((NORUN.EQ.3).OR.(NORUN.EQ.4))CALL DERV2
      RETURN
      END

      SUBROUTINE DERV1
      IMPLICIT DOUBLE PRECISION (A-H,O-Z)
      PARAMETER (NX=51)
      COMMON/T/         T,      NSTOP,      NORUN
     1      /Y/     E(NX)
     2      /F/    ET(NX)
     3      /S/    EX(NX),   EXX(NX)
     4      /C/       EPS,         MU,      SIGMA
     5      /I/         N,         IP
      DOUBLE PRECISION MU
C...
C...  BOUNDARY CONDITION AT X = 0
       E(1)=1.0D0
      ET(1)=0.0D0
C...
C...  BOUNDARY CONDITION AT X = 1
      EX(NX)=0.0D0
C...
C...  UXX
      IF(NORUN.EQ.1)THEN
         CALL DSS042(0.0D0,1.0D0,NX,E,EX,EXX,1,2)
      ELSE IF(NORUN.EQ.2)THEN
         CALL DSS044(0.0D0,1.0D0,NX,E,EX,EXX,1,2)
      END IF
C...
C...  PDE
      DO 1 I=2,NX
         ET(I)=EXX(I)
1     CONTINUE
      RETURN
      END

      SUBROUTINE DERV2
      IMPLICIT DOUBLE PRECISION (A-H,O-Z)
      PARAMETER (NX=101)
      COMMON/T/         T,      NSTOP,      NORUN
     1      /Y/    E1(NX),     E2(NX)
```

Program 5.2b Subroutines DERV, DERV1, and DERV2 for Equation (5.33), Case I and Case II. *Continued next page.*

```
     2      /F/   E1T(NX),    E2T(NX)
     3      /S/    EX(NX),    EXX(NX)
     4      /C/       EPS,         MU,      SIGMA
     5      /I/         N,         IP
      DOUBLE PRECISION MU
C...
C...  BOUNDARY CONDITION AT X = 0
      EX(1)=0.0D0
C...
C...  BOUNDARY CONDITION AT X = 1
      EX(NX)=0.0D0
C...
C...  UXX
      IF(NORUN.EQ.3)THEN
         CALL DSS042(0.0D0,1.0D0,NX,E1,EX,EXX,2,2)
      ELSE IF(NORUN.EQ.4)THEN
         CALL DSS044(0.0D0,1.0D0,NX,E1,EX,EXX,2,2)
      END IF
C...
C...  PDES
      DO 1 I=1,NX
         E1T(I)=E2(I)
         E2T(I)=EXX(I)-E2(I)
1     CONTINUE
      RETURN
      END
```

Program 5.2b *Continued.*

7. Next, the derivative $\partial^2 E/\partial x^2$ in equation (5.33) is computed in DERV2 either by second-order differences (NORUN=3 so DSS042 is called), or by fourth-order differences (NORUN=4 so DSS044 is called):

```
C...
C...  UXX
      IF(NORUN.EQ.3)THEN
         CALL DSS042(0.0D0,1.0D0,NX,E1,EX,EXX,2,2)
      ELSE IF(NORUN.EQ.4)THEN
         CALL DSS044(0.0D0,1.0D0,NX,E1,EX,EXX,2,2)
      END IF
```

   Note that both boundary conditions are Neumann, as specified by the last two arguments of the differentiation routines, i.e., 2 for $x = 0$ and $x = 1$.

8. Finally, equation (5.33) with $\epsilon = 1$ is programmed in DO loop 1 of DERV2:

```
C...
C...  PDES
      DO 1 I=1,NX
         E1T(I)=E2(I)
         E2T(I)=EXX(I)-E2(I)
1     CONTINUE
```

With $NX = 101$, 202 ODEs are programmed in DO loop 1.

Subroutine PRINT calls subroutine PRINT1 (for NORUN=1 and 2) and subroutine PRINT2 (for NORUN=3 and 4).

We can note the following points about these subroutines:

1. PRINT1 prints a heading for the numerical solution during the first call, then prints the numerical solution (for NORUN=1 and 2) during each call:

```
C...
C...  PRINT A HEADING FOR THE SOLUTION
      IP=IP+1
      IF(IP.EQ.1)WRITE(NO,1)
1     FORMAT(9X,'T',/,
     1 3X,'    E(0,T)',3X,' E(0.2,T)',3X,' E(0.4,T)',
     2 3X,' E(0.6,T)',3X,' E(0.8,T)',3X,' E(1.0,T)',/)
C...
C...  PRINT THE NUMERICAL SOLUTION
      WRITE(NO,2)T,(E(I),I=1,NX,10)
2     FORMAT(F9.2,/,6F12.5,/)
```

2. During each call, PRINT1 also stores the numerical solution for plotting, and by a call to PLOT1, writes the solution to a file for plotting at the end of the run:

```
C...
C...  STORE THE SOLUTION FOR PLOTTING
      DO 3 I=1,NX
         XP(I)=DFLOAT(I-1)/DFLOAT(NX-1)
         EP(IP,I)=E(I)
3     CONTINUE
      CALL PLOT1
```

3. PRINT1 also calls subroutine SPLOTS at the end of the solution for point plotting of the solution:

```
C...
C...  PLOT THE SOLUTION
      IF(IP.LT.6)RETURN
      CALL SPLOTS(IP,NX,XP,EP)
      WRITE(NO,4)
4     FORMAT(1H ,/,' E(X,T) VS X, T = 0, 0.2,..., 1.0')
```

Subroutine PRINT2 is essentially identical to PRINT1. Also, subroutines PLOT1 and PLOT2 are not listed as they are very similar to previous plotting routines, e.g., see Program 3.5d.

The data file for Program 5.2a to c is listed in Program 5.2d. Four runs are specified; for NORUN=1 and 2, 51 ODEs are integrated, and for NORUN=3 and 4, 202 ODEs are integrated. In each case, the error tolerance is 0.0001 (with integration by RKF45).

Abbreviated output from Programs 5.2a to d is listed in Table 5.2 and the numerical solution is plotted in Figures 5.2a and b.

We can note the following points about the output in Table 5.2:

1. The Case I problem (equations (5.33) to (5.36)) is relatively easy to solve as indicated by the continuous plot in Figure 5.2a, i.e., the solution is quite smooth. Further indication of this smoothness is evident from the close agreement of the

```
      SUBROUTINE PRINT(NI,NO)
      IMPLICIT DOUBLE PRECISION (A-H,O-Z)
      PARAMETER (NX=202)
      COMMON/T/      T,  NSTOP,  NORUN
     1      /Y/  U(NX)
     2      /F/ UT(NX)
C...
C...  PRINT AND PLOT THE SOLUTION
C...
C...     CASE 1
         IF((NORUN.EQ.1).OR.(NORUN.EQ.2))CALL PRINT1(NI,NO)
C...
C...     CASE 2
         IF((NORUN.EQ.3).OR.(NORUN.EQ.4))CALL PRINT2(NI,NO)
      RETURN
      END

      SUBROUTINE PRINT1(NI,NO)
      IMPLICIT DOUBLE PRECISION (A-H,O-Z)
      PARAMETER (NX=51)
      COMMON/T/         T,     NSTOP,     NORUN
     1      /Y/     E(NX)
     2      /F/    ET(NX)
     3      /S/    EX(NX),   EXX(NX)
     4      /C/       EPS,        MU,     SIGMA
     5      /I/         N,        IP
      DOUBLE PRECISION MU
C...
C...  DIMENSION THE ARRAYS FOR PLOTTING
      COMMON/P1/   XP(NX),  EP(6,NX)
C...
C...  PRINT A HEADING FOR THE SOLUTION
C...
C...  MONITOR THE OUTPUT
      WRITE(*,*)NORUN,T
C...
C...  PRINT A HEADING FOR THE SOLUTION
      IP=IP+1
      IF(IP.EQ.1)WRITE(NO,1)
1     FORMAT(9X,'T',/,
     1 3X,'   E(0,T)',3X,' E(0.2,T)',3X,' E(0.4,T)',
     2 3X,' E(0.6,T)',3X,' E(0.8,T)',3X,' E(1.0,T)',/)
C...
C...  PRINT THE NUMERICAL SOLUTION
      WRITE(NO,2)T,(E(I),I=1,NX,10)
2     FORMAT(F9.2,/,6F12.5,/)
C...
C...  STORE THE SOLUTION FOR PLOTTING
      DO 3 I=1,NX
         XP(I)=DFLOAT(I-1)/DFLOAT(NX-1)
         EP(IP,I)=E(I)
3     CONTINUE
```

Program 5.2c Subroutines PRINT, PRINT1, and PRINT2 for Equations (5.33) to (5.40). *Continued next page.*

```
      CALL PLOT1
C...
C...  PLOT THE SOLUTION
      IF(IP.LT.6)RETURN
      CALL SPLOTS(IP,NX,XP,EP)
      WRITE(NO,4)
4     FORMAT(1H ,/,' E(X,T) VS X, T = 0, 0.2,..., 1.0')
      RETURN
      END

      SUBROUTINE PRINT2(NI,NO)
      IMPLICIT DOUBLE PRECISION (A-H,O-Z)
      PARAMETER (NX=101)
      COMMON/T/          T,      NSTOP,      NORUN
     1      /Y/    E1(NX),    E2(NX)
     2      /F/   E1T(NX),   E2T(NX)
     3      /S/    EX(NX),   EXX(NX)
     4      /C/       EPS,        MU,      SIGMA
     5      /I/         N,        IP
      DOUBLE PRECISION MU
C...
C...  DIMENSION THE ARRAYS FOR PLOTTING
      COMMON/P2/   XP(NX),  EP(6,NX)
C...
C...  MONITOR THE OUTPUT
      WRITE(*,*)NORUN,T
C...
C...  PRINT A HEADING FOR THE SOLUTION
      IP=IP+1
      IF(IP.EQ.1)WRITE(NO,1)
1     FORMAT(9X,'T',/,
     1 3X,'   E(0,T)',3X,' E(0.2,T)',3X,' E(0.4,T)',
     2 3X,' E(0.6,T)',3X,' E(0.8,T)',3X,' E(1.0,T)',/)
C...
C...  PRINT THE NUMERICAL SOLUTION
      WRITE(NO,2)T,(E1(I),I=1,NX,20)
2     FORMAT(F9.2,/,6F12.5,/)
C...
C...  STORE THE SOLUTION FOR PLOTTING
      DO 3 I=1,NX
         XP(I)=DFLOAT(I-1)/DFLOAT(NX-1)
         EP(IP,I)=E1(I)
3     CONTINUE
      CALL PLOT2
C...
C...  PLOT THE SOLUTION
      IF(IP.LT.6)RETURN
      CALL SPLOTS(IP,NX,XP,EP)
      WRITE(NO,4)
4     FORMAT(1H ,/,' E(X,T) VS X, T = 0, 0.2,..., 1.0')
      RETURN
      END
```

Program 5.2c *Continued.*

```
MAXWELL'S EQUATION, EPS = 0, DSS042
0.         1.0       0.2
   51                      0.0001
MAXWELL'S EQUATION, EPS = 0, DSS044
0.         1.0       0.2
   51                      0.0001
MAXWELL'S EQUATION, EPS = 1, DSS042
0.         1.0       0.2
  202                      0.0001
MAXWELL'S EQUATION, EPS = 1, DSS044
0.         1.0       0.2
  202                      0.0001
END OF RUNS
```

Program 5.2d Data File for Programs 5.2a to d.

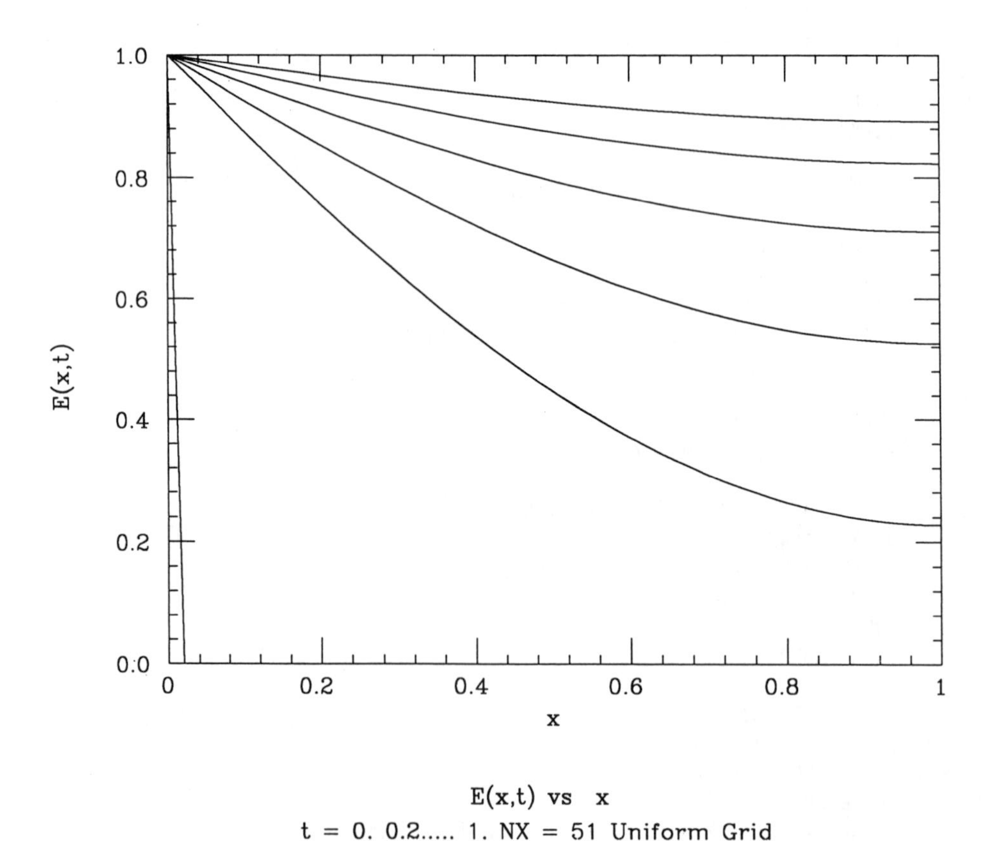

Figure 5.2 (a) Plotted Solution from Programs 5.2a to d for NORUN=1 and 2.

Table 5.2 Abbreviated Output from Programs 5.2a to d. *Continued next pages.*

```
RUN NO. -    1  MAXWELL'S EQUATION, EPS = 0, DSS042

INITIAL T -   0.000D+00

  FINAL T -   0.100D+01

  PRINT T -   0.200D+00

NUMBER OF DIFFERENTIAL EQUATIONS -  51

INTEGRATION ALGORITHM - RKF45

MAXIMUM INTEGRATION ERROR -  0.100D-03


        T
     E(0,T)    E(0.2,T)    E(0.4,T)    E(0.6,T)    E(0.8,T)    E(1.0,T)

    0.00
    1.00000     0.00000     0.00000     0.00000     0.00000     0.00000

    0.20
    1.00000     0.75575     0.53836     0.36962     0.26371     0.22773

    0.40
    1.00000     0.85331     0.72101     0.61607     0.54871     0.52551

    0.60
    1.00000     0.91047     0.82971     0.76561     0.72446     0.71028

    0.80
    1.00000     0.94534     0.89603     0.85690     0.83178     0.82312

    1.00
    1.00000     0.96663     0.93653     0.91263     0.89730     0.89202


RUN NO. -    2  MAXWELL'S EQUATION, EPS = 0, DSS044

INITIAL T -   0.000D+00

  FINAL T -   0.100D+01

  PRINT T -   0.200D+00

NUMBER OF DIFFERENTIAL EQUATIONS -  51

INTEGRATION ALGORITHM - RKF45

MAXIMUM INTEGRATION ERROR -  0.100D-03

        T
     E(0,T)    E(0.2,T)    E(0.4,T)    E(0.6,T)    E(0.8,T)    E(1.0,T)
```

Table 5.2 Continued.

```
   0.00
   1.00000     0.00000     0.00000     0.00000     0.00000     0.00000

   0.20
   1.00000     0.75575     0.53835     0.36960     0.26367     0.22769

   0.40
   1.00000     0.85331     0.72101     0.61607     0.54871     0.52551

   0.60
   1.00000     0.91047     0.82971     0.76562     0.72447     0.71032

   0.80
   1.00000     0.94535     0.89604     0.85691     0.83179     0.82311

   1.00
   1.00000     0.96663     0.93653     0.91264     0.89731     0.89204

RUN NO. -   3  MAXWELL'S EQUATION, EPS = 1, DSS042

INITIAL T -  0.000D+00

  FINAL T -  0.100D+01

  PRINT T -  0.200D+00

NUMBER OF DIFFERENTIAL EQUATIONS - 202

INTEGRATION ALGORITHM - RKF45

MAXIMUM INTEGRATION ERROR -  0.100D-03


     T
    E(0,T)    E(0.2,T)    E(0.4,T)    E(0.6,T)    E(0.8,T)    E(1.0,T)

   0.00
   0.00000     0.58779     0.95106     0.95106     0.58779     0.00000

   0.20
   0.55519     0.49292     0.78088     0.78088     0.49292     0.55519

   0.40
   0.88192     0.73845     0.38292     0.38292     0.73845     0.88192

   0.60
   0.89225     0.74154     0.38193     0.38193     0.74154     0.89225

   0.80
   0.63579     0.54782     0.70407     0.70407     0.54782     0.63579

   1.00
   0.27249     0.60853     0.82748     0.82748     0.60853     0.27249
```

***Table* 5.2** *Continued.*

```
RUN NO. -   4  MAXWELL'S EQUATION, EPS = 1, DSS044

INITIAL T -  0.000D+00

  FINAL T -  0.100D+01

  PRINT T -  0.200D+00

NUMBER OF DIFFERENTIAL EQUATIONS - 202

INTEGRATION ALGORITHM - RKF45

MAXIMUM INTEGRATION ERROR -  0.100D-03


        T
     E(0,T)    E(0.2,T)    E(0.4,T)    E(0.6,T)    E(0.8,T)    E(1.0,T)

    0.00
    0.00000     0.58779     0.95106     0.95106     0.58779     0.00000

    0.20
    0.57395     0.48893     0.78087     0.78087     0.48893     0.57395

    0.40
    0.86780     0.73803     0.37777     0.37777     0.73803     0.86780

    0.60
    0.89140     0.74166     0.37484     0.37484     0.74166     0.89140

    0.80
    0.62674     0.54182     0.70558     0.70558     0.54182     0.62674

    1.00
    0.26384     0.60722     0.82699     0.82699     0.60722     0.26384
```

two solutions for NORUN=1 and 2. For example, the solutions at $t = 0.20$ are (from Table 5.2):

```
Run No. 1 - DSS042

        T
     E(0,T)    E(0.2,T)    E(0.4,T)    E(0.6,T)    E(0.8,T)    E(1.0,T)

    0.20
    1.00000     0.75575     0.53836     0.36962     0.26371     0.22773
```

Table 5.2 *Continued.*

```
Plotted in run nos. 1 and 2

              ..1....1....1....1....1....1....1....1....1....1....1..
    0.100D+01+  CCCCCCCCCCCCC666666666                               +I
             -     22C33CC444CCC555555CCCCCC66666666666666666666666  -I
             -        222 3333  444444444   55555555555555555555555  -I
             -           22   33333      4444444444444               -I
             -             222     333333             4444444444444  -I
    0.667D+00+                222        3333333                     +I
             -                   2222           3333333333333        -I
             -                       222                     333333  -I
             -                          2222                         -I
             -                              22222                    -I
    0.333D+00+                                   2222222             +I
             -                                          22222222222  -I
             -                                                       -I
             -                                                       -I
             -                                                       -I
    0.000D+00+   11111111111111111111111111111111111111111111111111  +I
              ..1....1....1....1....1....1....1....1....1....1....1..
           0.000D+00  0.20D+00  0.40D+00  0.60D+00  0.80D+00  0.10D+01

 E(X,T) VS X, T = 0, 0.2,..., 1.0
```

```
Plotted in run nos. 3 and 4

              ..1....1....1....1....1....1....1....1....1....1....1..
    0.100D+01+                      CCCCCCCCCCC1                     +I
             -  CCCC             1CC           1CC             4CCC  -I
             -     3CCCC4      1C1 CCCCCCCCCCCCC6 CC       CCCCC     -I
             -          CCC   CCCCCCCC       2CCCCCCC1  CCC3         -I
             -             CCCC6CCCCCCCCCCCCCCCCCCC2CCCCC            -I
    0.667D+00+  5CCCCC5   6CCCCCC5                CCCCCCC    CCCCCC  +I
             -  2CCCCCCCCCCCCC CC                 CCC2CCCCCCCCCCCCC  -I
             -        6CCCCC     CC              CC    2CCCCC        -I
             -      6C611         CC           CC          C CC      -I
             -     C6 C1            CCCCCCCCCCCC            C1 C6    -I
    0.333D+00+  6CC  C                                       11 6CC  +I
             -      C                                         11     -I
             -     C                                           11    -I
             -    C                                             11   -I
             -  1C                                               11  -I
   -0.464D-07+                                                    1  +I
              ..1....1....1....1....1....1....1....1....1....1....1..
           0.000D+00  0.20D+00  0.40D+00  0.60D+00  0.80D+00  0.10D+01

 E(X,T) VS X, T = 0, 0.2,..., 1.0
```

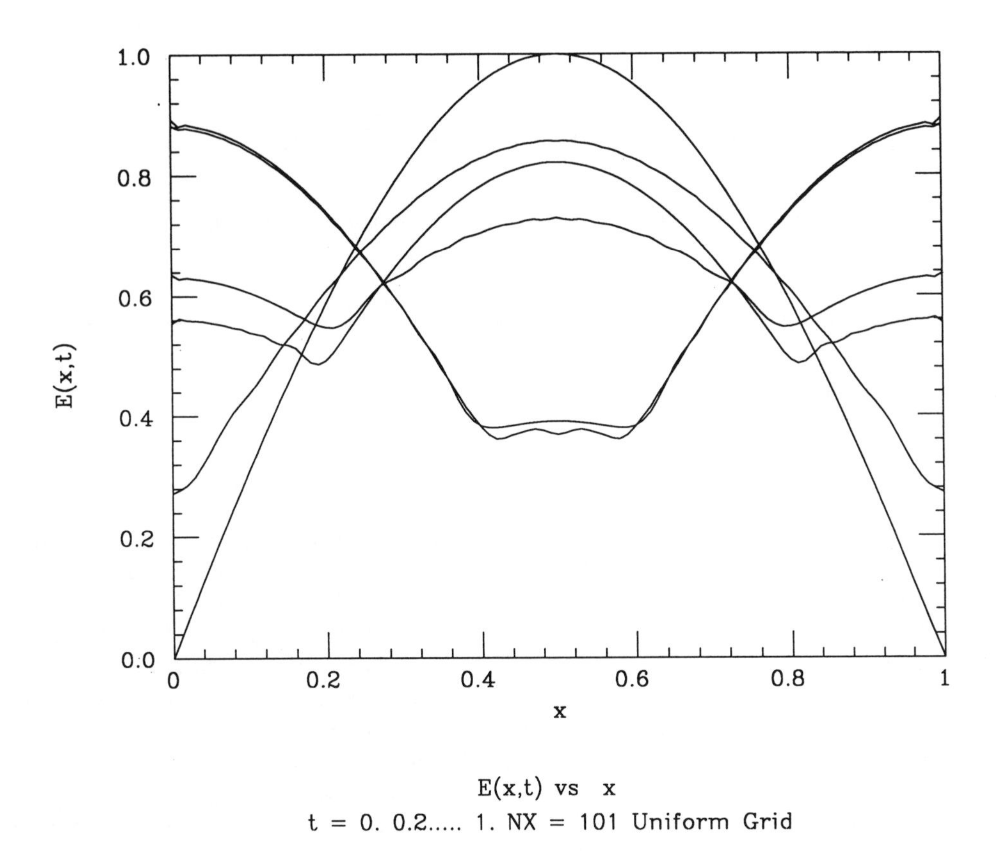

Figure 5.2 (b) Plotted Solution from Programs 5.2a to d for NORUN=3 and 4.

```
Run No. 2 - DSS044

       T
  E(0,T)      E(0.2,T)     E(0.4,T)     E(0.6,T)     E(0.8,T)     E(1.0,T)

  0.20
  1.00000     0.75575      0.53835      0.36960      0.26367      0.22769
```

This agreement to four figures suggests that the solution from the fourth-order approximations of DSS044 (from NORUN=2) is probably accurate to four figures. This could be confirmed, of course, with an analytical solution that could be derived using the Case II finite integral transform of equation (4.139).

2. The Case II problem (equations (5.33) and (5.37) to (5.40)) is relatively difficult to solve as indicated by the continuous plot in Figure 5.2b, i.e., the solution has relatively sharp spatial variations. In fact, when the solution was plotted from a 51-point grid, it clearly did not have adequate spatial resolution, and therefore the change to a 101-point grid was made. Further indication of this relative difficulty of computing an accurate solution is evident from the lack of

agreement of the two solutions for NORUN=3 and 4. For example, the solutions at $t = 0.20$ (from Table 5.2) are

```
Run No. 3 - DSS042

       T
     E(0,T)    E(0.2,T)    E(0.4,T)    E(0.6,T)    E(0.8,T)    E(1.0,T)

    0.20
    0.55519     0.49292     0.78088     0.78088     0.49292     0.55519


Run No. 4 - DSS044

       T
     E(0,T)    E(0.2,T)    E(0.4,T)    E(0.6,T)    E(0.8,T)    E(1.0,T)

    0.20
    0.57395     0.48893     0.78087     0.78087     0.48893     0.57395
```

Both numerical solutions have the symmetry required by initial conditions (5.37) and (5.38), and by boundary conditions (5.39) and (5.40), but they are substantially different numerically. For example, $E(0, 0.2) = 0.55519$ and 0.57395 for NORUN=3 and 4, respectively. Note also that the discrepancy between the two solutions is greatest near the boundaries at $x = 0$ and 1. This suggests that a nonuniform grid could be used to advantage for which the grid points are concentrated at the boundaries, e.g., the spline differentiator in subroutines DSS038 and DSS040 could be used.

This lack of agreement is not an untypical situation, and the question then naturally arises as to the accuracy of either solution. In other words, these results suggest that some additional investigation of the accuracy is required, perhaps by computing solutions with greater numbers of grid points, followed by comparison of the numerical results. For this particular problem, an analytical solution could also be derived, by using the Case III finite integral transform of equation (4.141), but in general, analytical solutions are not available to evaluate numerical solutions.

In fact, we offer the opinion that evaluating output from the numerical integration of differential systems is the most challenging aspect of these numerical studies. The required error analysis is not always straightforward and may consist of the use of a combination of special or limiting cases for which analytical solutions are available; changes in the number and spacing of the discrete points used in the numerical algorithm; changes in the order of the numerical algorithm; and even changes in the numerical algorithm itself (i.e., solutions from two or more different algorithms, which agree, give confidence in the validity of the numerical solutions).

For scientific and engineering applications, special knowledge of the problem can also be invaluable; certainly tentative solutions which violate established principles like conservation of energy, or contradict common sense, can be ruled out as invalid. Also, we can generally expect that each new problem that is being studied numerically for the first time will require a unique approach to an error analysis and evaluation of the solution. This process is not well defined in general, so even experience and judgment are important inputs to the process of evaluating numerical solutions. Certainly, critical evaluation of a numerical solution should be standard practice, since uninformed acceptance of computer output will most likely lead to incorrect conclusions.

As suggested previously, one approach to the evaluation of a numerical solution is to first compute a numerical solution to a closely related problem with a known analytical solution, that is, a special case, then compute a solution to the problem of interest and observe if it departs from the first solution in a reasonable and expected way. We illustrate this approach with a minor extension of equation (5.33):

$$a\frac{\partial^2 u}{\partial t^2} + b\frac{\partial u}{\partial t} = c\frac{\partial^2 u}{\partial x^2} + du \tag{5.41}$$

a, b, c, and d are given constants. Equation (5.41) differs from equation (5.33) (with V replaced by u) in only the additional term du. If $d = 0$, equation (5.41) reverts to equation (5.33), and is therefore Maxwell's equation in one dimension. If $d \neq 0$, equation (5.41) is known as the *telegraph equation.*

We choose the initial and boundary conditions of equation (5.41) as

$$u(x,0) = \sin(\pi x), \qquad u_t(x,0) = \{-b/(2a)\}\sin(\pi x) \qquad (5.42)(5.43)$$

$$u(0,t) = 0, \qquad u(1,t) = 0 \qquad (5.44)(5.45)$$

If we assume an analytical solution to equations (5.41) to (5.45) as

$$u(x,t) = e^{c_1 t}\cos(c_2 t)\sin(c_3 x) \tag{5.46}$$

then we can easily show that equation (5.46) is a solution to equations (5.41) to (5.45) if

$$c_1 = -b/(2a) \tag{5.47}$$

$$c_2 = \left\{(1/a)\left\{-b^2/(4a) + c{c_3}^2 - d\right\}\right\}^{1/2} \tag{5.48}$$

$$c_3 = \pi \tag{5.49}$$

Equation (5.46) can therefore be used to evaluate the numerical solution to equation (5.41).

Also, we will now extend equation (5.41) to a nonlinear case for which an analytical solution is not available to serve as a standard for the numerical solution. The nonlinearity introduced into equation (5.41) models damping which is common in physical systems; typically, nonlinear damping is modeled by a term proportional to the square of the velocity, $\partial u/\partial t$. Equation (5.41) then becomes

$$a\frac{\partial^2 u}{\partial t^2} + b\frac{\partial u}{\partial t} + e\left|\frac{\partial u}{\partial t}\right|\frac{\partial u}{\partial t} = c\frac{\partial^2 u}{\partial x^2} + du \tag{5.50}$$

where e is a coefficient for the nonlinear damping term. The absolute value is used in the velocity damping term, i.e., $|\partial u/\partial t|$, to ensure that this term has the correct sign (without the absolute value, if $u < 0$, equation (5.50) would be unstable).

The strategy in computing a method of lines solution to equation (5.50) is to program and test the linear problem first (with $e = 0$), using equation (5.46) to check the numerical solution. Then the nonlinear term in equation (5.50) can be added with one more line of code; thus by this *incremental approach,* the chances of a programming error are substantially reduced. Also, we can observe if the addition of the nonlinear term changes the linear solution in the direction we would expect.

Subroutines `INITAL`, `INIT1`, and `INIT2` are listed in Program 5.3a (we have programmed the linear and nonlinear cases separately in a set of two subroutines, starting with `INIT1` and `INIT2`).

We can note the following points about INITAL, INIT1, and INIT2:

1. Four runs are programmed as indicated in INITAL. For NORUN=1 and 2, equation (5.41) is programmed for the cases $d = 0$ (Maxwell's equation) and $d = 1$ (the telegraph equation). For NORUN=3 and 4, equation (5.50) is programmed with $e = 1$ and 100, respectively, corresponding to weak and strong nonlinear damping. In each run, the spatial grid has 21 points, thus the total of ODEs is 42.

2. In subroutine INIT1, the linear problem parameters are first defined:

```
C...
C...  PROBLEM PARAMETERS
      A=1.0D0
      B=1.0D0
      C=1.0D0
C...
C...  MAXWELL'S EQUATION
      IF(NORUN.EQ.1)THEN
         D=0.0D0
C...
C...  TELEGRAPH EQUATION
      ELSE IF(NORUN.EQ.2)THEN
         D=-1.0D0
      END IF
```

   Note the values $d = 0$ and -1 for NORUN=1 and 2, respectively.

3. Constants c_1, c_2, and c_3 are then computed in INIT1 for use in the analytical solution, equation (5.46):

```
C...
C...  PRECOMPUTE THE CONSTANTS IN THE ANALYTICAL SOLUTION FOR USE
C...  IN FUNCTION UANAL
      C3=4.0D0*DATAN(1.0D0)
      C1=-B/(2.0D0*A)
      C2=DSQRT((1.0D0/A)*(-(B**2)/(4.0D0*A)+C*(C3**2)-D))
```

4. Finally, in INIT1 the spatial grid and initial conditions (5.42) and (5.43) are defined (on a 51-point grid):

```
C...
C...  SPATIAL GRID
      DO 1 I=1,NX
         X(I)=DFLOAT(I-1)/DFLOAT(NX-1)
1     CONTINUE
C...
C...  INITIAL CONDITIONS
      DO 2 I=1,NX
         U1(I)=   DSIN(C3*X(I))
         U2(I)=C1*DSIN(C3*X(I))
2     CONTINUE
```

```
      SUBROUTINE INITAL
      IMPLICIT DOUBLE PRECISION (A-H,O-Z)
      PARAMETER (NX=21)
      COMMON/T/        T,   NSTOP,   NORUN
C...
C...  LINEAR PDE
      IF((NORUN.EQ.1).OR.(NORUN.EQ.2))THEN
         CALL INIT1
C...
C...  NONLINEAR PDE
      ELSE
     +IF((NORUN.EQ.3).OR.(NORUN.EQ.4))THEN
         CALL INIT2
      END IF
      RETURN
      END

      SUBROUTINE INIT1
      IMPLICIT DOUBLE PRECISION (A-H,O-Z)
      PARAMETER (NX=21)
      COMMON/T/        T,   NSTOP,   NORUN
     1      /Y/  U1(NX),  U2(NX)
     2      /F/ U1T(NX), U2T(NX)
     3      /S/ U1X(NX),U1XX(NX),   X(NX)
     4      /C/        A,        B,        C,        D,
     5                C1,       C2,       C3
     6      /I/       IP
C...
C...  PROBLEM PARAMETERS
      A=1.0D0
      B=1.0D0
      C=1.0D0
C...
C...  MAXWELL'S EQUATION
      IF(NORUN.EQ.1)THEN
         D=0.0D0
C...
C...  TELEGRAPH EQUATION
      ELSE IF(NORUN.EQ.2)THEN
         D=-1.0D0
      END IF
C...
C...  PRECOMPUTE THE CONSTANTS IN THE ANALYTICAL SOLUTION FOR USE IN
C...  FUNCTION UANAL
      C3=4.0D0*DATAN(1.0D0)
      C1=-B/(2.0D0*A)
      C2=DSQRT((1.0D0/A)*(-(B**2)/(4.0D0*A)+C*(C3**2)-D))
C...
```

Program 5.3a Subroutines `INITAL`, `INIT1`, and `INIT2` for Initial Conditions (5.42) and (5.43). *Continued next pages.*

```
C...  SPATIAL GRID
      DO 1 I=1,NX
         X(I)=DFLOAT(I-1)/DFLOAT(NX-1)
1     CONTINUE
C...
C...  INITIAL CONDITIONS
      DO 2 I=1,NX
         U1(I)=   DSIN(C3*X(I))
         U2(I)=C1*DSIN(C3*X(I))
2     CONTINUE
C...
C...  INITIAL DERIVATIVES
      CALL DERV1
      IP=0
      RETURN
      END

      SUBROUTINE INIT2
      IMPLICIT DOUBLE PRECISION (A-H,O-Z)
      PARAMETER (NX=21)
      COMMON/T/        T,   NSTOP,   NORUN
     1      /Y/  U1(NX),  U2(NX)
     2      /F/ U1T(NX), U2T(NX)
     3      /S/ U1X(NX),U1XX(NX),    X(NX)
     4      /C/        A,       B,        C,        D,
     5                C1,       E,       PI
     6      /I/       IP
C...
C...  PROBLEM PARAMETERS
      A= 1.0D0
      B= 1.0D0
      C= 1.0D0
      D=-1.0D0
C...
C...  WEAK NONLINEAR DAMPING
      IF(NORUN.EQ.3)THEN
         E=1.0D0
C...
C...  STRONG NONLINEAR DAMPING
      ELSE IF(NORUN.EQ.4)THEN
         E=100.0D0
      END IF
C...
C...  PRECOMPUTE SOME CONSTANTS FOR USE IN THE INITIAL CONDITIONS
      C3=4.0D0*DATAN(1.0D0)
      C1=-B/(2.0D0*A)
C...
```

Program 5.3a *Continued.*

```
C...   SPATIAL GRID
       DO 1 I=1,NX
          X(I)=DFLOAT(I-1)/DFLOAT(NX-1)
1      CONTINUE
C...
C...   INITIAL CONDITIONS
       DO 2 I=1,NX
          U1(I)=   DSIN(C3*X(I))
          U2(I)=C1*DSIN(C3*X(I))
2      CONTINUE
C...
C...   INITIAL DERIVATIVES
       CALL DERV2
       IP=0
       RETURN
       END
```

Program 5.3a *Continued.*

5. Subroutine INIT2 is similar to INIT1; thus we briefly consider just the essential differences. Constant e is defined for weak and strong nonlinear damping:

```
C...
C...   WEAK NONLINEAR DAMPING
       IF(NORUN.EQ.3)THEN
          E=1.0D0
C...
C...   STRONG NONLINEAR DAMPING
       ELSE IF(NORUN.EQ.4)THEN
          E=100.0D0
       END IF
```

6. Constants c_1 and c_3 are then computed as in INIT1. Finally, the spatial grid and initial conditions (5.42) and (5.43) are defined (on a 101-point grid):

```
C...
C...   PRECOMPUTE SOME CONSTANTS FOR USE IN THE INITIAL CONDITIONS
       C3=4.0D0*DATAN(1.0D0)
       C1=-B/(2.0D0*A)
C...
C...   SPATIAL GRID
       DO 1 I=1,NX
          X(I)=DFLOAT(I-1)/DFLOAT(NX-1)
1      CONTINUE
C...
C...   INITIAL CONDITIONS
       DO 2 I=1,NX
          U1(I)=   DSIN(C3*X(I))
          U2(I)=C1*DSIN(C3*X(I))
2      CONTINUE
```

To summarize, the execution of INITAL, INIT1, and INIT2 sets initial conditions (5.42) and (5.43) for the linear PDE, equation (5.41), and the nonlinear PDE, equation (5.50).

Subroutines DERV, DERV1, and DERV2 are listed in Program 5.3b. We can note the following points about subroutines DERV, DERV1, and DERV2:

1. DERV calls DERV1 for the linear problem, equation (5.41) (NORUN=1 and 2), and DERV2 for the nonlinear problem, equation (5.50) (NORUN=3 and 4).

2. In DERV1, boundary conditions (5.44) and (5.45) are first defined:

```
C...
C...  BOUNDARY CONDITION AT X = 0
      NL=1
      U1(1)=0.0D0
C...
C...  BOUNDARY CONDITION AT X = 1
      NU=1
      U1(NX)=0.0D0
```

3. The spatial derivative in equation (5.41), $\partial^2 u/\partial x^2$, is next computed in DERV1 by the fourth-order finite difference approximations in DSS044:

```
C...
C...  UXX
      CALL DSS044(0.0D0,1.0D0,NX,U1,U1X,U1XX,NL,NU)
```

4. Finally, the method of lines programming of equation (5.41) is in DO loop 1:

```
C...
C...  PDES
      DO 1 I=1,NX
      IF((I.EQ.1).OR.(I.EQ.NX))THEN
         U1T(I)=0.0D0
      ELSE
         U1T(I)=U2(I)
      END IF
         U2T(I)=(1.0D0/A)*(C*U1XX(I)-B*U1T(I)+D*U1(I))
1     CONTINUE
```

Again, the close resemblance of the PDE and the programming is evident.

5. Subroutine DERV2 is very similar to DERV1. The only difference is the addition of one line to include the nonlinear damping term, $e|\partial u/\partial t|\partial u/\partial t$, in equation (5.50):

```
    U2T(I)=(1.0D0/A)*(C*U1XX(I)-B*U1T(I)+D*U1(I))
1         -E*DABS(U1T(I))*U1T(I)
```

This example clearly indicates the power of numerical methods, that is, the straightforward additional programming required to go from a linear to a nonlinear problem.

Subroutines PRINT, PRINT1, and PRINT2 are listed in Program 5.3c. We can note the following points about subroutines PRINT, PRINT1, and PRINT2:

1. PRINT calls PRINT1 for NORUN=1 and 2 (the linear problem) and PRINT2 for NORUN=3 and 4 (the nonlinear problem).

```
      SUBROUTINE DERV
      IMPLICIT DOUBLE PRECISION (A-H,O-Z)
      PARAMETER (NX=21)
      COMMON/T/        T,   NSTOP,   NORUN
C...
C...  LINEAR PDE
      IF((NORUN.EQ.1).OR.(NORUN.EQ.2))THEN
         CALL DERV1
C...
C...  NONLINEAR PDE
      ELSE
     +IF((NORUN.EQ.3).OR.(NORUN.EQ.4))THEN
         CALL DERV2
      END IF
      RETURN
      END

      SUBROUTINE DERV1
      IMPLICIT DOUBLE PRECISION (A-H,O-Z)
      PARAMETER (NX=21)
      COMMON/T/        T,   NSTOP,   NORUN
     1      /Y/  U1(NX),  U2(NX)
     2      /F/ U1T(NX), U2T(NX)
     3      /S/ U1X(NX),U1XX(NX),    X(NX)
     4      /C/        A,        B,        C,        D,
     5                C1,       C2,       C3
     6      /I/       IP
C...
C...  BOUNDARY CONDITION AT X = 0
      NL=1
      U1(1)=0.0D0
C...
C...  BOUNDARY CONDITION AT X = 1
      NU=1
      U1(NX)=0.0D0
C...
C...  UXX
      CALL DSS044(0.0D0,1.0D0,NX,U1,U1X,U1XX,NL,NU)
C...
C...  PDES
      DO 1 I=1,NX
      IF((I.EQ.1).OR.(I.EQ.NX))THEN
         U1T(I)=0.0D0
      ELSE
         U1T(I)=U2(I)
      END IF
         U2T(I)=(1.0D0/A)*(C*U1XX(I)-B*U1T(I)+D*U1(I))
1     CONTINUE
      RETURN
      END
```

Program 5.3b Subroutines DERV, DERV1, and DERV2 for Equations (5.41) and (5.50). *Continued next page.*

```
      SUBROUTINE DERV2
      IMPLICIT DOUBLE PRECISION (A-H,O-Z)
      PARAMETER (NX=21)
      COMMON/T/          T,     NSTOP,     NORUN
     1      /Y/  U1(NX),  U2(NX)
     2      /F/ U1T(NX), U2T(NX)
     3      /S/ U1X(NX),U1XX(NX),      X(NX)
     4      /C/          A,          B,          C,          D,
     5                  C1,          E,         PI
     6      /I/         IP
C...
C...  BOUNDARY CONDITION AT X = 0
      NL=1
      U1(1)=0.0D0
C...
C...  BOUNDARY CONDITION AT X = 1
      NU=1
      U1(NX)=0.0D0
C...
C...  UXX
      CALL DSS044(0.0D0,1.0D0,NX,U1,U1X,U1XX,NL,NU)
C...
C...  PDES
      DO 1 I=1,NX
      IF((I.EQ.1).OR.(I.EQ.NX))THEN
         U1T(I)=0.0D0
      ELSE
         U1T(I)=U2(I)
      END IF
         U2T(I)=(1.0D0/A)*(C*U1XX(I)-B*U1T(I)+D*U1(I))
     1            -E*DABS(U1T(I))*U1T(I)
1     CONTINUE
      RETURN
      END
```

Program 5.3b *Continued.*

2. Within PRINT1, after an initial call (at $t = 0$) to subroutine MAP to map the ODE Jacobian matrix, the analytical solution, equation (5.46), is evaluated by a call to function UANAL (in Program 5.3d):

```
C...
C...  COMPUTE THE ANALYTICAL SOLUTION AT X = 0, 0.2, ..., 1
      DO 1 I=1,NX,4
         UA(I)=UANAL(X(I),T)
1     CONTINUE
```

3. The numerical and analytical solutions are then printed at $x = 0, 0.2, \ldots 1$:

```
C...
C...  PRINT THE NUMERICAL AND ANALYTICAL SOLUTIONS AT X = 0, 0.2,
C...  ..., 1
      IF(T.EQ.0.)WRITE(NO,4)
```

```
      SUBROUTINE PRINT(NI,NO)
      IMPLICIT DOUBLE PRECISION (A-H,O-Z)
      PARAMETER (NX=21)
      COMMON/T/       T,    NSTOP,    NORUN
C...
C...  LINEAR PDE
      IF((NORUN.EQ.1).OR.(NORUN.EQ.2))THEN
         CALL PRINT1(NI,NO)
C...
C...  NONLINEAR PDE
      ELSE
     +IF((NORUN.EQ.3).OR.(NORUN.EQ.4))THEN
         CALL PRINT2(NI,NO)
      END IF
      RETURN
      END

      SUBROUTINE PRINT1(NI,NO)
      IMPLICIT DOUBLE PRECISION (A-H,O-Z)
      PARAMETER (NX=21)
      COMMON/T/       T,    NSTOP,    NORUN
     1      /Y/  U1(NX),  U2(NX)
     2      /F/ U1T(NX), U2T(NX)
     3      /S/ U1X(NX),U1XX(NX),    X(NX)
     4      /C/       A,        B,        C,        D,
     5               C1,       C2,       C3
     6      /I/      IP
      DIMENSION  UA(NX)
C...
C...  INITIALIZE A COUNTER FOR THE PRINTING AND PLOTTING
      DATA IP/0/
      IP=IP+1
      IF(IP.EQ.1)THEN
C...
C...  MAP THE ODE JACOBIAN MATRIX AT THE BEGINNING OF THE SOLUTION
      CALL MAP
      END IF
C...
C...  MONITOR THE CALCULATIONS
      WRITE(*,*)NORUN,T
C...
C...  COMPUTE THE ANALYTICAL SOLUTION AT X = 0, 0.2, ..., 1
      DO 1 I=1,NX,4
         UA(I)=UANAL(X(I),T)
1     CONTINUE
C...
C...  PRINT THE NUMERICAL AND ANALYTICAL SOLUTIONS AT X = 0, 0.2,
C...  ..., 1
      IF(T.EQ.0.)WRITE(NO,4)
4     FORMAT(10X,'T',8X,'X=0',3X,'    X=0.2',3X,'    X=0.4',
     1                         3X,'    X=0.6',3X,'    X=0.8',
     2                         7X,' X=1')
```

Program 5.3c Subroutines PRINT, PRINT1, and PRINT2 to Print the Solutions to Equations (5.41) and (5.50) with Auxiliary Conditions (5.42) to (5.45). *Continued next page.*

```
      WRITE(NO,2)T,(U1(I),I=1,NX,4)
2     FORMAT(F11.2,/,'  NUMERICAL',6F11.5)
      WRITE(NO,3)(UA(I),I=1,NX,4)
3     FORMAT(' ANALYTICAL',6F11.5,/)
C...
C...  MAP THE ODE JACOBIAN MATRIX AT THE END OF THE SOLUTION
      IF(IP.LT.11)RETURN
      CALL MAP
C...
C...  RESET THE INTEGER COUNTER FOR THE NEXT RUN
      IP=0
      RETURN
      END

      SUBROUTINE PRINT2(NI,NO)
      IMPLICIT DOUBLE PRECISION (A-H,O-Z)
      PARAMETER (NX=21)
      COMMON/T/        T,   NSTOP,   NORUN
     1      /Y/  U1(NX),  U2(NX)
     2      /F/ U1T(NX), U2T(NX)
     3      /S/ U1X(NX),U1XX(NX),    X(NX)
     4      /C/        A,        B,        C,        D,
     5               C1,        E,       PI
     6      /I/       IP
C...
C...  INITIALIZE A COUNTER FOR THE PRINTING AND PLOTTING
      DATA IP/0/
      IP=IP+1
      IF(IP.EQ.1)THEN
C...
C...  MAP THE ODE JACOBIAN MATRIX AT THE BEGINNING OF THE SOLUTION
      CALL MAP
      END IF
C...
C...  MONITOR THE CALCULATIONS
      WRITE(*,*)NORUN,T
C...
C...  PRINT THE NUMERICAL SOLUTION AT X = 0, 0.2,..., 1
      IF(T.EQ.0.)WRITE(NO,4)
4     FORMAT(10X,'T',8X,'X=0',3X,'    X=0.2',3X,'    X=0.4',
     1                        3X,'    X=0.6',3X,'    X=0.8',
     2                        7X,' X=1')
      WRITE(NO,2)T,(U1(I),I=1,NX,4)
2     FORMAT(F11.2,/,'  NUMERICAL',6F11.5,/)
C...
C...  MAP THE ODE JACOBIAN MATRIX AT THE END OF THE SOLUTION
      IF(IP.LT.11)RETURN
      CALL MAP
C...
C...  RESET THE INTEGER COUNTER FOR THE NEXT RUN
      IP=0
      RETURN
      END
```

Program 5.3c *Continued.*

```
      DOUBLE PRECISION FUNCTION UANAL(X,T)
C...
C...  FUNCTION UANAL COMPUTES THE ANALYTICAL SOLUTION
C...
      IMPLICIT DOUBLE PRECISION (A-H,O-Z)
      COMMON/C/       A,          B,          C,          D,
     5                C1,         C2,         C3
      UANAL=DEXP(C1*T)*DCOS(C2*T)*DSIN(C3*X)
      RETURN
      END
```

Program 5.3d Function UANAL for Analytical Solution (5.46).

```
4     FORMAT(10X,'T',8X,'X=0',3X,'    X=0.2',3X,'    X=0.4',
     1                          3X,'    X=0.6',3X,'    X=0.8',
     2                          7X,' X=1')
      WRITE(NO,2)T,(U1(I),I=1,NX,4)
2     FORMAT(F11.2,/,'  NUMERICAL',6F11.5)
      WRITE(NO,3)(UA(I),I=1,NX,4)
3     FORMAT(' ANALYTICAL',6F11.5,/)
```

The ODE Jacobian matrix is also mapped by a call to MAP during the final call to PRINT1.

4. Subroutine PRINT2 is very similar to PRINT2. In fact, the only differences involve removal of the coding for the analytical solution from PRINT1 since an analytical solution is not available for the nonlinear problem, equation (5.50).

   Subroutines PLOT1, PLOT2, and MAP are similar to the earlier plotting and mapping subroutines, and therefore are not listed here (see Programs 3.1e and 3.5d).

Function UANAL for analytical solution (5.46) is listed in Program 5.3d. The one line of code for UANAL follows directly from equation (5.46). Of course, C1, C2, and C3 are first set by the call to INIT1.

The data file for Programs 5.3a to d is listed in Program 5.3e. Each set of data specifies 42 ODEs and integration over the interval $0 \leq t \leq 10$ with an accuracy of 0.00001. The integration was done with RKF45. The main program that calls RKF45 is similar to earlier main programs, e.g., Program 2.2a, and therefore is not listed here.

Abbreviated output from Programs 5.2a to e is given in Table 5.3a. We can note the following points about the output in Table 5.3a:

1. The numerical and analytical solutions in run nos. 1 and 2 agree to five figures, indicating that the spatial discretization on a 21-point grid, and the fourth-order accuracy of the finite difference approximations in subroutine DSS044 are adequate.

2. The Jacobian maps for the linear problem of equation (5.41) do not change with increasing t (this can be inferred by comparing the Jacobian maps from run no. 1 at $t = 0$ and $t = 10$). In other words, the Jacobian matrix for the linear case appears to be a constant matrix. This is confirmed by considering the following linear, constant coefficient ODE system (which is what the method

```
A*UTT + B*UT = C*UXX + D*U, D = 0
0.        10.0      1.0
  42                   0.00001
A*UTT + B*UT = C*UXX + D*U, D = -1
0.        10.0      1.0
  42                   0.00001
A*UTT + B*UT + E*ABS(UT)*UT = C*UXX + D*U, E=1
0.        10.0      1.0
  42                   0.00001
A*UTT + B*UT + E*ABS(UT)*UT = C*UXX + D*U, E=100
0.        10.0      1.0
  42                   0.00001
END OF RUNS
```

Program 5.3e Data File for Programs 5.3a to d.

of lines approximation of equation (5.41) produced, i.e., 21 first-order, linear constant coefficient ODEs)

$$dy_1/dt = f_1(y_1, y_2, \ldots, y_N) = a_{11}y_1 + a_{12}y_2 + \cdots + a_{1N}y_N$$

$$dy_2/dt = f_2(y_1, y_2, \ldots, y_N) = a_{21}y_1 + a_{22}y_2 + \cdots + a_{2N}y_N$$

$$\vdots \quad \vdots$$

$$dy_N/dt = f_N(y_1, y_2, \ldots, y_N) = a_{N1}y_1 + a_{N2}y_2 + \cdots + a_{NN}y_N \tag{5.51}$$

The Jacobian matrix is the $N \times N$ matrix of all first-order partial derivatives of the derivative functions, f_1 to f_N, of equation (5.51) with respect to the N dependent variables, y_1 to y_N. In other words, the ijth component of the Jacobian matrix is $\partial f_i/\partial y_j$. For example, for the linear system of equations (5.51), $\partial f_1/\partial y_1 = a_{11}$, $\partial f_1/\partial y_2 = a_{12}$, etc. so that the Jacobian matrix is just the right-hand side coefficient matrix of equations (5.51), which is a *constant matrix*.

3. Because of the programming in subroutine `DERV1` in terms of `U1` and `U2`, the main diagonal of the Jacobian matrix does not have other diagonals above and below it as we have observed before. Rather, the five-point approximations of `DSS044` produce a pentadiagonal band below the main diagonal.

4. The nonlinear damping term in equation (5.50) does dampen the oscillations of the solution to equation (5.50). This can be observed by comparing the solutions from run no. 2 (no damping since $e = 0$), run no. 3 (weak damping with $e = 1$), and run no. 4 (strong damping with $e = 100$). For example, the solution $u(0.4, t)$ from Table 5.3a is given in Table 5.3b.

The oscillation of the solution to equation (5.50) is reduced as the damping is increased (with increasing `E`). For `E=100`, the oscillation in the solution is essentially eliminated.

By this procedure of extending the analysis from a linear problem for which the solution is known (equation (5.41) with solution (5.46)), we have arrived at the solution to the nonlinear problem, equation (5.50) with some assurance that the programming is correct (since only one additional line of code was added in `DERV2` in Program 5.3b for

Table 5.3 (a) Abbreviated Output from Programs 5.3a to e. *Continued next pages.*

```
RUN NO. -    1  A*UTT + B*UT = C*UXX + D*U, D = 0

INITIAL T -  0.000D+00

  FINAL T -  0.100D+02

  PRINT T -  0.100D+01

NUMBER OF DIFFERENTIAL EQUATIONS -  42

INTEGRATION ALGORITHM - RKF45

MAXIMUM INTEGRATION ERROR -  0.100D-04


DEPENDENT VARIABLE COLUMN INDEX J (FOR YJ) IS PRINTED HORIZONTALLY

DERIVATIVE ROW INDEX I (FOR DYI/DT = FI(Y1,Y2,...,YJ,...,YN) IS
PRINTED VERTICALLY

JACOBIAN MATRIX ELEMENT IN THE MAP WITH INDICES I,J IS FOR PFI/PYJ
WHERE P DENOTES A PARTIAL DERIVATIVE


                 111111111122222222223333333333444
        123456789012345678901234567890123456789012
   1
   2                          4
   3                           4
   4                            4
   5                             4
   6                              4
   7                               4
   8                                4
   9                                 4
  10                                  4
  11                                   4
  12                                    4
  13                                     4
  14                                      4
  15                                       4
  16                                        4
  17                                         4
  18                                          4
  19                                           4
  20                                            4
  21
  22     88887
  23     77776                4
  24     7776                  4
  25     67776                  4
  26      67776                  4
  27       67776                  4
```

Table 5.3 (a) *Continued.*

```
28       67776                  4
29        67776                  4
30         67776                  4
31          67776                  4
32           67776                  4
33            67776                  4
34             67776                  4
35              67776                  4
36               67776                  4
37                67776                  4
38                 67776                  4
39                  67776                  4
40                   6777                   4
41                  67777                    4
42                  78888
```

T	X=0	X=0.2	X=0.4	X=0.6	X=0.8	X=1
0.00						
NUMERICAL	0.00000	0.58779	0.95106	0.95106	0.58779	0.00000
ANALYTICAL	0.00000	0.58779	0.95106	0.95106	0.58779	0.00000
1.00						
NUMERICAL	0.00000	-0.35622	-0.57638	-0.57638	-0.35622	0.00000
ANALYTICAL	0.00000	-0.35622	-0.57638	-0.57638	-0.35622	0.00000
2.00						
NUMERICAL	0.00000	0.21554	0.34875	0.34875	0.21554	0.00000
ANALYTICAL	0.00000	0.21554	0.34875	0.34875	0.21554	0.00000
3.00						
NUMERICAL	0.00000	-0.13021	-0.21068	-0.21068	-0.13021	0.00000
ANALYTICAL	0.00000	-0.13021	-0.21068	-0.21068	-0.13021	0.00000
4.00						
NUMERICAL	0.00000	0.07853	0.12706	0.12706	0.07853	0.00000
ANALYTICAL	0.00000	0.07853	0.12706	0.12706	0.07853	0.00000
5.00						
NUMERICAL	0.00000	-0.04728	-0.07651	-0.07651	-0.04728	0.00000
ANALYTICAL	0.00000	-0.04728	-0.07651	-0.07651	-0.04728	0.00000
6.00						
NUMERICAL	0.00000	0.02842	0.04599	0.04599	0.02842	0.00000
ANALYTICAL	0.00000	0.02842	0.04599	0.04599	0.02842	0.00000
7.00						
NUMERICAL	0.00000	-0.01706	-0.02760	-0.02760	-0.01706	0.00000
ANALYTICAL	0.00000	-0.01706	-0.02760	-0.02760	-0.01706	0.00000
8.00						
NUMERICAL	0.00000	0.01022	0.01653	0.01653	0.01022	0.00000
ANALYTICAL	0.00000	0.01022	0.01653	0.01653	0.01022	0.00000

Table 5.3 (a) *Continued.*

```
      9.00
 NUMERICAL    0.00000   -0.00611   -0.00989   -0.00989   -0.00611    0.00000
ANALYTICAL    0.00000   -0.00611   -0.00989   -0.00989   -0.00611    0.00000

     10.00
 NUMERICAL    0.00000    0.00365    0.00590    0.00590    0.00365    0.00000
ANALYTICAL    0.00000    0.00365    0.00590    0.00590    0.00365    0.00000


 DEPENDENT VARIABLE COLUMN INDEX J (FOR YJ) IS PRINTED HORIZONTALLY

 DERIVATIVE ROW INDEX I (FOR DYI/DT = FI(Y1,Y2,...,YJ,...,YN) IS PRINTED
 VERTICALLY

 JACOBIAN MATRIX ELEMENT IN THE MAP WITH INDICES I,J IS FOR PFI/PYJ
 WHERE P DENOTES A PARTIAL DERIVATIVE


                111111111122222222223333333333444
       123456789012345678901234567890123456789012
   1
   2                         4
   3                          4
   4                           4
   5                            4
   6                             4
   7                              4
   8                               4
   9                                4
  10                                 4
  11                                  4
  12                                   4
  13                                    4
  14                                     4
  15                                      4
  16                                       4
  17                                        4
  18                                         4
  19                                          4
  20                                           4
  21
  22    88887
  23    77776                4
  24    7776                  4
  25    67776                  4
  26     67776                  4
  27      67776                  4
  28       67776                  4
  29        67776                  4
  30         67776                  4
  31          67776                  4
```

Table 5.3 (a) *Continued.*

```
     32            67776                  4
     33             67776                  4
     34              67776                  4
     35               67776                  4
     36                67776                  4
     37                 67776                  4
     38                  67776                  4
     39                   67776                  4
     40                    6777                  4
     41                   67777                   4
     42                   78888

 RUN NO. -    2  A*UTT + B*UT = C*UXX + D*U, D = -1

 INITIAL T -  0.000D+00

   FINAL T -  0.100D+02

   PRINT T -  0.100D+01

 NUMBER OF DIFFERENTIAL EQUATIONS -  42

 INTEGRATION ALGORITHM - RKF45

 MAXIMUM INTEGRATION ERROR -  0.100D-04
                   .
                   .
                   .

         T       X=0      X=0.2      X=0.4      X=0.6      X=0.8        X=1
      0.00
 NUMERICAL   0.00000    0.58779    0.95106    0.95106    0.58779    0.00000
ANALYTICAL   0.00000    0.58779    0.95106    0.95106    0.58779    0.00000

      1.00
 NUMERICAL   0.00000   -0.35407   -0.57289   -0.57289   -0.35407    0.00000
ANALYTICAL   0.00000   -0.35406   -0.57289   -0.57289   -0.35406    0.00000

      2.00
 NUMERICAL   0.00000    0.21032    0.34031    0.34031    0.21032    0.00000
ANALYTICAL   0.00000    0.21032    0.34031    0.34031    0.21032    0.00000

      3.00
 NUMERICAL   0.00000   -0.12313   -0.19923   -0.19923   -0.12313    0.00000
ANALYTICAL   0.00000   -0.12313   -0.19923   -0.19923   -0.12313    0.00000

      4.00
 NUMERICAL   0.00000    0.07097    0.11483    0.11483    0.07097    0.00000
ANALYTICAL   0.00000    0.07097    0.11483    0.11483    0.07097    0.00000

      5.00
 NUMERICAL   0.00000   -0.04020   -0.06505   -0.06505   -0.04020    0.00000
ANALYTICAL   0.00000   -0.04020   -0.06505   -0.06505   -0.04020    0.00000
```

Table 5.3 (a) *Continued.*

```
      6.00
 NUMERICAL     0.00000    0.02233    0.03612    0.03612    0.02233    0.00000
ANALYTICAL     0.00000    0.02232    0.03612    0.03612    0.02232    0.00000

      7.00
 NUMERICAL     0.00000   -0.01211   -0.01959   -0.01959   -0.01211    0.00000
ANALYTICAL     0.00000   -0.01211   -0.01959   -0.01959   -0.01211    0.00000

      8.00
 NUMERICAL     0.00000    0.00637    0.01031    0.01031    0.00637    0.00000
ANALYTICAL     0.00000    0.00637    0.01031    0.01031    0.00637    0.00000

      9.00
 NUMERICAL     0.00000   -0.00322   -0.00522   -0.00522   -0.00322    0.00000
ANALYTICAL     0.00000   -0.00322   -0.00521   -0.00521   -0.00322    0.00000

     10.00
 NUMERICAL     0.00000    0.00154    0.00249    0.00249    0.00154    0.00000
ANALYTICAL     0.00000    0.00154    0.00249    0.00249    0.00154    0.00000
                  .
                  .
                  .

 RUN NO. -    3  A*UTT + B*UT + E*ABS(UT)*UT = C*UXX + D*U, E=1

 INITIAL T -  0.000D+00

   FINAL T -  0.100D+02

   PRINT T -  0.100D+01

 NUMBER OF DIFFERENTIAL EQUATIONS -  42

 INTEGRATION ALGORITHM - RKF45

 MAXIMUM INTEGRATION ERROR -  0.100D-04
                  .
                  .
                  .

         T         X=0      X=0.2      X=0.4      X=0.6      X=0.8        X=1
      0.00
 NUMERICAL     0.00000    0.58779    0.95106    0.95106    0.58779    0.00000

      1.00
 NUMERICAL     0.00000   -0.17594   -0.28734   -0.28734   -0.17594    0.00000

      2.00
 NUMERICAL     0.00000    0.08373    0.13769    0.13769    0.08373    0.00000

      3.00
 NUMERICAL     0.00000   -0.04514   -0.07444   -0.07444   -0.04514    0.00000
```

Table 5.3 (a) *Continued.*

```
     4.00
NUMERICAL     0.00000     0.02553     0.04213     0.04213     0.02553     0.00000

     5.00
NUMERICAL     0.00000    -0.01470    -0.02426    -0.02426    -0.01470     0.00000

     6.00
NUMERICAL     0.00000     0.00849     0.01400     0.01400     0.00849     0.00000

     7.00
NUMERICAL     0.00000    -0.00488    -0.00805    -0.00805    -0.00488     0.00000

     8.00
NUMERICAL     0.00000     0.00277     0.00457     0.00457     0.00277     0.00000

     9.00
NUMERICAL     0.00000    -0.00155    -0.00256    -0.00256    -0.00155     0.00000

    10.00
NUMERICAL     0.00000     0.00085     0.00141     0.00141     0.00085     0.00000
                  .
                  .
                  .

RUN NO. -   4  A*UTT + B*UT +E*ABS(UT)*UT = C*UXX + D*U, E=100

INITIAL T -  0.000D+00

  FINAL T -  0.100D+02

  PRINT T -  0.100D+01

NUMBER OF DIFFERENTIAL EQUATIONS -  42

INTEGRATION ALGORITHM - RKF45

MAXIMUM INTEGRATION ERROR -  0.100D-04


DEPENDENT VARIABLE COLUMN INDEX J (FOR YJ) IS PRINTED HORIZONTALLY

DERIVATIVE ROW INDEX I (FOR DYI/DT = FI(Y1,Y2,...,YJ,...,YN) IS PRINTED
VERTICALLY

JACOBIAN MATRIX ELEMENT IN THE MAP WITH INDICES I,J IS FOR PFI/PYJ
WHERE P DENOTES A PARTIAL DERIVATIVE


                  111111111122222222223333333333444
         123456789012345678901234567890123456789012
     1
     2                         4
     3                          4
```

Table 5.3 (a) *Continued.*

```
      4                           4
      5                            4
      6                             4
      7                              4
      8                               4
      9                                4
     10                                 4
     11                                  4
     12                                   4
     13                                    4
     14                                     4
     15                                      4
     16                                       4
     17                                        4
     18                                         4
     19                                          4
     20                                           4
     21
     22    88887
     23    77776                6
     24    7876                  6
     25    67876                  6
     26     67876                  6
     27      67876                  6
     28       67876                  6
     29        67876                  6
     30         67876                  6
     31          67876                  6
     32           67876                  7
     33            67876                  6
     34             67876                  6
     35              67876                  6
     36               67876                  6
     37                67876                  6
     38                 67876                  6
     39                  67876                  6
     40                   6787                   6
     41                  67777                    6
     42                  78888

        T        X=0      X=0.2      X=0.4      X=0.6      X=0.8        X=1
     0.00
NUMERICAL    0.00000    0.58779    0.95106    0.95106    0.58779    0.00000

     1.00
NUMERICAL    0.00000    0.38104    0.65064    0.65064    0.38104    0.00000

     2.00
NUMERICAL    0.00000    0.23422    0.40348    0.40348    0.23422    0.00000

     3.00
NUMERICAL    0.00000    0.12479    0.21506    0.21506    0.12479    0.00000
```

Table 5.3 (a) *Continued.*

```
     4.00
NUMERICAL    0.00000    0.04975    0.08571    0.08571    0.04975    0.00000

     5.00
NUMERICAL    0.00000    0.00824    0.01428    0.01428    0.00824    0.00000

     6.00
NUMERICAL    0.00000   -0.00206   -0.00333   -0.00333   -0.00206    0.00000

     7.00
NUMERICAL    0.00000    0.00091    0.00148    0.00148    0.00091    0.00000

     8.00
NUMERICAL    0.00000   -0.00047   -0.00075   -0.00075   -0.00047    0.00000

     9.00
NUMERICAL    0.00000    0.00025    0.00040    0.00040    0.00025    0.00000

    10.00
NUMERICAL    0.00000   -0.00014   -0.00022   -0.00022   -0.00014    0.00000


DEPENDENT VARIABLE COLUMN INDEX J (FOR YJ) IS PRINTED HORIZONTALLY

DERIVATIVE ROW INDEX I (FOR DYI/DT = FI(Y1,Y2,...,YJ,...,YN) IS PRINTED
VERTICALLY

JACOBIAN MATRIX ELEMENT IN THE MAP WITH INDICES I,J IS FOR PFI/PYJ
WHERE P DENOTES A PARTIAL DERIVATIVE


                   111111111122222222223333333333444
          123456789012345678901234567890123456789012
        1
        2                       4
        3                        4
        4                         4
        5                          4
        6                           4
        7                            4
        8                             4
        9                              4
       10                               4
       11                                4
       12                                 4
       13                                  4
       14                                   4
       15                                    4
       16                                     4
       17                                      4
       18                                       4
       19                                        4
       20                                         4
```

***Table* 5.3** (a) *Continued.*

```
21
22    88887
23    77776                5
24    7876                  5
25    67876                  5
26     67876                  5
27      67876                  5
28       67876                  5
29        67876                  5
30         67876                  5
31          67876                  5
32           67876                  5
33            67876                  5
34             67876                  5
35              67876                  5
36               67876                  5
37                67876                  5
38                 67876                  5
39                  67876                  5
40                   6787                   5
41                  67777                    5
42                  78888
```

the nonlinear case). However, this procedure of extending the analysis from a problem with a known solution to a problem for which we seek a solution can be considered more generally. For example, we can consider the coefficient e in equation (5.50) as a parameter for continuing the solution from the linear problem with $e = 0$ with a known solution, to the solution of the nonlinear problem with $e = 100$ for which we might seek a solution.

This process of embedding a parameter in the problem, then systematically varying the parameter from an initial value that defines a problem with a known solution to a final value that defines a problem for which we seek a solution is called *continuation*. Usually as the parameter is varied, the solution for one value of the parameter is used as the initial condition to compute the solution for the next value of the parameter; thus there is smooth transition through a series of problems as the parameter is varied from an initial value to a final value. This procedure of stepping through a series of problems to arrive at the final problem of interest is generally useful for the solution of nonlinear equations. General discussions of this procedure, which is also called *homotopy continuation*, are given by Rheinboldt (1980) and Watson (1986, 1987). Differential variation of the continuation parameter (through differential equations like equation (2.201)) is discussed by Boggs (1971), Davidenko (1953), Edelen (1976a,b), and Hachtel et al. (1974).

5.2 PDEs Zeroth Order in Time

We have, thus far, throughout this book considered problems in ODEs and PDEs which have an initial value variable, generally designated t. However, mathematical models in science and engineering are also formulated as ODEs and PDEs without an initial value variable, i.e., the differential equations contain derivatives with respect to only boundary value independent variables. We now consider two such problems, one in

Table 5.3 (b) Effect of the Nonlinear Damping Term in Equation (5.50).

	C2 = 0	C2 = 1	C2 = 100
T	X=0.4	X=0.4	X=0.4
0.00			
NUMERICAL	0.95106	0.95106	0.95106
ANALYTICAL	0.95106		
1.00			
NUMERICAL	-0.57289	-0.28734	0.65064
ANALYTICAL	-0.57289		
2.00			
NUMERICAL	0.34031	0.13769	0.40348
ANALYTICAL	0.34031		
3.00			
NUMERICAL	-0.19923	-0.07444	0.21506
ANALYTICAL	-0.19923		
4.00			
NUMERICAL	0.11483	0.04213	0.08571
ANALYTICAL	0.11483		
5.00			
NUMERICAL	-0.06505	-0.02426	0.01428
ANALYTICAL	-0.06505		
6.00			
NUMERICAL	0.03612	0.01400	-0.00333
ANALYTICAL	0.03612		
7.00			
NUMERICAL	-0.01959	-0.00805	0.00148
ANALYTICAL	-0.01959		
8.00			
NUMERICAL	0.01031	0.00457	-0.00075
ANALYTICAL	0.01031		
9.00			
NUMERICAL	-0.00522	-0.00256	0.00040
ANALYTICAL	-0.00521		
10.00			
NUMERICAL	0.00249	0.00141	-0.00022
ANALYTICAL	0.00249		

ODEs and one in PDEs. In particular, we consider how such boundary value problems can be solved numerically by reformulating them as related initial value problems (or initial and boundary value problems).

We start first with a boundary value ODE problem that comes from an analysis

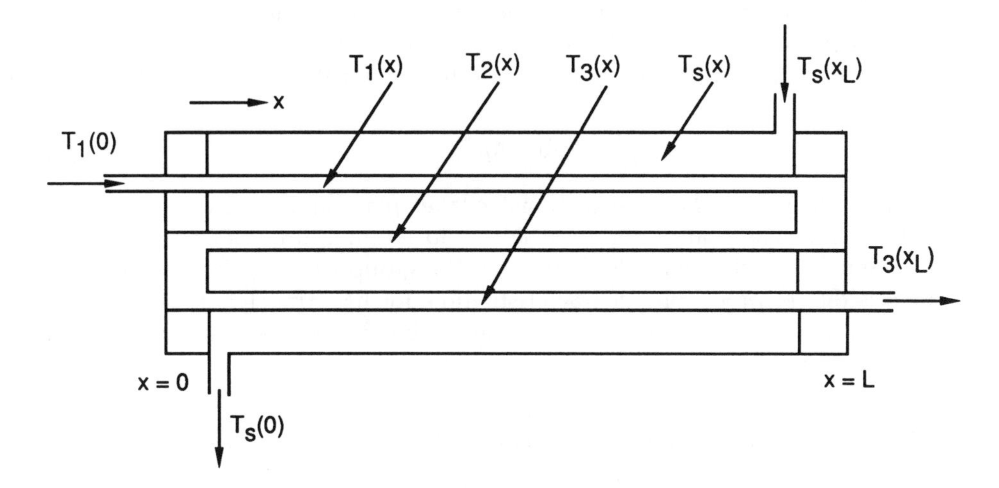

***Figure* 5.3** A Three-Pass Heat Exchanger.

of the heat transfer system illustrated in Figure 5.3. The heat exchanger of Figure 5.3 essentially provides a means for two fluid streams, flowing countercurrent through the exchanger, to exchange heat as a result of the difference in the temperatures of the two streams. Such heat exchangers are built of metal with a high thermal conductivity so that the difference in the stream temperatures can lead to a significant exchange of heat between the streams. The internal structure is typically a series of tubes within an outer shell. One stream flows through the tubes, and the other stream flows through the shell (and around the exterior of the tubes).

Usually the two entering fluid temperatures are specified, and the design of the exchanger is based on the specification of the flow rates through the exchanger and the two exiting fluid temperatures. The design of a heat exchanger requires the specification of the amount of area for heat transfer within the exchanger required to ensure that the specified exiting temperatures are achieved; this area is determined by the number and arrangement of the tubes within the shell. In the case of the exchanger in Figure 5.3, the tubes pass back and forth within the shell three times; thus this is a "three-pass, shell and tube heat exchanger."

We now develop a mathematical model for the exchanger of Figure 5.3 that can be used to calculate the two exiting fluid temperatures, when the two entering fluid temperatures and the fluid stream flow rates are specified. Because we are computing temperatures, we write energy balances for the two fluid streams over an incremental section of length Δx, and then take the limit as $\Delta x \to 0$ to arrive at the boundary value ODEs. For example, for the fluid in the first tube pass, with temperature $T_1(x)$ (here we use T to represent temperature rather that the usual u for the dependent variable of an ODE/PDE), the energy balance is

$$Av\rho C_p T_1|_x - Av\rho C_p T_1|_{x+\Delta x} = hA_h \Delta x(T_s - T_1) \tag{5.52}$$

where T_1 = fluid temperature in the first tube pass (*k*); x = position along the exchanger (*m*); A = total cross-sectional area for fluid in the tubes of the first pass (m^2); ρ = fluid density (kg/m^3); C_p = fluid specific heat (*j/kg-k*); v = fluid average linear velocity (*m/s*); h = heat transfer coefficient (*j/s-*m^2*-k*); A_h = heat transfer area per unit length of the exchanger ($m = m^2/m$); and T_s = shell-side fluid temperature (*k*).

Equation (5.52) states the the difference between the rate of energy flowing into and out of the incremental section in the tube-side fluid (the left-hand side of the energy balance) equals the rate at which energy is transferred to or from the tube-side fluid by heat transfer due to the temperature difference $(T_s - T_1)$.

Division of equation (5.52) by $A\rho C_p \Delta x$, followed by the limit $\Delta x \to 0$, gives the ODE

$$v\frac{dT_1}{dx} = hA_h/(A\rho C_p)(T_s - T_1) \tag{5.53}$$

As expected, when $T_1 = T_s$, no heat transfer takes place and equation (5.53) reduces to simply $dT_1/dx = 0$; that is, the tube-side fluid temperature does not change with position along the exchanger. Note also that the group $hA_h/(A\rho C_p)$ has the units of $1/s$, i.e., it is the reciprocal of a characteristic time for heat transfer.

Equation (5.53) is first order in x and therefore requires one boundary condition. We use

$$T_1(0) = T_{10} \tag{5.54}$$

where T_{10} is the specified entering temperature to the first tube pass.

We now require energy balances for the fluid in the second and third tube passes, as well as an energy balance for the shell-side fluid. These three additional energy balances are derived in essentially the same way as equation (5.53). The final results are

$$-v\frac{dT_2}{dx} = hA_h/(A\rho C_p)(T_s - T_2) \tag{5.55}$$

$$v\frac{dT_3}{dx} = hA_h/(A\rho C_p)(T_s - T_3) \tag{5.56}$$

$$-v_s\frac{dT_s}{dx} = hA_h/(A_s\rho_s C_{ps})\{(T_1 - T_s) + (T_2 - T_s) + (T_3 - T_s)\} \tag{5.57}$$

where a subscript s refers to the shell side, e.g., C_{ps} is the specific heat of the shell-side fluid. The velocities v and v_s are defined to be positive. Therefore, $-v$ and $-v_s$ are used in equations (5.55) and (5.57) to account for the flow in the second tube pass and the shell from right to left (in the negative x direction). Note that the right-hand side of equation (5.57) has three temperature differences, $(T_1 - T_s)$, $(T_2 - T_s)$, and $(T_3 - T_s)$, since the shell-side fluid exchanges heat with the fluid in all three tube passes.

Equations (5.55), (5.56), and (5.57) each require a boundary condition. We see from Figure 5.3 that the entering temperature for the second tube pass is the exiting temperature of the first tube pass, which we express as

$$T_2(L) = T_1(L) \tag{5.58}$$

Similarly, the entering temperature for the third tube pass is the exiting temperature of the second tube pass

$$T_3(0) = T_2(0) \tag{5.59}$$

Finally, the entering temperature of the shell-side fluid is specified at $x = L$

$$T_s(L) = T_{sL} \tag{5.60}$$

Equations (5.53) to (5.60) are the system of ODEs to be integrated numerically to obtain the four temperatures, $T_1(x)$, $T_2(x)$, $T_3(x)$, and $T_s(x)$. However, we cannot use one of the ODE integrators discussed previously such as `RKF45` or `LSODE` since equations (5.53) to (5.60) are boundary value ODEs, not initial value ODEs. This distinction was discussed in Sections 1.1.4, 1.1.5, and 1.1.6 (for the present problem, we have explicit boundary value ODEs as discussed in Section 1.1.4 since only one derivative appears in each ODE). We can then naturally ask how we can proceed to compute a solution to equations (5.53) to (5.60) since the numerical methods we have considered thus far do not seem to apply to this problem system.

The procedure we will follow is to add initial value derivatives, which we call time derivatives, to equations (5.53), (5.55), (5.56), and (5.57); that is, we convert the boundary values ODEs into first-order hyperbolic PDEs

$$\frac{\partial T_1}{\partial t} + v\frac{\partial T_1}{\partial x} = hA_h/(A\rho C_p)(T_s - T_1) \tag{5.61}$$

$$\frac{\partial T_2}{\partial t} - v\frac{\partial T_2}{\partial x} = hA_h/(A\rho C_p)(T_s - T_2) \tag{5.62}$$

$$\frac{\partial T_3}{\partial t} + v\frac{\partial T_3}{\partial x} = hA_h/(A\rho C_p)(T_s - T_3) \tag{5.63}$$

$$\frac{\partial T_s}{\partial t} - v_s\frac{\partial T_s}{\partial x} = hA_h/(A_s\rho_s C_{ps})\left\{(T_1 - T_s) + (T_2 - T_s) + (T_3 - T_s)\right\} \tag{5.64}$$

Boundary conditions (5.54) and (5.58) to (5.60) are then be restated as

$$T_1(0,t) = T_{10}(t) \tag{5.65}$$

$$T_2(L,t) = T_1(L,t) \tag{5.66}$$

$$T_3(0,t) = T_2(0,t) \tag{5.67}$$

$$T_s(L,t) = T_{sL}(t) \tag{5.68}$$

where $T_{10}(t)$ and $T_{sL}(t)$ are prescribed functions of t.

Since PDEs (5.61) to (5.64) now have initial value derivatives, they each require an initial condition

$$T_1(x,0) = T_{1i}(x) \tag{5.69}$$

$$T_2(x,0) = T_{2i}(x) \tag{5.70}$$

$$T_3(x,0) = T_{3i}(x) \tag{5.71}$$

$$T_s(x,0) = T_{si}(x) \tag{5.72}$$

where $T_{1i}(x)$, $T_{2i}(x)$, $T_{3i}(x)$, and $T_{si}(x)$ are functions of x, which can be prescribed arbitrarily since they will have no effect on the final solution of the boundary value ODE problem.

The solution procedure for the boundary value ODEs, equations (5.53) to (5.60), is to compute a method of lines solution of the PDEs, equations (5.61) to (5.72), and allow the solution to run long enough so that a steady-state condition is reached for which $\partial T_1/\partial t \approx 0$, $\partial T_2/\partial t \approx 0$, $\partial T_3/\partial t \approx 0$, and $\partial T_s/\partial t \approx 0$, in which case the PDEs, equations (5.61) to (5.64), revert to the boundary value ODEs, equations (5.53) and (5.55) to (5.57). Note that under these steady-state conditions, boundary conditions (5.65) to (5.68) are the same as boundary conditions (5.54) and (5.58) to (5.60). Also, initial conditions (5.69) to (5.72) have no effect on the steady-state solution. This approach to a boundary value problem by converting it into an initial value problem is the *method of false transients*. It can also be considered as an example of continuation in which the continuation parameter is the initial value variable t. In this case, we start with $t = 0$ and allow it to reach a final value for which the solution of the PDEs is also a solution to the problem of interest, the boundary value ODEs.

```
      SUBROUTINE INITAL
      IMPLICIT DOUBLE PRECISION (A-H,O-Z)
      PARAMETER (NX=11)
      COMMON/T/           T,      NSTOP,      NORUN
     1      /Y/   T1(NX),    T2(NX),    T3(NX),      TS(NX)
     2      /F/  T1T(NX),   T2T(NX),   T3T(NX),     TST(NX)
     3      /S/  T1X(NX),   T2X(NX),   T3X(NX),     TSX(NX)
     4      /C/       TO,        XL,        VT,          VS,
     5                CT,        CS,       TSO,           N,
     6            XP(NX),        IP,       ISP
C...
C...  MODEL PARAMETERS
      TO=25.0D0
      XL=10.0D0
      VT=1.0D0
      VS=0.5D0
      CT=0.1D0
      CS=0.1D0
      TSO=220.0D0
      N=NX
C...
C...  INITIALIZE THE SPATIAL GRID USED IN SUBROUTINE PRINT
      DO 1 I=1,NX
         XP(I)=XL*DFLOAT(I-1)/DFLOAT(NX-1)
1     CONTINUE
C...
C...  MODEL INITIAL CONDITIONS
      DO 2 I=1,NX
         T1(I)=TO
         T2(I)=TO
         T3(I)=TO
         TS(I)=TO
2     CONTINUE
C...
C...  INITIALIZE THE DERIVATIVES
      CALL DERV
      IP=0
      ISP=0
      RETURN
      END
```

Program 5.4a Subroutine `INITAL` for Initial Conditions (5.69) to (5.72).

We now consider the method of lines programming of equations (5.61) to (5.72) which is straightforward since we have previously developed all of the required numerical methods. Also, we obtain a second independent solution to equation (5.53) to (5.60) by applying finite differences directly to equations (5.53) and (5.55) to (5.57). Subroutine `INITAL` for initial conditions (5.69) to (5.72) is listed in Program 5.4a.

We can note the following details about subroutine `INITAL`:

1. The four temperatures of PDEs (5.61) to (5.65), $T_1(x,t)$, $T_2(x,t)$, $T_3(x,t)$, and $T_s(x,t)$, are defined on an 11-point spatial grid in `COMMON/Y/`; thus, the complete model consists of 44 ODEs to approximate the four PDEs. The temporal deriva-

tives of the four dependent variables, $\partial T_1/\partial t$, $\partial T_2/\partial t$, $\partial T_3/\partial t$, and $\partial T_s/\partial t$, are in COMMON/F/ and the spatial derivatives, $\partial T_1/\partial x, \partial T_2/\partial x, \partial T_3/\partial x$ and $\partial T_s/\partial x$, are in COMMON/S/. The problem parameters, including the spatial grid in XP(NX), are in COMMON/C/:

```
      PARAMETER (NX=11)
      COMMON/T/          T,      NSTOP,      NORUN
     1      /Y/   T1(NX),     T2(NX),     T3(NX),     TS(NX)
     2      /F/  T1T(NX),    T2T(NX),    T3T(NX),    TST(NX)
     3      /S/  T1X(NX),    T2X(NX),    T3X(NX),    TSX(NX)
     4      /C/       TO,         XL,         VT,         VS,
     5                CT,         CS,        TSO,          N,
     6            XP(NX),         IP,        ISP
```

2. The PDE model parameters are defined next:

```
C...
C...  MODEL PARAMETERS
      TO=25.0DO
      XL=10.0DO
      VT=1.0DO
      VS=0.5DO
      CT=0.1DO
      CS=0.1DO
      TSO=220.0DO
      N=NX
```

(2.1) The initial condition functions of equations (5.69) to (5.72) are set to a constant value of 25, i.e., $T_{1i}(x) = T_{2i}(x) = T_{3i}(x) = T_{4i}(x) = T_{si}(x) = 25$ (TO=25.0DO). This value is also used as the entering temperature to the first tube pass in boundary condition (5.65), $T_{10}(t) = 25$ (TO=25.0DO is passed from INITAL to DERV for this purpose). For the purpose of computing a solution to the boundary values ODEs, equations (5.53) to (5.60), the initial condition is immaterial; that is, the final steady-state solution does not depend on the initial condition.

(2.2) The length of the exchanger in Figure 5.3, x_L, is 10 (XL=10.0DO).

(2.3) The tube- and shell-side velocities, v and v_s, are 1 and 0.5 respectively (VT=1.0DO, VS=0.5DO).

(2.4) The reciprocals of the heat transfer characteristic times in equations (5.61) to (5.64) are $hA_h/(A\rho C_p) = 0.1$ (CT=0.1DO) and $hA_h/(A_s\rho_s C_{ps}) = 0.1$ (CS=0.1DO).

(2.5) The boundary condition function of equation (5.68) is set to a constant value of 220, i.e., $T_{sL}(t) = 25$ (TSO=220.0DO)

3. The 11-point spatial grid is defined in DO loop 1:

```
C...
C...  INITIALIZE THE SPATIAL GRID USED IN SUBROUTINE PRINT
      DO 1 I=1,NX
         XP(I)=XL*DFLOAT(I-1)/DFLOAT(NX-1)
1     CONTINUE
```

4. Finally, initial conditions (5.69) to (5.72) are programmed in DO loop 2, using TO (=25) set previously:

```
C...
C...  MODEL INITIAL CONDITIONS
      DO 2 I=1,NX
         T1(I)=TO
         T2(I)=TO
         T3(I)=TO
         TS(I)=TO
2     CONTINUE
```

Subroutine DERV is listed in Program 5.4b. We can note the following points about subroutine DERV:

1. Boundary conditions (5.65) to (5.68) are first programmed

```
C...
C...  BOUNDARY CONDITIONS
      T1( 1)=TO
      TS(NX)=TSO
      T2(NX)=T1(NX)
      T3( 1)=T2(1)
```

   The programming follows directly from equations (5.65) to (5.68).

2. The four spatial derivatives in PDEs (5.61) to (5.64), $\partial T_1/\partial x$, $\partial T_2/\partial x$, $\partial T_3/\partial x$, and $\partial T_s/\partial x$, are computed by the five-point biased upwind approximations in DSS020:

```
C...
C...  SPATIAL DERIVATIVES
      CALL DSS020(0.0D0,XL,NX,T1,T1X, 1.0D0)
      CALL DSS020(0.0D0,XL,NX,T2,T2X,-1.0D0)
      CALL DSS020(0.0D0,XL,NX,T3,T3X, 1.0D0)
      CALL DSS020(0.0D0,XL,NX,TS,TSX,-1.0D0)
```

   Note in particular the positive sixth arguments for PDEs (5.61) and (5.63) and the negative sixth arguments for PDEs (5.62) and (5.64), in accordance with the sign multiplying the velocities in these equations.

3. PDEs (5.61) to (5.64) are then programmed in DO loop 1, where again, the programming bears a close resemblance to the PDEs themselves:

```
C...
C...  MODEL PDES
      DO 1 I=1,N
         T1T(I)=-VT*T1X(I)+CT*(TS(I)-T1(I))
         T2T(I)= VT*T2X(I)+CT*(TS(I)-T2(I))
         T3T(I)=-VT*T3X(I)+CT*(TS(I)-T3(I))
         TST(I)= VS*TSX(I)+CS*((T1(I)-TS(I))
     1           +(T2(I)-TS(I))+(T3(I)-TS(I)))
1     CONTINUE
```

```
      SUBROUTINE DERV
      IMPLICIT DOUBLE PRECISION (A-H,O-Z)
      PARAMETER (NX=11)
      COMMON/T/          T,      NSTOP,      NORUN
     1      /Y/    T1(NX),     T2(NX),     T3(NX),     TS(NX)
     2      /F/   T1T(NX),    T2T(NX),    T3T(NX),    TST(NX)
     3      /S/   T1X(NX),    T2X(NX),    T3X(NX),    TSX(NX)
     4      /C/        TO,         XL,         VT,         VS,
     5                 CT,         CS,        TSO,          N,
     6             XP(NX),         IP,        ISP
C...
C...  BOUNDARY CONDITIONS
      T1( 1)=TO
      TS(NX)=TSO
      T2(NX)=T1(NX)
      T3( 1)=T2(1)
C...
C...  SPATIAL DERIVATIVES
      CALL DSS020(0.0D0,XL,NX,T1,T1X, 1.0D0)
      CALL DSS020(0.0D0,XL,NX,T2,T2X,-1.0D0)
      CALL DSS020(0.0D0,XL,NX,T3,T3X, 1.0D0)
      CALL DSS020(0.0D0,XL,NX,TS,TSX,-1.0D0)
C...
C...  MODEL PDES
      DO 1 I=1,N
         T1T(I)=-VT*T1X(I)+CT*(TS(I)-T1(I))
         T2T(I)= VT*T2X(I)+CT*(TS(I)-T2(I))
         T3T(I)=-VT*T3X(I)+CT*(TS(I)-T3(I))
         TST(I)= VS*TSX(I)+CS*((T1(I)-TS(I))
     1           +(T2(I)-TS(I))+(T3(I)-TS(I)))
1       CONTINUE
C...
C...  ZERO THE TEMPORAL DERIVATIVES AT THE BOUNDARIES
      T1T(1)=0.0D0
      T2T(N)=0.0D0
      T3T(1)=0.0D0
      TST(N)=0.0D0
      RETURN
      END
```

Program 5.4b Subroutine DERV for Equations (5.61) to (5.64).

4. Finally, the boundary temporal derivatives are set to zero in accordance with boundary conditions (5.65) to (5.68) programmed previously; this additional programming merely ensures the ODE integrator does not move the dependent variable boundary values away from the prescribed values of equations (5.65) to (5.68):

```
C...
C...  ZERO THE TEMPORAL DERIVATIVES AT THE BOUNDARIES
      T1T(1)=0.0D0
      T2T(N)=0.0D0
      T3T(1)=0.0D0
      TST(N)=0.0D0
```

Subroutine PRINT is listed in Program 5.4c. We can note the following points about subroutine PRINT:

1. A heading for the numerical solution is printed during the first call to PRINT:

```
C...
C...  PRINT A HEADING FOR THE SOLUTION
      IF(IP.EQ.0)WRITE(NO,1)
1     FORMAT(9X,1HT,4X,8H T1(0,T),4X,8HTS(XL,T),
     1                 4X,8HT3(XL,T),4X,8H TS(0,T))
```

2. Then the numerical solution is printed as a function of t every 10th call to PRINT (PRINT is called 101 times to provide enough points for plots of reasonable resolution, as produced by subroutine SPLOTS called subsequently). In particular, the entering and exiting temperatures of the exchanger of Figure 5.3 are printed as a function of time, that is $T_1(0,t)$ (= T1(1) = 25), $T_s(x_L,t)$ (= TS(NX) =220), $T_3(x_L,t)$ (=T3(NX)), and $T_s(0,t)$ (= TS(1)):

```
C...
C...  PRINT THE SOLUTION
      IP=IP+1
      IF((IP-1)/10*10.EQ.(IP-1))THEN
         WRITE(NO,2)T,T1(1),TS(NX),T3(NX),TS(1)
2        FORMAT(F10.1,4F12.1)
      END IF
```

3. The exiting temperatures, $T_3(x_L,t)$ (=T3(NX)) and $T_s(0,t)$ (= TS(1)), are then stored for plotting as a function of t

```
C...
C...  STORE THE SOLUTION FOR PLOTTING
      TP(IP)=T
      T3P(IP)=T3(NX)
      TSP(IP)=TS(1)
```

4. The temperatures are also stored as a function of x for plotting of the spatial temperature profiles every 20th call to PRINT:

```
C...
C...  THE SPATIAL PROFILES ARE STORED EVERY 20TH CALL TO PRINT
      IF((IP-1)/20*20.EQ.(IP-1))THEN
      ISP=ISP+1
      DO 5 I=1,NX
         T1SP(ISP,I)=T1(I)
         T2SP(ISP,I)=T2(I)
         T3SP(ISP,I)=T3(I)
         TSSP(ISP,I)=TS(I)
5     CONTINUE
      END IF
```

5. The maximum temperature, TS (=220), is stored in the arrays for the spatial

```
      SUBROUTINE PRINT(NI,NO)
      IMPLICIT DOUBLE PRECISION (A-H,O-Z)
      PARAMETER (NX=11)
      COMMON/T/          T,      NSTOP,      NORUN
     1      /Y/    T1(NX),    T2(NX),    T3(NX),      TS(NX)
     2      /F/   T1T(NX),   T2T(NX),   T3T(NX),     TST(NX)
     3      /S/   T1X(NX),   T2X(NX),   T3X(NX),     TSX(NX)
     4      /C/        TO,        XL,        VT,          VS,
     5                 CT,        CS,       TS0,           N,
     6             XP(NX),        IP,       ISP
C...
C...  DIMENSION THE ARRAYS FOR PLOTTING
      DIMENSION   T3P(101), TSP(101),   TP(101),
     1          T1SP(7,NX),          T2SP(7,NX),
     2          T3SP(7,NX),          TSSP(7,NX)
C...
C...  MONITOR THE CALCULATION
      WRITE(*,*)NORUN,T
C...
C...  PRINT A HEADING FOR THE SOLUTION
      IF(IP.EQ.0)WRITE(NO,1)
1     FORMAT(9X,1HT,4X,8H T1(0,T),4X,8HTS(XL,T),
     1               4X,8HT3(XL,T),4X,8H TS(0,T))
C...
C...  PRINT THE SOLUTION
      IP=IP+1
      IF((IP-1)/10*10.EQ.(IP-1))THEN
         WRITE(NO,2)T,T1(1),TS(NX),T3(NX),TS(1)
2        FORMAT(F10.1,4F12.1)
      END IF
C...
C...  STORE THE SOLUTION FOR PLOTTING
      TP(IP)=T
      T3P(IP)=T3(NX)
      TSP(IP)=TS(1)
C...
C...  THE SPATIAL PROFILES ARE STORED EVERY 20TH CALL TO PRINT
      IF((IP-1)/20*20.EQ.(IP-1))THEN
      ISP=ISP+1
      DO 5 I=1,NX
         T1SP(ISP,I)=T1(I)
         T2SP(ISP,I)=T2(I)
         T3SP(ISP,I)=T3(I)
         TSSP(ISP,I)=TS(I)
5     CONTINUE
      END IF
C...
C...  PLOT THE SOLUTION AT THE END OF THE RUN.  THE ENTERING SHELL
C...  FLUID TEMPERATURE IS FIRST STORED TO SCALE THE TEMPERATURE PLOTS
4     IF(IP.LT.101)RETURN
      DO 6 I=1,NX
```

Program 5.4c Subroutine PRINT to Print the Solutions to Equations (5.61) to (5.72). *Continued next page.*

```
         T1SP(7,I)=TSO
         T2SP(7,I)=TSO
         T3SP(7,I)=TSO
         TSSP(7,I)=TSO
6     CONTINUE
      CALL SPLOTS(1,IP,TP,T3P)
      WRITE(NO,10)
10    FORMAT(1H ,//,31H T3(XL,T) VS T                  )
      CALL SPLOTS(1,IP,TP,TSP)
      WRITE(NO,11)
11    FORMAT(1H ,//,31H TS(0,T) VS T                   )
      CALL SPLOTS(7,NX,XP,T1SP)
      WRITE(NO,12)
12    FORMAT(1H ,//,31H T1(X,T) VS X, T = 0, 10,... 50)
      CALL SPLOTS(7,NX,XP,T2SP)
      WRITE(NO,13)
13    FORMAT(1H ,//,31H T2(X,T) VS X, T = 0, 10,... 50)
      CALL SPLOTS(7,NX,XP,T3SP)
      WRITE(NO,14)
14    FORMAT(1H ,//,31H T3(X,T) VS X, T = 0, 10,... 50)
      CALL SPLOTS(7,NX,XP,TSSP)
      WRITE(NO,15)
15    FORMAT(1H ,//,31H TS(X,T) VS X, T = 0, 10,... 50)
C...
C...  THE FOLLOWING CALLS TO SUBROUTINES BVODE1 AND BVODE2 (BOUNDARY-
C...  VALUE ORDINARY DIFFERENTIAL EQUATIONS, VERSIONS 1 AND 2) SOLVE
C...  THE STEADY-STATE PROBLEM (T1  = T2  = T3  = TS  = 0) DIRECTLY, FOR
C...                               T     T     T     T
C...  COMPARISON WITH THE DYNAMIC SOLUTION AT ESSENTIALLY INFINITE TIME
C...  (STEADY STATE).  SUBROUTINES BVODE1 AND BVODE2 DIFFER IN THE WAY
C...  IN WHICH THE BOUNDARY CONDITIONS OF THE STEADY-STATE BVODE PROBLEM
C...  ARE IMPLEMENTED.  THE COMMENTS IN BVODE1 AND BVODE2 GIVE ALL OF THE
C...  DETAILS
      CALL BVODE1
      CALL BVODE2
      RETURN
      END
```

Program 5.4c *Continued.*

profiles to ensure that the four plots are scaled from 25 to 220 in each case:

```
C...
C...  PLOT THE SOLUTION AT THE END OF THE RUN.  THE ENTERING SHELL
C...  FLUID TEMPERATURE IS FIRST STORED TO SCALE THE TEMPERATURE PLOTS
4     IF(IP.LT.101)RETURN
      DO 6 I=1,NX
         T1SP(7,I)=TSO
         T2SP(7,I)=TSO
         T3SP(7,I)=TSO
         TSSP(7,I)=TSO
6     CONTINUE
```

6. Subroutine SPLOTS is next called to plot the exiting temperatures vs. time:

```
      CALL SPLOTS(1,IP,TP,T3P)
      WRITE(NO,10)
10    FORMAT(1H ,//,31H T3(XL,T) VS T                          )
      CALL SPLOTS(1,IP,TP,TSP)
      WRITE(NO,11)
11    FORMAT(1H ,//,31H TS(0,T) VS T                           )
```

7. Subroutine SPLOTS is then called to plot the four sets of spatial profiles (for the three tube passes and the shell):

```
      CALL SPLOTS(7,NX,XP,T1SP)
      WRITE(NO,12)
12    FORMAT(1H ,//,31H T1(X,T) VS X, T = 0, 20,... 100)
      CALL SPLOTS(7,NX,XP,T2SP)
      WRITE(NO,13)
13    FORMAT(1H ,//,31H T2(X,T) VS X, T = 0, 20,... 100)
      CALL SPLOTS(7,NX,XP,T3SP)
      WRITE(NO,14)
14    FORMAT(1H ,//,31H T3(X,T) VS X, T = 0, 20,... 100)
      CALL SPLOTS(7,NX,XP,TSSP)
      WRITE(NO,15)
15    FORMAT(1H ,//,31H TS(X,T) VS X, T = 0, 20,... 100)
```

8. Finally, two subroutines, BVODE1 and BVODE2, are called to directly compute a solution to the boundary value ODE problem, equations (5.53) to (5.60). This is done by replacing the spatial derivatives in equations (5.53) and (5.55) to (5.57) with three-point centered approximations. For example, equation (5.53) becomes

$$v\frac{T_{1,i+1} - T_{1,i-1}}{2\Delta x} = hA_h/(A\rho C_p)(T_{s,i} - T_{1,i}) \tag{5.73}$$

Equation (5.73) is now entirely algebraic. Also, because the original ODEs, equations (5.53) and (5.55) to (5.57), are linear, the algebraic equations that result by replacing the derivatives in x with finite difference approximations are also linear. We therefore arrive at a system of $4 \times 11 = 44$ linear algebraic equations, which can be solved with standard library routines such as DECOMP and SOLVE [Forsythe et al. (1977)].

9. Subroutines BVODE1 and BVODE2 differ only in the way that the boundary values defined by equations (5.54) and (5.58) to (5.60) are included in the system of algebraic equations. In BVODE1, the boundary values are included directly in the algebraic equations, thus algebraic equations at the boundaries are not used. In BVODE2, noncentered, three-point finite difference approximations are used at the boundaries, and then the boundary values specified by equations (5.54) and (5.58) to (5.60) are used to eliminate unknowns in the resulting algebraic equations. The detailes are given in a series of comments in BVODE1 and BVODE2. Also, a subroutine, MAP, is called from BVODE1 and BVODE2 to map the coefficient matrix of the linear algebraic system. Since this approach of a full finite difference solution of a system of differential equations is outside the scope of this book, and since the code for BVODE1, BVODE2, and MAP is relatively long, these subroutines are not listed here (they are on the diskette available from the author). Only abbreviated output from these subroutines is presented as an independent check

```
METHOD OF LINES INTEGRATION OF BOUNDARY VALUE ODES
0.          100.       1.0
  44                         0.00001
END OF RUNS
```

Program 5.4d Data File for Programs 5.4a to c.

of the method of lines solutions of equations (5.61) to (5.72) (and therefore, as a check of the solution of ODEs (5.53) to (5.60)). We might note parenthetically that the coding of equations (5.53) to (5.60) by finite differences in `BVODE1` and `BVODE2` was considerably more complicated than the programming of equations (5.61) to (5.72) in subroutine `DERV`. This complexity was due in large part to the lack of correspondence between the approximating algebraic equations and the PDEs, which is usually the situation when differential equations are replaced entirely by algebraic equations. On the other hand, the method of lines generally provides a close correspondence between the coding and the PDEs which facilitates the programming of PDE problems.

Program 5.4d is the data file for Programs 5.4a to c. One run is programmed for 44 ODEs integrated with an error tolerance of 0.00001. The integrator is `RKF45`, called by a main program similar to Program 2.2a, and therefore the main program is not listed here.

Abbreviated output from Programs 5.4a to d, produced by subroutine `PRINT`, is listed in Table 5.4. We can note the following points about the output in Table 5.4:

1. The system of PDEs, equations (5.61) to (5.72), approaches a steady state which is the solution to the boundary value ODE system, equations (5.53) to (5.60):

T	T1(0,T)	TS(XL,T)	T3(XL,T)	TS(0,T)
0.0	25.0	220.0	25.0	25.0
10.0	25.0	220.0	53.4	25.2
	.		.	
	.		.	
	.		.	
90.0	25.0	220.0	100.0	68.3
100.0	25.0	220.0	100.3	68.5

In particular, the exiting temperatures are $T_3(x_L, 100) = 100.3$ and $T_s(0, 100) = 68.5$.

2. This approach to steady state is also apparent in the point plots of $T_3(x_L, t)$ and $T_s(0, t)$ vs. t; note in particular how these plots approach the steady-state values, 100.3 and 68.5.

3. The spatial profiles also obviously approach a steady state. For example, the profile of $T_1(x, t)$ at $t = 100$ goes from $T_1(0, 100) = 25$ to $T_1(x_L, 100) = 78.2$. Similarly, the profile of $T_s(x, t)$ at $t = 100$ goes from $T_s(x_L, 100) = 220$ to $T_s(0, 100) = 68.5$

Table 5.4 Abbreviated Output From Programs 5.4a to d.

```
RUN NO. -   1  METHOD OF LINES INTEGRATION OF BOUNDARY VALUE ODES

INITIAL T -  0.000D+00

  FINAL T -  0.100D+03

  PRINT T -  0.100D+01

NUMBER OF DIFFERENTIAL EQUATIONS -  44

INTEGRATION ALGORITHM - RKF45

MAXIMUM INTEGRATION ERROR -  0.100D-04


         T      T1(0,T)    TS(XL,T)    T3(XL,T)     TS(0,T)
       0.0        25.0       220.0        25.0        25.0
      10.0        25.0       220.0        53.4        25.2
      20.0        25.0       220.0        62.7        45.9
      30.0        25.0       220.0        82.1        55.2
      40.0        25.0       220.0        89.4        60.8
      50.0        25.0       220.0        94.2        64.2
      60.0        25.0       220.0        97.0        66.1
      70.0        25.0       220.0        98.5        67.2
      80.0        25.0       220.0        99.5        67.9
      90.0        25.0       220.0       100.0        68.3
     100.0        25.0       220.0       100.3        68.5
```

```
             ..1....1....1....1....1....1....1....1....1....1....1..
 0.100D+03+                              1CCCCCCCCCCCCCCCCCCCCCCCC  +I
          -                        1CCCCC1                          -I
          -                    1CCC1                                -I
          -                  CC1                                    -I
          -                CC                                       -I
 0.752D+02+              C                                          +I
          -             C                                           -I
          -            C                                            -I
          -        1CCC                                             -I
          -      1C1                                                -I
 0.501D+02+     11                                                  +I
          -    11                                                   -I
          -   11                                                    -I
          -   1                                                     -I
          -                                                         -I
 0.250D+02+  1                                                      +I
             ..1....1....1....1....1....1....1....1....1....1....1..
          0.000D+00  0.20D+02  0.40D+02  0.60D+02  0.80D+02  0.10D+03

T3(XL,T) VS T
```

```
            ..1....1....1....1....1....1....1....1....1....1....1..
0.685D+02+                                CCCCCCCCCCCCCCCCCCCCCC  +I
         -                          CCCCCC                         -I
         -                      1CCC                               -I
         -                    CC1                                  -I
         -                  CC                                     -I
0.539D+02+                CC                                       +I
         -              1C                                         -I
         -             11                                          -I
         -            11                                           -I
         -            1                                            -I
0.394D+02+          1                                              +I
         -         11                                              -I
         -         1                                               -I
         -        1                                                -I
         -  1CCCC11                                                -I
0.248D+02+      1                                                  +I
            ..1....1....1....1....1....1....1....1....1....1....1..
        0.000D+00  0.20D+02  0.40D+02  0.60D+02  0.80D+02  0.10D+03

TS(0,T) VS T
```

```
            ..1....1....1....1....1....1....1....1....1....1....1..
0.220D+03+  7    7    7    7    7    7    7    7    7    7    7  +I
         -                                                       -I
         -                                                       -I
         -                                                       -I
         -                                                       -I
0.155D+03+                                                       +I
         -                                                       -I
         -                                                       -I
         -                                                       -I
         -                                                       -I
0.900D+02+                                                    C  +I
         -                                               C    C  -I
         -                                     C    C    3    2  -I
         -                      C    C    C    C    2    2       -I
         -       C    C    C    C    2    2                      -I
0.250D+02+  C    1    1    1    1    1    1    1    1    1    1  +I
            ..1....1....1....1....1....1....1....1....1....1....1..
        0.000D+00  0.20D+01  0.40D+01  0.60D+01  0.80D+01  0.10D+02

T1(X,T) VS X, T = 0, 20,... 100
                  .
                  .
                  .
```

(the spatial profiles for $T_2(x,t)$ and $T_3(x,t)$ were deleted to conserve space):

```
           ..1....1....1....1....1....1....1....1....1....1....1..
 0.220D+03+  7    7    7    7    7    7    7    7    7    7    C  +I
          -                                                      -I
          -                                                      -I
          -                                                      -I
          -                                                 C    -I
 0.155D+03+                                                 2    +I
          -                                                      -I
          -                                            C         -I
          -                                            2         -I
          -                                  C    C              -I
 0.900D+02+                        C    C    C    2              +I
          -  C    C    C    C      C    3    2                   -I
          -  3    3                2    2                        -I
          -  2    2    2    2                                    -I
          -                                                      -I
 0.250D+02+  1    1    1    1    1    1    1    1    1    1      +I
           ..1....1....1....1....1....1....1....1....1....1....1..
         0.000D+00  0.20D+01  0.40D+01  0.60D+01  0.80D+01  0.10D+02

TS(X,T) VS X, T = 0, 20,... 100
```

4. Subroutines BVODE1 and BVODE2, called at the end of subroutine PRINT in Program 5.4c, also print out the steady-state profiles. For example, BVODE1 gives $T_3(x_L) = 100.7$ and $T_s(0) = 68.6$; the discrepancies with the method of lines solution, 100.3 and 68.5, respectively, are probably due to the coarse grid of 11 points.

In conclusion, we have found the solution of boundary value problems, like equations (5.53) to (5.60), by the method of false transients, in which temporal derivatives are added to the original equations, generally to be a straightforward, reliable, and accurate procedure. There is the requirement, however, to add the time derivatives in such a way that the resulting transient problem is stable and approaches a steady state for which the time derivatives essentially vanish. The addition of the time derivatives is frequently guided by physical insight and understanding of the problem system. Alternatively, the time derivatives can be added in accordance with the Davidenko differential equation (2.201) [Boggs (1971), Davidenko (1953), Edelen (1976a,b), and Hachtel et al. (1974)].

There is also the matter of the choice of an initial condition, such as equations (5.69) to (5.72). For linear problems like equations (5.53) to (5.60), the initial condition will not affect the final steady-state solution. For nonlinear problems, which can have multiple solutions, different choices of initial conditions might lead to different steady-state solutions. In general, the initial condition should be selected to be as close as possible to the final steady-state solution that is to be computed.

We conclude this chapter with a PDE zeroth order in time, which models the aquifer illustrated in Figure 5.4 [Prasher, (1987)]. The hydraulic head (height of water in the aquifer which produces a flow of water), $h(x, y)$, is given by *Poisson's equation* in Cartesian coordinates:

$$T_r \left\{ \frac{\partial^2 h}{\partial x^2} + \frac{\partial^2 h}{\partial y^2} \right\} = R_r(x, y) \tag{5.74}$$

where h = liquid hydraulic head (m); x, y = spatial coordinates of the aquifer (m); T_r = transmissivity of the aquifer (m/day); and R_r = recharge rate ($1/day$).

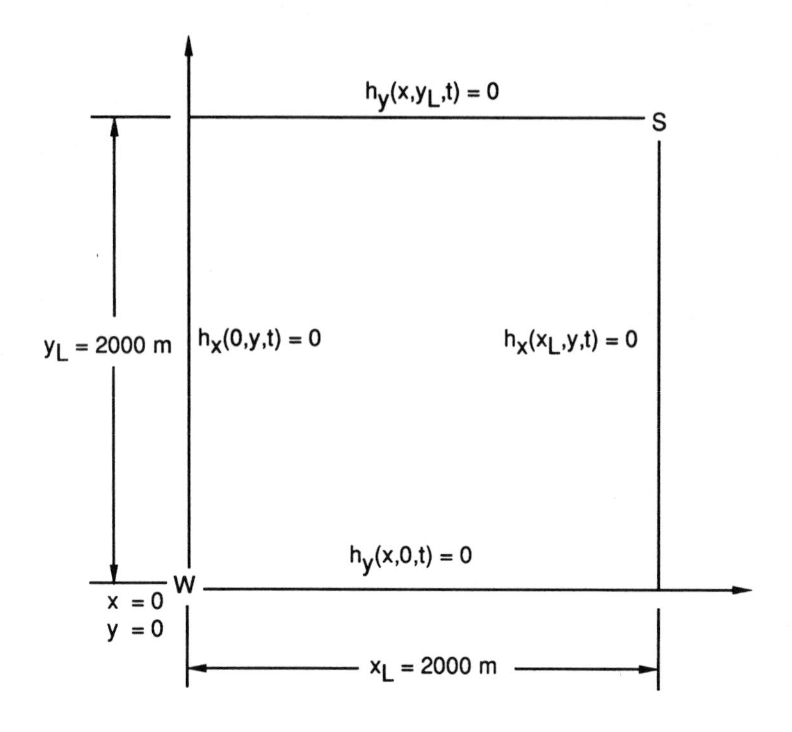

Figure 5.4 A Simplified Aquifer.

Equation (5.74) is derived from an analog to Fourier's first law

$$q_x = T_r \frac{\partial h}{\partial x}$$

where q_x is the flux of water in the x direction (with the units of m^3/m^2-day).

Note that equation (5.75) does not have an initial value independent variable, i.e., x and y are boundary value independent variables. It therefore requires two boundary conditions in x and y

$$\frac{\partial h(0,y)}{\partial x} = \frac{\partial h(x_L,y)}{\partial x} = 0 \tag{5.75)(5.76}$$

$$\frac{\partial h(x,0)}{\partial y} = \frac{\partial h(x,y_L)}{\partial y} = 0 \tag{5.77)(5.78}$$

Equations (5.75) to (5.78) are homogeneous Neumann boundary conditions which state that there is no flow into or out of the aquifer at its boundaries; that is, there are no flow boundary conditions at $x = 0$ and x_L, $y = 0$ and y_L. We should also note that equation (5.74) is an elliptic PDE since it has only boundary value independent variables; in fact, equation (5.74) was discussed earlier (see equation (1.51)).

We will compute a solution to equations (5.74) to (5.78) for two cases:

Case I:

$$R_r(x,y) = (Q_s/(\Delta x \Delta y))\delta(x)\delta(y) \tag{5.79}$$

Case II:

$$R_r(x,y) = (Q_s/(\Delta x \Delta y))\delta(x)\delta(y) - (Q_s/(\Delta x \Delta y))\delta(x - x_L)\delta(y - y_L) \tag{5.80}$$

where Q_s is the flow rate of the source and sink (m^3/day), Δx and Δy are the increments in x and y used subsequently in the finite difference approximation of equation (5.74), and $\delta(x)$ is a *Dirac delta (impulse) function*.

Case I corresponds to a *sink* (well) located at $x = 0$ and $y = 0$ which withdraws water from the aquifer until the water is depleted; thus the solution to equations (5.74) to (5.79) is $h(x, y) = 0$ (the sink is denoted as W at $x = 0$ and $y = 0$ in Figure 5.4). Note that $h(x, y) = 0$ is a solution to equations (5.74) to (5.79) except for $x = 0$ and $y = 0$.

Case II corresponds to a sink (well) located at $x = 0$ and $y = 0$ which withdraws water from the aquifer, and a source located at $x = x_L$ and $y = y_L$ which recharges water at the same rate as it is withdrawn by the sink so that a steady-state distribution of the hydraulic head develops (the source is denoted as S at $x = x_L$ and $y = y_L$ in Figure 5.4).

In order to compute the solution to equation (5.74) to (5.80), we again add a derivative to equation (5.74) with respect to an initial value variable, t. In other words, we convert elliptic PDE (5.74) into a parabolic PDE (Fourier's second law in two dimensions)

$$S_c \frac{\partial h}{\partial t} = T_r \left\{ \frac{\partial^2 h}{\partial x^2} + \frac{\partial^2 h}{\partial y^2} \right\} - R_r(x, y) \tag{5.81}$$

The temporal derivative, $\partial h/\partial t$, can be considered as an accumulation term in a material balance written for a differential section of the aquifer (or it can be considered as just a mathematical term added to equation (5.74) to produce a solution to equation (5.74)). If the temporal derivative is given this physical interpretation of an accumulation term, then the multiplying coefficient, S_c, is the *storage coefficient* [Prasher (1987)].

To develop a method of lines solution to equations (5.75) to (5.81), we also require an initial condition for equation (5.81). Again, as in the case of boundary value ODEs (5.53) and (5.55) to (5.57), we can select the initial condition arbitrarily since it will have no effect on the solution of equation (5.74). Thus we simply use

$$h(x, y, 0) = h_0 \tag{5.82}$$

where h_0 is a constant.

We now have a complete parabolic problem, equations (5.75) to (5.82), and we can proceed with the programming of a method of lines solution. One approach, which we have used repeatedly in previous examples, is to compute the boundary value derivatives in equation (5.81) by calling library spatial differentiation routines, as we did, for example, in programming the two-dimensional advection equation (4.195). However, we can also expect to encounter PDE problems for which an appropriate spatial differentiation routine is not available, and we therefore are required to select and program algebraic approximations for the spatial derivatives. To illustrate this approach to the method of lines solution of a PDE problem system, we now approximate the spatial derivatives of equation (5.81) with three-point centered approximations (this, of course, could also be done by calling soubroutine `DSS002` over the two-dimensional x–y domain, in a way similar to Programs 4.11a to d). The *Laplacian group* can be approximated as

$$\frac{\partial^2 h}{\partial x^2} + \frac{\partial^2 h}{\partial y^2} \approx \frac{h_{i+1,j} - 2h_{i,j} + h_{i-1,j}}{\Delta x^2} + \frac{h_{i,j+1} - 2h_{i,j} + h_{i,j-1}}{\Delta y^2} \tag{5.83}$$

where i and j are indices for the position within a two-dimensional x–y grid. Thus, i defines the position in x

$$x = (i-1)\Delta x, \qquad \Delta x = x_L/(n-1), \qquad i = 1, 2, \ldots, n \tag{5.84}$$

and j defines the position in y

$$y = (j-1)\Delta y, \qquad \Delta y = y_L/(n-1), \qquad j = 1, 2, \ldots, n \tag{5.85}$$

The method of lines approximation of equation (5.81) then becomes

$$S_c \frac{dh_{i,j}}{dt} = T_r \left\{ \frac{h_{i+1,j} - 2h_{i,j} + h_{i-1,j}}{\Delta x^2} + \frac{h_{i,j+1} - 2h_{i,j} + h_{i,j-1}}{\Delta y^2} \right\} - R_r(x_i, y_j) \tag{5.86}$$

which is a system of $n \times n$ ODEs.

Equation (5.86) must also be applied at the boundaries $x = 0$ and x_L, $y = 0$ and y_L. According to boundary conditions (5.75) to (5.78), expressed in terms of second-order, centered finite difference approximations, we have

$$\frac{h_{2,j} - h_{0,j}}{2\Delta x} = 0, \qquad \frac{h_{n+1,j} - h_{n-1,j}}{2\Delta x} = 0 \qquad (5.87)(5.88)$$

$$\frac{h_{i,2} - h_{i,0}}{2\Delta y} = 0, \qquad \frac{h_{i,n+1} - h_{i,n-1}}{2\Delta y} = 0 \qquad (5.89)(5.90)$$

Each of these four equations can be solved for the fictitious values which fall outside the spatial grid defined by $1 \le i \le n$, $1 \le j \le n$. From equations (5.87) to (5.90) we have

$$h_{0,j} = h_{2,j}, \quad h_{n+1,j} = h_{n-1,j}, \quad h_{i,0} = h_{i,2}, \quad h_{i,n+1} = h_{i,n-1} \tag{5.91}$$

Then equation (5.86) applied at the boundary ($x = 0$, $y \neq 0$, $y \neq y_L$) becomes

$$S_c \frac{dh_{i,j}}{dt} = T_r \left\{ \frac{2\left\{h_{2,j} - 2h_{1,j}\right\}}{\Delta x^2} + \frac{h_{1,j+1} - 2h_{1,j} + h_{1,j-1}}{\Delta y^2} \right\} - R_r(x_1, y_j) \tag{5.92}$$

Similar equations follow from equation (5.86) for the boundaries ($x = x_L$, $y \neq 0$, $y \neq x_L$), ($x \neq 0$, $x \neq x_L$, $y = 0$), and ($x \neq 0$, $x \neq x_L$, $y = y_L$).

At the four corners, we can also apply equations (5.91). For example, for ($x = 0$, $y = 0$), equation (5.86) becomes

$$S_c \frac{dh_{1,1}}{dt} = T_r \left\{ \frac{2\left\{h_{2,1} - 2h_{1,1}\right\}}{\Delta x^2} + \frac{2\left\{h_{1,2} - 2h_{1,1}\right\}}{\Delta y^2} \right\} - R_r(x_1, y_1) \tag{5.93}$$

Similar equations follow from equation (5.86) for the other three-corner points ($x = 0$, $y = y_L$), ($x = x_L$, $y = 0$), ($x = x_L$, $y = y_L$).

We can now consider the method of lines programming of the system of $n \times n$ ODEs. Subroutine `INITAL` is listed in Program 5.5a. We can note the following points about subroutine `INITAL`:

1. The dependent variable of equation (5.81) (as approximated by equation (5.86)), $h(x, y, t)$, is defined on an 11×11 spatial grid in `COMMON/Y/`, and its temporal derivative, $\partial h(x, y, t)/\partial t$, is in `COMMON/F/`.

```
      PARAMETER (N=11)
      COMMON/T/          T,      NSTOP,      NORUN
     1      /Y/     H(N,N)
     2      /F/    HT(N,N)
     3      /C/         QS,         TR,         SC,
     4                  XL,         YL,          A,
     5                  DX,         DY,         RR,
     6                  HO
     7      /I/         IP
```

```
      SUBROUTINE INITAL
      IMPLICIT DOUBLE PRECISION (A-H,O-Z)
      PARAMETER (N=11)
      COMMON/T/          T,      NSTOP,      NORUN
     1      /Y/    H(N,N)
     2      /F/   HT(N,N)
     3      /C/         QS,         TR,         SC,
     4                  XL,         YL,          A,
     5                  DX,         DY,         RR,
     6                  HO
     7      /I/         IP
C...
C...  MODEL PARAMETERS
C...
C...     SOURCE AND SINK FLOW RATES (M**3/DAY)
         QS=2000.0D0
C...
C...     TRANSMISSIVITY (M/DAY)
         TR=300.0D0
C...
C...     STORAGE COEFFICIENT (DIMENSIONLESS)
         SC=0.02D0
C...
C...     LENGTH OF THE AQUIFER IN THE X DIRECTION (M)
         XL=2000.0D0
C...
C...     LENGTH OF THE AQUIFER IN THE Y DIRECTION (M)
         YL=2000.0D0
C...
C...     GRID SPACING IN THE X DIRECTION
         DX=XL/DFLOAT(N-1)
C...
C...     GRID SPACING IN THE Y DIRECTION
         DY=YL/DFLOAT(N-1)
C...
C...     AREA OF A GRID UNIT (M**2)
         A=DX*DY
C...
C...     RECHARGE RATE (M/DAY)
         RR=QS/A
C...
C...     INITIAL HYDRAULIC HEAD (M)
         HO=10.0D0
C...
C...  INITIAL CONDITIONS
      DO 1 I=1,N
      DO 1 J=1,N
         H(I,J)=HO
```

Program 5.5a Subroutine `INITAL` for Initial Condition (5.82). *Continued next page.*

```
1     CONTINUE
C...
C...  COMPUTE THE INITIAL DERIVATIVES
      CALL DERV
      IP=0
      RETURN
      END
```

Program 5.5a *Continued.*

2. The parameters of equations (5.75) to (5.82) are then defined. Note in particular the length of the aquifer in the x- and y-directions is 2000 m:

```
C...
C...      LENGTH OF THE AQUIFER IN THE X-DIRECTION (M)
          XL=2000.0D0
C...
C...      LENGTH OF THE AQUIFER IN THE Y-DIRECTION (M)
          YL=2000.0D0
```

   Of course, the numerical method of lines formulation does not require equal lengths or numbers of grid points in the two directions.

3. Initial condition (5.82) is then set in DO loop 1:

```
C...
C...      INITIAL HYDRAULIC HEAD (M)
          H0=10.0D0
C...
C...   INITIAL CONDITIONS
       DO 1 I=1,N
       DO 1 J=1,N
          H(I,J)=H0
1      CONTINUE
```

Subroutine DERV is listed in Program 5.5b. We can note the following points about subroutine DERV:

1. A physical constraint is first imposed on the solution, i.e., the hydraulic head must be nonnegative:

```
C...
C...   SET CONSTRAINTS ON THE SOLUTION
       DO 7 I=1,N
       DO 8 J=1,N
          IF(H(I,J).LT.0.0D0)H(I,J)=0.0D0
8      CONTINUE
7      CONTINUE
```

   This constraint is required in general since the recharge rate, $R_r(x, y)$, in equation (5.81), when negative, as it is in Case I, can cause the temporal derivative,

```
      SUBROUTINE DERV
      IMPLICIT DOUBLE PRECISION (A-H,O-Z)
      PARAMETER (N=11)
      COMMON/T/          T,      NSTOP,      NORUN
     1      /Y/    H(N,N)
     2      /F/   HT(N,N)
     3      /C/         QS,         TR,         SC,
     4                  XL,         YL,          A,
     5                  DX,         DY,         RR,
     6                  HO
     7      /I/         IP
C...
C...  SET CONSTRAINTS ON THE SOLUTION
      DO 7 I=1,N
      DO 8 J=1,N
         IF(H(I,J).LT.0.0D0)H(I,J)=0.0D0
8     CONTINUE
7     CONTINUE
C...
C...  BOUNDARY POINTS FROM I = 2 TO N-1
      DO 1 I=2,N-1
C...
C...  Y = 0
      HT(I,1)=
     1  (TR/SC)*(      H(I+1,  1)-2.0D0*H(I   ,  1)+   H(I-1,  1))/DX**2
     2 +(TR/SC)*(2.0D0*H(I   ,  2)-2.0D0*H(I   ,  1)               )/DY**2
C...
C...  Y = YL
      HT(I,N)=
     1  (TR/SC)*(      H(I+1,N  )-2.0D0*H(I   ,N  )+   H(I-1,N  ))/DX**2
     2 +(TR/SC)*(2.0D0*H(I   ,N-1)-2.0D0*H(I   ,N  )               )/DY**2
1     CONTINUE
C...
C...  BOUNDARY POINTS FROM J = 2 TO N-1
      DO 2 J=2,N-1
C...
C...  X = 0
      HT(1,J)=
     1  (TR/SC)*(2.0D0*H(  2,J  )-2.0D0*H(  1,J  )               )/DX**2
     2 +(TR/SC)*(      H(  1,J+1)-2.0D0*H(  1,J  )+   H(  1,J-1))/DY**2
C...
C...  X = XL
      HT(N,J)=
     1  (TR/SC)*(2.0D0*H(N-1,J  )-2.0D0*H(N  ,J  )               )/DX**2
     2 +(TR/SC)*(      H(N  ,J+1)-2.0D0*H(N  ,J  )+   H(N  ,J-1))/DY**2
2     CONTINUE
C...
C...  CORNER POINTS
C...
C...  X = 0, Y = 0
      HT(1,1)=
     1  (TR/SC)*(2.0D0*H(  2,  1)-2.0D0*H(  1,  1)               )/DX**2
```

Program 5.5b Subroutine DERV for Equations (5.75) to (5.81). *Continued next page.*

```
      2 +(TR/SC)*(2.0D0*H(  1,  2)-2.0D0*H(  1,  1)              )/DY**2
C...
C...    X = XL, Y = 0
        HT(N,1)=
      1  (TR/SC)*(2.0D0*H(N-1,  1)-2.0D0*H(N  ,  1)              )/DX**2
      2 +(TR/SC)*(2.0D0*H(N  ,  2)-2.0D0*H(N  ,  1)              )/DY**2
C...
C...    X = 0, Y = YL
        HT(1,N)=
      1  (TR/SC)*(2.0D0*H(  2,N  )-2.0D0*H(  1,N  )              )/DX**2
      2 +(TR/SC)*(2.0D0*H(  1,N-1)-2.0D0*H(  1,N  )              )/DY**2
C...
C...    X = XL, Y = YL
        HT(N,N)=
      1  (TR/SC)*(2.0D0*H(N-1,N  )-2.0D0*H(N  ,N  )              )/DX**2
      2 +(TR/SC)*(2.0D0*H(N  ,N-1)-2.0D0*H(N  ,N  )              )/DY**2
C...
C...    INTERIOR POINTS
        DO 4 I=2,N-1
        DO 4 J=2,N-1
        HT(I,J)=
      1  (TR/SC)*(     H(I+1,J  )-2.0D0*H(I  ,J  )+   H(I-1,J  ))/DX**2
      2 +(TR/SC)*(     H(I  ,J+1)-2.0D0*H(I  ,J  )+   H(I  ,J-1))/DY**2
4       CONTINUE
C...
C...    INCLUDE THE SOURCE AND SINK
        IF(NORUN.EQ.1)THEN
           HT(1,1)=HT(1,1)-RR/SC
        ELSE IF(NORUN.EQ.2)THEN
           HT(1,1)=HT(1,1)-RR/SC
           HT(N,N)=HT(N,N)+RR/SC
        END IF
        RETURN
        END
```

Program 5.5b *Continued.*

$\partial h(x, y, t)/\partial t$, to be negative long enough that when integrated in time, the hydraulic head, $h(x, y, t)$, will also become negative. Of course, the recharge rate could be modified so that if $h(x, y, t) < 0$, $R_r(x, y, t) = 0$. Another argument might be made that, depending on the definition of the hydraulic head, particularly its reference level, it could actually be negative. In any case, the constraint in DO loops 7 and 8 is one possibility for including a limit on the solution to conform to a physical constraint.

2. In DO loops 1 and 2, equation (5.86) is programmed for the boundaries of the aquifer. For example, at $x = 0$ and $y \neq 0$, $y \neq y_L$, equation (5.86) becomes equation (5.92):

$$S_c \frac{dh_{1,j}}{dt} = T_r \left\{ \frac{2\left\{h_{2,j} - 2h_{1,j}\right\}}{\Delta x^2} + \frac{h_{1,j+1} - 2h_{1,j} + h_{1,j-1}}{\Delta y^2} \right\} - R_r(x_1, y_j) \tag{5.92}$$

which is programmed in DO loop 2 as

```
C...
C...  X = 0
      HT(1,J)=
     1  (TR/SC)*(2.0D0*H(  2,J  )-2.0D0*H(  1,J  )          )/DX**2
     2 +(TR/SC)*(        H(  1,J+1)-2.0D0*H(  1,J  )+   H(  1,J-1))/DY**2
```

3. Equation (5.86) is next programmed for the four corners of the aquifer. For example, for $x = 0$ and $y = 0$, equation (5.93)

$$S_c \frac{dh_{1,1}}{dt} = T_r \left\{ \frac{2\{h_{2,1} - 2h_{1,1}\}}{\Delta x^2} + \frac{2\{h_{1,2} - 2h_{1,1}\}}{\Delta y^2} \right\} - R_r(x_1, y_1) \tag{5.93}$$

is programmed as

```
C...
C...  X = 0, Y = 0
      HT(1,1)=
     1  (TR/SC)*(2.0D0*H(  2,  1)-2.0D0*H(  1,  1)          )/DX**2
     2 +(TR/SC)*(2.0D0*H(  1,  2)-2.0D0*H(  1,  1)          )/DY**2
```

4. Then equation (5.86)

$$S_c \frac{dh_{i,j}}{dt} = T_r \left\{ \frac{h_{i+1,j} - 2h_{i,j} + h_{i-1,j}}{\Delta x^2} + \frac{h_{i,j+1} - 2h_{i,j} + h_{i,j-1}}{\Delta y^2} \right\} - R_r(x_i, y_j) \tag{5.86}$$

is programmed in the interior of the aquifer in DO loops 4 as

```
C...
C...  INTERIOR POINTS
      DO 4 I=2,N-1
      DO 4 J=2,N-1
      HT(I,J)=
     1  (TR/SC)*(        H(I+1,J  )-2.0D0*H(I  ,J  )+   H(I-1,J  ))/DX**2
     2 +(TR/SC)*(        H(I  ,J+1)-2.0D0*H(I  ,J  )+   H(I  ,J-1))/DY**2
4     CONTINUE
```

5. In the preceding programming of the PDE, the nonhomogeneous term, the recharge rate, $R_r(x, y)$, was not included since it is nonzero only at two points in the grid of 121 points, that is at $(x = 0, y = 0)$ and $(x = x_L, y = y_L)$, or in terms of the grid indices, $(i = 1, j = 1)$ and $(i = 11, j = 11)$. This term is then included in the final coding of the PDE, equation (5.81):

```
C...
C...  INCLUDE THE SOURCE AND SINK
      IF(NORUN.EQ.1)THEN
         HT(1,1)=HT(1,1)-RR/SC
      ELSE IF(NORUN.EQ.2)THEN
         HT(1,1)=HT(1,1)-RR/SC
         HT(N,N)=HT(N,N)+RR/SC
      END IF
```

Note the Case I and Case II programming (for NORUN=1 and 2, respectively).

Subroutine PRINT is listed in Program 5.5c. We can note the following points about subroutine PRINT:

1. The solution is printed over the 11×11-point grid:

```
C...
C...  PRINT THE SOLUTION
      NS=2
      WRITE(NO,1)T
1     FORMAT(1H ,//,5H T = ,F6.1,//,11H    H(X,Y,T))
      WRITE(NO,2)((  H(I,J),I=1,N,NS),J=1,N,NS)
      WRITE(NO,3)
3     FORMAT(11H  HT(X,Y,T))
      WRITE(NO,2)(( HT(I,J),I=1,N,NS),J=1,N,NS)
2     FORMAT(
     +                9X,10H      X = 0,10H    X = 400,10H    X = 800,
     +                   10H  X = 1200,10H  X = 1600,10H  X = 2000,/,
     +                    9H Y = 0    ,6F10.4,/,9H Y = 400 ,6F10.4,/,
     +                    9H Y = 800 ,6F10.4,/,9H Y = 1200,6F10.4,/,
     +                    9H Y = 1600,6F10.4,/,9H Y = 2000,6F10.4,//)
```

2. Next, the solution and the temporal derivative at ($x = 0$, $y = 0$), ($x = x_L/2$, $y = y_L/2$), and ($x = x_L$, $y = y_L$) are stored for subsequent printing and plotting:

```
C...
C...  STORE THE SOLUTION AND THE TEMPORAL DERIVATIVE FOR PLOTTING
      IP=IP+1
      TP(IP)=T
      HP(1,IP)=H(1,1)
      HP(2,IP)=H(6,6)
      HP(3,IP)=H(N,N)
      HTP(1,IP)=HT(1,1)
      HTP(2,IP)=HT(6,6)
      HTP(3,IP)=HT(N,N)
```

3. The solution and the temporal derivative at the three points of (2) are then printed:

```
C...
C...  PRINT THE SOLUTION AND THE TEMPORAL DERIVATIVE AT THE END OF THE
C...  RUN FOR THE POINTS (X=0,Y=0), (X=XL/2,Y=YL/2), (X=XL,Y=YL)
      IF(IP.EQ.11)THEN
         DO 5 J=1,IP
         WRITE(NO,4)TP(J),(HP(I,J),I=1,3),(HTP(I,J),I=1,3)
4        FORMAT(1X,'T = ',F6.1,/,3F12.4,/,3F12.4,/)
5        CONTINUE
```

4. Finally, the solution and the temporal derivative at the three points of (2) are plotted:

```
C...
C...  PLOT THE SOLUTION AND THE TEMPORAL DERIVATIVE AT THE END OF THE
C...  RUN FOR THE POINTS (X=0,Y=0), (X=XL/2,Y=YL/2), (X=XL,Y=YL)
         CALL SPLOTS(3,IP,TP,HP)
```

```
      SUBROUTINE PRINT(NI,NO)
      IMPLICIT DOUBLE PRECISION (A-H,O-Z)
      PARAMETER (N=11)
      COMMON/T/          T,      NSTOP,     NORUN
     1      /Y/    H(N,N)
     2      /F/   HT(N,N)
     3      /C/        QS,         TR,         SC,
     4                 XL,         YL,          A,
     5                 DX,         DY,         RR,
     6                 HO
     7      /I/        IP
C...
C...  ARRAYS FOR PLOTTING
      DIMENSION TP(11), HP(3,11), HTP(3,11)
C...

C...  MONITOR THE CALCULATION
      WRITE(*,*)NORUN,T
C...
C...  PRINT THE SOLUTION
      NS=2
      WRITE(NO,1)T
1     FORMAT(1H ,//,5H T = ,F6.1,//,11H   H(X,Y,T))
      WRITE(NO,2)((  H(I,J),I=1,N,NS),J=1,N,NS)
      WRITE(NO,3)
3     FORMAT(11H  HT(X,Y,T))
      WRITE(NO,2)(( HT(I,J),I=1,N,NS),J=1,N,NS)
2     FORMAT(
     +           9X,10H     X = 0,10H   X = 400,10H   X = 800,
     +              10H  X = 1200,10H  X = 1600,10H  X = 2000,/,
     +               9H Y = 0    ,6F10.4,/,9H Y = 400  ,6F10.4,/,
     +               9H Y = 800  ,6F10.4,/,9H Y = 1200,6F10.4,/,
     +               9H Y = 1600,6F10.4,/,9H Y = 2000,6F10.4,//)
C...
C...  STORE THE SOLUTION AND THE TEMPORAL DERIVATIVE FOR PLOTTING
      IP=IP+1
      TP(IP)=T
      HP(1,IP)=H(1,1)
      HP(2,IP)=H(6,6)
      HP(3,IP)=H(N,N)
      HTP(1,IP)=HT(1,1)
      HTP(2,IP)=HT(6,6)
      HTP(3,IP)=HT(N,N)
C...
C...  PRINT THE SOLUTION AND THE TEMPORAL DERIVATIVE AT THE END OF THE
C...  RUN FOR THE POINTS (X=0,Y=0), (X=XL/2,Y=YL/2), (X=XL,Y=YL)
      IF(IP.EQ.11)THEN
         DO 5 J=1,IP
         WRITE(NO,4)TP(J),(HP(I,J),I=1,3),(HTP(I,J),I=1,3)
4        FORMAT(1X,'T = ',F6.1,/,3F12.4,/,3F12.4,/)
```

Program 5.5c Subroutine PRINT to Print and Plot the Solution to Equations (5.75) to (5.82). *Continued next page.*

```
5         CONTINUE
C...
C...   PLOT THE SOLUTION AND THE TEMPORAL DERIVATIVE AT THE END OF THE
C...   RUN FOR THE POINTS (X=0,Y=0), (X=XL/2,Y=YL/2), (X=XL,Y=YL)
          CALL SPLOTS(3,IP,TP,HP)
          WRITE(NO,6),NORUN
6         FORMAT(/,' RUN ',I2,3X,
     1    ' 1 - H(0,0), 2 - H(XL/2,YL/2), 3 - H(XL,YL)',/)
          CALL SPLOTS(3,IP,TP,HTP)
          WRITE(NO,7),NORUN
7         FORMAT(/,' RUN ',I2,3X,
     1    ' 1 - HT(0,0), 2 - HT(XL/2,YL/2), 3 - HT(XL,YL)',/)
       END IF
       RETURN
       END
```

Program 5.5c *Continued.*

```
AQUIFER SIMULATION, SINK ONLY
0.         100.      10.
  121                    0.00001
AQUIFER SIMULATION, SOURCE AND SINK
0.         10.       1.0
  121                    0.00001
END OF RUNS
```

Program 5.5d Data for Programs 5.5a to c.

```
          WRITE(NO,6),NORUN
6         FORMAT(/,' RUN ',I2,3X,
     1    ' 1 - H(0,0), 2 - H(XL/2,YL/2), 3 - H(XL,YL)',/)
          CALL SPLOTS(3,IP,TP,HTP)
          WRITE(NO,7),NORUN
7         FORMAT(/,' RUN ',I2,3X,
     1    ' 1 - HT(0,0), 2 - HT(XL/2,YL/2), 3 - HT(XL,YL)',/)
       END IF
```

The data file for Program 5.5a to c is listed in Program 5.5d. In each of two runs, 121 ODEs (for the $11 \times 11 = 121$-point spatial grid) are integrated with an accuracy of 0.00001. The integration is by RKF45. Again, the main program that calls RKF45 is not listed since it is similar to earlier main programs (see Program 2.2a). Note also that the time scale for the first run is 10 times that for the second run. This longer time scale was selected for the solution to more closely approach a steady state.

Abbreviated output from Programs 5.5a to d is listed in Table 5.5. We can note the following points about the output in Table 5.5:

1. For Case I (a sink or well at $(x = 0, y = 0)$), the drawdown of the water in the aquifer is relatively slow. Thus, at $t = 100$ days, the hydraulic head does not depart significantly from its initial value of 10 m except in the immediate area

Table 5.5 Abbreviated Output from Programs 5.5a to d. *Continued next pages.*

```
RUN NO. -   1  AQUIFER SIMULATION, SINK ONLY

INITIAL T -  0.000D+00

  FINAL T -  0.100D+03

  PRINT T -  0.100D+02

NUMBER OF DIFFERENTIAL EQUATIONS - 121

INTEGRATION ALGORITHM - RKF45

MAXIMUM INTEGRATION ERROR -  0.100D-04


T =    0.0

  H(X,Y,T)
                X = 0    X = 400    X = 800   X = 1200   X = 1600   X = 2000
Y = 0        10.0000    10.0000    10.0000    10.0000    10.0000    10.0000
Y = 400      10.0000    10.0000    10.0000    10.0000    10.0000    10.0000
Y = 800      10.0000    10.0000    10.0000    10.0000    10.0000    10.0000
Y = 1200     10.0000    10.0000    10.0000    10.0000    10.0000    10.0000
Y = 1600     10.0000    10.0000    10.0000    10.0000    10.0000    10.0000
Y = 2000     10.0000    10.0000    10.0000    10.0000    10.0000    10.0000


 HT(X,Y,T)
                X = 0    X = 400    X = 800   X = 1200   X = 1600   X = 2000
Y = 0        -2.5000     0.0000     0.0000     0.0000     0.0000     0.0000
Y = 400       0.0000     0.0000     0.0000     0.0000     0.0000     0.0000
Y = 800       0.0000     0.0000     0.0000     0.0000     0.0000     0.0000
Y = 1200      0.0000     0.0000     0.0000     0.0000     0.0000     0.0000
Y = 1600      0.0000     0.0000     0.0000     0.0000     0.0000     0.0000
Y = 2000      0.0000     0.0000     0.0000     0.0000     0.0000     0.0000


T =   10.0

  H(X,Y,T)
                X = 0    X = 400    X = 800   X = 1200   X = 1600   X = 2000
Y = 0         7.1724     9.4559     9.8926     9.9831     9.9980     9.9996
Y = 400       9.4559     9.7414     9.9344     9.9887     9.9986     9.9997
Y = 800       9.8926     9.9344     9.9798     9.9961     9.9995     9.9999
Y = 1200      9.9831     9.9887     9.9961     9.9992     9.9999    10.0000
Y = 1600      9.9980     9.9986     9.9995     9.9999    10.0000    10.0000
Y = 2000      9.9996     9.9997     9.9999    10.0000    10.0000    10.0000


 HT(X,Y,T)
                X = 0    X = 400    X = 800   X = 1200   X = 1600   X = 2000
Y = 0        -0.0550    -0.0413    -0.0180    -0.0048    -0.0009    -0.0002
Y = 400      -0.0413    -0.0311    -0.0136    -0.0036    -0.0007    -0.0002
```

Table 5.5 *Continued.*

Y = 800	-0.0180	-0.0136	-0.0059	-0.0016	-0.0003	-0.0001
Y = 1200	-0.0048	-0.0036	-0.0016	-0.0004	-0.0001	0.0000
Y = 1600	-0.0009	-0.0007	-0.0003	-0.0001	0.0000	0.0000
Y = 2000	-0.0002	-0.0002	-0.0001	0.0000	0.0000	0.0000

T = 20.0

H(X,Y,T)

	X = 0	X = 400	X = 800	X = 1200	X = 1600	X = 2000
Y = 0	6.7952	9.1476	9.7208	9.9145	9.9765	9.9899
Y = 400	9.1476	9.4889	9.7931	9.9319	9.9806	9.9915
Y = 800	9.7208	9.7931	9.8996	9.9632	9.9888	9.9949
Y = 1200	9.9145	9.9319	9.9632	9.9853	9.9953	9.9978
Y = 1600	9.9765	9.9806	9.9888	9.9953	9.9984	9.9992
Y = 2000	9.9899	9.9915	9.9949	9.9978	9.9992	9.9996

HT(X,Y,T)

	X = 0	X = 400	X = 800	X = 1200	X = 1600	X = 2000
Y = 0	-0.0270	-0.0235	-0.0156	-0.0080	-0.0034	-0.0019
Y = 400	-0.0235	-0.0205	-0.0136	-0.0070	-0.0029	-0.0017
Y = 800	-0.0156	-0.0136	-0.0090	-0.0046	-0.0020	-0.0011
Y = 1200	-0.0080	-0.0070	-0.0046	-0.0024	-0.0010	-0.0006
Y = 1600	-0.0034	-0.0029	-0.0020	-0.0010	-0.0004	-0.0002
Y = 2000	-0.0019	-0.0017	-0.0011	-0.0006	-0.0002	-0.0001

. . .

T = 100.0

H(X,Y,T)

	X = 0	X = 400	X = 800	X = 1200	X = 1600	X = 2000
Y = 0	5.8894	8.2935	8.9952	9.3351	9.5054	9.5580
Y = 400	8.2935	8.6825	9.1053	9.3794	9.5287	9.5760
Y = 800	8.9952	9.1053	9.3064	9.4785	9.5855	9.6212
Y = 1200	9.3351	9.3794	9.4785	9.5794	9.6491	9.6736
Y = 1600	9.5054	9.5287	9.5855	9.6491	9.6964	9.7136
Y = 2000	9.5580	9.5760	9.6212	9.6736	9.7136	9.7283

HT(X,Y,T)

	X = 0	X = 400	X = 800	X = 1200	X = 1600	X = 2000
Y = 0	-0.0069	-0.0069	-0.0067	-0.0065	-0.0063	-0.0062
Y = 400	-0.0069	-0.0068	-0.0066	-0.0064	-0.0063	-0.0062
Y = 800	-0.0067	-0.0066	-0.0065	-0.0063	-0.0061	-0.0060
Y = 1200	-0.0065	-0.0064	-0.0063	-0.0061	-0.0059	-0.0059
Y = 1600	-0.0063	-0.0063	-0.0061	-0.0059	-0.0058	-0.0057
Y = 2000	-0.0062	-0.0062	-0.0060	-0.0059	-0.0057	-0.0056

Table 5.5 *Continued.*

```
 T =     0.0
     10.0000       10.0000       10.0000
     -2.5000        0.0000        0.0000

 T =    10.0
      7.1724        9.9956       10.0000
     -0.0550       -0.0018        0.0000

 T =    20.0
      6.7952        9.9600        9.9996
     -0.0270       -0.0049       -0.0001
                      .
                      .
                      .

 T =    90.0
      5.9598        9.5321        9.7834
     -0.0072       -0.0062       -0.0054

 T =   100.0
      5.8894        9.4696        9.7283
     -0.0069       -0.0063       -0.0056


              ..1....1....1....1....1....1....1....1....1....1....1..
   0.100D+02+  C    C    C    C    C    C    3    3    3    3    3   +I
            -                                2    2    2    2    2   -I
            -                                                         -I
            -                                                         -I
            -                                                         -I
   0.863D+01+                                                         +I
            -                                                         -I
            -                                                         -I
            -                                                         -I
            -                                                         -I
   0.726D+01+       1                                                 +I
            -            1                                            -I
            -                 1                                       -I
            -                      1    1    1                        -I
            -                                     1    1    1         -I
   0.589D+01+                                                    1   +I
              ..1....1....1....1....1....1....1....1....1....1....1..
           0.000D+00  0.20D+02  0.40D+02  0.60D+02  0.80D+02  0.10D+03

RUN  1     1 - H(0,0), 2 - H(XL/2,YL/2), 3 - H(XL,YL)
           .                              .
           .                              .
           .                              .

RUN NO. -   2  AQUIFER SIMULATION, SOURCE AND SINK

INITIAL T -  0.000D+00
```

Table 5.5 *Continued.*

```
 FINAL T -  0.100D+02

 PRINT T -  0.100D+01

NUMBER OF DIFFERENTIAL EQUATIONS - 121

INTEGRATION ALGORITHM - RKF45

MAXIMUM INTEGRATION ERROR -  0.100D-04


T =     0.0

  H(X,Y,T)
               X = 0    X = 400   X = 800   X = 1200   X = 1600   X = 2000
Y = 0        10.0000   10.0000   10.0000   10.0000   10.0000   10.0000
Y = 400      10.0000   10.0000   10.0000   10.0000   10.0000   10.0000
Y = 800      10.0000   10.0000   10.0000   10.0000   10.0000   10.0000
Y = 1200     10.0000   10.0000   10.0000   10.0000   10.0000   10.0000
Y = 1600     10.0000   10.0000   10.0000   10.0000   10.0000   10.0000
Y = 2000     10.0000   10.0000   10.0000   10.0000   10.0000   10.0000


 HT(X,Y,T)
               X = 0    X = 400   X = 800   X = 1200   X = 1600   X = 2000
Y = 0        -2.5000    0.0000    0.0000    0.0000    0.0000    0.0000
Y = 400       0.0000    0.0000    0.0000    0.0000    0.0000    0.0000
Y = 800       0.0000    0.0000    0.0000    0.0000    0.0000    0.0000
Y = 1200      0.0000    0.0000    0.0000    0.0000    0.0000    0.0000
Y = 1600      0.0000    0.0000    0.0000    0.0000    0.0000    0.0000
Y = 2000      0.0000    0.0000    0.0000    0.0000    0.0000    2.5000


T =     1.0

  H(X,Y,T)
               X = 0    X = 400   X = 800   X = 1200   X = 1600   X = 2000
Y = 0         8.6211    9.9781    9.9999   10.0000   10.0000   10.0000
Y = 400       9.9781    9.9992   10.0000   10.0000   10.0000   10.0000
Y = 800       9.9999   10.0000   10.0000   10.0000   10.0000   10.0000
Y = 1200     10.0000   10.0000   10.0000   10.0000   10.0000   10.0001
Y = 1600     10.0000   10.0000   10.0000   10.0000   10.0008   10.0219
Y = 2000     10.0000   10.0000   10.0000   10.0001   10.0219   11.3789


 HT(X,Y,T)
               X = 0    X = 400   X = 800   X = 1200   X = 1600   X = 2000
Y = 0        -0.7321   -0.0471   -0.0005    0.0000    0.0000    0.0000
Y = 400      -0.0471   -0.0030    0.0000    0.0000    0.0000    0.0000
Y = 800      -0.0005    0.0000    0.0000    0.0000    0.0000    0.0000
Y = 1200      0.0000    0.0000    0.0000    0.0000    0.0000    0.0005
Y = 1600      0.0000    0.0000    0.0000    0.0000    0.0030    0.0471
Y = 2000      0.0000    0.0000    0.0000    0.0005    0.0471    0.7321
```

Table 5.5 *Continued.*

T = 2.0

H(X,Y,T)

	X = 0	X = 400	X = 800	X = 1200	X = 1600	X = 2000
Y = 0	8.1267	9.9173	9.9982	10.0000	10.0000	10.0000
Y = 400	9.9173	9.9909	9.9997	10.0000	10.0000	10.0000
Y = 800	9.9982	9.9997	10.0000	10.0000	10.0000	10.0000
Y = 1200	10.0000	10.0000	10.0000	10.0000	10.0003	10.0018
Y = 1600	10.0000	10.0000	10.0000	10.0003	10.0091	10.0827
Y = 2000	10.0000	10.0000	10.0000	10.0018	10.0827	11.8733

HT(X,Y,T)

	X = 0	X = 400	X = 800	X = 1200	X = 1600	X = 2000
Y = 0	-0.3375	-0.0692	-0.0030	-0.0001	0.0000	0.0000
Y = 400	-0.0692	-0.0142	-0.0006	0.0000	0.0000	0.0000
Y = 800	-0.0030	-0.0006	0.0000	0.0000	0.0000	0.0001
Y = 1200	-0.0001	0.0000	0.0000	0.0000	0.0006	0.0030
Y = 1600	0.0000	0.0000	0.0000	0.0006	0.0142	0.0692
Y = 2000	0.0000	0.0000	0.0001	0.0030	0.0692	0.3375

. . .

T = 9.0

H(X,Y,T)

	X = 0	X = 400	X = 800	X = 1200	X = 1600	X = 2000
Y = 0	7.2305	9.4989	9.9105	9.9877	9.9989	10.0000
Y = 400	9.4989	9.7731	9.9477	9.9924	10.0000	10.0011
Y = 800	9.9105	9.9477	9.9858	10.0000	10.0076	10.0123
Y = 1200	9.9877	9.9924	10.0000	10.0142	10.0523	10.0895
Y = 1600	9.9989	10.0000	10.0076	10.0523	10.2269	10.5011
Y = 2000	10.0000	10.0011	10.0123	10.0895	10.5011	12.7695

HT(X,Y,T)

	X = 0	X = 400	X = 800	X = 1200	X = 1600	X = 2000
Y = 0	-0.0614	-0.0446	-0.0177	-0.0041	-0.0005	0.0000
Y = 400	-0.0446	-0.0324	-0.0129	-0.0029	0.0000	0.0005
Y = 800	-0.0177	-0.0129	-0.0048	0.0000	0.0029	0.0041
Y = 1200	-0.0041	-0.0029	0.0000	0.0048	0.0129	0.0177
Y = 1600	-0.0005	0.0000	0.0029	0.0129	0.0324	0.0446
Y = 2000	0.0000	0.0005	0.0041	0.0177	0.0446	0.0614

T = 10.0

H(X,Y,T)

	X = 0	X = 400	X = 800	X = 1200	X = 1600	X = 2000
Y = 0	7.1724	9.4559	9.8926	9.9832	9.9982	10.0000
Y = 400	9.4559	9.7414	9.9346	9.9892	10.0000	10.0018

Table 5.5 *Continued.*

```
Y = 800      9.8926     9.9346     9.9806    10.0000    10.0108    10.0168
Y = 1200     9.9832     9.9892    10.0000    10.0194    10.0654    10.1074
Y = 1600     9.9982    10.0000    10.0108    10.0654    10.2586    10.5441
Y = 2000    10.0000    10.0018    10.0168    10.1074    10.5441    12.8276


 HT(X,Y,T)
               X = 0    X = 400    X = 800   X = 1200   X = 1600   X = 2000
Y = 0       -0.0550    -0.0413    -0.0180    -0.0048    -0.0007     0.0000
Y = 400     -0.0413    -0.0311    -0.0135    -0.0034     0.0000     0.0007
Y = 800     -0.0180    -0.0135    -0.0055     0.0000     0.0034     0.0048
Y = 1200    -0.0048    -0.0034     0.0000     0.0055     0.0135     0.0180
Y = 1600    -0.0007     0.0000     0.0034     0.0135     0.0311     0.0413
Y = 2000     0.0000     0.0007     0.0048     0.0180     0.0413     0.0550


T =     0.0
    10.0000     10.0000     10.0000
    -2.5000      0.0000      2.5000

T =     1.0
     8.6211     10.0000     11.3789
    -0.7321      0.0000      0.7321

T =     2.0
     8.1267     10.0000     11.8733
    -0.3375      0.0000      0.3375
        .                       .
        .                       .
        .                       .

T =     9.0
     7.2305     10.0000     12.7695
    -0.0614      0.0000      0.0614

T =    10.0
     7.1724     10.0000     12.8276
    -0.0550      0.0000      0.0550
```

around the sink. This conclusion is supported by $h(x, y, 100)$ at ($x = 0$, $y = 0$), ($x = x_L/2$, $y = y_L/2$), and ($x = x_L$, $y = y_L$):

```
T =   100.0
      5.8894      9.4696      9.7283
     -0.0069     -0.0063     -0.0056
```

That is, $h(0, 0, 100) = 5.8894$, $h(x_L/2, y_L/2, 100) = 9.4696$, and $h(x_L, y_L, 100) = 9.7283$; note also the small values of $\partial h(x, y, 100)/\partial t$. This slow response is also apparent from the point plot in which the head $h(0, 0, t)$ (with points denoted by

Table 5.5 *Continued.*

```
            ..1....1....1....1....1....1....1....1....1....1....1..
   0.128D+02+                                3    3    3    3    3  +I
            -                  3    3    3                           -I
            -             3                                          -I
            -        3                                               -I
            -                                                        -I
   0.109D+02+                                                        +I
            -                                                        -I
            -   C    2    2    2    2    2    2    2    2    2    2  -I
            -                                                        -I
            -                                                        -I
   0.906D+01+                                                        +I
            -        1                                               -I
            -             1                                          -I
            -                  1    1    1                           -I
            -                                1    1    1    1        -I
   0.717D+01+                                                     1  +I
            ..1....1....1....1....1....1....1....1....1....1....1..
         0.000D+00  0.20D+01  0.40D+01  0.60D+01  0.80D+01  0.10D+02

RUN  2     1 - H(0,0), 2 - H(XL/2,YL/2), 3 - H(XL,YL)
           .                             .
           .                             .
           .                             .
```

"1") changes significantly with t, but $h(x_L/2, y_L/2, 100)$ (denoted with "2"), and $h(x_L, y_L, 100)$ (denoted with "3") change very little over $0 \le t \le 100$ days. Thus, the Case I solution does not demonstrate the steady-state solution $h(x, y, \infty) = 0$; obviously a substantially longer problem time would be required to demonstrate this solution.

2. For the same parameters, the Case II solution (a sink or well at $(x = 0, y = 0)$ and a source at $(x = x_L, y = y_L)$), reaches a steady-state solution relatively quickly. Thus, after 9 and 10 days, the values $h(x, y, 9)$ and $h(x, y, 10)$ at $(x = 0, y = 0)$, $(x = x_L/2, y = y_L/2)$, and $(x = x_L, y = y_L)$ are

```
T =     9.0
       7.2305     10.0000     12.7695
      -0.0614      0.0000      0.0614

T =    10.0
       7.1724     10.0000     12.8276
      -0.0550      0.0000      0.0550
```

This fast response is also apparent from the point plot.

3. A line of symmetry is established running from $(x = 0, y = 0)$ to $(x = x_L, y = y_L)$. This is demonstrated, for example, by the solution at $t = 10$:

```
T =    10.0

  H(X,Y,T)
                X = 0    X = 400    X = 800   X = 1200   X = 1600   X = 2000
Y = 0          7.1724     9.4559     9.8926     9.9832     9.9982    10.0000
Y = 400        9.4559     9.7414     9.9346     9.9892    10.0000    10.0018
Y = 800        9.8926     9.9346     9.9806    10.0000    10.0108    10.0168
Y = 1200       9.9832     9.9892    10.0000    10.0194    10.0654    10.1074
Y = 1600       9.9982    10.0000    10.0108    10.0654    10.2586    10.5441
Y = 2000      10.0000    10.0018    10.0168    10.1074    10.5441    12.8276
```

Of course, this symmetry would not occur if the source and sink are of different strengths, or if the source and sink are not located symmetrically along the diagonal from $(x = 0, y = 0)$ to $(x = x_L, y = y_L)$.

We have presented this final example to demonstrate two ideas:

1. The solution of elliptic PDEs by converting them to parabolic PDEs (they could also be converted to hyperbolic PDEs, which might have a relatively faster rate of convergence to the final steady-state solution).

2. The explicit programming of the spatial derivatives in PDEs (rather than the use of spatial differentiation routines). This approach obviously leads to code which is not as compact and closely related in appearance to the problem system PDEs as we observed in previous examples, but it might provide greater flexibility in the sense that, at least in principle, any spatial derivatives and their associated boundary conditions can be approximated in developing a method of lines solution.

5.3 Conclusions

A discussion of numerical methods for the solution of ODE/DAE/PDE problems in science and engineering can be organized in at least two ways:

1. The numerical methods are the major topics and the equation types are the subtopics. For example, we could have a chapter on finite differences, with possibly applications to elliptic, parabolic, and hyperbolic PDEs. Then we could have a chapter on finite elements, with possibly applications to elliptic, parabolic, and hyperbolic PDEs, and then a chapter on finite volumes, with applications to elliptic, parabolic, and hyperbolic PDEs, etc. This organization might be the choice of numerical analysts and applied mathematicians who intend to emphasize the theoretical and computational aspects of the numerical methods. Of course, this is not the organization of this book.

2. The classification of equations is the origin of the major topics, with the numerical methods as the subtopics. For example, in this book, Chapter 1 is a general survey of the types of ODEs/DAEs/PDEs that serves as the basis for most of the subsequent discussion. Chapter 2 is a discussion of numerical methods for ODEs and rather briefly, DAEs. Chapters 3 and 4 treat PDEs first order in time, which generally are classified as first-order hyperbolic and parabolic PDEs. Then

Chapter 5 covers ODEs second and zeroth order in time that generally are classified as second-order hyperbolic and elliptic PDEs. The length of discussion of PDEs first order in time (Chapters 3 and 4) might suggest that they are considered the most important or useful PDEs. This is not the case; rather, we have chosen PDEs first order in time to illustrate a series of numerical methods, e.g., finite differences, elements, and volumes, but these numerical methods can, in principle, be applied to the other major classes of PDEs as well, like the PDEs second and zeroth order in time covered in Chapter 5 (for the latter, as well as boundary value ODEs, a first-order derivative in time is generally added to the original problem equations).

In other words, we have decided to emphasize the equation types as the key to applications and then consider the numerical methods as a means to an end, that is, the means to the computer generation of solutions. This might be the choice of organization of a scientist or engineer who is interested primarily in applications and views the equations as the fundamental mode of expression of a new problem. The numerical methods are then the path to a solution once the problem is stated mathematically. To express this approach in other words, a scientist or engineer typically writes down a mathematical statement of a problem as a system of equations, then looks for a method(s) to compute a solution. This is the preferred approach. The alternative, perhaps, is to modify the problem equations so that they fit within the framework of an established numerical method, which typically is implemented in existing library software. This approach, of course, means that the numerical methods and existing software are dictating the form of the problem statement, which should be viewed as generally unacceptable; the justification for this approach is typically one of economy of time and effort, but, of course, leaves open the question of whether the final results are really useful in the sense of providing a meaningful answer to the problem of interest.

This discussion suggests three additional aspects of the computer solution of ODE/DAE/PDE systems:

3. Within the present state of knowledge, general algorithms which can reliably provide solutions to all of the major classes of differential equations are not available. What we have discussed in this book are generally accepted methods that have been found to be useful in a variety of applications. However, each new problem should be viewed as a challenge, with some research required before an acceptable solution is computed; in other words, there can be no guarantees in advance that a numerical solution can be produced with a given investment of time and effort. Of course, there are cases for which the effort to compute a solution can be anticipated, but usually these are for minor extensions of existing problems. New problems which have not been investigated previously, and even problems which appear to be minor extensions of problems already solved, might have unexpected research requirements.

4. The method of lines discussed in this book is an approach to PDEs which at least in principle can be applied to all of the major classes of PDEs, elliptic, parabolic and hyperbolic, linear and nonlinear, in one, two, and three spatial dimensions. Although its performance cannot be guaranteed in advance for any particular problem, the method of lines at least provides a framework for the solution of broad classes of PDEs; also, it can be implemented through the use of exisiting, quality initial value integrators, e.g., `RKF45`, `LSODE`, and `DASSL`.

5. The evaluation of numerical results is perhaps the most difficult aspect of the computer solution of ODE/DAE/PDE systems. In other words, we must always

ask the question: When do we believe what comes out of the computer? An error analysis is an essential part of any numerical solution.

In any case, we have now considered some numerical methods and associated software for initial and boundary value ODEs, linear and nonlinear, and elliptic, parabolic and hyperbolic PDEs, in one and two dimensions. The extension to three dimensions follows directly from the approach to two-dimensional PDEs, and in fact, a spatial differentiation routine for three-dimensional PDEs, `DSS036`, is available which is used in the same way as the two-dimensional routine, `DSS034`, called in Program 4.11c.

In Appendix A, we have listed an ODE integrator based on the fixed-step, second-order Adams Bashforth (SOAB) method. Although this ODE integrator is rudimentary since it does not provide error monitoring and control through automatic step size adjustment, it does illustrate how the equations for an ODE integrator are coded, and it has been applied to ODE problems (starting with equation (2.126) in Chapter 2). Also, the coding for the SOAB integrator is modest in size and can therefore be studied with minimum effort. Other more sophisticated integrators like `RKF45`, `LSODE`, and `DASSL` are too lengthy to be included in this book.

In Appendix B, we have listed some spatial differentiation routines which can provide the second essential component of a method of lines code; that is, we require an ODE integrator, like the SOAB integrator in Appendix A, and a means for spatial differentation, such as the routines in Appendix B.

Finally, in Appendix C we list some ODE/DAE/PDE applications in various fields of science and engineering that illustrate the applications of the ideas presented in this book. We hope that these applications serve a useful teaching purpose by illustrating various approaches to the solution of ODE/DAE/PDE problems; they are all coded in the same format as the programs discussed previously (i.e., using the basic routines `INITAL`, `DERV`, and `PRINT`, plus subordinate routines). Certainly the techniques discussed in this book are open ended, and have as a significant requirement the application of the experience and ingenuity of the analyst. We hope that the methods discussed in this book are generally useful; the author welcomes inquiries concerning the solution of new problems.

References

Boggs, G. T. (1971), "The Solution of Nonlinear Systems of Equations by A-stable Integration Techniques", *Siam J. Numer. Anal.*, **8**, pp. 767–785.

Davidenko, D. F. (1953), "On a New Method for Numerical Solution of Systems of Nonlinear Equations" (Russian), *Dokaldy Akad. Nauk. SSSR*, **88**, pp. 601–602 (Journal translated as *Soviet Math. Dokl.*); also, *Ukr. Mat. Z.5*.

Edelen, D. G. B. (1976a), "On the Construction of Differential Systems for the Solution of Nonlinear Algebraic and Transcendental Systems of Equations," *Numerical Methods for Differential Systems* (L. Lapidus and W. E. Schiesser (eds.), Academic Press, New York, pp. 67–84.

Edelen, D. G. B. (1976b), "Differential Procedures for Systems of Implicit Relations and Implicitly Coupled Nonlinear Boundary-value Problems," *Numerical Methods for Differential Systems* (L. Lapidus and W. E. Schiesser, eds.), Academic Press, New York, pp. 85–95.

Forsythe, G. E., M. A. Malcolm and C. B. Moler (1977), *Computer Methods for Mathematical Computations*, Prentice-Hall, Englewood Cliffs, NJ.

Hachtel, G. and M. Mack (1974), "A Pseudo Dynamic Method for Solving Nonlinear Algebraic Equations," *Stiff Differential Systems*, R. A. Willoughby (ed.), Plenum Press, New York, pp. 135–150.

Myint-U, T. and L. Debnath (1987), *Partial Differential Equations for Scientists and Engineers*, Third Edition, North Holland, New York.

Prasher, S. O. (1987), Private communication.

Rheinboldt, W. C. (1980), "Solution Fields on Nonlinear Equations and Continuation Methods," *SIAM J. Numer. Anal.*, **17**(2), pp. 221–237, April.

Watson, L. T. (1986), "Numerical Linear Algebra Aspects of Globally Convergent Homotopy Methods," *SIAM Review*, **28**(4).

Watson, L. T., S. C. Billups and A. P. Morgan, (1987), "Algorithm 652: HOMPACK: A Suite of Codes for Globally Convergent Homotopy Algorithms," *ACM Trans. Math. Software*, **13**(3), pp. 281–310.

appendix a

A Second-Order Adams-Bashforth ODE Integrator

```
      PROGRAM ADAMS
C...
C...  PROGRAM ADAMS IMPLEMENTS A FIXED-STEP, SECOND-ORDER ADAMS-
C...  BASHFORTH ODE INTEGRATION
C...
C...  THE PROBLEM INITIAL CONDITIONS ARE SET IN SUBROUTINE INITAL, AND
C...  THE PROBLEM DERIVATIVES ARE PROGRAMMED IN SUBROUTINE DERV.  THE
C...  NUMERICAL SOLUTION IS PRINTED AND PLOTTED IN SUBROUTINE PRINT.
C...
C...  COMMON AREA TO LINK THIS MAIN PROGRAM WITH THE USER-SUPPLIED
C...  SUBROUTINES INITAL, DERV, AND PRINT.  THE FOLLOWING CODING IS FOR
C...  250 ORDINARY DIFFERENTIAL EQUATIONS (ODES).  IF MORE ODES ARE TO
C...  BE INTEGRATED, ALL OF THE 250*S SHOULD BE CHANGED TO THE REQUIRED
C...  NUMBER
      COMMON/T/          T,      NSTOP,      NORUN
     1      /Y/    Y(250)
     2      /F/    F(250)
     3      /H/        DT
C...
C...  ARRAYS REQUIRED BY THE INTEGRATION ALGORITHM
      COMMON/R/ FNM2(250), FNM1(250)
C...
C...  COMMON AREA TO PROVIDE THE INPUT/OUTPUT UNIT NUMBERS TO OTHER
C...  SUBROUTINES
      COMMON/IO/       NI,        NO
C...
C...  ARRAY FOR THE TITLE (FIRST LINE OF DATA), CHARACTERS  END OF RUNS
      CHARACTER TITLE(20)*4, ENDRUN(3)*4
C...
C...  VARIABLE FOR THE TYPE OF ERROR
      CHARACTER*3 ABSREL
C...
C...  DEFINE THE CHARACTERS  END OF RUNS
      DATA ENDRUN/'END ','OF R','UNS '/
C...
C...  DEFINE THE INPUT/OUTPUT UNIT NUMBERS
      NI=5
      NO=6
C...
C...  OPEN INPUT AND OUTPUT FILES
      OPEN(NI,FILE='DATA'  ,STATUS='OLD')
```

```
      OPEN(NO,FILE='OUTPUT',STATUS='NEW')
C...
C...  INITIALIZE THE RUN COUNTER
      NORUN=0
C...
C...  BEGIN A RUN
1     NORUN=NORUN+1
C...
C...  INITIALIZE THE RUN TERMINATION VARIABLE
      NSTOP=0
C...
C...  READ THE FIRST LINE OF DATA
      READ(NI,1000,END=999)(TITLE(I),I=1,20)
C...
C...  TEST FOR  END OF RUNS  IN THE DATA
      DO 2 I=1,3
      IF(TITLE(I).NE.ENDRUN(I))GO TO 3
2     CONTINUE
C...
C...  AN END OF RUNS HAS BEEN READ, SO TERMINATE EXECUTION
999   STOP
C...
C...  READ THE SECOND LINE OF DATA
3     READ(NI,1001,END=999)TO,TF,TP
C...
C...  READ THE THIRD LINE OF DATA
      READ(NI,1002,END=999)NEQN,NMAX,NSTART,NCONT
C...
C...  PRINT A DATA SUMMARY
      WRITE(NO,1003)NORUN,(TITLE(I),I=1,20),
     1      TO,TF,TP,NEQN,NMAX,NSTART,NCONT
C...
C...  INITIALIZE TIME
      T=TO
C...
C...  SET THE INTEGRATION INTERVAL
      DT=TP/FLOAT(NMAX)
C...
C...  SET THE INITIAL CONDITIONS
      CALL INITAL
C...
C...  PRINT THE INITIAL CONDITIONS
      CALL PRINT(NI,NO)
C...
C...  BEGIN THE TIME STEPPING
         NSTEPS=NMAX
C...
C...     NSTART READ FROM THE DATA FILE CAN BE USED TO SELECT THE
C...     INITIAL INTEGRATION ALGORITHM (THIS SHOULD BE 0 OR 1)
C...
C...        NSTART= 0 - FIRST-ORDER, RUNGE KUTTA (EULER)
C...
C...        NSTART = 1 - SECOND-ORDER, RUNGE KUTTA (MODIFIED EULER)
4     DO 5 I=1,NSTEPS
C...
C...  TAKE A FIRST-ORDER (NSTART = 0) OR SECOND-ORDER (NSTART = 1)
```

```
C...  RUNGE-KUTTA STEP AT THE BEGINNING
      IF((NSTART.EQ.0).OR.(NSTART.EQ.1))THEN
C...
C...     DERIVATIVE AT N-1 = 1-1 = 0
         CALL DERV
C...
C...     FIRST-ORDER, RUNGE-KUTTA (EULER) STEP FROM N-1 TO N = 1-1 TO 1
         DO 6 J=1,NEQN
         FNM2(J)=F(J)
         Y(J)=Y(J)+F(J)*DT
6        CONTINUE
         T=T+DT
C...
C...     DERIVATIVE AT N = 1
         CALL DERV
C...
C...     THE DERIVATIVE AT N = 1 BECOMES THE DERIVATIVE AT N-1 FOR
C...     THE FIRST STEP OF THE ADAMS-BASHFORTH INTEGRATION
         DO 7 J=1,NEQN
         FNM1(J)=F(J)
7        CONTINUE
C...
C...     SECOND-ORDER, RUNGE-KUTTA STEP
         IF(NSTART.EQ.1)THEN
         DO 8 J=1,NEQN
         Y(J)=Y(J)+0.5*(FNM1(J)-FNM2(J))*DT
8        CONTINUE
         END IF
      END IF
C...
C...  TAKE A SECOND-ORDER, ADAMS-BASHFORTH STEP FROM N-1 TO N
      IF(NSTART.EQ.2)THEN
         DO 9 J=1,NEQN
         Y(J)=Y(J)+DT*(1.5*FNM1(J)-0.5*FNM2(J))
9        CONTINUE
         T=T+DT
C...
C...     DERIVATIVE AT N
         CALL DERV
C...
C...     DERIVATIVES AT N-1 AND N-2 FOR NEXT STEP
         DO 10 J=1,NEQN
         FNM2(J)=FNM1(J)
         FNM1(J)=F(J)
10       CONTINUE
         END IF
C...
C...  ALL NEQN DEPENDENT VARIABLES HAVE BEEN ADVANCED, SO TAKE THE
C...  NEXT STEP IN TIME.  NSTART SET AT THIS POINT (NSTART = NCONT,
C...  WHERE NCONT IS READ FROM THE DATA FILE) CAN BE USED TO CHANGE
C...  THE INTEGRATION ALGORITHM
C...
C...     NSTART = 0 - FIRST-ORDER, RUNGE KUTTA (EULER)
C...
C...     NSTART = 1 - SECOND-ORDER, RUNGE KUTTA (MODIFIED EULER)
C...
```

```
C...      NSTART = 2 - SECOND-ORDER, ADAMS BASHFORTH
          NSTART=NCONT
5      CONTINUE
C...
C...   ONE PRINT INTERVAL IS COMPLETE, SO PRINT THE SOLUTION
       CALL PRINT(NI,NO)
C...
C...   CHECK FOR A RUN TERMINATION
       IF(NSTOP.NE.0)GO TO 1
C...
C...   CHECK FOR THE END OF THE RUN
       IF(T.LT.(TF-0.5*TP))GO TO 4
C...
C...   THE CURRENT RUN IS COMPLETE, SO GO ON TO THE NEXT RUN
       GO TO 1
C...
C...   ******************************************************************
C...
C...   FORMATS
C...
1000   FORMAT(20A4)
1001   FORMAT(3E10.0)
1002   FORMAT(4I5)
1003   FORMAT(1H1,//,
     1 ' RUN NO. - ',I3,2X,20A4,//,
     2 ' INITIAL T - ',E10.3,//,
     3 '   FINAL T - ',E10.3,//,
     4 '   PRINT T - ',E10.3,//,
     5 ' NUMBER OF DIFFERENTIAL EQUATIONS - ',I3,//,
     6 ' PRINT INTERVAL/INTEGRATION INTERVAL - ',I5,//,
     7 ' STARTING ALGORITHM - ',I2,/,
     8 '   0 -  FIRST-ORDER, RUNGE KUTTA (EULER METHOD)',/,
     9 '   1 - SECOND-ORDER, RUNGE KUTTA (MODIFIED EULER METHOD)',//,
     A ' CONTINUING ALGORITHM - ',I2,/,
     B '   0 -  FIRST-ORDER, RUNGE KUTTA (EULER METHOD)',/,
     C '   1 - SECOND-ORDER, RUNGE KUTTA (MODIFIED EULER METHOD)',/,
     D '   2 - SECOND-ORDER, ADAMS BASHFORTH')
      END
```

appendix b

Spatial Differentiation Routines

Several spatial differentiation routines, with names beginning with DSS, e.g., DSS002, DSS020, have been discussed in some detail. They are part of a library of routines for which a brief description follows:

Routine	Description
DSS002	One-dimensional, three-point centered approximations for first-order derivatives
DSS004	One-dimensional, five-point centered approximations for first-order derivatives
DSS006	One-dimensional, seven-point centered approximations for first-order derivatives
DSS008	One-dimensional, nine-point centered approximations for first-order derivatives
DSS010	One-dimensional, eleven-point centered approximations for first-order derivatives
DSS012	One-dimensional, two-point upwind approximations for first-order derivatives
DSS014	One-dimensional, three-point upwind approximations for first-order derivatives
DSS016	One-dimensional, five-point upwind approximations for first-order derivatives
DSS018	One-dimensional, four-point biased upwind approximations for first-order derivatives
DSS020	One-dimensional, five-point biased upwind approximations for first-order derivatives
DSS022	One-dimensional, combined two-point and four-point biased upwind approximations for first-order derivatives
DSS024	One-dimensional, combined two-point and five-point biased upwind approximations for first-order derivatives
DSS026	Two-dimensional approximations of first-order derivatives by calls to DSS004 and DSS024 over a two-dimensional domain
DSS028	Two-dimensional approximations of second-order derivatives by calls to DSS044 over a two-dimensional domain
DSS030	One-dimensional, five-point centered approximations for fourth-order derivatives
DSS032	One-dimensional, five-point centered and biased upwind approximation for first-order derivatives on a user-defined nonuniform grid
DSS034	Two-dimensional, five-point centered and biased upwind approximations for first-order derivatives

- `DSS036` Three-dimensional, five-point centered and biased upwind approximations for first-order derivatives
- `DSS038` One-dimensional, cubic spline differentiator for first-order derivatives on a user-defined nonuniform grid
- `DSS040` One-dimensional, cubic spline differentiator for first- and second-order derivatives on a user-defined nonuniform grid
- `DSS042` One-dimensional, three-point centered approximations for second-order derivatives with Dirichlet and Neumann boundary conditions
- `DSS044` One-dimensional, five-point centered approximations for second-order derivatives with Dirichlet and Neumann boundary conditions
- `DSS046` One-dimensional, orthogonal collocation approximations for first-order derivatives with user-specified finite element collocation and user-selected second-, fourth-,... or tenth-order orthogonal polynomials
- `DSS048` Two-dimensional, orthogonal collocation approximations for first-order derivatives with user-specified finite element collocation and user selected second-, fourth-,... or tenth-order orthogonal polynomials
- `DSS050` Three-dimensional, orthogonal collocation approximations for first-order derivatives with user-specified finite element collocation and user-selected second-, fourth-,... or tenth-order orthogonal polynomials

The increasing numbers of these differentiators, e.g., `DSS002` then `DSS004`, etc., reflect their historical development. All of these routines are available from the author, with example applications, in single- and double-precision Fortran 77 formats.

appendix c

Library of ODE/DAE/PDE Applications

Physics

1. Solution of the steady-state and transient Sine-Gordon equations by quasilinearization and the method of lines
2. $1 + 1$ Sine-Gordon equation
3. Liouville's equation
4. Transient response of a resonant RCL circuit
5. Dynamic response of a manometer
6. Simulation of a spacecraft modeled as a triangular truss
7. Dynamics of a discrete-distributed mass-spring system
8. Numerical integration of a Fokker-Planck equation (van der Meer antiproton accumulator equation)
9. Einstein equation for the scale factor of the universe
10. Time-dependent Schrödinger equation in two dimensions
11. Dynamics of a rocket
12. Euler integration of the differential equations for the dynamics of a rocket
13. Dynamic analysis by difference and differential equations
14. Integration of the Gibbs-Duhem equation
15. Integration of Bessel's equation of order zero and one
16. Bessel's equation of order zero (second approach)
17. Numerical integration of the normal probability distribution function
18. Integration of the solar-radiation spectrum
19. Coupled linear harmonic oscillators
20. Shuttle launch simulation
21. Settling of a particle in a fluid

22. A good test problem in partial differential equations for the numerical method of lines
23. Laplace transform solution of simultaneous ODEs
24. Davidenko's method (a differential form of Newton's method)
25. Flow-through porous media
26. Dynamics of a double pendulum
27. One-dimensional, time-dependent Maxwell's equations
28. Damping in a second-order linear system
29. Effect of damping on a mass-spring system
30. Eigenvalue and transient response analysis of a resistance-inductance circuit
31. Linear stability analysis of a system of nonlinear, ordinary differential equations
32. One-dimensional, time-dependent Maxwell's equations—explicit programming of the spatial derivatives

Mathematics

1. Alternate implementation of time-dependent Dirichlet boundary conditions in the numerical method of lines
2. The quasi steady-state approximation
3. Multiple steady states of a nonlinear distributed (PDE) system
4. Frequency of the Jacobian matrix updating in Davidenko's method
5. Automatic updating of the Jacobian matrix in Davidenko's method
6. A new finite difference approximation of second-order derivatives with Neumann boundary conditions
7. Ordering of operations in the numerical method of lines
8. Investigation of the order of the Euler and modified Euler methods
9. One-dimensional, parabolic partial differential equation with a time-dependent nonhomogeneous term
10. Euler integration applied to integrals
11. Comparison of the analytical and numerical solutions and eigenvalues of a second-order, nonoscillatory ordinary differential equation
12. Comparison of the analytical and numerical solutions and eigenvalues of a second-order, oscillatory ordinary differential equation
13. Testing energy conservation of one-dimensional, finite difference operators
14. A comparison of stagewise differentiation and direct differentiation in the numerical method of lines—linear case

15. A comparison of stagewise differentiation and direct differentiation in the numerical method of lines—nonlinear case

16. Explicit and implicit ordinary differential equations

17. Even when the analytical solution to an ODE is available, the numerical solution may be much easier to compute and use

18. Effect of ODE stiffness on explicit integrator performance

19. Linear stability analysis of a nonlinear, second-order ordinary differential equation

20. PDE application executed under DASSL with specification of the ODE Jacobian matrix bandwidth

21. One-dimensional, nonlinear, parabolic PDE by second- and fourth-order formulas for second-order spatial derivatives

22. The logistic equation

23. A differential form of the Henon "strange attractor" equations

24. An analysis of the nonlinear structure of the Henon strange attractor equations

25. Crisis transition in the chaotic Lorenz equations

26. Intermittency in the choatic Lorenz equations

27. One-dimensional Burgers' equation with Dirichlet and Neumann boundary conditions, illustrating the numerical method of lines integration of one-dimensional, elliptic, hyperbolic, and parabolic partial differential equations—explicit time integration

28. One-dimensional Burgers' equation with Dirichlet and Neumann boundary conditions, illustrating the numerical method of lines integration of one-dimensional, elliptic, hyperbolic, and parabolic partial differential equations—implicit (stiff) time integration by LSODE

29. Two-dimensional Burgers' equation with Dirichlet and Neumann boundary conditions, illustrating the numerical method of lines integration of one-dimensional, elliptic, hyperbolic, and parabolic partial differential equations

30. Three-dimensional Burgers' equation with Dirichlet and Neumann boundary conditions, illustrating the numerical method of lines integration of one-dimensional, elliptic, hyperbolic, and parabolic partial differential equations

31. Comparison of the analytical and numerical solutions for a first- and second-order ODE system

32. A parabolic partial differential equation with inconsistent initial and boundary conditions

33. Numerical integration of a partial differential equation over a semi-infinite domain

34. An introduction to finite integral transforms

35. LSODE solution of two linear ODEs with a stiffness ratio of 10^6—printing of the numerical and exact solutions
36. LSODE solution of two linear ODEs with a stiffness ratio of 10^6—plotting of the computational statistics
37. DASSL solution of two linear ODEs with a stiffness ratio of 10^6—printing of the numerical and exact solutions
38. Krogh's ODE problem—solution of a nonstiff case by Runge Kutta
39. Krogh's ODE problem—solution of a stiff case by LSODE
40. Bandwidth reduction in the numerical method of lines integration of partial differential equations
41. Two-dimensional partial differential equation with a mixed partial derivative
42. Effect of discontinuous initial/boundary conditions on the numerical method of lines solution of a two-dimensional, parabolic partial differential equation
43. Two ODEs with a stiffness ratio of 10^6 integrated by IMSL routine IVPAG using the DSS/2 format—printing of the numerical and exact solutions
44. Two ODEs with a stiffness ratio of 10^6 integrated by IMSL routine IVPAG using the DSS/2 format—printing of the computational statistics
45. Solution of two fourth-order PDEs
46. Adaptive grid solution of the one-dimensional Burgers' equation with BDF time integration via LSODE
47. DIFFPACK—a collection of problems illustrating the NUMOL solution of initial and boundary value ODEs, and elliptic, hyperbolic, and parabolic PDEs in one, two, and three dimensions

Biochemical, Biomedical, and Environmental Systems

1. Solution of steady-state equations of human cardiovascular mechanics by Davidenko's method
2. Oxygen transfer in a red blood cell
3. Biosynthesis of the antibiotic erythromycin
4. Glucose tolerance test
5. Solution of dynamic equations of human cardiovascular mechanics
6. River pollution model
7. Diffusion and reaction in an immobolized enzyme catalyst
8. Kinetics of biochemical reactions when two species compete for the same substrate
9. An analysis of population growth in Asia

10. Forrester World 2 model
11. Boyd modification of the Forrester World 2 model
12. Two-sector Globe 6 model
13. Nomenclature for the Globe 6 model
14. A low-order, nonlinear, generic model for the dynamics of HIV transmission

Separation Systems

1. Dynamics of a binary batch distillation column
2. Dynamic simulation of a deisobutanizer
3. Diffusion and nonlinear adsorption in a pore
4. Dynamics of a countercurrent mixer-settler system
5. Dynamics of a decanter
6. Dynamic model of a three-stage mixer-settler system
7. Multicomponent chromatographic separation
8. Start-up and control of a plate absorber
9. Packed column absorber response
10. Multicomponent chromatographic separation with cosine pulse input
11. Performance of a moving-bed adsorption column
12. PI control of a binary batch distillation column
13. PI control of a double-effect evaporator
14. Jacobian matrix map and temporal eigenvalues of the model for a batch distillation column
15. Nonlinear diffusion
16. Proportional-integral control of a humidifier
17. Evaporation from a tank
18. Countercurrent liquid-liquid extractor for a partially miscible system
19. Nonequilibrium model for fixed bed, multicomponent adiabatic adsorption (chromatography)
20. Diffusion in a porous solid—linear case
21. Diffusion in a porous solid—nonlinear case
22. Control of a triple-effect evaporator
23. Solution of the Beattie-Bridgeman equation of state via the Davidenko algorithm
24. Solution of the Beattie-Bridgeman equation of state via minimum squared error

25. Batch separation of *n*-heptane and *n*-octane without automatic control
26. Batch separation of *n*-heptane and *n*-octane with automatic control
27. Relative effects of surface mass transfer resistance, internal diffusion, and internal reaction in the performance of a spherical catalyst particle
28. Dynamic simulation of a packed column absorber

Kinetics and Reactor Models

1. Nonisothermal catalytic tubular reactor
2. Dynamic two-dimensional tubular reactor
3. Dynamics of an ethylene oxide reactor
4. Ethane pyrolysis in a tubular reactor
5. Hydrodealkylation of toluene to benzene
6. First-order kinetics or consecutive, irreversible reactions
7. Decay of Plutonium 239
8. Three continuous stirred tank reactors (CSTRs) with PI control
9. Three CSTRs with proportional-integral control
10. Transient response of two CSTRs with variable holdup
11. Polymerization with two initiators
12. PI control of a CSTR with competing reactions
13. Oxidation of naphthalene to phthalic anhydride: model No. 1—resistance to heat and mass transfer at the catalyst surface is neglected
14. Oxidation of naphthalene to phthalic anhydride: model No. 2—resistance to heat and mass transfer at the catalyst surface is included
15. CSTR with proportional temperature control
16. Three-component CSTR with temperature control
17. Adsorption with chemical reaction in a tray column
18. Nonlinear kinetics of consecutive, irreversible reactions
19. Effect of the axial Peclet number on the solution of convective-diffusion (hyperbolic-parabolic) partial differential equations
20. Recycle in dynamic simulation
21. Hydrolysis of acetic anhydride
22. Moving boundary problem
23. Dynamics of a tubular reactor with axial dispersion

24. Kinetics of catalytic cracking selectivity
25. Conduction, diffusion, and reaction in a catalyst particle
26. Temperature profiles in a tubular reactor
27. Relative effects of surface mass transfer resistance, internal diffusion, and internal reaction on the performance of a spherical catalyst particle
28. Solution of a second-order, nonlinear two-point boundary value ordinary differential equation by quasilinearization—application in chemical kinetics
29. Esterification of acetic acid and ethyl alcohol
30. Semi-batch production of hexamethylenetetramine

Heat Transfer

1. Method of lines integration of boundary value ODEs
2. Solution of the one-dimensional heat conduction equation with Neumann boundary conditions and boundary conditions of the third type
3. Sizing of a vapor-condensing heat exchanger
4. Laminar flow fluid temperature profiles in a cylindrical tube
5. One-dimensional heat conduction with plotted temperature profiles
6. Simplified transient analysis of a LOCA for Three Mile Island reactor no. 2
7. Air temperature within a large city
8. Dynamics of a nonisothermal holding tank
9. Integration of the Planck distribution law over the visible spectrum
10. Integration of the Planck distribution law over the visible spectrum—second approach
11. Regenerator simulated by CSTRs in series and centered differences
12. Dynamics of a double pipe heat exchanger
13. Thermal bonding of solids
14. Temperature rise from mechanical agitation of a fluid
15. Dynamics of a regenerator with plotting of the spatial profiles
16. Temperature distribution in a nuclear fuel rod assembly
17. Dynamics of a three-pass shell and tube heat exchanger
18. Improvement in the performance of a multipass heat exchanger from adding another pass
19. PI control of a three-pass shell and tube heat exchanger
20. Parameter estimation in heat transfer systems

21. Heat conduction in a sphere
22. Convective cooling of a moving polymer sheet
23. Dynamics of a cross-flow heat exchanger
24. Temperature control of a nuclear fuel rod assembly
25. Control of a triple-effect evaporator
26. Dynamics of a two-stage heat exchanger system
27. PI control of a three-pass shell and tube heat exchanger (case 2)
28. Temperature control of a nuclear fuel rod
29. Validation of dynamic models through conservation principles
30. A comparison of two-point upwind and five-point biased upwind approximations in the numerical method of lines integration of first-order hyperbolic partial differential equations
31. Effect of heat loss from a heat exchanger
32. Water cooling of sulfuric acid
33. Cooling of sulfuric acid with limited cooling water
34. Heat conduction in a finite cylinder
35. Heat conduction with internal heat generation
36. Heating a liquid in a tube with axial conduction

Fluid Flow

1. Dynamics of liquid transfer from a tank car
2. Dynamics of a holding tank with two long lines and a pump
3. Drainage from a vertical cylindrical tank
4. Drainage from a horizontal cylindrical tank
5. Steady-state simulation via dynamic analysis
6. Liquid transfer between tanks via a centrifugal pump
7. Isentropic discharge of a perfect gas from a duct—first approach
8. Isentropic discharge of a perfect gas from a duct—second approach
9. Dynamics of a gravity flow tank
10. Transient response of a first-order hydraulic system
11. Transient response of a pneumatic system with a centrifugal compressor
12. Dynamics of two tanks connected by a long line with feed and discharge lines
13. Transients in a hydraulic system with a long line

14. Advection equation with variable velocity
15. Numerical instability due to incorrect modeling of flow reversals
16. Nonlinear advection equation
17. Dynamics of two tanks connected by a long line
18. Dynamics of two tanks connected by a long line with proportional control of the liquid height in the second tank
19. Liquid level control of two cascaded tanks
20. Pressures and flows in a pump-pipe-tank network by the multidimensional Davidenko algorithm (differential form of the Newton-Raphson method)
21. Aquifer simulation in Cartesian coordinates
22. Euler integration of three differential equations for the dynamics of two tanks connected by a long line
23. Euler integration of the differential equation for draining of a tank
24. Hydraulic system dynamics
25. Euler integration of the differential equation for a manometer
26. Eigenvalue analysis of a fluid flow system with a fast momentum balance
27. Least squares analysis of friction factor/Reynolds number data—case 1
28. Least squares analysis of friction factor/Reynolds number data—case 2
29. Time to drain a tank through a vertical pipe
30. Liquid level dynamics of two tanks with a long discharge line

Automatic Control

1. Simulation of a linear system with a time delay
2. PI control of the liquid levels in two interacting holding tanks
3. Comparative evaluation of controlling the inlet vs. outlet flows of a dynamic unit
4. Sinusoidal response of a nonlinear system
5. Ultimate control of a feedback control system
6. Effect of saturation on control system performance
7. Eigenvalue stability analysis of feedback control systems
8. Effect of time delays on the performance of feedback control systems
9. Root locus plotting for an nth-order system
10. Ultimate gain and tuning of the PI control of a plate absorber

11. Direct frequency response testing
12. Root locus stability analysis of a fourth-order system
13. Frequency response stability analysis of a fourth-order system
14. Offset/stability tradeoff in the proportional control of dynamic systems
15. Root locus stability analysis
16. Feedback control with sensor dynamics
17. Damping in a second-order linear system
18. Simulation and root locus analysis of a third-order system with PI control

index